D1516896

# GASEOUS POLLUTANTS

# Volume

# 24

### in the Wiley Series in

## Advances in Environmental
## Science and Technology

## JEROME O. NRIAGU, Series Editor

# GASEOUS POLLUTANTS

## Characterization and Cycling

Edited by

## Jerome O. Nriagu
**National Water Research Institute**
**Burlington, Ontario, Canada**

**A WILEY-INTERSCIENCE PUBLICATION**

## JOHN WILEY & SONS, INC.
**New York • Chichester • Brisbane • Toronto • Singapore**

In recognition of the importance of preserving what has been
written, it is a policy of John Wiley & Sons, Inc., to have books
of enduring value published in the United States printed on
acid-free paper, and we exert our best efforts to that end.

*Library of Congress Cataloging in Publication Data:*
Gaseous pollutants: characterization and cycling / edited by Jerome
  O. Nriagu
    p.   cm.—(Wiley series in advances in environmental science
    and technology, ISSN 0065-2563; v. 24)

    "A Wiley-Interscience publication."
    Includes bibliographical references and index.
    ISBN 0-471-54898-7
    1. Pollutants—Analysis.   2. Gases—Analysis.   3. Environmental
    chemistry.   I. Nriagu, Jerome O.   II. Series: Advances in
    environmental science and technology; v. 24

TD180.A36 vol. 24
[TD193]
628.5′3—dc20                                                    91-14369
                                                               CIP

Printed and bound in the United States of America by Braun-Brumfield, Inc.

10  9  8  7  6  5  4  3  2  1

# CONTRIBUTORS

BETTERTON, ERIC A., Department of Atmospheric Sciences, University of Arizona, Tucscon, Arizona

CECINATO, A., Istituto Sull'inquinamento Atmosferic, Consiglio Nazionale delle Ricerche, Area della Ricerca di Roma, Rome, Italy

CICCIOLI, P., Istituto Sull'inquinamento Atmosferic, Consiglio Nazionale delle Ricerche, Area della Ricerca di Roma, Rome, Italy

FELLIN, PHILLIP, Concord Scientific Corporation, Toronto, Ontario, Canada

GRUMM, D. M., Department of Physics, New Mexico Institute of Mining & Technology, Socorro, New Mexico

JALLIFFIER-MERLON, E., Observatoire Oceanologie de Villefranche-sur-Mer, Laboratoire de Physique et Chimie Marines, Villefranche-sur-Mer, France

KAPLAN, ISAAC R., Institute of Geophysics and Planetary Physics, University of California, Los Angeles, California

MARTY, J. C., Observatoire Oceanologie de Villefranche-sur-Mer, Laboratoire de Physique et Chimie Marines, Villefranche-sur-Mer, France

MATZING, H., Laboratorium fur Aerosolphysik und Filtertechnik I, Kernforschungszentrum Karlsruhe, Karlsruhe, Germany

OTSON, REIN, Bureau of Chemical Hazards, Environmental Health Directorate, Health Protection Branch, Health & Welfare Canada, Ottawa, Ontario, Canada

PAUR, H.-R., Laboratorium fur Aerosolphysik und Filtertechnik I, Kernforschungszentrum Karlsruhe, Karlsruhe, Germany

SAKUGAWA, H., Institute of Geophysics and Planetary Physics, University of California, Los Angeles, California

SCHERY, STEPHEN D., Department of Physics, New Mexico Institute of Mining & Technology, Socorro, New Mexico

SICKLES, JOSEPH E., II, Centre for Environmental Measurements, Research Triangle Institute, Research Triangle Park, North Carolina

SINGH, HANWANT B., NASA Ames Research Center, Moffett Field, California

SISKOS, Laboratory of Analytical Chemistry, Department of Chemistry, University of Athens, Greece

SLANINA, J., Netherlands Energy Research Foundation ECN, ZG Petten, The Netherlands

v

VIRAS, LOISOS G., Directory of Air and Noise Pollution Control, Ministry of Environment, Physical Planning and Public Works, Athens, Greece

WILD, P. J., Netherlands Energy Research Foundation ECN, ZG Petten, The Netherlands

WYERS, G. P., Netherlands Energy Research Foundation ECN, ZG Petten, The Netherlands

ZIMMERMAN, PATRICK B., National Center for Atmospheric Research, Boulder, Colorado

# INTRODUCTION
# TO THE SERIES

The deterioration of environmental quality, which began when mankind first congregated into villages, has existed as a serious problem since the industrial revolution. In the second half of the twentieth century, under the ever increasing impacts of exponentially growing population and of industrializing society, environmental contamination of the air, water, soil, and food has become a threat to the continued existence of many plant and animal communities of various ecosystems and may ultimately threaten the very survival of the human race. Understandably, many scientific, industrial, and governmental communities have recently committed large resources of money and human power to the problems of environmental pollution and pollution abatement by effective control measures.

*Advances in Environmental Sciences and Technology* deals with creative reviews and critical assessments of all studies pertaining to the quality of the environment and to the technology of its conservation. The volumes published in the series are expected to service several objectives: (1) stimulate interdisciplinary cooperation and understanding among the environmental scientists; (2) provide the scientists with a periodic overview of environmental developments that are of general concern or that are of relevance to their own work or interests; (3) provide the graduate student with a critical assessment of past accomplishment, which may help stimulate him or her toward the career opportunities in this vital area; and (4) provide the research manager and the legislative or administrative official with an assured awareness of newly developing research work on the critical pollutants and with the background information important to their responsibility.

As the skills and techniques of many scientific disciplines are brought to bear on the fundamental and applied aspects of the environmental issues, there is a heightened need to draw together the numerous threads and to present a coherent picture of the various research endeavors. This need and the recent tremendous growth in the field of environmental studies have clearly made some editorial adjustments necessary. Apart from the changes in style and format, each future volume in the series will focus on one particular theme or timely topic, starting with Volume 12. The author(s) of each pertinent section will be expected to critically review the literature and the most important recent developments in the particular field; to critically evaluate new concepts, methods, and data; and to focus attention on important unresolved or controversial questions and on probable future trends. Monographs embodying the results of unusually extensive and well-rounded investigations will also be published in the series. The net result of the new editorial policy should be more integrative and comprehensive volumes on key environmental issues and pollutants. Indeed, the development of realistic standards of environmental quality for many pollutants often entails such a holistic treatment.

JEROME O. NRIAGU, Series Editor

# PREFACE

At the onset of the environmental movement, many countries established ambient air quality standards designed to mitigate the conventional problems associated with the release of particulates, sulfur, nitrogen, and carbon oxides from automobiles and industrial point sources. The legislative and technological efforts have generally been successful in reducing the problems that they were set up to address in so far as the air in many urban areas of the developed countries is now perceived to be less polluted. Since the late 1970s, however, concern has shifted to the regional and global effects of the common pollutants and to the ever-increasing number of trace contaminants being emitted to the atmosphere. Besides the numerous inorganic constituents, the 1000 or so organic pollutants that have been identified in the atmosphere represent only a tiny fraction of what potentially can be there. Acid rain, the greenhouse effect, depletion of the ozone layer, and the Arctic haze have become familiar environmental problems linked to longer-range transport of gaseous pollutants with either large sources or long lifetime in the troposphere. Despite many years of research, the picture on gaseous pollutants in the atmosphere remains hazy because of the complexity of the sources, the large variety of species involved, and the lack of suitable analytical techniques. Our understanding of the potential effects of the gaseous contaminants on terrestrial and aquatic ecosystems is further constrained by the inadequacy of the measurement protocols.

This volume deals with the conventional and sophisticated techniques used in the sampling and analysis of gaseous contaminants in the atmosphere. A book on the theory and practice of air analysis would be of limited use to many environmental scientists. The purposes of the measurements in terms of material fluxes and the understanding of chemical and physical processes in the atmosphere have therefore been emphasized. The sources, occurrence, and characterization of volatile organic compounds in the indoor environment have also been reviewed authoritatively in the book, since this topic has become a public health issue. It should be noted that a comprehensive coverage of the available information on all gaseous pollutants is neither possible nor intended. Rather, this volume evaluates the present state of the measurement technology, provides a flavor of the current research results, and draws attention to the challenging research opportunities available in this important field of study. The chapters should be of particular use to scientists and engineers engaged in the monitoring of air pollutants, atmospheric chemists interested in deciphering the behavior and fate of atmospheric constituents, and regulatory agencies concerned with air pollution control strategies. The book itself should be read by anyone concerned about the quality of the air we breathe.

JEROME O. NRIAGU

*Burlington, Ontario, Canada*
*February 1992*

# CONTENTS

# 1

# HENRY'S LAW CONSTANTS OF SOLUBLE AND MODERATELY SOLUBLE ORGANIC GASES: EFFECTS ON AQUEOUS PHASE CHEMISTRY

*E. A. Betterton*

*Department of Atmospheric Sciences, University of Arizona, Tucson, Arizona*

*Gaseous Pollutants: Characterization and Cycling,* Edited by Jerome O. Nriagu.
ISBN 0-471-54898-7 © 1992 John Wiley & Sons, Inc.

## 1.  INTRODUCTION

The exchange of chemical compounds between the gas—and aqueous—phases is an important flux process in the environment for a wide range of inorganic and organic species. Often the compound, referred to here as a solute, is emitted directly into one phase but then partitions into a second phase where it is exposed to a different set of conditions that determine its environmental fate. For example, $SO_2$ is commonly emitted into the atmosphere where a major loss process is oxidation by the hydroxyl radical, $OH_g^.$. However, $SO_2$ is very soluble in most cloud and fog droplets where the major oxidants are dissolved $H_2O_2$, $O_3$, and $O_2$ (in the presence of trace metal catalysts). The lifetime of $SO_2$ with respect to oxidation by $OH_g^.$ is approximately 10 days, but in the aqueous phase its lifetime may be as short as a few hours depending on pH and oxidant concentration. It is therefore important to be able to predict the rate, direction, and extent of phase transfer in order to be able to assess the fate and impact of various solutes on the environment. All of these processes are affected by the Henry's law constant, which therefore becomes an important parameter in modeling the fate of natural and anthropogenic species.

A two-film model is commonly used to describe the exchange of gases between the atmosphere and water bodies (Stumm and Morgan, 1981; Liss and Slater, 1974) (although other models such as the surface renewal (Danckwerts, 1951) and penetration models (Higbie, 1935) are also available). The two-film model invokes two thin stagnant fluid films, one on each side of the phase boundary. In the bulk, the two fluids are turbulent and well mixed to within a small distance of the interface (see Fig. 1). All resistance to phase transfer is assumed to take place in one or both films where transport occurs solely by molecular diffusion. From Fick's first law the steady flux of solute across the gas film into the liquid is given by

$$F_g = \frac{D_g}{\delta_g} \Delta C \tag{1}$$

where $F_g$ is the gas phase flux (with units such as mol cm$^{-2}$ s$^{-1}$), $D_g$ is the gas phase diffusion coefficient, $\delta_g$ is the thickness of the gas film, and $\Delta C$ is the concentration gradient across the distance $\delta_g$. For a steady flux, $F = F_g = F_L$, where $F_L$ is the flux into the liquid phase, and, therefore,

$$F = k_g(C_g - C_g^i) = k_L(C_L^i - C_L) \tag{2}$$

where $k_g = D_g/\delta_g$, $k_L = D_L/\delta_L$, $C_g$ and $C_L$ are the bulk concentrations in the gas and liquid phases, respectively, and $C_g^i$ and $C_L^i$ are the respective interfacial concentrations. The constants $k_g$ and $k_L$ have units of velocity (e.g., cm s$^{-1}$) and are known as exchange rate constants or mass transfer coefficients. The troublesome interfacial concentrations are eliminated using Hen-

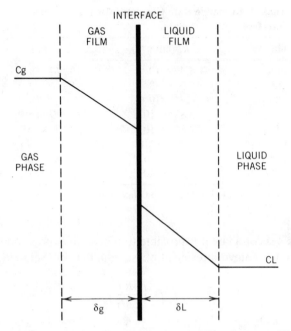

**Figure 1.** Two-film model for the transfer of solutes between the gas and liquid phases.

ry's Law (discussed in detail below), $C_g^i/C_L^i = H$, and, therefore,

$$F = K_L\left(\frac{C_g}{H} - C_L\right) \tag{3}$$

where $K_L = 1/k_L + 1/Hk_g$. Often the overall mass transfer coefficient $K_L$ is written as the sum of two resistances in series,

$$R_L = r_L + r_g \tag{4}$$

Similar expressions can be derived for the overall mass transfer coefficient based on the gas phase: $K_g = 1/k_g + H/k_L$. When the liquid film resistance is large, that is, when $r_L \gg r_g$, the flux is controlled by the small liquid exchange rate constant, while when $r_g \gg r_L$, the flux is limited by the small product $HK_g$ and is said to be gas phase controlled. In short, the flux of soluble gases (small $H$) is limited by gas film resistance, whereas the flux of insoluble gases (large $H$) is limited by liquid film resistance. It is difficult to obtain numerical values of the individual exchange rate constants but Liss and Slater (1974) were able to obtain estimates for the gases shown in Table 1 (for the atmosphere–ocean interface). Based on this work, Mackay and Leinonen (1975) have suggested that $r_g \approx r_L$ when $H \approx 1.6 \times 10^{-4}$ atm m$^3$ mol$^{-1}$ (6.25 M atm$^{-1}$).

**Table 1  Exchange Rate Constants for the Air–Sea Interface**

| Gas | $k_g$ (cm s$^{-1}$) | $k_L$ (cm s$^{-1}$) |
|---|---|---|
| SO$_2$ | $4.4 \times 10^{-1}$ | 9.561 |
| N$_2$O | $5.3 \times 10^{-1}$ | $5.6 \times 10^{-3}$ |
| CO | $6.7 \times 10^{-1}$ | $5.6 \times 10^{-3}$ |
| CH$_4$ | $8.83 \times 10^{-1}$ | $5.6 \times 10^{-3}$ |
| CCl$_4$ | $2.86 \times 10^{-1}$ | $2.97 \times 10^{-3}$ |
| CCl$_3$F | $3.013 \times 10^{-1}$ | $3.14 \times 10^{-3}$ |
| CH$_3$I | $2.97 \times 10^{-1}$ | $3.08 \times 10^{-3}$ |
| (CH$_3$)$_2$S | $4.50 \times 10^{-1}$ | $5.6 \times 10^{-3}$ |

*Source.* Adapted from Liss and Slater (1974).

Using Fick's second law it is possible to derive an expression for the characteristic time to achieve interfacial phase equilibrium (Seinfeld, 1986):

$$\tau = D_L \left( \frac{4RT}{\alpha \bar{c} H} \right)^2 \tag{5}$$

where $\tau$ is the characteristic time, $R$ is the gas constant, $\alpha$ is the accommodation coefficient, and $\bar{c}$ is the average speed of gas molecules, which is given by the kinetic theory of gases:

$$\bar{c} = \left( \frac{8RT}{\pi M} \right)^{0.5} \tag{6}$$

where $M$ is the molecular weight of the gas. For many solutes $D_L \approx 10^{-5}$ cm$^2$ s$^{-1}$ and so, assuming that $\alpha = 1$ and $T = 298$ K, characteristic times to achieve interfacial equilibrium can be calculated. Table 2 shows the results for several important atmospheric gases. It can be seen that the characteristic times are strongly dependent on the Henry's law constant. Very soluble gases require longer to establish equilibrium because more gas is required to cross the interface to achieve equilibrium.

This brief introduction should make it clear that the Henry's law constant is a key parameter is estimating the magnitude, rate, and direction of the flux of solutes between the gas and aqueous phases.

## 2.  HENRY'S LAW

### 2.1.  Background

The French chemist François Raoult found in a series of experiments with mixtures of closely related liquids such as benzene and toluene that the vapor

**Table 2 Characteristic Times Required to Achieve Interfacial Equilibrium**

| Solute | $H$ (atm $M^{-1}$) | $\tau$ (s) |
|---|---|---|
| $HNO_3$[a] | $4.8 \times 10^{-7}$ | $5.7 \times 10^2$ |
| $H_2O_2$ | $1.4 \times 10^{-5}$ | $2.6 \times 10^{-1}$ |
| $HCHO$[a] | $1.6 \times 10^{-4}$ | $1.8 \times 10^{-3}$ |
| $SO_2$ | $8.1 \times 10^{-1}$ | $1.5 \times 10^{-10}$ |
| $O_3$ | $1.1 \times 10^2$ | $6.4 \times 10^{-15}$ |

[a]This (effective) Henry's law constant includes the enhancement due to chemical reaction in water.

*Source*. Adapted from Seinfeld (1986).

pressure of a given component, A, is proportional to the mole fraction of that component in the liquid mixture. This is known as Raoult's law:

$$P_A = X_A P_A^\circ \tag{7}$$

$P_A$ = vapor pressure of component A above the liquid mixture, $P_A^\circ$ = vapor pressure of the pure liquid A, and $X_A$ = mole fraction of A in the liquid mixture. Equation 7 is based on experimental observation and states that a straight-line plot of $P_A$ against $X_A$ has a slope that can be identified as being equal to the vapor pressure of pure A. Mixtures of two components A and B that are chemically similar often obey Raoult's law very well; those that obey the law from a composition of pure A through to pure B are known as ideal solutions. For mixtures that deviate strongly from Raoult's law (e.g., a mixture of 2-propanol and water (Pierotti et al., 1959)) and no longer show a single straight-line relationship, it is found that the deviation becomes smaller as the mole fraction of the *solvent* approaches unity. Therefore, Raoult's law generally holds for the solvent (even if it is not an ideal solution) as long as it is nearly pure.

For ideal solutions, the vapor pressure of the *solute* is also governed by Raoult's law, but for dilute real solutions it is found that although there is a straight-line relationship between $P_A$ and $X_A$, the slope of the line is no longer equal to $P_A^\circ$. This new behavior was discovered experimentally by the English chemist William Henry and is known as Henry's law:

$$P_A = X_A H_A \tag{8}$$

where $P_A$ and $X_A$ have the same definitions as previously, and $H_A$ is the Henry's law constant. Henry's law states that there is a straight-line relationship between the partial pressure of a gas and its concentration in dilute solution, and that the line has a slope of $H_A$. (As will be seen later, $H_A$ can be identified as the product of $P_A^\circ$ and an activity coefficient.) We assume for the present that the solute does not interact chemically in solution and

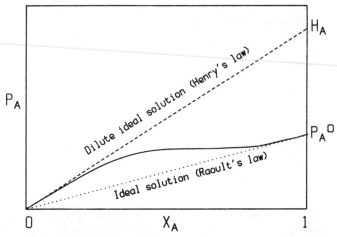

**Figure 2.** The relationship between Raoult's law, which is valid for all ideal solutions, and Henry's law, which is valid for dilute ideal solutions. Component A follows Raoult's law when it is nearly pure solvent and it follows Henry's law when it is a solute at low concentration.

that we are therefore concerned only with the intrinsic, or physical, solubility of the solute. Later, chemical interactions will be taken into account to define an effective Henry's law constant.

The relationship between Raoult's law and Henry's law is shown in Figure 2. Raoult's law is obeyed for both the solvent and the solute in ideal solutions at any concentration, and also for dilute mixtures that are nearly pure solvent; but for the solute in dilute mixtures only Henry's law is obeyed. For real solutions that cannot be approximated as pure solvent or dilute solute, neither law necessarily holds. When considering the behavior of atmospheric gases at normal concentrations in equilibrium with hydrometeors or large natural water bodies such as lakes and oceans, Henry's law is generally accepted as being a good approximation.

Henry's law constants are reported in the literature in several different unit systems and this can give rise to a certain amount of confusion. From equation 8 it can be seen that the units involve the ratio of gas phase to aqueous phase concentrations (for the purposes of this discussion the liquid phase is always aqueous), and the units of $H_A$ therefore depend on those selected for $P_A$ and $X_A$. For example, if the gas phase concentration unit is atm and the aqueous phase concentration unit is mol $L^{-1}$, then $H_A$ has units of atm mol$^{-1}$ L. However, if $P_A$ is expressed as mol $L^{-1}$, then $H_A$ becomes dimensionless. The dimensionless Henry's law constant is also called the air/water (or water/air) partition coefficient. It should be noted that the gas phase concentration is always in the numerator for the present definition of $H_A$ and therefore highly soluble gases are characterized by numerically small $H_A$ values.

In atmospheric studies it is often convenient to consider the equilibrium of species A between the atmosphere and water, in which case one writes

$$A_g \rightleftharpoons A_{aq} \tag{9}$$

and the equilibrium constant is then, by convention, written as

$$K_{H,A} = \frac{[A_{aq}]}{[A_g]} \tag{10}$$

The aqueous phase concentration is now in the numerator and $K_{H,A}$ is the inverse of $H_A$. Both $H_A$ and $K_{H,A}$ are called Henry's law constants (or coefficients) and so it is necessary, when making use of published Henry's law data, to determine which relationship (equation 8 or 10) the author used to define Henry's law. $K_{H,A}$ is found in the literature in a variety of units but unlike $H_A$, the numerical value of $K_{H,A}$ is large for soluble gases and small for insoluble gases. For this reason we shall tabulate Henry's law constants as $K_{H,A}$ values. We choose the units of mol $L^{-1}$ atm$^{-1}$, that is, M atm$^{-1}$, because they are commonly used. The correct units in the SI system are mol $m^{-3}$ Pa$^{-1}$.

$K_{H,A}$ depends on temperature as any other equilibrium constant does. The van't Hoff expression, which shows the variation of the equilibrium constant with temperature at constant pressure, is

$$\frac{d(\ln K_{H,A})}{dT} = \frac{-\Delta H°}{RT^2} \tag{11}$$

Assuming that the standard enthalpy of the reaction, $\Delta H°$, is essentially independent of temperature, equation 11 can be integrated to give the useful expression

$$\ln\frac{(K_{H,A})_{T_2}}{(K_{H,A})_{T_1}} = \frac{\Delta H°}{R}\left(\frac{1}{T_1} - \frac{1}{T_2}\right) \tag{12}$$

Since nearly all reported $\Delta H°$ values for the process defined by equation 9 are negative, $K_{H,A}$ increases as the temperature decreases. Using the data of Table 2 is can readily be shown that $K_H^*$ for formaldehyde increases by nearly an order of magnitude from $2.97 \times 10^3$ M atm$^{-1}$ at 25°C to $2.71 \times 10^4$ M atm$^{-1}$ at 0°C.

Until now, we have implicitly dealt only with binary systems, that is, we have been concerned only with the behavior of a single solute A in a solvent B (water). Real systems, however, are rarely this simple since they usually contain many cosolutes. Multicomponent systems, especially in polar solvents, are difficult to adequately describe in thermodynamic terms, but, fortunately, it is not usually necessary to do so for environmentally relevant systems. Concentrations of individual solutes are generally so low that the solute molecules behave independently, collision with a solvent molecule being much more frequent than with another solute molecule. This is borne out in experimental studies such as those of Munz and Roberts (1982) who found no effect of added methanol on the Henry's law constants of chlorinated aliphatics over a mole fraction range of at least $0-1.8 \times 10^{-4}$ $CH_3OH$. Similarly, Bomben et al. (1973) found no effect of added methanol on the activity coefficients (which are directly proportional to the Henry's law constants) of 1-octanol or 2-heptanone with up to $10^{-2}$ mole fraction $CH_3OH$.

## 2.2. Experimental Determination of Henry's Law Constants

There are two commonly used techniques for the experimental determination of Henry's law constants, although others exist (Battino and Clever, 1964). They are the head-space and bubble column techniques. In the head-space method (see, for example, Ioffe and Vitenberg, 1984; Buttery et al., 1969, 1971; Friant and Suffet, 1979) a stirred solution of the solute in a sealed container is allowed to equilibrate with the head-space vapor at a fixed temperature, and a small sample of the head-space vapor is then analyzed. The aqueous phase is similarly analyzed, and the ratio of the aqueous to gas phase concentrations gives the Henry's law constant. The head-space technique is simple and accurate and requires only a small sample volume, but it is limited to solutes that are sufficiently volatile to provide a usable analytical signal in the gas phase. Gas chromatography is by far the most common analytical tool since it displays high sensitivity and good resolution, and the signal is often linear over many orders of magnitude. It is not always necessary to calibrate the gas chromatograph absolutely since it is the ratio of the gas to aqueous phase peak areas that is required (assuming that injection of aqueous and gas samples yields the same response). A potential source of error includes sorption losses to the syringe walls and needle, but techniques have been described by Kieckbusch and King (1979) and McNally and Grob (1985) that overcome these problems by continuous circulation of the head-space vapor through the syringe followed by needleless injection. Precision is reported to be better than 0.5%.

Henry's law constants can also be obtained through gas chromatography of solutes over thin aqueous films on conventional solid supports and subsequent analysis of retention volume data (Martin, 1961; Karger et al., 1971).

In the bubble column technique an inert carrier gas is bubbled through a dilute solution of the solute at a known flow rate (see, for example, Burnett, 1963; Mackay et al., 1979; Betterton and Hoffmann, 1988). The rate of loss of the solute from solution then yields the Henry's law constant (see below). The loss rate can be determined by directly analyzing the aqueous or gas phases, or, alternatively, the solute can be stripped from the carrier gas by passing it through a second bubble column containing a suitable stripping solution that is subsequently analyzed. If only a small volume of stripping solution is used this option provides a useful concentration step. The bubble column technique is more complicated than the head-space technique and it relies on accurate determination of the carrier gas flow rate, but it offers a degree of flexibility in the choice of solute concentration and analytical tools that the head-space technique does not. The possibility of unwanted physical loss of solute by carry over of aerosol particles generated by bubble bursting is an inherent weakness. The relative error of this technique is estimated to be ≈5% except for highly insoluble gases where the error may be as high as 10% due to sorption losses to the container walls (Mackay and Shiu, 1981).

A variation of the bubble column technique avoids the problem of aerosol carry over by continuously recirculating the test solution through a column that is loosely packed with porcelain saddles or some other inert packing, and passing the carrier gas through the packed column in a counter current direction. This experimental setup entirely avoids bubble formation yet provides a large surface area for gas–liquid exchange (Johnson, 1990).

The theory behind the bubble column technique has been described by Burnett (1963) and by Mackay et al. (1979), among others. For an isothermal, well-mixed system, a mass balance for the solute gives

$$\frac{-V\,dC}{dt} = \frac{P_A}{RT} = \frac{H_A CG}{RT} \tag{13}$$

where the carrier gas flow rate is $G$, and the (fixed) solution volume is $V$. Integration from $t_0$ to $t$ while the solute concentration varies from $C_0$ to $C$ yields

$$\ln\frac{C}{C_0} = -\left(\frac{H_A G}{VRT}\right)_t \tag{14}$$

A plot of $\ln(C/C_0)$ against time yields a straight line, and $H_A$ can be obtained from the slope.

The validity of equation 14 hinges on the assumption that the vapor that exits the column is in equilibrium with the aqueous phase, that is, that the bubble has sufficient time to equilibrate during its rise time. Burnett (1963) found that for aqueous acetone, a 3-cm-diameter sintered glass disk (porosity 20–30 μm) when used at carrier gas flow rates of 60–120 mL min$^{-1}$ produced a fully equilibrated exit vapor even though the liquid depth was only ≈2 cm; while Betterton and Hoffmann (1988) showed that for some simple aldehydes, the use of a medium porosity sintered glass frit resulted in an equilibrated vapor for liquid depths that ranged from 20 to 50 cm and carrier gas flow rates of 110–390 mL min$^{-1}$. For less soluble gases equilibrium is achieved only after the transfer of relatively large quantities of solute to the bubble and therefore equilibration takes longer. Mackay et al. (1979) studied aqueous benzene solutions ($K_H$ = 0.182 M atm$^{-1}$) and found that there was 80% approach to equilibrium for every 10-cm increase in liquid depth in their bubble column fitted with a medium porosity frit.

Following the treatment of Mackay et al. (1979), the partial pressure of the solute in the exit vapor is dependent on the interfacial area for mass transfer:

$$P_A = H_A C \left[ 1 - \exp\left( -\frac{K_L A R T}{G H_A} \right) \right] \qquad (15)$$

where $K_L$ is the overall liquid phase mass transfer coefficient and $A$ is the total interfacial area in the bubble column. As the exponential term in equation 15 tends to zero, so $P_A$ tends to the Henry's law equilibrium value. For $P_A$ to reach >99% of the equilibrium value, the term $\exp(-K_L A R T / G H_A)$ must be <0.01. A reasonable value of $K_L$ is 5.6 × 10$^{-3}$ cm s$^{-1}$ for a moderately soluble gas (Liss and Slater, 1974), and this allows one to estimate the minimum surface area required for different values of $H_A$ and $G$. As an example, when $H_A$ = 1 atm mol L$^{-1}$ and $G$ = 200 mL min$^{-1}$, the minimum value of $A$ required to achieve 99% approach to equilibrium is 11 cm$^2$, which corresponds to just 14 bubbles of 5 mm diameter. For a swarm of noninteracting, independently rising bubbles that are assumed to maintain a constant size, the total number of bubbles and hence the interfacial area increases with carrier gas flow rate; the ratio $A/G$ therefore remains fairly constant for a given experimental setup. The magnitude of $A$ is varied by changing the liquid depth.

A third technique for determining Henry's law constants employs microporous membranes. The carrier gas is passed at a known flow rate through a microporous membrane tube that is immersed in a dilute thermostatted solution of known concentration. At the low flow rates normally employed, the solute equilibrates with the carrier gas across the membrane, and the carrier gas is then analyzed directly or stripped of the solute using a stripping solution. Dasgupta has successfully used this approach to accurately determine the Henry's law constants of formaldehyde (Dong and Dasgupta, 1986) and ammonia (Dasgupta and Dong, 1986).

## 2.3.  Calculation of Henry's Law Constants Through Thermodynamic Cycles

In many instances an experimentally determined Henry's law constant is unavailable for the solute of interest. In this case a thermodynamic calculation will sometimes yield the desired result as shown by Schwartz (1984) for HO$_2^{\cdot}$. As an example, take the case of formic acid for which no experimentally determined Henry's law constant has been published. We write the following equations which represent a thermodynamic cycle. The thermodynamic data are taken from Bard et al. (1985).

$$HCOOH_g \rightleftharpoons H_{2g} + CO_{2g} \qquad \Delta G° = -58.68 \text{ kJ/mol} \qquad (16)$$

$$H_{2g} \rightleftharpoons 2H_{aq}^+ + 2e^- \qquad \Delta G° = 0 \qquad (17)$$

$$CO_{2g} + 2H_{aq}^+ + 2e^- \rightleftharpoons HCOOH_{aq} \qquad \Delta G° = +38.4 \text{ kJ/mol} \qquad (18)$$

$$HCOOH_g \rightleftharpoons HCOOH_{aq} \qquad \Delta G° = -20.28 \text{ kJ/mol} \qquad (19)$$

$K_H$ is calculated from $K_H = \exp(-\Delta G°/RT)$, which yields $K_H = 3.6 \times 10^3$ M atm$^{-1}$ at 25°C. This calculated value is reasonably close to an experimentally determined (but unpublished) value of $7.6 \times 10^3$ M atm$^{-1}$ ($\approx$24°C) (Johnson, 1990). It should be noted that the calculation is sensitive to the accuracy of the thermochemical data. Similar calculations by Chameides and Davis (1983), Jacob (1986), and Keene and Galloway (1986) (using different sources for the thermodynamic data) yield $K_H = 3.7 \times 10^3$ M atm$^{-1}$ and 5.6 $\times 10^3$ M atm$^{-1}$, respectively.

The Henry's law constant of 56.4 M atm$^{-1}$ calculated for ammonia by Dasgupta and Dong (1986) is in good agreement with their measured value of 56.0 M atm$^{-1}$ (25°C).

## 2.4. Calculation of Henry's Law Constants from Vapor Pressure and Solubility Data

At equilibrium (at constant temperature and pressure) the fugacity of a solute in the liquid phase and the vapor phase are equal. The fugacity (from the Latin word *fugare*, meaning to flee) may be regarded as the tendency of a substance to escape from a phase (Mackay and Paterson, 1981, 1982; Mackay et al., 1983):

$$f_{V,A} = f_{L,A} \qquad (20)$$

Here, $f_{V,A}$ is the fugacity of the solute A in the gas phase and $f_{L,A}$ is the fugacity of the solute A in the liquid phase. We write

$$f_{L,A} = X_A \gamma_A P_A° \qquad (21)$$

while in the gas phase,

$$f_{V,A} = Y_A \phi_A P_T \qquad (22)$$

where $X_A$ is the solute mole fraction in the liquid phase, $Y_A$ is the solute mole fraction in the gas phase, $\gamma_A$ is the solute activity coefficient in the liquid phase ($\gamma_A = 1$ when $X_A = 1$), $\phi_A$ is the solute fugacity coefficient in the gas phase, $P_A°$ is the vapor pressure of pure solute A, and $P_T$ is the total system pressure (atmospheric pressure in the environment). For a nonassociating solute at the relatively low pressure of the atmosphere $\phi_A = 1$, and the gas phase fugacity therefore becomes equal to the partial pressure of solute, $P_A$:

$$f_{V,A} = Y_A P_T = P_A \qquad (23)$$

Substituting into equation 13 gives

$$P_A = X_A \gamma_A P_A^o \qquad (24)$$

Comparing equation 24 with equation 8 shows that

$$H_A = \gamma_A P_A^o \qquad (25)$$

and therefore $H_A$ can in principle be calculated from a knowledge of the vapor pressure of pure A and its activity coefficient in water. Both parameters are often available from tables. Activity coefficients cannot yet be reliably calculated from first principles using only a knowledge of molecular structure, so experimental measurements of $\gamma_A$ are often needed. Alternatively, several correlation methods exist that can be used to estimate $\gamma_A$ from such properties as molar volume (Leinonen et al., 1971), molecular surface area (Yalkowsky and Valvani, 1979), carbon number (Tsonopoulos and Prausnitz, 1971; Mackay and Shiu, 1977; Kabadi and Danner, 1979), and so on. Pierotti et al. (1959) tabulate semiempirical parameters based on molecular interactions that can be used to calculate the activity coefficients of homologous series of n-carboxylic acids, n-alcohols, n-carbonyls, n-ethers, n-acetals, n-nitriles, n-esters, n-paraffins, n-alkylbenzenes, and n-monoalkyl chlorides in water (and organic solvents).

A related method for calculating Henry's law constants relies on a knowledge of the vapor pressure of the pure solute and the solubility of the solute (in water) (Mackay and Shiu, 1981). In equation 8, $P_A$ becomes the saturated vapor pressure of pure A and $X_A$ becomes the solubility of A in the solvent water. As an example, for benzene, vapor pressure = 0.127 atm, solubility = 1780 g m$^{-3}$, and therefore $K_{H,C6H6} = 0.18$ M atm$^{-1}$. It should be noted that this type of calculation is applicable only to solutes that are barely miscible with water. If a significant amount of water dissolves in the solute and dilutes it, then the solute vapor pressure becomes significantly lower than that of the pure solute. It should also be noted that it is difficult to accurately measure the solubility of a nearly insoluble compound and that great inaccuracies can therefore be introduced. It is also necessary to ensure that the vapor pressure and solubility data are for the same phase (solid or liquid) of the solute. It is erroneous, for example, to use liquid vapor pressure data with solid solubility data (Mackay and Shiu, 1981).

## 2.5.  Using Linear Correlations to Obtain Henry's Law Constants

A third method for obtaining missing Henry's law constants is to use correlations of experimentally determined values with some property of a closely related set of molecules. For example, simple correlations of log $H_A$ with carbon number for homologous series of alkanols, alkan-2-ones, alkanals, or methyl alkanoates are linear from $C_2$ to $C_9$, as can be seen in Fig. 3 (Buttery et al., 1969). Other simple correlations for nonhomologous series of aldehydes and ketones based on the calculated Taft $\sigma^*$ parameter of the molecule ($\sigma^*$ is a measure of the ability of a substituent to withdraw electron density from the carbonyl group) show good linearity but also show that anomalies can be expected as can be seen in Fig. 4 (Betterton and Hoffmann, 1988).

More sophisticated correlations, such as those of Hine and Mookerjee (1975) and the UNIFAC method of Gmehling et al. (1982), also are available. Hine and Mookerjee (1975) provide bond contribution and group contribution data that have been independently applied to predict activity coefficients of nearly 30 organic compounds including hydrocarbons, halocarbons, ethers, sulfides, alcohols, mercaptans, phenols, thiophenols, aldehydes, ketones, carboxylic acids, esters, amines, nitriles, nitro compounds, pyridines, and pyrazines. However, it should be noted that the correlations can sometimes break down for the first member of the series. For acetaldehyde, the bond contribution correlation of Hine and Mookerjee (1975) is in reasonably good agreement with experiment ($1.8 \times 10^1$ versus $1.3 \times 10^1$ M atm$^{-1}$, respectively), but the agreement between calculation and experiment is not so good for the

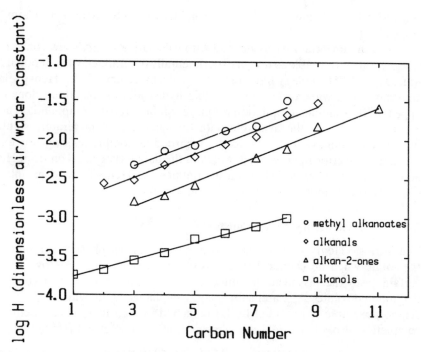

**Figure 3.** Correlation of log $H$ with carbon number for a series of esters, aldehydes, ketones, and alcohols. (Adapted from Buttery et al. (1969).)

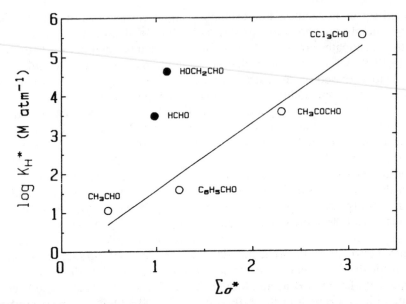

**Figure 4.** Correlation of log $K_H^*$ with Taft's $\sigma^*$ parameter, a measure of the electronegativity of the substituent R, for a series of aldehydes RCHO. Data points denoted by filled circles were not included in the linear least squares analysis. (Adapted from Betterton and Hoffmann (1988).)

(intrinsic) Henry's law constant of formaldehyde (9.8 versus 1.3 M atm$^{-1}$, respectively).

Given the uncertainties associated with calculating Henry's law constants, it is generally preferable to use an experimentally determined value if possible. Mackay and Shiu (1981) have recently critically reviewed the Henry's law constants for a wide variety of insoluble hydrocarbons, but there does not appear to be a similar compilation for more soluble organic compounds. Table 3 contains a list of the available data for soluble and moderately soluble organic compounds. It should not be regarded as comprehensive; nor have the data been critically reviewed. Much of the data has been taken from Hine and Mookerjee (1975). Table 3 also contains the Henry's law constants of some important inorganic solutes for comparison.

## 2.6.  Apparent Henry's Law Constant

The intrinsic Henry's law constant is a measure of the physical solubility of the solute when no chemical reactions occur after dissolution. However, in certain cases the dissolved gas undergoes chemical reactions that consume the aquated solute, $A_{aq}$, and shift the gas–aqueous equilibrium to the right. A good example is provided by formaldehyde which undergoes essentially complete hydrolysis to yield the *gem–diol*, methylene glycol (Bell, 1966):

$$HCHO_{aq} + H_2O \rightleftharpoons HCH(OH)_2 \tag{26}$$

Table 3 Henry's Law Constants of Some Soluble and Moderately Soluble Gases in Water at 25 °C (unless otherwise stated)

| Compound | Formula | $K_H$ (M atm$^{-1}$) | $\Delta H_{298}$ (kJ mol$^{-1}$) | Reference |
|---|---|---|---|---|
| Alcohols | | | | |
| 4-Bromophenol | $C_6H_4(OH)Br$ | $6.6 \times 10^3$ | | Parsons et al. (1971) |
| 1-Butanol | $C_4H_9OH$ | $1.1 \times 10^2$ | | Buttery et al. (1969), Butler et al. (1933) |
| | | $1.4 \times 10^2$ | | Burnett (1963) |
| | | $1.3 \times 10^2$ | $-61.9$ | Snider and Dawson (1985) |
| | | $5.48 \times 10^1$ (30 °C) | | Friant and Suffet (1979) |
| 2-Butanol | $C_2H_5CH(OH)CH_3$ | $1.1 \times 10^2$ | $-62.8$ | Snider and Dawson (1985) |
| 4-tert-Butylphenol | $C_6H_4(OH)C(CH_3)_3$ | $1.5 \times 10^1$ | | Buttery et al. (1969) |
| 2-Cresol | $C_6H_4(OH)CH_3$ | $8.2 \times 10^2$ | | Parsons et al. (1972) |
| 4-Cresol | $C_6H_4(OH)CH_3$ | $1.3 \times 10^3$ | | Parsons et al. (1972) |
| Cyclohexanol | $C_6H_{11}OH$ | $1.74 \times 10^2$ | | Hansch et al. (1968) |
| 2,3-Dimethylbutanol | $CH_3CHCH_3CHCH_3CH_2OH$ | $3.0 \times 10^1$ | | Hansch et al. (1968) |
| Ethanol | $C_2H_5OH$ | $2.0 \times 10^2$ | | Buttery et al. (1971) |
| | | $1.9 \times 10^2$ | $-52.3$ | Snider and Dawson (1985) |
| | | $1.6 \times 10^2$ | | Hine and Weimar (1965) |
| Ethylene glycol | $HOCH_2CH_2OH$ | $1.7 \times 10^4$ | | Butler and Ramachandani (1935) |
| Glycerol | $CH_2OHCHOHCH_2OH$ | $5.8 \times 10^4$ | | Butler and Ramachandani (1935) |
| Hexafluoro-2-propanol | $CF_3CH(OH)CF_3$ | $2.4 \times 10^1$ | | Rochester and Symonds (1973) |
| 1-Heptanol | $C_7H_{15}OH$ | $5.4 \times 10^1$ | | Butler et al. (1935) |
| 1-Hexanol | $C_6H_{13}OH$ | $5.84 \times 10^1$ | | Hansch et al. (1968) |
| 3-Hexanol | $C_3H_7CH(OH)C_2H_5$ | $2.0 \times 10^1$ | | Hansch et al. (1968) |
| 2-Methyl-2-pentanol | $C_3H_7C(OH)(CH_3)CH_3$ | $3.1 \times 10^1$ | | Hansch et al. (1968) |

**Table 3** (*Continued*)

| Compound | Formula | $K_H$ (M atm$^{-1}$) | $\Delta H_{298}$ (kJ mol$^{-1}$) | Reference |
|---|---|---|---|---|
| 2-Methyl-3-pentanol | $C_2H_5CHOHCHCH_3CH_3$ | $2.9 \times 10^1$ | | Hansch et al. (1968) |
| 4-Methyl-2-pentanol | $CH_3CHCH_3CH_2CHOHCH_3$ | $2.2 \times 10^1$ | | Hansch et al. (1968) |
| 4-Nitrophenol | $C_6H_4(OH)NO_2$ | $2.6 \times 10^6$ | | Parsons et al. (1971) |
| 1-Octanol | $C_8H_{17}OH$ | $4.1 \times 10^1$ | | Buttery et al. (1969) |
| 1-Pentanol | $C_5H_{11}OH$ | $8.0 \times 10^1$ | | Butler et al. (1935) |
| 2-Pentanol | $C_3H_7CHOHCH_3$ | $6.8 \times 10^1$ | | Butler et al. (1935) |
| Phenol | $C_6H_5OH$ | $2.5 \times 10^3$ | | Parsons et al. (1971), Hine and Weimar (1965) |
| Methanol | $CH_3OH$ | $7.8 \times 10^{-2}$ | | Howe et al. (1987) |
| | | $2.2 \times 10^2$ | | Buttery et al. (1971), Butler et al. (1933) |
| | | $2.2 \times 10^2$ | $-44.8$ | Snider and Dawson (1985) |
| 2-Methyl-1-butanol | $C_2H_5CHCH_3CH_2OH$ | $7.1 \times 10^1$ | | Butler et al. (1935) |
| 2-Methyl-2-butanol | $C_2H_5C(OH)(CH_3)CH_3$ | $7.3 \times 10^1$ | | Butler et al. (1935) |
| 2-Methyl-1-propanol (*i*BuOH) | $(CH_3)_2CHCH_2OH$ | $1.0 \times 10^2$ | $-64.4$ | Snider and Dawson (1985) |
| 2-Methyl-2-propanol (*t*BuOH) | $(CH_3)_3COH$ | $6.9 \times 10^1$ | $-64.4$ | Snider and Dawson (1985) |
| 2,2,3,3,3-Pentafluoropropanol | $CF_3CF_2CH_2OH$ | $4.5 \times 10^1$ | | Rochester and Symonds (1973) |
| 1-Propanol | $C_3H_7OH$ | $1.5 \times 10^2$ | | Buttery et al. (1971) |
| | | $1.3 \times 10^2$ | $-57.7$ | Snider and Dawson (1985) |
| | | $1.6 \times 10^2$ | | Burnett (1963) |
| 2,2,3,3-Tetrafluoropropanol | $CHF_2CF_2CH_2OH$ | $1.59 \times 10^2$ | | Rochester and Symonds (1973) |

| | | | | |
|---|---|---|---|---|
| 1,1,1-Trifluoro-2-propanol | $CH_3CHOHCF_3$ | $4.6 \times 10^1$ | | Rochester and Symonds (1973) |
| Esters | | | | |
| Amyl acetate | $CH_3COOC_5H_7$ | 2.6 | | Deno and Berkheimer (1960) |
| Amyl propionate | $C_2H_5COOC_5H_7$ | 1.2 | | Deno and Berkheimer (1960) |
| Butyl acetate | $CH_3COOC_4H_9$ | 3.55 | −47.6 | Kieckbusch and King (1979) |
| Ethyl acetate | $CH_3COOC_2H_5$ | 7.4 | | Butler and Ramchandani (1935) |
| Ethyl butyrate | $C_3H_7COOC_2H_5$ | 2.8 | | Deno and Berkheimer (1960), Hansch et al. (1968) |
| Ethyl formate | $HCOOC_2H_5$ | 3.6 | | Hansch et al. (1968) |
| Ethyl heptanoate | $C_6H_{13}COOC_2H_5$ | 2.0 | | Deno and Berkheimer (1960), Hansch et al. (1968) |
| Ethyl pentanoate | $C_4H_9COOC_2H_5$ | 2.9 | | Deno and Berkheimer (1960), Hansch et al. (1968) |
| Ethyl propionate | $C_2H_5COOC_2H_5$ | 4.6 | | Deno and Berkheimer (1960) |
| Hexyl acetate | $CH_3COOC_6H_{13}$ | 1.9 | | Deno and Berkheimer (1960) |
| Isoamyl acetate | $CH_3COO(CH_2)_2CH(CH_3)_2$ | 1.7 | | Deno and Berkheimer (1960) |
| Isobutyl acetate | $CH_3COOCH_2CH(CH_3)_2$ | 2.2 | | Deno and Berkheimer (1960), Hansch et al. (1968) |

**Table 3** (*Continued*)

| Compound | Formula | $K_H$ (M atm$^{-1}$) | $\Delta H_{298}$ (kJ mol$^{-1}$) | Reference |
|---|---|---|---|---|
| Isoamyl formate | HCOO(CH$_2$)$_2$CH(CH$_3$)$_2$ | 1.5 | | Deno and Berkheimer (1960) |
| Isobutyl acetate | CH$_3$COOCH$_2$CH(CH$_3$)$_2$ | 2.2 | | Deno and Berkheimer (1960), Hansch et al. (1968) |
| Isobutyl formate | HCOOCH$_2$CH(CH$_3$)$_2$ | 1.7 | | Deno and Berkheimer (1960) |
| Isopropyl acetate | CH$_3$COOCH(CH$_3$)$_2$ | 3.6 | | Deno and Berkheimer (1960), Hansch et al. (1968) |
| Isopropyl formate | HCOOCH(CH$_3$)$_2$ | 1.2 | | Deno and Berkheimer (1960) |
| Isopropyl propionate | C$_2$H$_5$COOCH(CH$_3$)$_2$ | 1.7 | | Deno and Berkheimer (1960), Hansch et al. (1968) |
| Methyl acetate | CH$_3$COOCH$_3$ | 8.7 7.76 | $-39.2$ | Buttery et al. (1969) Kieckbusch and King (1979) |
| Methyl benzoate | C$_6$H$_5$COOCH$_3$ | $5.6 \times 10^1$ | | Butler and Ramchandani (1935) Deno and Berkheimer (1960) |
| Methyl formate | HCOOCH$_3$ | 4.5 | | Pecsar and Martin (1966) |
| Methyl propionate | C$_2$H$_5$COOCH$_3$ | 5.8 5.89 | $-41.4$ | Buttery et al. (1969) Kieckbusch and King (1979) |
| | | 6.2 | | Hansch et al. (1968) |

| Compound | Formula | | | Reference |
|---|---|---|---|---|
| Methyl butyrate | $C_3H_7COOCH_3$ | 4.9 | | Buttery et al. (1969) |
| Methyl pentanoate | $C_4H_9COOCH_3$ | 3.1 | | Buttery et al. (1969) |
| Methyl hexanoate | $C_5H_{11}COOCH_3$ | 2.7 | | Buttery et al. (1969) |
| Methyl octanoate | $C_7H_{15}COOCH_3$ | 1.3 | | Buttery et al. (1969) |
| Pentyl acetate | $CH_3COOC_5H_{11}$ | 2.81 | $-51.4$ | Kieckbusch and King (1979) |
| Propyl acetate | $CH_3COOC_3H_7$ | 4.59 | $-43.2$ | Kieckbusch and King (1979) |
| | | 5.0 | | Butler and Ramchandani, (1935) |
| | | | | Hansch et al. (1968) |
| Propyl butyrate | $C_3H_7COOC_3H_7$ | 1.9 | | Deno and Berkheimer (1960), Hansch et al. (1968) |
| Propyl formate | $HCOOC_3H_7$ | 2.7 | | Deno and Berkheimer (1960), Hansch et al. (1968) |
| Propyl propionate | $C_2H_5COOC_3H_7$ | 2.6 | | Deno and Berkheimer (1960), Hansch et al. (1968) |
| **Amines** | | | | |
| n-Butylamine | $C_4H_9NH_2$ | $6.6 \times 10^1$ | | Butler and Ramchandani (1935) |
| n-Dibutylamine | $(C_4H_9)_2NH$ | $1.1 \times 10^1$ | | Christie and Crisp (1967) |
| Diethylamine | $(C_2H_5)_2NH$ | $3.9 \times 10^1$ | | Christie and Crisp (1967) |
| Dimethylamine | $(CH_3)_2NH$ | $5.6 \times 10^1$ | | Christie and Crisp (1967) |
| n-Dipropylamine | $(C_3H_7)_2NH$ | $2.9 \times 10^1$ | | Christie and Crisp (1967) |
| Ethylamine | $C_2H_5NH_2$ | $9.8 \times 10^1$ | | Butler and Ramchandani (1935) |

**Table 3** (*Continued*)

| Compound | Formula | $K_H$ (M atm$^{-1}$) | $\Delta H_{298}$ (kJ mol$^{-1}$) | Reference |
|---|---|---|---|---|
| Ethylenediamine | $H_2NCH_2CH_2NH_2$ | $5.8 \times 10^5$ | | Westheimer and Ingraham (1956) |
| n-Hexylamine | $C_6H_{13}NH_2$ | $3.7 \times 10^1$ | | Christie and Crisp (1967) |
| N-Methylpiperidine | $C_5H_{10}NCH_3$ | $2.9 \times 10^1$ | | Cabani et al. (1971a) |
| N-Methylpyrrolidine | $C_4H_8NCH_3$ | $3.3 \times 10^1$ | | Cabani et al. (1971a) |
| n-Pentylamine | $C_5H_{11}NH_2$ | $4.1 \times 10^1$ | | Christie and Crisp (1967) |
| n-Propylamine | $C_3H_7NH_2$ | $8.2 \times 10^1$ | | Butler and Ramchandani (1935) |
| Triethylamine | $(C_2H_5)_3N$ | 6.8 | | Christie and Crisp (1967) |
| Trimethylamine | $(CH_3)_3N$ | 9.6 | | Christie and Crisp (1967) |
| **Nitrogen Heterocycles** | | | | |
| 2-Isobutyl-3-methoxypyrazine | $C_4N_2H_3(C_4H_9)OCH_3$ | $2.0 \times 10^1$ | | Buttery et al. (1971) |
| 2-Isobutylpyrazine | $C_4N_2H_3C_4H_9$ | $2.04 \times 10^2$ | | Buttery et al. (1971) |
| 2,3-Dimethylpyridine | $C_5H_3N(CH_3)_2$ | $1.39 \times 10^2$ | | Andon et al. (1954) |
| 2,4-Dimethylpyridine | $C_5H_3N(CH_3)_2$ | $1.48 \times 10^2$ | | Andon et al. (1954) |
| 2,5-Dimethylpyridine | $C_5H_3N(CH_3)_2$ | $1.15 \times 10^2$ | | Andon et al. (1954) |
| 2,6-Dimethylpyridine | $C_5H_3N(CH_3)_2$ | $9.6 \times 10^1$ | | Andon et al. (1954) |
| 3,4-Dimethylpyridine | $C_5H_3N(CH_3)_2$ | $2.70 \times 10^2$ | | Andon et al. (1954) |
| 3,5-Dimethylpyridine | $C_5H_3N(CH_3)_2$ | $1.46 \times 10^2$ | | Andon et al. (1954) |
| 2-Ethyl-3-methoxypyrazine | $C_4N_2H_3(C_2H_5)OCH_3$ | $6.8 \times 10^1$ | | Buttery et al. (1971) |
| 2-Ethylpyrazine | $C_4N_2H_3C_2H_5$ | $4.09 \times 10^2$ | | Buttery et al. (1971) |
| 2-Ethylpyridine | $C_5H_4NC_2H_5$ | $6.0 \times 10^1$ | | Andon et al. (1954) |
| 3-Ethylpyridine | $C_5H_4NC_2H_5$ | $9.6 \times 10^1$ | | Andon et al. (1954) |
| 4-Ethylpyridine | $C_5H_4NC_2H_5$ | $1.18 \times 10^2$ | | Andon et al. (1954) |

| | | | |
|---|---|---|---|
| 2-Methylpyrazine | $C_4N_2H_3CH_3$ | $4.54 \times 10^2$ | Buttery et al. (1971) |
| 2-Methylpyridine | $C_5H_4NCH_3$ | $1.00 \times 10^2$ | Andon et al (1954) |
| 3-Methylpyridine | $C_5H_4NCH_3$ | $1.29 \times 10^2$ | Andon et al. (1954) |
| 4-Methylpyridine | $C_5H_4NCH_3$ | $1.67 \times 10^2$ | Andon et al. (1954) |
| Piperidine | $C_5H_{10}NH$ | $2.25 \times 10^2$ | Cabani et al. (1971a) |
| Pyridine | $C_5H_5N$ | $1.13 \times 10^2$ | Andon et al (1954), Irmann (1965) |
| Pyrolidine | $C_4H_8NH$ | $4.18 \times 10^2$ | Cabani et al. (1971a) |

## Nitriles, Nitrates, Nitro Compounds

| | | | | |
|---|---|---|---|---|
| Acetonitrile | $CH_3CN$ | $4.91 \times 10^1$ | $-30.5$ | Snider and Dawson (1984) |
| n-Butyl nitrate | $C_4H_9ONO_2$ | $1.26 \ (\approx 22 \ ^\circ C)$ | | Luke et al. (1989) |
| 2-Butyl nitrate | $C_2H_5CH(ONO_2)CH_3$ | $7.4 \times 10^{-1} \ (\approx 22 \ ^\circ C)$ | | Luke et al. (1989) |
| Butyronitrile | $C_3H_7CN$ | $1.9 \times 10^1$ | | Butler and Ramchandani (1935) |
| Nitroethane | $C_2H_5NO_2$ | $1.4 \times 10^2 \ (30 \ ^\circ C)$ | | Friant and Suffet (1979) |
| Nitrobenzene | $C_6H_5NO_2$ | $4.3 \times 10^1$ | | Deno and Berkheimer (1960) |
| 1-Nitropropane | $C_2H_5CH_2NO_2$ | $1.15 \times 10^1$ | | Deno and Berkheimer (1960) |
| 2-Nitropropane | $CH_3CHNO_2CH_3$ | $8.2$ | | Deno and Berkheimer (1960) |
| 2-Nitrotoluene | $C_6H_4(NO_2)CH_3$ | $1.7 \times 10^1$ | | Deno and Berkheimer (1960) |
| 3-Nitrotoluene | $C_6H_4(NO_2)CH_3$ | $1.4 \times 10^1$ | | Deno and Berkheimer (1960) |
| Peroxyacetyl nitrate (PAN) | $CH_3C(O)OONO_2$ | $5$ | | Holdren et al. (1984) |
| Propionitrile | $C_2H_5CN$ | $2.7 \times 10^1$ | | Butler and Ramchandani (1935) |

**Table 3** (*Continued*)

| Compound | Formula | $K_H$ (M atm$^{-1}$) | $\Delta H_{298}$ (kJ mol$^{-1}$) | Reference |
|---|---|---|---|---|
| Aldehydes[a] | | | | |
| Acetaldehyde | $CH_3CHO$ | $1.14 \times 10^1$ | $-52.1$ | Betterton and Hoffmann (1988) |
| | | $1.3 \times 10^1$ | $-46.0$ | Snider and Dawson (1985) |
| | | $1.5 \times 10^1$ | | Buttery et al. (1969) |
| Acrolein | $CH_2{=}CHCHO$ | $7.4$ | $-40.6$ | Snider and Dawson (1985) |
| Benzaldehyde | $C_6H_5CHO$ | $3.74 \times 10^1$ | $-42.2$ | Betterton and Hoffmann (1988) |
| Butanal | $C_3H_7CHO$ | $8.7$ | | Buttery et al. (1969) |
| But-*trans*-2-enal | $C_2H_5CH{=}CHO$ | $5.1 \times 10^1$ | | Buttery et al. (1971) |
| Formaldehyde | $HCHO$ | $2.97 \times 10^3$ | $-59.8$ | Betterton and Hoffmann (1988) |
| | | $8.5 \times 10^3$ (20 °C) | | Blair and Ledbury (1925) |
| | | $3.16 \times 10^3$ | | Dong and Dasgupta (1986) |
| Glyoxal | $CHOCHO$ | $\geq 3 \times 10^5$ | | Betterton and Hoffmann (1988) |
| Heptanal | $C_6H_{13}CHO$ | $3.7$ | | Buttery et al. (1969) |
| Hexanal | $C_5H_{11}CHO$ | $4.7$ | | Buttery et al. (1969) |
| Hexa-*trans-trans*-2,4-dienal | $CH_3CH{=}(CH)_2{=}CHCHO$ | $1.02 \times 10^2$ | | Buttery et al (1971) |
| Hex-*trans*-2-enal | $C_3H_7CH{=}CHCHO$ | $2.0 \times 10^1$ | $-38.5$ | Buttery et al. (1971) |
| Hydroxy-acetaldehyde | $CH_2(OH)CHO$ | $4.14 \times 10^4$ | | Betterton and Hoffmann (1988) |
| 4-Hydroxy-benzaldehyde | $C_6H_4(OH)CHO$ | $2.0 \times 10^6$ | | Parsons et al (1971) |

| Compound | Formula | Value | | Reference |
|---|---|---|---|---|
| Methylglyoxal | $CH_3COCHO$ | $3.71 \times 10^5$ | $-62.7$ | Betterton and Hoffmann (1988) |
| Nonanal | $C_8H_{17}CHO$ | 1.4 | | Buttery et al. (1969) |
| Octanal | $C_7H_{15}CHO$ | 2.0 | | Buttery et al. (1969) |
| Oct-*trans*-2-enal | $C_5H_{11}CHCH{=}CHO$ | 4 | | Buttery et al. (1971) |
| Pentanal | $C_4H_9CHO$ | 6.8 | | Buttery et al. (1969) |
| Propanal | $C_2H_5CHO$ | $1.36 \times 10^1$ | | Buttery et al. (1969) |
| Trichloro-acetaldehyde | $CCl_3CHO$ | $3.44 \times 10^5$ | $-29.1$ | Betterton and Hoffmann (1988) |
| **Ketones** | | | | |
| Acetone | $CH_3COCH_3$ | $2.56 \times 10^1$ | | Buttery et al. (1969) |
| | | $2.58 \times 10^1$ | $-37.2$ | Snider and Dawson (1985) |
| | | $3.2 \times 10^1$ | $-48$ | Betterton (1991) |
| | | $2.8 \times 10^1$ | | Burnett (1963) |
| Acetophenone | $C_6H_5COCH_3$ | $1.10 \times 10^2$ (25.1 °C) | $-50$ | Betterton (1991) |
| 2,3-Butanedione (biacetyl) | $(CH_3CO)_2$ | $5.7 \times 10^1$ | | Snider and Dawson (1985) |
| 2-Butanone (MEK) | $C_2H_5COCH_3$ | $7.4 \times 10^1$ | $-47$ | Betterton (1991) |
| | | $1.05 \times 10^1$ (30 °C) | | Friant and Suffet (1979) |
| | | $2.2 \times 10^1$ | | Buttery et al. (1969) |
| | | 4.1–7.7 | $-45$ | Howe et al. (1987) |
| Chloroacetone | $CH_2ClCOCH_3$ | $5.9 \times 10^1$ | | Betterton (1991) |
| 2-Heptanone | $C_5H_{11}COCH_3$ | 6.9 | | Buttery et al. (1969) |
| 4-Methyl-2-pentanone (MIBK) | $(CH_3)_2CHCH_2COCH_3$ | 2.6–5.2 | | Howe et al. (1987) |
| 2-Nonanone | $C_7H_{15}COCH_3$ | 2.7 | | Buttery et al. (1969) |
| 2-Octanone | $C_6H_{13}COCH_3$ | 5.3 | | Buttery et al. (1969) |
| 2-Pentanone | $C_3H_7COCH_3$ | $1.57 \times 10^1$ | | Buttery et al. (1969) |
| 1,1,1-Trifluoroacetone | $CCl_3COCH_3$ | $1.38 \times 10^2$ (25.1 °C) | $-74$ | Betterton (1991) |

**Table 3** (*Continued*)

| Compound | Formula | $K_H$ (M atm$^{-1}$) | $\Delta H_{298}$ (kJ mol$^{-1}$) | Reference |
|---|---|---|---|---|
| 2-Undecanone | $C_9H_{19}COCH_3$ | 1.6 | | Buttery et al. (1969) |
| **Carboxylic Acids** | | | | |
| Acetic acid | $CH_3COOH$ | $8.8 \times 10^3$ | | Keene and Galloway (1986) |
| Formic acid | $HCOOH$ | $5.6 \times 10^3$ | | Keene and Galloway (1986) |
| | | $3.7 \times 10^3$ | | Chameides and Davis (1983), Jacob (1986) |
| | | $3.6 \times 10^3$ | | This work |
| | | $7.6 \times 10^3$ ($\approx$24 °C) | | Johnson (1990) |
| **Sulfur-Containing Compounds** | | | | |
| Diethyl sulfide | $C_2H_5SC_2H_5$ | $4.6 \times 10^{-1}$ | | Deno and Berkheimer (1960) |
| Dimethyl sulfide | $CH_3SCH_3$ | $5.6 \times 10^{-1}$ | | Dacey et al. (1984) |
| | | $5.5 \times 10^{-1}$ | | Hine and Weimar (1965) |
| Dimethyl sulfoxide | $(CH_3)_2SO$ | $1.4 \times 10^3$ | | Gmehling et al. (1981) |
| Ethanethiol | $C_2H_5SH$ | $3.6 \times 10^{-1}$ | | Deno and Berkheimer (1960) |
| Methanethiol | $CH_3SH$ | $3.3 \times 10^{-1}$ | | Hine and Weimar (1965) |
| Thioanisole | $C_6H_5SCH_3$ | 4.1 | | Hine and Weimar (1965) |
| Thiophenol | $C_6H_5SH$ | 3.0 | | Hine and Weimar (1965) |
| **Oxygen Heterocycles** | | | | |
| 1,4-Dioxane | $C_4O_2H_8$ | $1.47 \times 10^2$ (28 °C; 4M $Na_2SO_4$) | | Friant and Suffet (1979) |
| | | $2.05 \times 10^2$ | | Hine and Weimar (1965) |
| | | $2.21 \times 10^2$ | | Vitenburg et al. (1975) |
| | | $2.1 \times 10^2$ | | Cabani et al. (1971b) |

Inorganic Gases

| | | | |
|---|---|---|---|
| Ammonia | $NH_3$ | $6.2 \times 10^1$ | Van Krevelen et al. (1949) |
| | | $\exp(4092/T - 9.7)$ | Dasgupta and Dong (1986) |
| Hydrogen bromide | HBr | $7.6 \times 10^1$ | Hales and Drewes (1979) |
| | | $1.32 \times 10^9$ | Brimblecombe and Clegg (1989) |
| Hydrogen chloride | HCl | $2.04 \times 10^6$ | Brimblecombe and Clegg (1989) |
| Hydrogen fluoride | HF | $9.6$ | Brimblecombe and Clegg (1989) |
| Hydrogen iodide | HI | $2.5 \times 10^9$ | Brimblecombe and Clegg (1989) |
| Hydrogen peroxide | $H_2O_2$ | $1.42 \times 10^5$ (20 °C) | Yoshizumi et al. (1984) |
| | | $0.7–1.0 \times 10^5$ | Martin and Damschen (1981) |
| | | $6.9 \times 10^4$ | Hwang and Dasgupta (1985) |
| Hydrogen sulfide | $H_2S$ | $1.0 \times 10^{-1}$ | Stumm and Morgan (1981) |
| Hydroperoxyl radical | $HO_2$ | $1–3 \times 10^3$ | Schwartz (1984) |
| Nitric acid | $HNO_3$ | $2.1 \times 10^5$ | Schwartz and White (1981) |
| | | $2.45 \times 10^6$ | Brimblecombe and Clegg (1989) |
| Nitrous acid | $HNO_2$ | $4.9 \times 10^1$ | Schwartz and White (1981) |
| Sulfur dioxide | $SO_2$ | $1.24$ | Maahs (1982) |

[a]These are apparent Henry's law constants and include the hydration constant.

25

The hydration constant is defined as

$$K_{HYD,HCHO} = \frac{[HCH(OH)_2]}{[HCHO]_{aq}} = 2 \times 10^3 \tag{27}$$

The apparent (or effective) Henry's law constant $K_H^*$ is defined as the ratio of the concentration of total dissolved solute to the gas phase concentration:

$$K_{H,HCHO}^* = \frac{[HCHO]_{aq} + [HCH(OH)_2]}{[HCHO]_g} \tag{28}$$

while, as we have seen, the intrinsic Henry's law constant is given by

$$K_{H,HCHO} = \frac{[HCHO]_{aq}}{[HCHO]_g} \tag{29}$$

The two Henry's law constants are therefore related by

$$K_{H,HCHO}^* = K_{H,HCHO}(1 + K_{HYD,HCHO}) \tag{30}$$

If $K_{HYD} \ll 1$, then $K_H^* \approx K_H$ and the distinction between the two constants becomes somewhat superfluous from a practical point of view; but if $K_{HYD} \gg 1$, as is the case with formaldehyde, then the distinction between the intrinsic Henry's law constant $K_H$ and the apparent Henry's law constant $K_H^*$ is crucial. Since $K_{HYD} = 2 \times 10^3$ for formaldehyde (Bell, 1966), $K_H^*$ is nearly three orders of magnitude larger than $K_H$ in this case. An experimentally determined Henry's law constant is invariably the apparent Henry's law constant, which may or may not be distinguishable from the intrinsic Henry's law constant depending on the chemical nature of the solute. Other chemical reactions can further complicate the expression for $K_H^*$, and again formaldehyde serves as a useful example.

Formaldehyde forms a stable bisulfite adduct, hydroxymethanesulfonate (HMSA), in the presence of $SO_{2aq}$ (Deister et al., 1986), and this provides an example of a ternary system that requires the introduction of an additional term in the expression for $K_H^*$ when $SO_2$ is present (Munger et al., 1984):

$$K_{H,HCHO}^* = K_{H,HCHO}(1 + K_{HYD,HCHO})\left(1 + \frac{P_{SO2}K_{H,SO2}K_{HSO3}K_{a1}}{[H^+]}\right) \tag{31}$$

where $P_{SO2}$ is the partial pressure of $SO_2$, $K_{HSO3}$ is the stability constant for reaction of bisulfite with formaldehyde ($[CH_2OHSO_3^-]/[HSO_3^-][HCHO]_T$) and $K_{a1}$ is the first acid dissociation constant of $H_2O \cdot SO_2$.

Solutes that undergo acid/base reactions in solution also show an apparent Henry's law constant. Taking oxalic acid as a diprotic example, we find

$$K_H = \frac{[(COOH)_2]_{aq}}{[(COOH)_2]_g} \tag{32}$$

$$K_{a1} = \frac{[H^+][COOHCOO^-]}{[(COOH)_2]_{aq}} \tag{33}$$

$$K_{a2} = \frac{[H^+][(COO^-)_2]}{[COOHCOO^-]} \tag{34}$$

We define the total dissolved oxalic acid to be $[(COOH)_2]_T$, and then

$$[(COOH)_2]_T = [(COOH)_2]_{aq} + [COOHCOO^-] + [(COO^-)_2] \tag{35}$$

and

$$[(COOH)_2]_T = K_H \left\{ 1 + \frac{K_{a1}}{[H^+]} + \frac{K_{a1}K_{a2}}{[H^+]^2} \right\} [(COOH)_2]_g \tag{36}$$

or

$$[(COOH)_2]_T = K_H^*[(COOH)_2]_g \tag{37}$$

where

$$K_H^* = K_H \left\{ 1 + \frac{K_{a1}}{[H^+]} + \frac{K_{a1}K_{a2}}{[H^+]^2} \right\} \tag{38}$$

The similarity between equations 30 and 38 is self-evident; in the case of the latter, however, a pH dependence has been introduced. For monoprotic acids the term involving $K_{a2}$ is dropped.

### 2.7. What Constitutes a Soluble Gas?

It is impossible to provide a definition for a soluble gas that is universally applicable, since a gas may be regarded as soluble in one context but insoluble in another. For example, acetone ($K_H^* = 25.6$ M atm$^{-1}$) is very soluble compared to most halogenated aliphatics, pesticides, and so on, but it is insoluble compared to formaldehyde ($K_H^* = 2.97 \times 10^3$ M atm$^{-1}$). Seinfeld (1986) provides a useful bench mark for Henry's law constants with respect to the distribution of a solute between the gas and aqueous phases in a liquid water cloud. The ratio of the molar concentration of solute A in the two phases per unit volume of air is

$$\frac{[A_{liquid}]}{[A_{gas}]} = \frac{K_{H,A}P_A L}{P_A/RT} = K_{H,A}RTL \tag{39}$$

where $L$ is the dimensionless liquid water content of the cloud (e.g., $m^3$ liquid water/$m^3$ air). $K_{H,A}$ is replaced by $K^*_{H,A}$ for those solutes that undergo chemical reactions in solution. If $K_{H,A} \ll (RTL)^{-1}$, then species A will be found mainly in the gas phase. A typical value for $L$ is $10^{-6}$ and therefore $(RTL)^{-1}$ is approximately $10^4$ M atm$^{-1}$. This implies that those species with $K_H \ll 10^4$ M atm$^{-1}$ will partition strongly to the gas phase and could reasonably be regarded as being insoluble in the cloud water, while species with $K_H \gg 10^4$ M atm$^{-1}$ will be highly soluble. As can be seen from Table 3, remarkably few gases meet this criterion even when the effective Henry's law constant is taken into account. This does not imply, however, that only those gases with $K_{H,A}$ (or $K^*_{H,A}$) $\geq 10^4$ M atm$^{-1}$ play an important role in cloud water chemistry.

Farmer and Dawson's (1982) definition of solubility is even more stringent for the collection of atmospheric gases by co-condensation. By their definition, a poorly soluble gas has $K_H < 10^4$ M atm$^{-1}$, a moderately soluble gas has $K_H \approx 10^4$ to $10^5$ M atm$^{-1}$, and a highly soluble gas has $K_H > 10^5$ M atm$^{-1}$.

For the purposes of this discussion, however, any gas that is more soluble than those in Mackay and Shiu's (1981) compilation, that is, greater than $\approx 0.1$ M atm$^{-1}$, is regarded as being moderately soluble while those with $K_H$ greater than $\approx 10^4$ M atm$^{-1}$ are regarded as being highly soluble.

## 3.   EFFECTS ON AQUEOUS PHASE CHEMISTRY

### 3.1.   Alcohols and Esters

There are numerous natural and anthropogenic sources of alcohols, including starch fermenting bacteria (Cavanagh et al., 1969), plant emissions, forest fires, hydrocarbon photooxidation products, and auto emissions. Graedel (1978) lists some 150 alkyl, olefinic, cyclic, and aromatic alcohols that have been identified in the atmosphere. Methanol and ethanol are often the most abundant alcohols in both urban (0.5–10 ppbv) and rural atmospheres (0.5–5 ppbv). Snider and Dawson (1985) identified $C_1$ to $C_4$ aliphatic alcohols in rural and urban atmospheres. They also found methanol, ethanol, and $n$-butanol in precipitation samples (0.9–22 ppbm). The introduction of methanol- and ethanol-containing automobile fuels is likely to increase atmospheric alcohol (and aldehyde) concentrations.

The moderate solubility of the lower alcohols ($K_H \approx 2 \times 10^2$ M atm$^{-1}$) implies that wet deposition is probably not a large sink. $OH^._g$ oxidation is likely to be the most important loss mechanism, the lifetime of ethanol and methanol being approximately 4 and 10 days, respectively, for this process (Finlayson-Pitts and Pitts, 1986). In solution, oxidation by $OH^._{aq}$ yields aldehydes or ketones (Graedel and Weschler, 1981).

Approximately 100 esters have been identified in the atmosphere (Graedel, 1978). Most of them probably originate from plants, while methyl formate and methyl and ethyl acetate have small anthropogenic sources as well. The ubiquitous phthalate esters such as di-2(ethylhexyl)-phthalate (DEHP), di-*n*-butylphthalate (DBP), and diisodecyl phthalate, are common plasticizers in PVC resins, while the carbamic acid esters 2-isopropylphenyl *N*-methylcarbamate (Baygon) and 1-naphthyl *N*-methylcarbamate (Carbaryl), and the organophosphate esters parathion and methyl parathion are used as insecticides.

The low $K_H$ of most esters probably precludes wet deposition from being a large sink although atmospheric deposition of DEHP and DBP to the Great Lakes is estimated to be 100 metric tons per year (Eisenreich et al., 1981).

Rapid hydrolysis leads to the corresponding carboxylic acid and alcohol and could markedly increase $K_H^*$:

$$
\begin{array}{c}
\text{O} \\
\parallel \\
\text{RCOR}' + \text{H}_2\text{O}
\end{array}
\underset{}{\overset{\text{H}^+ \text{ or OH}^-}{\rightleftharpoons}}
\text{RCOOH} + \text{R}'\text{OH}
\tag{40}
$$

The reaction is catalyzed by acids and bases (both general and specific), and by trace metals (see, for example, Bender, 1963; Buckingham, 1977; Martell, 1963). A general rate law is

$$
\frac{-d[\text{ester}]}{dt} = k_{\text{obsd}}[\text{ester}]
\tag{41}
$$

where

$$
k_{\text{obsd}} = k_{\text{H2O}}[\text{H}_2\text{O}] + k_{\text{H}^+}[\text{H}^+] + k_{\text{OH}^-}[\text{OH}^-]
$$
$$
+ k_A[\text{A}] + k_B[\text{B}] + \frac{k_M K_{a1}[\text{M}]_T}{[\text{H}^+] + K_{a1}}
\tag{42}
$$

and $k_{\text{H}_2\text{O}}$ is the rate constant for hydrolysis due to water alone, $k_{\text{H}^+}$ and $k_{\text{OH}^-}$ are due to specific acid and base catalysis, $k_A$ and $k_B$ are due to general acid and base catalysis, $k_M$ is due to specific metal catalysis, and $K_{a1}$ is the first acid dissociation constant of the aquated metal ion. The half-life for hydrolysis thus depends strongly on the nature of the ester, on the nature of the catalysts, and on the prevailing conditions, and may range from minutes to years.

## 3.2. Amines and Nitriles

Alkylamines have natural sources such as the microbial degradation of natural organic matter, while anthropogenic sources include releases from feed lot operations, sewage treatment, and various industrial activities (Graedel, 1978; Finlayson-Pitts and Pitts, 1986), although only a few atmospheric measurements appear to have been made (Mosier *et al.*, 1973). A major atmospheric sink for simple amines is thought to be reaction with $OH_g^{\bullet}$, with lifetimes ranging from 4 to 13 h (Finlayson-Pitts and Pitts, 1986). Photochemical oxidation in $NO_x$–hydrocarbon mixtures may also be a rapid loss process (Finlayson-Pitts and Pitts, 1986). When the typical intrinsic $K_H$ for alkylamines is taken to be $\approx 50$ M atm$^{-1}$ (see Table 3) and the typical $pK_a$ is $\approx 9$ (Smith and Martell, 1975), then the effective $K_H^*$ with respect to a hydrometeor at pH 6 becomes $5 \times 10^4$ M atm$^{-1}$, suggesting that even mildly acidic clouds and fogs could be major sinks for amines. Aqueous amines are expected to display much of the chemical behavior of ammonia, such as acid-neutralizing capacity and the ability to form complexes with formaldehyde (see Section 3.3) and certain metal ions (Smith and Martell, 1975).

The major sources of the heterocyclic amine, pyridine, are thought to be coal coking and combustion processes (Graedel, 1978; Schulte and Arnold, 1990). It is found in cigarette smoke and therefore it may also be formed during biomass burning. Pyridine is miscible with water. It is a weak base ($pK_a = 5.2$) that forms stable pyridinium salts with aqueous acids, for example,

$$\tag{43}$$

and it is also readily mercurated by mercuric salts ($\log K = 5.1$) (Smith and Martell, 1975). These properties make pyridine a potentially interesting atmospheric component, but its $K_H$ value is only moderately high ($1.13 \times 10^2$ M atm$^{-1}$), suggesting that wet deposition is likely to be an unimportant loss processes in all but the most acidic hydrometeors.

The introduction of the nitrogen atom into the benzene ring lowers its reactivity so that, in the gas phase, the lifetime of pyridine is on the order of 23–46 days with respect to loss by $OH_g^{\bullet}$ attack. Reaction with ozone is also an unlikely sink for atmospheric pyridine since lifetimes are expected to be $>1$ year (Atkinson *et al.*, 1987). It has been suggested that reaction with $HNO_{3g}$ may be the most important sink since the lifetime is calculated to be $\leq 3$ days for 1 ppb $HNO_3$. Pyridinium nitrate has been identified in the laboratory as a product of reaction between pyridine and $HNO_{3g}$ ($k \geq 1.5 \times 10^{16}$ cm$^3$ molecule$^{-1}$S$^{-1}$) (Atkinson *et al.*, 1987).

The aquated pyridinium ion $C_5H_5NH^+(H_2O)_n$ has been shown to be an abundant positive ion in the free troposphere (3000–6000 m) over central

Europe (Schulte and Arnold, 1990). It is thought to be formed by reaction of ambient $NH_4^+$ ions with free pyridine (with an abundance of 1.6 pptv). Little is known about pyridine concentrations at ground level.

Acetonitrile is a polar molecule that is miscible with water. Sources are thought to include biomass burning (Lobert et al., 1990) and industrial solvent emissions (Graedel, 1978) but few measurements of ambient levels have been reported. Snider and Dawson (1984) found acetonitrile at rural sites in the southwestern United States with a mean concentration of 56 pptv, suggesting that natural sources exist, while Hamm and Warneck (1990) found acetonitrile at approximately 50–150 pptv over the North Atlantic, the Bay of Helgoland, and in the stratosphere over Northern Europe. These authors also found aetonitrile at levels of approximately 300–800 pptv in urban air at Mainz. The moderate $K_H$ of $1.65 \times 10^2$ M atm$^{-1}$ (0 °C) implies that clouds and precipitation are unlikely to be important sinks. The relatively slow rate of $OH_g^\cdot$ attack implies that gas phase scavenging is also likely to be a poor sink and the strong C≡N bond is not expected to undergo significant photolysis in the troposphere. The ground-level measurements (56 pptv) together with stratospheric measurements (3 pptv at 25 km decreasing to 0.5 pptv at 40 km) strengthen the argument that there is only a weak source of acetonitrile and that atmospheric sinks such as $OH_g^\cdot$ scavenging and wet deposition are commensurately small (Brasseur et al., 1983; Snider and Dawson, 1984).

Acetonitrile hydrolyzes in the presence of strong base to yield acetate and ammonia, but the rate of this reaction is too slow to be of significance under environmental conditions (Linetskii and Serebryakov, 1965):

$$CH_3C\equiv N \underset{60\ °C}{\overset{2\text{--}5\% \text{ NaOH}}{\rightleftharpoons}} CH_3COO^- + NH_3 \tag{44}$$

$$k = 5.25 \times 10^{-3} \text{ min}^{-1}$$

## 3.3. Carbonyl Compounds

Formaldehyde is both a primary anthropogenic pollutant and a secondary pollutant that arises from the photochemical oxidation of hydrocarbons, both anthropogenic and natural (Seinfeld, 1986; Finlayson-Pitts and Pitts, 1986). It has been described as the "funnel" through which hydrocarbons pass on their way to being photochemically oxidized to carbon monoxide and carbon dioxide (G. A. Dawson, personal communication, 1990). Formaldehyde is therefore the most important and ubiquitous carbonyl in the atmosphere and we focus mainly on its chemistry here. Typical concentrations are 0.5 ppbv in the clean troposphere and 50 ppbv in polluted air. Many other carbonyls have been found in the atmosphere, including saturated and unsaturated aliphatics, dicarbonyls (Grosjean, 1990), and aromatic carbonyls (see Grosjean et al., 1990, for example). The concentrations of the α-dicarbonyls, glyoxal (CHOCHO) and methylglyoxal ($CH_3COCHO$), have on occasion even exceeded that of formaldehyde in urban fogs (Munger et al., 1989). Due to almost complete hydration, $K_H^*$ values of formaldehyde, glyoxal, and methylglyoxal are particularly high. $C_5-C_{11}$ aldehydes, particularly nonanal, have recently been found at ppbv levels in remote Japanese islands and may be the predominant organic species in these areas (Yokouchi et al., 1990). $C_1-C_{18}$ alkanals, acetone, n-butanone, benzaldehyde, glyoxal, methylglyoxal, acrolein, and crotonal have been detected at sub-ppbv levels in clean marine air (Zhou and Mopper, 1990).

In aqueous solution, carbonyls can undergo a wide variety of reactions. They may act as reductants; they are well known to undergo addition reactions with a variety of nucleophiles that include $H_2O$ (see Section 2.6), $SO_{2aq}$ and $NH_{3aq}$ (see Fig. 5); certain carbonyls may undergo photochemical reactions in solution (Graedel and Weschler, 1981; Betterton, 1990); and they may also be oxidized to carboxylic acids by strong oxidants such as $OH^\cdot_{aq}$ (Chameides and Davis, 1983). Each class of reaction is briefly discussed in turn.

Formaldehyde is a good reductant especially under alkaline conditions (Bard et al., 1985):

$$HCOO^- + 2H_2O + 2e^- \rightleftharpoons HCHO + 3OH^- \qquad E° = -1.07 \text{ V}$$

Formaldehyde is able to reduce the salts, oxides, and hydroxides of metals such as iron, copper, nickel, bismuth, mercury, gold, silver, and platinum. It also reduces $H_2O_2$ at high pH, yielding formate and hydrogen (Walker, 1964). The Cannizzaro reaction, which is catalyzed by hydroxide, involves the reduction of one molecule of formaldehyde by another:

$$2HCHO + H_2O \xrightarrow{OH^-} CH_3OH + HCOOH \qquad (45)$$

The reversible hydration of aldehydes and ketones is represented by equation 46. Here the planar $sp^2$ carbonyl carbon atom becomes tetrahedral $sp^3$:

$$RR'CO \underset{-H_2O}{\overset{H_2O}{\rightleftharpoons}} RR'C(OH)_2 \qquad (46)$$

The hydration constants for a wide variety of carbonyls have been measured (Bell, 1966), often by nuclear magnetic resonance techniques (Greenzaid et al., 1967, 1968; Hooper, 1967; Buschmann et al., 1980), and correlated with

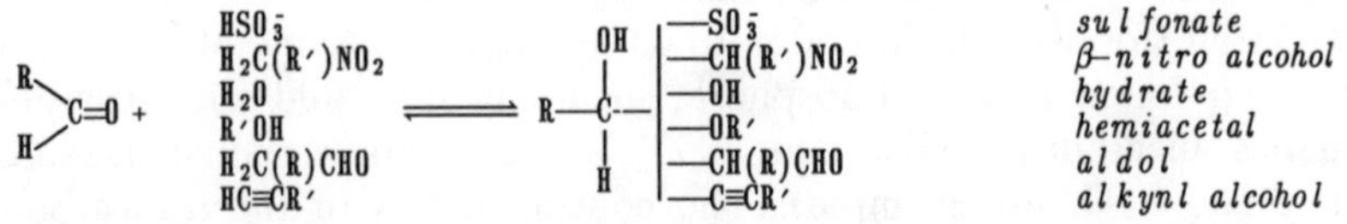

**Figure 5.** The reactions of aldehydes with a variety of potential nucleophiles.

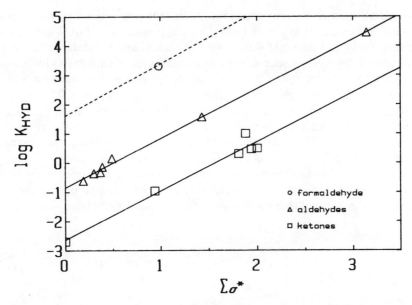

**Figure 6.** The correlation of the hydration constant of carbonyls with Taft's $\sigma^*$ parameter showing the existence of two independent series for aldehydes and ketones and a hypothetical third series with formaldehyde as the sole member.

the Taft $\sigma^*$ parameter (Taft, 1956; Perrin *et al.*, 1981). Figure 6 shows such a correlation. It is apparent that ketones and aldehydes form separate well-defined series and that formaldehyde could be regarded as the sole member of a third series. The progressive substitution of H atoms by C atoms on the carbonyl carbon as one moves from $H_2CO$ through HRCO to RR'CO, is thought to account for the existence of the three series, although a simple increase in steric repulsion is probably not the sole reason. In general, there is a strong positive correlation between $K_{HYD}$ and $K_H^*$.

It should be noted that glyoxal, CHOCHO, is able to form both the hemihydrate, $(HO)_2CHCHO$, and the dihydrate, $(HO)_2CHCH(OH)_2$ (Wasa and Musha, 1970).

The magnitude of the hydration constant for HCHO indicates that it is almost completely (>99.9%) hydrated in solution, and therefore the carbonyl chromophore is lost. This makes it impossible for formaldehyde to undergo significant direct photochemical reactions in tropospheric hydrometeors. Other carbonyls, particularly the ketones, are not so strongly hydrated and the chromophore is thus retained, allowing the possibility of photoinitiated reactions. Ketones begin to absorb strongly at wavelengths below $\approx 310$ nm so there is a partial overlap of the chromophore with the solar spectrum, which begins to decrease rapidly at about 320 nm and becomes almost negligible below about 290 nm (a few high-energy photons will always reach ground level but the flux is so low that we ignore it). Carbonyl photochemistry in aqueous solution under environmentally relevant conditions does not appear to have been studied in detail (although gas phase photolysis is well documented (Finlayson-Pitts and Pitts, 1986)). Graedel and Weschler (1981) have summarized the likely aqueous photolysis reactions of carbonyls. Photolysis of acetone at 300 nm $\leq \lambda \leq$ 350 nm leads to hydrogen atom abstraction but does not give rise to $\alpha$-cleavage (equation 47). Higher ketones (C $\geq$ 4) may undergo both H-atom abstraction and $\alpha$-cleavage depending on the photolysis conditions. Acetaldehyde, the simplest aldehyde to largely retain its chromophore in solution, likewise undergoes H-atom abstraction but not $\alpha$-cleavage, whereas propanal is subject to both H-atom abstraction and $\alpha$-cleavage. Cleavage yields an alkoxyl radical that is expected to abstract a hydrogen atom from an available source to form the corresponding aldehyde (equation 48); radicals are thereby generated that could initiate radical-chain reactions.

$$\begin{array}{c} R \\ \diagdown \\ \phantom{xx} C{=}O + h\nu \longrightarrow RCO^{\cdot} + R''^{\cdot} \\ \diagup \\ R' \end{array} \qquad (47)$$

$$RCO^{\cdot} + R''H \rightarrow RCHO + R'''^{\cdot} \qquad (48)$$

The ability of aldehydes to form bisulfite adducts has been known for many years (organic chemists routinely use this reaction to recover and purify aldehydes), but it is only recently that the kinetics and mechanism of the formation of bisulfite adducts has been studied in detail. Olson and Hoffmann (1989) have reviewed the formation of bisulfite adducts in atmospheric water droplets. As an example we consider the reaction of formaldehyde with $SO_{2aq}$ to yield hydroxymethanesulfonate (HMSA) (Boyce and Hoffmann, 1984) which has been identified in polluted urban fogs at concentrations exceeding 100 $\mu$M (Munger et al., 1986). Other aldehydes undergo similar reactions (Betterton et al., 1988). The formation of HMSA stabilizes $SO_{2aq}$ and can have important effects on the generation and transport of atmospheric acids. The decomposition of HMSA is slow except at high pH (Sorensen and Andersen, 1970) and consequently HMSA is a relatively long-lived sink for $SO_{2aq}$. The reaction pathways for HMSA formation are summarized in Fig. 7.

The free, unhydrated aldehyde reacts with both $HSO_3^-$ and $SO_3^{2-}$ to yield the S(IV) adducts, $CH_2(OH)SO_3^-$ and $CH_2(O^-)SO_3^-$, which are related by a simple acid/base equilibrium. The relative magnitudes of $K_{a0}$ and $K_{a3}$ are such that $H_2C(OH)SO_3^-$ is the sole product at equilibrium in nearly all atmospheric droplets. The rate law for this reaction at 25 °C is

$$\frac{d[HMSA]}{dt} = k_1[HSO_3^-][H_2CO]_T + k_2[SO_3^{2-}][H_2CO]_T \qquad (49)$$

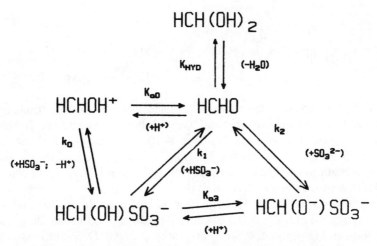

**Figure 7.** The reaction of aqueous formaldehyde with dissolved $SO_2$.

with $k_1 = 7.90 \times 10^2$ $M^{-1}s^{-1}$ and $k_2 = 2.48 \times 10^7$ $M^{-1}s^{-1}$. $[H_2CO]_T$ is the sum of the free and hydrated aldehyde (Boyce and Hoffmann, 1984). Comparison of the rate constants shows that sulfite is a much stronger nucleophile than bisulfite. Only below about pH 3 does a significant portion of the reaction proceed through the $k_1$ pathway.

Direct oxidation of HMSA by $OH_{aq}^{\bullet}$ is a possible sink for the adduct, but the rate of reaction is relatively slow (second order rate constant = $2.77 \times 10^8$ $M^{-1}s^{-1}$) and is likely to be unimportant in all but the most polluted urban fogs (Olson, 1990). Direct oxidation of HMSA by common oxidants such as dissolved $O_3$ and $H_2O_2$ does not appear to occur.

The acid/base equilibria shown in Fig. 7 are diffusion controlled and are therefore "instantaneous" on the time scale of the other chemical reactions, but the same is not true of the carbonyl hydration equilibrium. Since the formation of the adduct proceeds only through the free aldehyde, the rate of dehydration can become rate limiting under certain conditions (Olson and Hoffmann, 1986; Betterton and Hofmann, 1987). The rate of dehydration is subject to general and specific acid and base catalysis. Bell and Evans (1966) studied the reaction

$$CH_2(OH)_2 \underset{k_{hyd}}{\overset{k_{-hyd}}{\rightleftharpoons}} CH_2O_{aq} + H_2O \tag{50}$$

in the presence of free-formaldehyde scavengers such as sulfite that prevented the reverse reaction from occurring. Under these conditions the rate law for dehydration is

$$\frac{d[CH_2O]}{dt} = k_{-hyd}[CH_2(OH)_2] \tag{51}$$

with

$$k_{-hyd} = k_0 + k_H[H^+] + k_{OH}[OH^-] + k_{SH}[SH]$$
$$+ k_S[S] + k_A[A] + k_B[B] \quad (52)$$

SH and S represent the scavenger in the acid and base forms, respectively, and A and B are the general acid and base catalyst S, respectively. Numerical values of the rate constants are $k_0 = 5.1 \times 10^{-3}\ s^{-1}$; $k_H = 2.7\ M^{-1}s^{-1}$; $k_{OH} \approx 1.6 \times 10^3\ M^{-1}s^{-1}$. Values for $k_{SH}$, $k_S$, $k_A$, and $k_B$ depend on the nature of the scavenger or catalyst concerned. Bell and Evans (1966) report constants for 30 different scavengers and catalysts.

By equating equation 49 with equation 51 is it possible to show, on a plot of [S(IV)] against pH, the domain where dehydration is the rate determining step (Olson and Hoffmann, 1986). This is seen in Fig. 8. Dehydration becomes the rate-limiting step for the formation of HMSA only when $[S(IV)] > 10^{-5}$ M at pH 6, and when $[S(IV)] > 10^{-4}$ M at pH 5. Dehydration is never rate limiting below $2 \times 10^{-6}$ M S(IV).

The reactions of S(IV) with other aldehydes, including glyoxal (Olson and Hoffmann, 1988a), methylglyoxal (Betterton and Hoffmann, 1987), glyoxylic acid (Olson and Hoffmann, 1988b), and benzaldehyde (Olson et al., 1986),

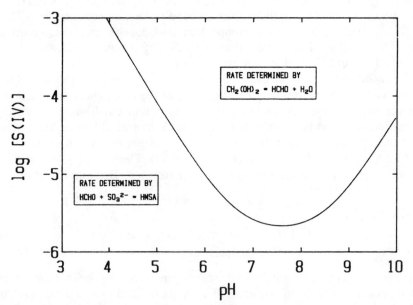

**Figure 8.** The rate of formation of HMSA may be limited by the rate of dehydration of methylene glycol or by the rate of addition of sulfite to formaldehyde depending on pH and S(IV) concentration [S(IV) = $H_2O \cdot SO_2$ + $HSO_3^-$ + $SO_3^{2-}$].

have also been studied in detail. In general, the same pattern is seen as with formaldehyde. The important equilibria can be summarized as follows:

$$RCHO + H_2O \overset{K_{HYD}}{\rightleftharpoons} RCH(OH)_2 \qquad\qquad K_{HYD} = \frac{k_{hyd}}{k_{-hyd}} \qquad (53)$$

$$RC^+HOH + HSO_3^- \overset{K_0}{\rightleftharpoons} RCH(OH)SO_3^- + H^+ \qquad\qquad (54)$$

$$RCHO + HSO_3^- \overset{K_1}{\rightleftharpoons} RCH(OH)SO_3^- \qquad\qquad K_1 = \frac{k_1}{k_{-1}} \qquad (55)$$

$$RCHO + SO_3^{2-} \overset{K_2}{\rightleftharpoons} RCH(O^-)SO_3^- \qquad\qquad K_2 = \frac{k_2}{k_{-2}} \qquad (56)$$

$$RCH(OH)SO_3H \overset{K_{a0}}{\rightleftharpoons} RCH(OH)SO_3^- + H^+ \qquad\qquad (57)$$

$$H_2O \cdot SO_2 \overset{K_{a1}}{\rightleftharpoons} HSO_3^- + H^+ \qquad\qquad (58)$$

$$HSO_3^- \overset{K_{a2}}{\rightleftharpoons} SO_3^{2-} + H^+ \qquad\qquad (59)$$

$$RCH(OH)SO_3^- \overset{K_{a3}}{\rightleftharpoons} RCH(O^-)SO_3^- + H^+ \qquad\qquad (60)$$

Rate constants and equilibrium constants for different aldehydes are given in Table 4.

**Table 4 Rate Constants and Equilibrium Constants for the Reaction of Certain Aldehydes, R—CHO, with Water and S(IV) at 25 °C[a]**

| R— | $\log K_{HYD}$ | $\log K_1$ | $\log K_2$ | $pK_{a3}$ | $\log k_1$ | $\log k_2$ |
|---|---|---|---|---|---|---|
| $CCl_3$— | 4.4 | | | 7.2 | | |
| $CH_3CO$— | 3.4 | 8.91 | 4.73 | 10.3 | 3.54 | 8.56 |
| $C_6H_5$— | | 3.98 | 2.08 | 10.3 | −0.15 | 4.34 |
| $HOCH_2$— | 1.0 | 6.30 | | 10.3 | | |
| H— | 3.3 | 9.82 | 5.34 | 11.2 | 2.90 | 7.40 |
| $CH_3$— | 0.2 | 5.84 | | 11.4 | | |
| Iso-Pr— | −0.4 | 4.68 | | 11.3 | | 4.15 |

[a]The constants are defined in Fig. 5 and equations 53–60.

*Source.* Data taken from Betterton et al. (1988), Olson and Hoffmann (1989), and references therein.

Aldehydes, particularly formaldehyde, also undergo addition reactions with a variety of nitrogen-containing compounds including ammonia, alkylamines, hydrazine, hydroxylamine, and hydrogen cyanide (Walker, 1964). Formaldehyde is also known to interact strongly with RNA, DNA, and DNA bases with potential toxicological effects (McGhee and von Hippel, 1975, 1977; IARC, 1982, 1985), but we focus here on the reaction with $NH_3$, the most important basic gas in the atmosphere. The reversible addition of $NH_3$ leads first to the formation of a tetrahedral carbinolamine (**1**) which, being unstable, forms an imine (**2**) by elimination of water (Walker, 1964):

$$RCHO + NH_3 \rightleftharpoons RCH(NH_2)OH \tag{61}$$
$$(\mathbf{1})$$

$$RCH(NH_2)OH \rightleftharpoons RCH{=}NH + H_2O \tag{62}$$
$$(\mathbf{2})$$

For simple aldehydes and amines, the reaction does not stop at this stage, however; polymers are the ultimate products. In the case of formaldehyde, tricyclic hexamethylenetetramine (**3**), also called hexamine, is formed. Hexamine is a crystalline solid and is an important commercial product used in the manufacture of resins (Walker, 1964):

$$6HCHO_{aq} + 4NH_{3g} \longrightarrow (\mathbf{3}) + 6H_2O + 339 \text{ kJ} \tag{63}$$

The mechanistic details of the formation of this polymer in aqueous solution are not well understood. The emphasis of kinetic studies has been on the rate of consumption of $NH_3$ or HCHO, or on the rate of production of hexamine. Most studies conclude that the consumption of $NH_3$ and HCHO is governed by a third-order rate law (Baur and Rüetschi, 1941; Richmond et al., 1948; Bose, 1957; Ogata and Kawasaki, 1964; Wood and Stevens, 1964):

$$\text{rate} = k[\text{HCHO}]_T^2[\text{NH}_3] \tag{64}$$

The rate constant is strongly pH dependent with a maximum of $k = 25.2$ $M^{-2}s^{-1}$ at pH 9.8 (20 °C) (Ogata and Kawasaki, 1964). The reaction is therefore quite slow: The half-time for a solution containing $10^{-2}$ M $NH_3$ and $10^{-4}$ M HCHO at pH 9.8 would be 11 h.

The kinetics of formation of hexamine leads to a more complex rate law. The interested reader is referred to the original work of Nielsen et al. (1979) for further details.

The thermodynamics of hexamine formation also appears to minimize its potential environmental importance. Hexamine is stable at high pH but not at low pH. The formation constant, defined by equation 65, is $6.3 \times 10^{-11}$ $M^{-5}$ at 25 °C (Baur and Rüetschi, 1941):

$$K = \frac{[(CH_2)_6N_4][H^+]^4}{[CH_2O]_T^6[NH_4^+]_T^4} \tag{65}$$

The mechanism of formation of acetaldehyde–ammonia adducts is somewhat better characterized. The primary product is the trimer, 2,4,6-trimethylhexahydro-s-triazine (**4**) (Hull et al., 1973).

$$
\begin{array}{c}
\overset{\displaystyle H}{\underset{\displaystyle N}{}} \\
CH_3CH \qquad CHCH_3 \\
\\
HN \qquad NH \\
\\
CHCH_3 \\
(\mathbf{4})
\end{array}
$$

The rate of consumption of acetaldehyde obeys a second-order rate law (Ogata and Kawasaki, 1964),

$$\text{rate} = k[NH_3]_T[CH_3CHO]_T \tag{66}$$

but the exact value of the intrinsic rate constant $k$ does not appear to be known.

Aminoalkanesulfonates have been known for many years, and synthetic procedures are well documented (Neelakantan and Hartung, 1959). They do not appear to have been detected in the atmosphere. These compounds are thought to form by the addition of sulfite to the protonated amine:

$$RC\overset{+}{=}NH_2 + SO_3^{2-} \rightleftharpoons RCH\begin{array}{l} {}^{\nearrow NH_2} \\ {}_{\searrow SO_3^{-4}} \end{array} \tag{67}$$

Equimolar mixtures of a wide variety of aldehydes, primary amines (or ammonia), and bisulfite yield the aminoalkanesulfonate at room temperature, but the reaction is facilitated by warming. The adducts are stable in acid solution and can be isolated at pH 2. LeHenaff (1962) reported the following stability constant for the HMSA−NH$_3$ system:

$$K = \frac{[H_2NCH_2SO_3^-]}{[NH_3][CH_2OHSO_3^-]} = 735 \text{ M}^{-1}$$

## 3.4. Carboxylic Acids

The most abundant carboxylic acids in the atmosphere are formic and acetic acid, although Graedel (1978) lists some 80 aliphatic, olefinic, and aromatic acids that have been detected. Keene and Galloway (1986) have recently reviewed the sources of formic and acetic acids. Concentrations are generally less than 1 ppbv in rural areas, but can rise to 2–10 ppbv in urban areas. Typical gas phase levels in rural and remote areas can be found in the work of Dawson et al. (1980), Goldman et al. (1984), Andreae et al. (1988), and Talbot et al. (1988), while Grosjean (1988, 1989) provides data for urban atmospheres. Direct emissions from vegetation (Keene and Galloway, 1986; Dawson et al., 1980) and formicine ants (Graedel and Eisner, 1988) together with photochemical oxidation of plant emissions such as isoprene (Jacob and Wofsy, 1988) probably account for most of the formic and acetic acid in remote areas, while direct emissions from automobiles and the reaction of ozone with olefins probably account for most urban light carboxylic acids.

Natural sources appear to be characterized by a formate to acetate ratio slightly greater than unity, the ratio being highest during the growing season and dropping to slightly below one during the nongrowing season. In the greater Los Angeles urban air shed the formate to acetate ratio is also typically 1.0–1.2 (Grosjean, 1989), in spite of the much higher levels of acetic acid in auto exhaust.

Acid dissociation constants (25 °C) are $1.8 \times 10^{-4}$ M for formic acid and $1.7 \times 10^{-5}$ for acetic acid (Smith and Martell, 1977), so that $K_H^*$ in hydrometeors at pH 6 will be approximately $10^6$ and $10^5$ M atm$^{-1}$, respectively, making these highly soluble species. Indeed, formate and acetate are important anions in rain in many parts of the world and are important natural sources of acidity (Keene and Galloway, 1986; Chapman et al., 1986). The primary effect of these acids on the aqueous phase is to lower the pH and/or to lower the buffering capacity.

Dry deposition can also be an important loss process for light carboxylic acids. Grosjean (1989) estimates that 92% of the total organic acid deposition in the California South Coast air basin is due to dry deposition.

Since the proton is a master variable in aquatic systems, the uptake of carboxylic acids (and other acids) can have important ramifications that affect all pH-dependent equilibria and reaction kinetics. Take, for example, the effect of pH on the rate of oxidation of dissolved $SO_2$ by a variety of oxidants (Seinfeld, 1986). A change in pH not only has a strong effect on the rate of reaction with a given oxidant, but can also determine which of several potential oxidants is the most reactive.

A second possible role for aqueous carboxylic acids is metal complexation. The stability constants for reaction of formic and acetic acids with some metal ions are given in Table 5. It can be seen that while $Fe^{3+}$ and $Pb^{2+}$ complexation are only moderately strong, very stable $Hg^{2+}$ complexes could be formed. Stability constants of the simplest dicarboxylic acid, oxalic acid, are included in Table 5. Oxalic and other dicarboxylic acids have much higher stability constants than the monocarboxylic acids, but their low abundance probably make them less important except in areas characterized by high local anthropogenic releases.

Finally, carboxylic acids may act as metal ion reducants. Formic acid is widely used in the laboratory for this purpose. As can be seen from equation 68, formic acid is thermodynamically capable of reducing all metals with standard reduction potentials more positive than $-0.199$ V (Bard et al., 1985):

$$CO_2 + 2H^+ + 2e^- \rightleftharpoons HCOOH \qquad E° = -0.199 \text{ V} \qquad (68)$$

The reduction of $Fe^{3+}$, for example, is highly favorable.

$$HCOOH + 2Fe^{3+} \rightleftharpoons CO_2 + 2Fe^{2+} + 2H^+$$

$$\Delta G° = -187 \text{ kJ mol}^{-1} \qquad (69)$$

**Table 5 Carboxylic Acid–Metal Ion Stability Constants (25 °C)[a]**

| Metal Ion | log $K_{1HCOOH}$ | log $K_{1CH3COOH}$ | log $K_{1(COOH)2}$ |
|---|---|---|---|
| $H^+$ | 3.745 | 4.757 | 4.266 |
| $Fe^{3+}$ | 3.1 | 3.38 | 7.74 |
| $Pb^{2+}$ | 1.65 | 2.15 | 4.91 |
| $Hg^{2+}$ | 5.43[b] | 5.89 | 9.66 |
| $CH_3Hg^+$ | 2.67 | 3.18 | |
| $Cu^{2+}$ | 2.00 | 2.22 | 6.23 |
| $Cd^{2+}$ | 1.15 | 1.93 | 3.89 |
| $Co^{2+}$ | 0.68 | 1.46 | 4.72 |

[a]Data quoted are for the first stability constant defined by the metal (M) to ligand (L) ratio [ML]/[M][L].
[b]The value for this constant is uncertain.

*Source.* From Smith and Martell (1977).

Other metal ions that could theoretically also be reduced include Pb(II), $CrO_4^{2-}$, Sn(IV), Cu(II), and Hg(II).

### 3.5.  Organic Sulfur Compounds

Anthropogenic sources of organic sulfur compounds include odorants for natural gas (alkyl sulfides and thiols), and wood pulping operations that employ the Kraft process. Graedel (1978) lists more than 20 organic sulfur species that have been identified in the atmosphere, including thiols, sulfides, and heterocyclic compounds. Perhaps most important on a global scale are the reduced sulfur species of natural origin (Aneja and Cooper, 1989; Andreae, 1985). The oxidation of biogenic sulfur compounds such as dimethylsulfide (DMS), methanethiol ($CH_3SH$), dimethyldisulfide ($CH_3SSCH_3$), carbon disulfide ($CS_2$), carbonyl sulfide (OCS), and $H_2S$ is thought to be a potentially important source of sulfate aerosol in the remote atmosphere. DMS alone accounts for approximately one-third of the total natural sulfur flux, while OCS is the largest reservoir of reduced sulfur in the atmosphere. Particulate sulfate can act as a cloud condensation nucleus and therefore could influence cloud microphysics, planetary albedo, and global change phenomena. Clouds, including marine stratiform clouds, reflect more incident sunlight if their liquid water is distributed among a larger number of smaller drops— the Twomey effect (Twomey, 1977). There is therefore the possibility that marine phytoplankton, a strong source of DMS, could provide a feedback loop if their sulfur production is climate sensitive (Charlson et al., 1987; Erickson et al., 1990; Ghan et al., 1990).

Most work to date on the fate of organic sulfur compounds has focused on the gas phase oxidation of sulfides, disulfides, thiols, and so on, by $OH_g^{•}$ and, to a lesser extent, by $NO_{3g}^{•}$ and $IO_g^{•}$. Comparatively little work has been done on the aqueous oxidation, partly because of the low Henry's law constant. DMS is oxidized by $OH_g^{•}$ to yield mainly $SO_2$, by an H-atom abstraction route, with smaller quantities of methanesulfonic acid ($CH_3SO_3H$) and dimethylsulfone ($CH_3SO_2CH_3$) produced by an $OH^{•}$ addition pathway (Saltzman and Cooper, 1989). The ultimate fate of most organic sulfur is oxidation to sulfate.

Organic sulfides and mercaptans are characterized by low $K_H$ values and thus they are unlikely to play a significant direct role in the aqueous chemistry of hydrometeors. However, a knowledge of the Henry's law constant of DMS is important in assessing the magnitude of the flux of DMS from the sea to the atmosphere. In the remote marine environment the average DMS concentration in the photic zone of the sea is approximately two orders of magnitude higher than the Henry's law equilibrium value, indicating that the sea surface is supersaturated with respect to DMS and that a net outward flux to the atmosphere must exist. Andreae (1985) estimates the global average DMS concentration in surface seawater to be $\approx 100$ ng S $L^{-1}$ and calculates a global sea to air flux of $\approx 0.11$ g S $m^{-2}$ $yr^{-1}$ (40 Tg S $yr^{-1}$).

## REFERENCES

Andon, R. J. L., Cox, J. D., and Herington, E. F. G. (1954). "Phase relationships in the pyridine series. Part V. The thermodynamic properties of dilute solutions of pyridine bases in water at 25° and 40°." *J. Chem. Soc.*, 3188.

Andreae, M. O. (1985). "The emission of sulfur to the remote atmosphere: Background paper." In J. N. Galloway, R. J. Carlson, M. O. Andreae, and H. Rodhe, eds.), *The Biogeochemical Cycling of Sulfur and Nitrogen in the Remote Atmosphere*. Reidel, Dordrecht.

Andreae, M. O., Talbot, R. W., Andreae, T. W., and Harriss, R. C. (1988). "Formic and acetic acid over the central Amazon region, Brazil. 1. Dry season." *J. Geophys. Res.* **93**, 1616.

Aneja, V. P., and Cooper, W. J. (1989). "Biogenic sulfur emissions: A review." *ACS Symp. Ser.* **393**, 2.

Atkinson, R., Tuazon, E. C., Wallington, J. J., Aschmann, S. M., Arey, T., Winer, A. M., and Pitts, J. N. (1987). "Atmospheric chemistry of aniline, n-n-dimethylaniline, pyridine, 1,3,5-triazine and nitrobenzene." *Environ. Sci. Technol.* **21**, 64.

Bard, A. J., Parsons, R., and Jordan, J., eds. (1985). *Standard Potentials in Aqueous Solution*. Dekker, New York.

Battino, R., and Clever, H. L. (1964). "The solubility of gases in liquids." *Chem. Rev.* **66**, 395.

Baur, E., and Rüetschi, W. (1941). "Über Bildung und Zerfall von Hexamethyletetramin." *Helv. Chim. Acta* **24**, 754.

Bell, R. P. (1966). "The reversible hydration of carbonyl compounds." *Adv. Phys. Org. Chem.* **4**, 1.

Bell, R. P., and Evans, P. G. (1966). "Kinetics of the dehydration of methylene glycol in aqueous solution." *Proc. R. Soc. London, Ser. A* **291**, 297.

Bender, M. L. (1963). "Metal ion catalysis of nucleophilic organic reactions in solution." *Adv. Chem. Ser.* **37**, 19.

Betterton, E. A. (1991). "The partitioning of ketones between the gas and aqueous phases." *Atmos. Environ.* **25A**, 1473.

Betterton, E. A., and Hoffmann, M. R. (1987). "Kinetics, mechanism, and thermodynamics of the reversible reaction of methylglyoxal ($CH_3COCHO$) with S(IV)." *J. Phys. Chem.* **91**, 3011.

Betterton, E. A., and Hoffmann, M. R. (1988). "Henry's law constants of some environmentally important aldehydes." *Environ. Sci. Technol.* **22**, 1415.

Betterton, E. A., Erel, Y., and Hoffmann, M. R. (1988). "Aldehyde-bisulfite adducts: Prediction of some of their thermodynamic and kinetic properties." *Environ. Sci. Technol.* **22**, 92.

Blair, E. W., and Ledbury, W. (1925). "Partial formaldehyde vapor pressures of aqueous solutions of formaldehyde." *J. Chem. Soc.* **127**, 26.

Bomben, J. L., Briun, S., Thijssen, H. A. C., and Merson, R. L. (1973). "Aroma recovery and retention in concentration and drying of foods." *Adv. Food. Res.* **20**, 2.

Bose, S. (1957). "Kinetics of the reaction between formaldehyde and ammonia." *J. Indian Chem. Soc.* **34**, 663.

Boyce, S. D., and Hoffmann, M. R. (1984). "Kinetics and mechanism of the formation of hydroxymethanesulfonic acid at low pH." *J. Phys. Chem.* **88**, 4740.

Brasseur, G., Arijs, E., De Rudder, A., Nevejans, D., and Ingels, J. (1983). "Acetonitrile in the atmosphere." *Geophys. Res. Lett.* **10**, 725.

Brimblecombe, P., and Clegg, S. L. (1989). "Erratum." *J. Atmos. Chem.* **8**, 95.

Buckingham, D. A. (1977). "Metal-OH and its ability to hydrolyze (or hydrate) substrates of biological interest." In A. W. Addison, W. R. Cullen, D. Dolphin, and B. R. James, eds., *Biological Aspects of Inorganic Chemistry*. Wiley (Interscience), New York.

Burnett, M. G. (1963). "Determination of partition coefficients at infinite dilution by the gas chromatographic analysis of the vapor above dilute solutions." *Anal. Chem.* **35**, 1567.

Buschmann, H.-J., Fuldner, H.-H., and Knoche, W. (1980). "The reversible hydration of carbonyl compounds in aqueous solution. Part I. The keto/gem-diol equilibrium." *Ber. Bunsenges. Phys. Chem.* **84**, 41.

Butler, J. A. V., and Ramchandani, C. N. (1935). "The solubility of non-electrolytes. Part II. The influence of the polar group on the energy of hydration of aliphatic compounds." *J. Chem. Soc.*, 952.

Butler, J. A. V., Thomson, D. W., and Maclennan, W. H. (1933). "The free energy of the normal aliphatic alcohols in aqueous solution. Part I. The partial vapour pressures of aqueous solutions of methyl,n-propyl and n-butyl alcohols. Part II. The solubilities of some normal aliphatic alcohols in water. Part III. The theory of binary solutions, and its application to aqueous-alcoholic solutions." *J. Chem. Soc.*, 674.

Butler, J. A. V., Ramchandani, C. N., and Thomson, D. W. (1935). "The solubility of non-electrolytes. Part I. The free energy of hydration of some aliphatic alcohols." *J. Chem. Soc.*, 280.

Buttery, R. G., Ling, L. C., and Guadagni, D. G. (1969). "Food volatiles: Volatilities of aldehydes, ketones, and esters in dilute water solution." *J. Agric. Food Chem.* **17**, 385.

Buttery, R. G., Bomben, J. L., Guadagni, D. G., and Ling, L. C. (1971). "Some considerations of the volatilities of organic flavor compounds in foods." *J. Agric. Food Chem.* **19**, 1045.

Cabani, S., Conti, G., and Lepori, L. (1971a). "Thermodynamic studies of aqueous dilute solutions of organic compounds. Part 1. Cyclic amines." *Trans. Faraday Soc.* **67**, 1933.

Cabani, S., Conti, G., and Lepori, L. (1971b). "Thermodynamic studies of aqueous dilute solutions of organic compounds. Part 2. Cyclic ethers." *Trans. Faraday. Soc.* **67**, 1943.

Cavanagh, L. A., Schadt, C. F., Conrad, F., and Robinson, E. (1969). "Atmospheric hydrocarbon and carbon monoxide measurements at Point Barrow, Alaska." *Environ. Sci. Technol.* **3**, 251.

Chameides, W. L., and Davis, D. D. (1983). "Aqueous-phase source for formic acid in clouds." *Nature (London)* **304**, 427.

Chapman, E. G., Sklarew, D. S., and Flickinger, J. S. (1986). "Organic acids in springtime Wisconsin precipitation samples." *Atmos. Environ.* **20**, 1717.

Charlson, R. J., Lovelock, J. E., Andreae, M. O., and Warren, S. G. (1987). "Ocean phytoplankton, atmospheric sulfur, cloud albedo and climate." *Nature (London)* **326**, 655.

Christie, A. O., and Crisp, D. J. (1967). "Activity coefficients of the n-primary, secondary and tertiary aliphatic amines in aqueous solution." *J. Appl. Chem.* **17**, 11.

Dacey, J. W. H., Wakeham, S. G., and Howes, B. L. (1984). "Henry's law constants for dimethylsulfide in freshwater and seawater." *Geophys. Res. Lett.* **11**, 991.

Danckwerts, P. V. (1951). "Significance of liquid-film coefficients in gas absorption." *Ind. Eng. Chem.* **43**, 1460.

Dasgupta, P. K., and Dong, S. (1986). "Solubility of ammonia in liquid water and generation of trace levels of standard gaseous ammonia." *Atmos. Environ.* **20**, 565.

Dawson, G. A., Farmer, J. C., and Moyers, J. L. (1980). "Formic and acetic acids in the atmosphere of the southwest U.S.A." *Geophys. Res. Lett.* **7**, 725.

Deister, U., Neeb, R., Helas, G., and Warneck, P. (1986). "Temperature dependence of the equilibrium $CH_2(OH)_2 + HSO_3^- = CH_2(OH)SO_3^- + H_2O$ in aqueous solution." *J. Phys. Chem.* **90**, 3213.

Deno, N. C., and Berkheimer, H. E. (1960). "Activity coefficients as a function of structure and media." *J. Chem. Eng. Data* **5**, 1.

Dong, S., and Dasgupta, P. K. (1986). "Solubility of gaseous formaldehyde in water and generation of trace standard gaseous formaldehyde." *Environ. Sci. Technol.* **20**, 637.

Eisenreich, S. J., Looney, B. B., and Thornton, J. D. (1981). "Airborne organic contaminants in the Great Lakes ecosystem." *Environ. Sci. Technol.* **15**, 30.

Erickson, D. J., Ghan, S. J., and Penner, J. E. (1990). "Global ocean-to-atmosphere dimethyl sulfide flux." *J. Geophys. Res.* **95**, 7543.

Farmer, J. C., and Dawson, G. A. (1982). "Condensation sampling of soluble atmospheric trace gases." *J. Geophys. Res.* **87**, 8931.

Finlayson-Pitts, B. J., and Pitts, J. N. (1986). *Atmospheric Chemistry: Fundamentals and Experimental Techniques.* Wiley (Interscience), New York.

Friant, S. L., and Suffet, I. H. (1979). "Interactive effects of temperature, salt concentration, and pH on head space analysis for isolating volatile trace organics in aqueous environmental samples." *Anal. Chem.* **51**, 2167.

Ghan, S. J., Taylor, K. E., Penner, J. E., and Erickson, D. J. (1990). "Model test of CCN-cloud albedo climate forcing." *Geophys. Res. Lett.* **17**, 607.

Gmehling, J., Onken, U., and Arlt, W. (1981). *Vapor-Liquid Equilibrium Data Collection.* Behrens, D., Eckermann, R., eds. Dechema, Frankfurt/Main Vol 1a.

Gmehling, J., Rasmussen, P., and Fredenslund, A. (1982). "Vapor-liquid equilibria by UNIFAC group contribution. Revision and extension. 2." *Ind. Eng. Chem. Process Des. Dev.* **21**, 118.

Goldman, A., Murcray, F. H., Murcray, D. G., and Rinsland, C. P. (1984). "A search for formic acid in the upper troposphere: A tentative identification of the 1105-$cm^{-1}$ $v_6$ band Q branch in high-resolution balloon-borne solar absorption spectra." *Geophys. Res. Lett.* **11**, 307.

Graedel, T. E. (1978). *Chemical Compounds in the Atmosphere.* Academic Press, New York.

Graedel, T. E., and Eisner, T. (1988). "Atmospheric formic acid from formicine ants: A preliminary assessment." *Tellus* **40B**, 335.

Graedel, T. E., and Weschler, C. J. (1981). "Chemistry within aqueous aerosols and raindrops." *Rev. Geophys.* **19**, 505.

Greenzaid, P., Luz, Z., and Samuel, D. (1967). "A nuclear magnetic resonance study of the reversible hydration of aliphatic aldehydes and ketones. I. Oxygen-17 and proton spectra and equilibrium constants." *J. Am. Chem. Soc.* **89**, 749.

Greenzaid, P., Luz, Z., and Samuel, D. (1968). "N.M.R. study of reversible hydration of aliphatic aldehydes and ketones." *Trans. Faraday Soc.* **64**, 2780.

Grosjean, D. (1988). "Aldehydes, carboxylic acids and inorganic nitrate during NSMCS." *Atmos. Environ.* **22**, 1637.

Grosjean, D. (1989). "Organic acids in southern California air: Ambient concentrations, mobile source emissions, in situ formation and removal processes." *Environ. Sci. Technol.* **23**, 1506.

Grosjean, D. (1990). "Gas-phase reaction of ozone with 2-methyl-2-butene: Dicarbonyl formation from Criegee biradicals." *Environ. Sci. Technol.* **24**, 1428.

Grosjean, D., Miguel, A. H., and Tavares, T. M. (1990). "Urban air pollution in Brazil: Acetaldehyde and other carbonyls." *Atmos. Environ.* **24B**, 101.

Hales, J. M., and Drewes, D. R. (1979). "Solubility of ammonia in water at low concentrations." *Atmos. Environ.* **13**, 1133.

Hamm, S., and Warneck, P. (1990). "Gas chromatographic determination of acetonitrile in air using a thermionic detector." *Anal. Chem.* **62**, 1876.

Hansch, C., Quinlan, J. E., and Lawrence, G. L. (1968). "Linear free-energy relationship between partition coefficients and aqueous solubilities of organic liquids." *J. Org. Chem.* **33**, 347.

Higbie, R. (1935). "The rate of absorption of a pure gas into a still liquid during short periods of exposure." *Trans. Am. Inst. Chem. Eng.* **31**, 365.

Hine, J., and Mookerjee, P. K. (1975). "The intrinsic hydrophilic character of organic compounds. Correlations in terms of structural contributions." *J. Org. Chem.* **40**, 292.

Hine, J., and Weimar, R. D. (1965). "Carbon basicity." *J. Am. Chem. Soc.* **87**, 3387.

Holdren, M. W., Spicer, C. W., and Hales, J. M. (1984). "Peroxyacetyl nitrate solubility and decomposition rate in acidic water." *Atmos. Environ.* **18**, 1171.

Hooper, D. L. (1967). "Nuclear magnetic resonance measurements of equilibria involving hydration and hemiacetal formation from some carbonyl compounds." *J. Chem. Soc. B*, 169.

Howe, G. B., Mullins, M. E., and Rogers, T. N. (1987). *Evaluation and Prediction of Henry's Law Constants and Aqueous Solubilities for Solvents and Hydrocarbon Fuel Components.* Vol. 2. *Experimental Henry's Law Data.* Research Triangle Institute for Engineering and Services Lab., Air Force Engineering and Services Center, Research Triangle Park, NC, NTIS ADA-202-262/3.

Hull, W. E., Sykes, B. D., and Babior, B. M. (1973). "A proton nuclear magnetic resonance study of the aqueous chemistry of acetaldehyde and ammonia. The formation of 2,4,6-trimethylhexahydro-s-triazine." *J. Org. Chem.* **38**, 2931.

Hwang, H., and Dasgupta, P. K. (1985). "Thermodynamics of the hydrogen peroxide-water system." *Environ. Sci. Technol.* **19**, 255.

IARC (1982). *IARC Monographs on the Evaluation of the Carcinogenic Risk of Chemicals to Humans. Some Industrial Chemicals and Dyestuffs.* IARC, Lyon, Vol. 29.

IARC (1985). *IARC Monographs on the Evaluation of the Carcinogenic Risk of Chemicals to Humans. Allyl Compounds, Aldehydes, Epoxides and Peroxides.* IARC, Lyon, Vol. 36.

Ioffe, B. V., and Vitenberg, A. G. (1984). *Head-Space Analysis and Related Methods in Gas Chromatography.* Wiley (Interscience), New York.

Irmann, F. (1965). "Eine einfache Korrelation Zwischen Wasserlöslichkeit und Struktur von Kohlenwasserstoffen und Halogenkohlenwasseerstoffen." *Chem.-Ing.-Tech.* **37**, 789.

Jacob, D. J. (1986). "Chemistry of OH in remote clouds and its role in the production of formic acid and peroxymonosulfate." *J. Geophys. Res.* **91**, 9807.

Jacob, D. J., and Wofsy, S. C. (1988). "Photochemistry of biogenic emissions over the Amazon forest." *J. Geophys. Res.* **93**, 1477.

Johnson, B. J. (1990). "The carbon isotope content and concentration of ambient formic acid and acetic acid." Ph.D. dissertation, University of Arizona, Tucson.

Kabadi, V. N., and Danner, R. P. (1979). "Nomograph solves for solubilities of hydrocarbons in water." *Hydrocarbon Process.* **58**, 245.

Karger, B. L., Sewell, P. A., Castells, R. C., and Hartkopf, A. (1971). "Gas chromatographic study of the adsorption of insoluble vapors on water." *J. Colloid Interface Sci.* **35**, 328.

Keene, W. C., and Galloway, J. N. (1986). "Considerations regarding sources for formic and acetic acids in the troposphere." *J. Geophys. Res.* **91**, 14,466.

Kieckbusch, T. G., and King, C. J. (1979). "An improved method of determining vapor-liquid equilibria for dilute organics in aqueous solution." *J. Chromatogr. Sci.* **17**, 273.

LeHenaff, P. (1962). "Étude de quelques réactions des composés quaternaires de l'hexaméthylènetétramine." *Ann. Chim. (Paris)* [13] **7**, 367.

Leinonen, P. J., Mackay, D., and Phillips, C. R. (1971). "A correlation for the solubility of hydrocarbons in water." *Can. J. Chem. Eng.* **49**, 288.

Linetskii, V. A., and Serebryakov, B. R. (1965). "The effect of temperature on the reaction rate of the alkaline hydrolysis of nitriles." *Izv. Vyssh. Uchebn. Zaved., Neft. Gaz* **8**, 62. (*CA* **62**, 15745c.)

Liss, P. S., and Slater, P. G. (1974). "Flux of gases across the air-sea interface." *Nature (London)* **247**, 181.

Lobert, J. M., Scharffe, D. H., Hao, W. M., and Crutzen, P. J. (1990). "Importance of biomass burning in the atmospheric budgets of nitrogen containing gases." *Nature (London)* **346**, 552.

Luke, W. T., Dickerson, R. R., and Nunnermacker, L. J. (1989). "Direct measurements of the photolysis rate coefficients and Henry's law constants of several alkyl nitrates." *J. Geophys. Res.* **94**, 14,905.

Maahs, H. G. (1982). "Sulfur dioxide/water equilibria between 0° and 50°C: An examination of data at low concentrations." In D. R. Schryer, ed., *Heterogeneous Atmospheric Chemistry.* Am. Geophys. Union, Washington, DC, Geophys. Monog. 26.

Mackay, D., and Leinonen, P. J. (1975). "Rate of evaporation of low-solubility contaminants from water bodies to atmosphere." *Environ. Sci. Technol.* **9**, 1178.

Mackay, D., and Paterson, S. (1981). "Calculating fugacity." *Environ. Sci. Technol.* **15**, 1006.

Mackay, D., and Paterson, S. (1982). "Fugacity revisited." *Environ. Sci. Technol.* **16**, 645A.

Mackay, D., and Shiu, W. Y. (1977). "Aqueous solubility of polynuclear aromatic hydrocarbons." *J. Chem. Eng. Data* **22**, 399.

Mackay, D., and Shiu, W. Y. (1981). "A critical review of Henry's law constants for chemicals of environmental interest." *J. Phys. Chem. Ref. Data* **10**, 1175.

Mackay, D., Shiu, W. Y., and Sutherland, R. P. (1979). "Determination of air-water Henry's law constants for hydrophobic pollutants." *Environ. Sci. Technol.* **13**, 333.

Mackay, D., Joy, M., and Paterson, S. (1983). "A quantitative water, air, sediment interaction (QWASI) fugacity model for describing the fate of chemicals in lakes." *Chemosphere* **12**, 981.

Martell, A. E. (1963). "Metal chelate compounds as acid catalysts in solvolysis reactions." *Adv. Chem. Ser.* **37**, 161.

Martin, R. L. (1961). "Adsorption on the liquid phase in gas chromatography." *Anal. Chem.* **33**, 347.

Martin, L. R., and Damschen, D. E. (1981). "Aqueous oxidation of sulfur dioxide by hydrogen peroxide at low pH." *Atmos. Environ.* **15**, 1615.

McGhee, J. D., and von Hippel, P. H. (1975). "Formaldehyde as a probe of DNA structure. 1. Reaction with Exocyclic Amino Groups of DNA bases." *Biochemistry* **14**, 1281.

McGhee, J. D., and von Hippel, P. H. (1977). "Formaldehyde as a probe of DNA structure. 4. Mechanism of the initial reaction for formaldehyde with DNA." *Biochemistry* **16**, 3276.

McNally, M. E., and Grob, R. L. (1985). "Static and dynamic headspace analysis. Part one. Environmental applications." *Am. Lab.* **17**, 20.

Mosier, A. R., Andre, C. E., and Viets, F. G. (1973). "Identification of aliphatic amines volatalized from cattle feedyards." *Environ. Sci. Technol.* **7**, 642.

Munger, J. W., Jacob, D. J., and Hoffmann, M. R. (1984). "The occurrence of bisulfite-aldehyde addition products in fog and cloudwater." *J. Atmos. Chem.* **1**, 335.

Munger, J. W., Tiller, C., and Hoffmann, M. R. (1986). "Identification of hydroxymethane-sulfonate in fog water." *Science* **231**, 247.

Munger, J. W., Collett, J., Daube, B., and Hoffmann, M. R. (1989). "Fogwater chemistry at Riverside, California." *Atmos. Environ.* **24B**, 185.

Munz, C., and Roberts, P. V. (1982). *Transfer of Volatile Organic Contaminants into a Gas Phase during Bubble Aeration*. Dept. Civil Eng., Stanford University, Stanford, CA, Tech. Rep. No. 262.

Neelakantan, L., and Hartung, W. H. (1959). "α-Aminoalkanesulfonic acids." *J. Org. Chem.* **24**, 1943.

Nielsen, A. T., Moore, D. W., Ogan, M. D., and Atkins, R. K. (1979). "Structure and chemistry of the aldehyde ammonias. 3. Formaldehyde-ammonia reaction. 1,3,5-hexahydrotriazine." *J. Org. Chem.* **44**, 1678.

Ogata, Y., and Kawasaki, A. (1964). "The kinetics of the reaction of formaldehyde with ammonia." *Bull. Chem. Soc. Jpn.* **37**, 514.

Olson, T. M. (1990). "Kinetics of the oxidation of sulfonic acids by hydroxyl radicals in aqueous solution." *Environ. Eng. Proc., EE Div./ASCE*, July 8–11.

Olson, T. M., and Hoffmann, M. R. (1986). "On the kinetics of formaldehyde S(IV) adduct formation in slightly acidic solution." *Atmos. Environ.* **20**, 2277.

Olson, T. M., and Hoffmann, M. R. (1988a). "The kinetics mechanism and thermodynamics of glyoxal-S(IV) adduct formation." *J. Phys. Chem.* **92**, 533.

Olson, T. M., and Hoffmann, M. R. (1988b). "Formation, kinetics, mechanism and thermodynamics of glyoxylic acid-S(IV) adduct formation." *J. Phys. Chem.* **92**, 4246.

Olson, T. M., and Hoffmann, M. R. (1989). "Hydroxyalkylsulfonate formation: Its role as a S(IV) reservoir in atmospheric water droplets." *Atmos. Environ.* **23**, 985.

Olson, T. M., Boyce, S. D., and Hoffmann, M. R. (1986). "Kinetics thermodynamics and mechanism of the formation of benzaldehyde-S(IV) adducts." *J. Phys. Chem.* **90**, 2482.

Parsons, G. H., Rochester, C. H., and Wood, C. E. C. (1971). "Effect of 4-substitution on the thermodynamics of hydration of phenol and phenoxide anion." *J. Chem. Soc. B*, 533.

Parsons, G. H., Rochester, C H., Rostron, A., and Sykes, P. C. (1972). "The thermodynamics of hydratioin of phenols." *J. Chem. Soc., Perkin Trans. 2*, 136.

Pecsar, R. E., and Martin, J. J. (1966). "Solution thermodynamics from gas-liquid chromatography." *Anal. Chem.* **38**, 1661.

Perrin, D. D., Dempsey, B., and Serjeant, E. P. (1981). *pKa Prediction for Organic Acids and Bases*. Chapman & Hall, London.

Pierotti, G. J., Deal, C. H., and Derr, E. L. (1959). "Activity coefficients and molecular structure." *Ind. Eng. Chem.* **51**, 95.

Richmond, H. H., Myers, G. S., and Wright, G. F. (1948). "The reaction between formaldehyde and ammonia." *J. Am. Chem. Soc.* **70**, 3659.

Rochester, C. H., and Symonds, J. R. (1973). "Thermodynamic studies of fluoroalcohols." *J. Chem. Soc., Faraday Trans. I*, 1577.

Saltzman, E. S., and Cooper, W. J., eds. (1989). *Biogenic Sulfur in the Environment*, Am. Chem. Soc., Washington, DC, ACS Symp. Ser. 393.

Schulte, P., and Arnold, F. (1990). "Pyridinium ions and pyridine in the free troposphere." *Geophys. Res. Lett.* **17**, 1077.

Schwartz, S. E. (1984). "Gas- and aqueous-phase chemistry of $HO_2$ in liquid water clouds." *J. Geophys. Res.* **89**, 11,589.

Schwartz, S. E., and White, W. H. (1981). "Solubility equilibria of the nitrogen oxides and oxyacids in dilute aqueous solution." *Adv. Environ. Sci. Eng.* **4**, 1.

Seinfeld, J. H. (1986). *Atmospheric Chemistry and Physics of Air Pollution*. Wiley (Interscience), New York.

Smith, R. M., and Martell, A. E. (1975). *Critical Stability Constants*. Plenum, New York, Vol. 2.

Smith, R. M., and Martell, A. E. (1977). *Critical Stability Constants*. Plenum, New York, Vol. 3.

Snider, J. R., and Dawson, G. A. (1984). "Surface acetonitrile near Tucson, Arizona." *Geophys. Res. Lett.* **11**, 241.

Snider, J. R., and Dawson, G. A. (1985). "Tropospheric light alcohols, carbonyls, and acetonitrile: Concentrations in the Southwestern United States and Henry's law data." *J. Geophys. Res.* **90**, 3797.

Sørensen, P. E., and Andersen, V. S. (1970). "The formaldehyde-hydrogen sulfite system in alkaline solution. Kinetics, mechanisms and equilibria." *Acta Chem. Scand.* **24**, 1301.

Stumm, W., and Morgan, J. J. (1981). *Aquatic Chemistry.* Wiley (Interscience), New York.

Taft, R. W. (1956). "Separation of polar, steric, and resonance effects in reactivity." In M. S. Newman, ed., *Steric Effects in Organic Chemistry.* Wiley, New York.

Talbot, R. W., Beecher, K. M., Harriss, R. C., and Cofer, W. R. (1988). "Atmospheric geochemistry of formic and acetic acids at a mid-latitude temperate site." *J. Geophys. Res.* **93**, 1638.

Tsonopoulos, C., and Prausnitz, J. M. (1971). "Activity coefficients of aromatic solutes in dilute aqueous solutions." *Ind. Eng. Chem. Fundam.* **10**, 593.

Twomey, S. (1977). "The influence of pollution on the shortwave albedo of clouds." *J. Atmos. Sci.* **34**, 1149.

Van Krevelen, D. W., Hoftijzer, P. J., and Huntjens, F. J. (1949). "Composition and vapor pressures of aqueous solutions of ammonia, carbon dioxide and hydrogen sulfide." *Recl. Trav. Chim. Pays-Bas* **68**, 191.

Vitenburg, A. G., Toffe, B. V., St. Dimitrova, Z., and Butaeva, I. L. (1975). "Determination of gas-liquid partition coefficients by means of gas chromatographic analysis." *J. Chromatogr.* **112**, 319.

Walker, J. F. (1964). *Formaldehyde.* Reinhold, New York, 3rd ed.

Wasa, T., and Musha, S. (1970). "Polarographic behavior of glyoxal and its related compounds." *Univ. Osaka Prefect., Ser. A* **19**, 169.

Westheimer, F. H., and Ingraham, L. L. (1956). "The entropy of chelation." *J. Phys. Chem.* **60**, 1668.

Wood, R. K., and Stevens, W. F. (1964). "Reaction kinetics of the formation of hexamethylenetetramine." *J. Appl. Chem.* **14**, 325.

Yalkowsky, S. H., and Valvani, S. (1979). "Solubilities and partitioning. 2. Relationships between aqueous solubilities, partition coefficients, and molecular surface areas of rigid aromatic hydrocarbons." *J. Chem. Eng. Data* **24**, 127.

Yokouchi, Y., Mukai, H., Nakajima, K., and Ambe, Y. (1990). "Semi-volatile aldehydes as predominant organic gases in remote areas." *Atmos. Environ.* **24A**, 439.

Yoshizumi, K., Aoki, K., Nouchi, I., Okita, T., Kobayashi, T., Kamakura, S., and Tajima, M. (1984). "Measurements of the concentration of rainwater and of the Henry's law constant of hydrogen peroxide." *Atmos. Environ.* **18**, 395.

Zhou, X., and Mopper, K. (1990). "Measurement of sub-parts-per-billion levels of carbonyl compounds in marine air by a simple cartridge trapping procedure followed by liquid chromatography." *Environ. Sci. Technol.* **24**, 1482.

# 2

# SAMPLING AND ANALYSIS FOR AMBIENT OXIDES OF NITROGEN AND RELATED SPECIES

## J. E. Sickles II

*Research Triangle Institute, Research Triangle Park, North Carolina*

*Gaseous Pollutants: Characterization and Cycling,* Edited by Jerome O. Nriagu.
ISBN 0-471-54898-7 © 1992 John Wiley & Sons, Inc.

## 1.  INTRODUCTION

This chapter addresses various methods to measure selected airborne nitrogen-containing species. The focus is on methodologies currently available or in general use for in situ monitoring of airborne concentrations in both ambient and indoor environments. Methods for measuring the species of interest at their respective sources are not considered, and remote sensing technologies are mentioned in only a few cases.

Although the primary focus in this document is nitrogen dioxide, other nitrogen-containing species are also considered. This chapter is organized into several sections, with each section devoted to a different species. The species under consideration are nitric oxide (NO), nitrogen dioxide ($NO_2$), oxides of nitrogen ($NO_x$), total reactive odd nitrogen ($NO_y$), peroxyacetyl nitrate (PAN) and other organic nitrates, nitric acid ($HNO_3$), nitrous acid ($HNO_2$), dinitrogen pentoxide ($N_2O_5$), the nitrate radical ($NO_3$), particulate nitrate ($NO_3^-$), and nitrous oxide ($N_2O$).

Where possible, discussions of sampling and analysis methods for each species address the pertinent characteristics for each method. These topics include method type (i.e., in situ, remote, active, passive), description, status (i.e., concept, laboratory prototype, commercially available), interferences, time resolution, sensitivity, and precision and accuracy. A good overview of many of the currently available methods for measuring nitrogen-containing species is the proceedings of a recent National Aeronautics and Space Administration (NASA) workshop (NASA, 1983).

Methods development usually progresses through several stages: concepts, laboratory prototypes, laboratory evaluations, field tests, field evaluations and comparisons against other "proven" methods, and, finally, consensus acceptance by the user community. At each stage, modifications may be implemented to improve or resolve weaknesses that have been revealed. This is usually a winnowing process. As a result of limitations discovered during this process, many candidate methods may be abandoned in favor of other methods. At some stage near the end of the process, commercialization may occur. In the current document, those methods that have successfully progressed to the final stages of development are emphasized.

## 2. NITRIC OXIDE (NO)

Although $NO_2$ rather than NO is the primary focus of this document, the most commonly used method of measuring $NO_2$ does not detect the $NO_2$ molecule directly. Instead, the method relies on a chemiluminescent reaction of the NO molecule after $NO_2$ has been converted to NO. Thus, to provide a background for subsequent discussions of measurement methods for $NO_2$ and other nitrogen-containing species, NO rather than $NO_2$ is the first species that is addressed.

Airborne concentrations of NO can be determined by various methods. As noted previously, the most commonly used method is chemiluminescence (CLM). Other methods include laser-induced fluorescence (LIF), absorption spectroscopy, and passive collection with subsequent wet chemical analysis.

## 2.1.  Chemiluminescence

CLM can be used to detect several airborne nitrogen-containing species (i.e., NO, $NO_2$, $HNO_2$, $HNO_3$, $N_2O_5$, PAN, $NO_x$, $NO_y$, $NH_3$, and $NO_3^-$). Among these compounds, only NO is detected directly, while the other compounds must be converted in some manner to NO prior to detection.

The principle is based on the detection of the light emitted following the reaction of NO with ozone ($O_3$). Excess $O_3$ is added to an air sample containing NO, which is passing through a darkened reaction vessel with infrared-reflective walls and a window for viewing by a photomultiplier (PM) tube. The light-emitting species is electronically excited nitrogen dioxide, $NO_2^*$, a product of the reaction of NO and $O_3$. The $NO_2^*$ relaxes by photon emission that ranges in wavelength well beyond 600 nm and is centered near 1200 nm. Light is detected by a red-sensitive PM tube fitted with optical filters to prevent radiation below 600 nm produced by ozonalysis of other materials from interfering. The intensity of the measured light is proportional to the concentration of NO in the air sample, and the concentration can be determined by calibration with atmospheres of known composition.

In applications to detect other airborne nitrogen-containing species, the air sample is preconditioned prior to entering the reaction vessel to convert some or all of these species to NO, and that signal is compared to the signal for an unconditioned sample. The signal from the unconditioned sample represents NO, while that from the preconditioned sample represents the sum of the originally present NO along with the NO resulting from conversion of the other nitrogen species. Signal differencing permits determination of the other nitrogen-containing species. The specificity of the preconditioning process may be controversial and is discussed in subsequent sections.

CLM is designated by the U.S. Environmental Protection Agency (EPA) as the Reference Method for determining $NO_2$ in ambient air (see Section 3.1). As a result, commercial instruments for measuring NO and $NO_2$ are available. Detection limits of approximately 5 ppb with response times on the order of minutes are claimed by suppliers. While these performance parameters are adequate for monitoring NO and $NO_2$ in relatively polluted urban and suburban environments, they may be inadequate in less polluted remote areas. Efforts have been reported by several researchers to improve the sensitivity and response of CLM NO measurement technology to permit deployment in remote locations both on ground-based and airborne platforms. Delany et al. (1982), Dickerson et al. (1984), Tanner et al. (1983), and Kelly (1986) reported techniques for modifying commercially available oxides of nitrogen detectors to achieve improved sensitivity and response times. Modifications that can be employed include (1) operation at a low pressure, high flow rate, and increased $O_3$ supply; (2) addition of a prereactor where sample air and $O_3$ flows are mixed out of view of the PM tube to obtain a more stable background signal; (3) use of a larger more efficient reaction vessel with highly reflective walls that promotes reaction close to the PM tube; (4)

use of pure oxygen as the $O_3$ source; (5) cooling the PM tube to reduce noise in the dark current; and (6) change of the electronics to employ photon counting techniques rather than analog signal processing.

Kelly (1986) has provided instructions for application of the first three modifications to the Thermo Electron Model 14-B and the Monitor Labs Model 8840. Postmodification detection limits of 0.1–0.2 ppb and 90% response times of 5–10 s were claimed for these instruments. Dickerson et al. (1984) applied modifications 1–3 and 5 noted above to a Thermo Electron Model 14 B/E and report minimum detection limits (MDLs) for NO of 10 ppt with an 1/*e* response time of 20 s.

Other workers have developed and built highly sensitive research grade instruments for the CLM determination of NO (Ridley and Howlett, 1974; Kley and McFarland, 1980; Helas et al., 1981; Drummond et al., 1985; Carroll et al., 1985; Torres, 1985; Kondo et al., 1987). Such devices have been used to measure NO at the earth's surface, from airborne platforms in the troposphere and from balloon-borne platforms in the stratosphere. These instruments generally employ those features listed above. MDLs of 5 ppt or less, response times of 2–60 s, and accuracy of 10–20% have been claimed.

CLM NO instruments appear to be specific for NO. Water vapor may act to quench excited $NO_2$ efficiently (Matthews et al., 1977; Folsom and Courtney, 1979). Operation at reduced pressure reduces this problem. With a commercial analyzer, a 7% reduction in the NO signal was reported for 81% RH versus dry air (MacPhee et al., 1976). Recent tests of eight commercial analyzers have not shown a water vapor interference with the $NO_2$ signal (Michie et al., 1983). With various research grade instruments, interference due to varying humidities has been reported to be negligible below 20 ppm $H_2O$, increase by less than 10% with relative humidity (RH) up to 2.5% RH, and show no change between 2 and 100% RH (Fahey et al., 1985a; Drummond et al., 1985). Using commercial CLM instruments, no or very small (i.e., less than 2%) interferences have been reported for six chlorine-containing (Joshi and Bufalini, 1978), 14 sulfur-containing (Sickles and Wright, 1979), seven nitrogen-containing, and three sulfur-containing species (Grosjean and Harrison, 1985b). Zafiriou and True (1986), however, do report interferences from $H_2S$ and from gases purged from anoxic waters that may have contained sulfur compounds. Using research grade CLM instruments, Fahey et al. (1985a) found no NO interference for $NO_2$, $HNO_3$, $N_2O_5$, and PAN and negligible responses for $NH_3$, HCN, $N_2O$, $CH_4$, nine chlorine-containing and three sulfur-containing compounds. These findings are consistent with those of Drummond et al. (1985) who report no or negligible NO interferences from $NO_2$, $HNO_3$, PAN, $HO_2NO_2$, $H_2O_2$, $C_3H_6$, $H_2O$, and aerosols using a research grade instrument with a humidified $O_3$ source.

From an operational perspective, aerosols can accumulate on the glass filter separating the reaction chamber from the PM tube, causing a reduction in sensitivity (Klapheck and Winkler, 1985). Cleaning the filter is reported to restore the original sensitivity.

Whereas most of the CLM methods discussed above are continuous, CLM has also been used to analyze nitrogen species collected in integrated samples. Gallagher et al. (1985) have taken cryosamples (4 K) of whole stratospheric air. Samples were analyzed following desorption in the laboratory for NO and $NO_2$ using modified commercial CLM instruments. Braman et al. (1986) have employed a series of hollow denuder tubes coated with chemicals chosen to preconcentrate various oxides of nitrogen. The collected nitrogen species are thermally desorbed and detected as NO with a commercial CLM instrument. The coating materials used to preconcentrate the various target species in sequence are as follows: tungstic acid removes $HNO_3$; potassium–iron oxide removes $HNO_2$; copper(I) iodide removes $NO_2$; and cobalt(III) oxide removes NO. Future field testing is needed to demonstrate the adequacy of this method.

## 2.2.   Laser-Induced Fluorescence

LIF techniques may incorporate single-photon (SP), two-photon (TP), or photofragmentation (PF) schemes. Although SP-LIF has been used for NO (Bradshaw et al., 1982), TP-LIF represents an advancement in the state of the art (Bradshaw et al., 1985) and is discussed below. The TP-LIF detection principle requires that a molecule have more than one bonding excited state and can be sequentially pumped into the highest state. If the lifetime of the excited state is short compared to collisional deactivation, then the excited molecule will decay to a more stable state by a fluorescence process. The fluorescence wavelength is shifted relative to that of the pumping wavelengths and thus overcomes noise problems associated with background nonresonant fluorescence. For application to NO, pulsed ultraviolet (UV) and infrared (IR) laser light sources are used. Ground-state $X^2\Pi$ NO is excited to the $A^2\Sigma$ electronic level using UV of 226-nm wavelength. Then, using IR wavelengths of 1.06–1.15 μm, the molecule is further pulsed to the $D^2\Sigma$ level. The fluorescence resulting from the $D^2\Sigma$ to $X^2\Pi$ transition is monitored at 187–220 nm. By using long-wavelength blocking filters with solar blind PM tubes, this type of detector discriminates against noise and becomes signal, rather than signal-to-noise, limited. Photon counting and grated-charge integrators are used for signal processing. The intensity of the light is related to the concentration of NO in the air sample by calibration with atmospheres of known concentration.

Since the TP-LIF instrument is signal limited, the sensitivity is defined by the integration time (e.g., 1 ppt for 5 min and 10 ppt for 30 s) (Davis et al., 1987). Propagation of error analysis has been used to place 90% confidence limits on the accuracy of TP-LIF NO field measurements performed on an aircraft at ±16%.

The TP-LIF technique is expected to be highly specific, since it has two levels of spectroscopic selectivity. If some trace atmospheric compound were to produce an NO molecule by interaction with the 226-nm beam, then the opportunity for interference exists. The potential interference from $HNO_3$, $CH_3NO_2$, $C_2H_5NO_2$, $CH_3ONO_2$, $NO_2$, PAN, $HNO_2$, $SO_2$, and $CH_3ONO$ has been evaluated. Only the last compound was found to show potential interference, and arguments have been given to neglect its influence when sampling tropospheric air (Davis et al., 1987).

## 2.3.   Absorption Spectroscopy

Absorption spectroscopy encompasses techniques that measure the change in radiance from a source that occurs as a result of absorption by analyte molecules over a known path length. Several techniques, including Fourier transform infrared spectroscopy (FTIR), long-path absorption, and infrared tunable-diode laser spectroscopy (TDLAS), have been employed for measuring the concentration of various nitrogen oxides in the atmosphere (NASA, 1983). Among these techniques, the TDLAS is a well-developed technique that has been applied to NO as well as $NO_2$ and $HNO_3$. Similar sensitivities have been reported for both remote sensing applications using open air path lengths and in situ application using multipass cells (Cassidy and Reid, 1982). The latter configuration has found broader application for ambient measurements of oxides of nitrogen, and the use of the White cell avoids atmospheric turbulence-related errors that can affect open air application. As a result, in situ TDLAS is the primary focus of this section.

TDLAS employs a tunable diode laser to scan over a narrow wavelength region around a particular absorption line or feature of the gas of interest. High sensitivity is achieved by the high spectral radiance of the diodes and the rapid tunability of the laser. With rapid scanning back and forth across an absorption line, the absorption appears as an ac signal at twice the tuning frequency and can be sensitively detected by sychronous demodulation. System sensitivities sufficient to measure signal changes of $10^{-5}$ permit the detection of concentrations of less than 10 ppt with a 1-km path length. For analyte molecules that have resolved absorption spectra that are not coincident with other atmospheric constituents, TDLAS is highly specific. Additional information on the operating principles and hardware for TDLAS is provided by Schiff et al. (1983) and NASA (1983).

For optically thin systems, the Beer–Lambert law suggests that the fraction of the power transmitted through an absorbing medium is proportional to the concentration of the absorbing molecule. However, since the total laser power is not measured, it is usually necessary to calibrate the TDLAS by introducing a known concentration of the target gas and determining the proportionality between signal and concentration.

Using a 40-m path length near 1850 $cm^{-1}$ the MDL for NO is 0.5 ppb (Schiff et al., 1983). At a sampling rate giving a 4-s residence time in the White cell, stable NO signals are achieved in approximately 1 min. Linearity has been demonstrated between the signal and NO concentration at levels between 7 and 175 ppb. Since the NO calibration gas is introduced directly into the sampling line, surface losses are compensated for automatically. As noted previously, the measurement of NO using TDLAS is highly specific.

A newly developed method, two-tone frequency modulated spectroscopy (TTFMS), has shown great promise in the laboratory for the measurement of NO, $NO_2$, PAN, $HNO_3$, $N_2O$, and other atmospheric trace gases (Hansen, 1989). TTFMS uses a diode laser light source that is modulated simultaneously at two arbitrary but closely spaced frequencies. The beat tone between these two frequencies is monitored as the laser carrier and associated sidebands are tuned through an absorption line. The method is fast, specific, and extremely sensitive. Using a low-pressure (20-torr), multiple-reflection optical cell with a 100-m path length and 1-min signal averaging time, the projected MDL for NO is 4 ppt, and the projected MDLs for $NO_2$, PAN, $HNO_3$, and $N_2O$ are lower. Additional development of this laboratory prototype is needed to demonstrate its performance in the field.

## 2.4.  Passive Samplers

Whereas the previous methods are focused primarily on low NO concentrations representative of ambient air, passive samplers are focused on atmospheres having higher concentrations, such as those found indoors or in the workplace. The Palmes tube is a passive sampler that relies on diffusion of an analyte molecule through a quiescent diffusion path of known length to a reactive surface where the molecule is captured by chemical reaction (Palmes et al., 1976). After exposure durations ranging from hours to days, the reactive surface is analyzed and the integrated loading of the reaction product is used to infer the average gas concentration. A quiescent diffusion path is required to ensure that sampling is diffusion controlled and as a result is relatively constant. This permits the average ambient concentration to be related directly to the ratio of the reaction product loading to the exposure time. This proportionality factor is analogous to the reciprocal of a sampling rate and is the product of the diffusivity of the analyte gas and the area of the opening through which the molecules diffuse divided by the distance they must travel to be collected.

Palmes tubes are fabricated from 1.27-cm (½ in.) o.d., 0.95-cm (⅜ in.) i.d. acrylic tubing that has been cut into 7.1-cm (2.8 in.) segments (Palmes and Tomczyk, 1979). The dimensions are chosen to provide a ratio of sampling area to diffusion distance of 0.1 and thus ensure diffusion-controlled sampling. Reactive grids are secured and sealed at one end of the tubing segment using a plastic cap. The opposite end of the tube is sealed with a similar cap. The capped sampler is stored until the sample is to be collected. A sample is collected by placing the tube in the appropriate location (e.g., for personal sampling the tube may be attached to a worker's lapel), removing the end cap opposite the gridded end with the open end facing down, sampling for the appropriate period, recording the time, recapping the tube, and returning the sampler to the laboratory for analysis.

The Palmes tube passive sampler does not measure NO directly. Two tubes are required: One has reactive grids coated with triethanolamine (TEA) to collect $NO_2$. The second tube is similar but has an additional reactive surface coated with chromic acid to convert NO to $NO_2$ which is in turn collected by the TEA-coated grids. The NO concentration is determined by subtraction after correction for differences in sampling rates caused by differences in diffusivities of the two molecules. To ensure reliable results, contact between the chromic acid-coated surface and the TEA-coated grids for longer than 24 h must be avoided.

Analysis is accomplished by extracting the grids in solution and analyzing the extract for nitrite ion, $NO_2^-$. This analysis may be performed by adding a solution of water, sulfanilamide, and $N$-1-naphthylethylenediamine dihydrochloride (NEDA) reagent directly into the tube and determining the concentration using a spectrophotometer at 540 nm (Palmes et al., 1976). Increased sensitivities are claimed by analyzing the solution using ion chromatography (IC) with a concentrator column (Miller, 1984). The analytical finish is calibrated by dilution of gravimetrically prepared nitrite solutions.

The sampling rate (i.e., 0.02 cm$^3$ s$^{-1}$) for NO is reported to be independent of pressure, to increase approximately 1% for each 5.5 °C increase in temperature, and to increase by 3% for each 5 cm s$^{-1}$ increase in wind velocity (Palmes and Tomczyk, 1979). Linear response was found for dosages between 2 and 120 ppm-h. This method was proposed for sampling occupational exposures where the dosage is not to exceed 25 ppm for 8 h (i.e., 200 ppm-h.). This method cannot be used for sampling periods longer than 24 h. The reliability of this method in the field at both ppb and ppm levels remains to be demonstrated.

A badge-type sampler similar to the Palmes tube has been devised by Yanagisawa and Nishimura (1982). Their device uses a series of 12 layers of $CrO_3$-impregnated glass fiber to oxidize NO to $NO_2$. The filters also act as a diffusion barrier between the ambient air and a TEA-coated cellulose fiber filter. Nitric oxide is oxidized to $NO_2$ on the oxidizing filters and collected along with $NO_2$ that has diffused from ambient air through the filters to the TEA-coated collection surface. The TEA-coated filter is extracted and analyzed for $NO_2^-$. Either a colorimetric or IC analytical finish may be employed. The analytical finish is calibrated by dilution of gravimetrically prepared nitrite solutions. An effective proportionality factor (i.e., calibration factor) for the badge is provided by the supplier. This technique is claimed to be more sensitive than the Palmes tube and to have a lower limit dosage of 0.07 ppm-h.

### 2.5. Calibration

CLM, TP-LIF, and TDLAS NO measurement systems all employ calibration cylinders containing known concentrations of NO in $N_2$ at nominal concentrations of 1 to 50 ppm (Carroll et al., 1985; Bradshaw et al., 1985; Schiff et al., 1983). Calibrations are performed using dynamic dilution with air.

In the calibration procedure for the measurement of $NO_2$ by CLM, the EPA specifies the use of an NO concentration standard (Code of Federal Regulations, 1987a). This standard is a cylinder of compressed gas containing between 50 and 100 ppm NO in nitrogen. The concentration must be traceable according to a certification protocol to a National Bureau of Standards (NBS) Standard Reference Material (SRM) or an NBS/EPA-approved, commercially available Certified Reference Material (CRM). NBS provides 10 NO SRMs at nominal concentrations between 5 and 3000 ppm (NBS, 1988). Shores and Smith (1984) have demonstrated that aluminum calibration cylinders containing 10–150 ppm NO in nitrogen were stable over time, and for 103 cylinders, the average change was less than 1% over an 18-month period. Commercially supplied cylinders from 11 producers containing certified concentrations at nominal values of 70 and 400 ppm were evaluated for accuracy (Wright et al., 1987). In all cases, the certified and auditor-measured concentrations were within 5%, and in over two-thirds of the cases the agreement was within 2%.

Passive NO samplers do not employ full calibration of sampling and analysis operations (Palmes et al., 1976). Only the analysis portion of the procedure is calibrated. Calibration standards for colorimetric or IC determination of nitrite are prepared similarly. Dilution of gravimetrically prepared liquid solutions of nitrite is used to produce calibration standards that cover the working range of analysis.

## 2.6. Intercomparisons

Several intercomparisons of the performance of research grade NO instrumentation have been conducted recently (Walega et al., 1984; Hoell et al., 1985, 1987; Fehsenfeld et al., 1987; Gregory et al., 1990d). Walega et al. (1984) have reported comparisons of NO measurements made with a highly sensitive CLM instrument and a TDLAS system. Measurements of NO-spiked synthetic air were made in the laboratory and field. In addition, measurements were made of ambient and captive downtown Los Angeles air. Good agreement was found for all test conditions.

Tests to compare the performance of several instruments at measuring trace gases in the troposphere have been performed as part of NASA's Global Tropospheric Experiment/Chemical Instrumentation Test and Evaluation 1 and 2 (GTE/CITE 1 and 2). Hoell et al. (1985, 1987) and Gregory et al. (1990d) have reported comparisons of NO measurements made with two highly sensitive CLM instruments and a TP-LIF instrument. The first intercomparison was a ground-based study performed at Wallops Island, Virginia. The second intercomparison was an airborne study comprising two missions performed on a Convair CV-990 flown out of California and Hawaii. The third intercomparison involved 13 flights sampling tropical, nontropical, maritime, and continental air masses at altitudes between 150 and 5000 m. The two CLM instruments were of similar design (Kley and McFarland, 1980), with the main differences being the injection of water vapor to the airstream entering the reaction chamber of one instrument to minimize the background variability caused by changing ambient humidity and to suppress an $O_3$-related background signal. In the first study, measurements of ambient NO concentrations ranged from 10 to 60 ppt and of NO-enriched ambient air ranged from 20 to 170 ppt. Agreement among the techniques at the 95% confidence level was ±30%, and no artifact or species interferences were identified. In the second and third studies, NO concentrations ranged from below 5 ppt to above 100 ppt, with the majority below 20 ppt. At NO concentrations below 20 ppt, measurements agreed to within stated instrument precision and accuracy (i.e., to within 15 to 20 ppt). Good correlation was observed between CLM and TP-LIF measurements. The authors concluded that equally valid measurements of ambient NO can be expected from either instrument.

A field intercomparison of instruments designed to measure NO, $NO_x$, and $NO_y$ was conducted near Boulder, Colorado (Fehsenfeld et al., 1987). The study was performed to compare the performance of instruments that employed different approaches to reduce $NO_x$ or $NO_y$ to NO prior to detection by CLM. In several tests, both zero air and ambient air spiked with NO were measured. Excellent agreement was found among the measurements of the three tested instruments. These results confirm the equivalence of CLM NO detection systems.

### 2.7. Sampling Considerations for NO and Other Nitrogen-Containing Species

Nitric oxide reacts rapidly with $O_3$ to form $NO_2$. In the presence of sunlight, $NO_2$ will photolyze to form NO and atomic oxygen (O), which will add with atmospheric oxygen ($O_2$) to form $O_3$. Thus, under daylight conditions NO, $O_3$, and $NO_2$ can exist simultaneously in ambient air in a condition known as photostationary state where the rate of photolysis of $NO_2$ is nearly equal to the rate of reaction between NO and $O_3$ to form $NO_2$. The relative amounts of the three species at any time are influenced by the intensity of the sunlight present at that moment. When a sample is drawn into a dark sampling line, photolysis ceases, while NO continues to react with $O_3$ to form $NO_2$. As a result, long residence times in sampling lines must be avoided to ensure a representative sample. Sampling requirements for a given error in tolerance were discussed by Butcher and Ruff (1971). Figure 1 shows the absolute error in $NO_2$ introduced for a 10-s residence time in a dark sampling line in the presence of NO and $O_3$ at various concentrations.

In addition to sampling time considerations, sampling surfaces should be considered. Oxides of nitrogen are, in general, reactive species. As a result, the most nearly inert materials (i.e., glass and Teflon) are recommended for use in sampling trains. If water molecules accumulate on sampling train sur-

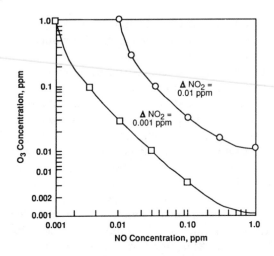

Source: Butcher and Ruff, 1971.

**Figure 1.** Absolute error in $NO_2$ and $\Delta NO_2$ for 10 s in dark sampling line.

faces and influence sample integrity, then species solubility may be one in-dicator of the susceptibility of a species to surface effects. Solubilities ex-pressed as Henry's law coefficients (M atm$^{-1}$) for selected nitrogen-containing species are NO, $2 \times 10^{-3}$; $NO_2$, $1 \times 10^{-2}$; $N_2O$, $3 \times 10^{-2}$; PAN, 4; $HNO_2$, 50; $NH_3$, 60; and $HNO_3$, $2 \times 10^5$ (Schwartz, 1983). This suggests that of the oxides or nitrogen, NO may be the least susceptible to surface effects, while surface effects may be very important in the sampling of $HNO_3$.

## 3.   NITROGEN DIOXIDE ($NO_2$)

Among the oxides of nitrogen, $NO_2$ is the only criteria pollutant and the only species to have sampling and analysis methodologies specified by the EPA for determining ambient airborne concentrations. As a result, methods for sampling and analysis of $NO_2$ are emphasized in this document. Airborne concentrations of $NO_2$ can be determined by several methods including CLM, LIF, absorption spectroscopy, and bubbler and passive collection with sub-sequent wet chemical analysis.

### 3.1.   CLM (NO + $O_3$)

Instruments discussed in this section sample continuously, employ the CLM reaction of NO and $O_3$, but do not detect $NO_2$ directly. Instead, they rely on the direct detection of NO, the conversion of some or all of the $NO_2$ in the air sample to NO, reaction of the resulting NO with $O_3$, and the appropriate signal processing to infer the $NO_2$ concentration. The CLM NO detection principle and hardware are described in Section 2.1. To measure $NO_2$, a CLM NO detector, a converter, plumbing modifications, and changes in signal processing are required.

Several methods have been employed to reduce $NO_2$ to NO (Kelly, 1986). They include catalytic reduction using heated molybdenum or stainless steel, reaction with CO over a gold catalyst surface, reaction with $FeSO_4$ at room temperature, reaction with carbon at 200 °C, and photolysis of $NO_2$ to NO at 320–400 nm.

Since CLM is designated by the EPA as the Reference Method for $NO_2$ in ambient air (Code of Federal Regulations, 1987a), CLM instruments for the determination of $NO_2$ are available commercially. As noted previously, these instruments are used to measure both NO and $NO_2$. Nominal detection limits of approximately 5 ppb and response times on the order of minutes are claimed by suppliers. Field evaluation of nine instruments has shown the MDLs to range from 5 to 13 ppb (Michie et al., 1983; Holland and McElroy, 1986). Recent field and laboratory evaluation of two commercial instruments operated on a 0.00- to 0.05-ppm range revealed detection limits of between 0.5 and 1.0 ppb and operating precision estimates of $\pm 0.3$ ppb (Rickman et al., 1989). While these performance parameters are adequate for monitoring NO and $NO_2$ in urban and suburban environments, they may be inadequate in less polluted remote areas. As noted in Section 2.1, efforts have been reported by several researchers to improve sensitivity and response of CLM NO measurement technology to permit deployment in remote locations on ground-based and airborne platforms. Since the research grade instruments employed by these workers included $NO_2$ to NO converters and were designed to measure both NO and $NO_2$, instrument performance for the determination of $NO_2$ is also improved substantially over that of commercially available instruments. Typically reported performance parameters for $NO_2$ response using research grade CLM instruments are MDLs of 10–25 ppt, response times of 1–100 s, and accuracy of 30–40% (Helas et al., 1987; Fehsenfeld et al., 1987).

Different converters may not be specific for $NO_2$ and may convert several nitrogen-containing compounds to NO, giving rise to artificially high values for $NO_2$. Using commercial instruments, Winer et al. (1974) found over 90% conversion of PAN, ethyl nitrate, and ethyl nitrite to NO with a molybdenum converter and similar responses to PAN and $n$-propyl nitrate with a carbon converter. With a stainless steel converter at 650 °C, Matthews et al. (1977) reported 100% conversion for $NO_2$, 86% for $NH_3$, 82% for $CH_3NH_2$, 68% for HCN, 1% for $N_2O$ and 0 for $N_2$. Using a commercial instrument, Joseph and Spicer (1978) found quantitative conversion of $HNO_3$ to NO with a molybdenum converter at 350 °C. Similar responses to PAN, methyl nitrate, $n$-propyl nitrate, $n$-butyl nitrate, and $HNO_3$, substantial response to nitrocresol, and no response to PBzN were reported with a commercial instrument using a molybdenum converter at 450 °C (Grosjean and Harrison, 1985b). These results were confirmed for PAN and $HNO_3$ by Rickman and Wright

(1986) using commercial instruments with a molybdenum converter at 375 °C and a carbon converter at 285 °C.

Interferences from species that do not contain nitrogen have also been reported. Joshi and Bufalini (1978) using a commercial instrument with a carbon converter found significant apparent $NO_2$ responses to phosgene, trichloroacetyl chloride, chloroform, chlorine, HCl, and photochemical reaction products of a perchloroethylene–$NO_x$ mixture. Grosjean and Harrison (1985b) reported substantial responses to photochemical reaction products of $Cl_2$–$NO_x$ and $Cl_2$–methanethiol mixtures and small negative responses to methanethiol, methyl sulfide, and ethyl sulfide. Sickles and Wright (1979), using a commercial instrument with a molybdenum converter at 450 °C, found small negative responses to 3-methylthiophene, methanethiol, ethanethiol, ethyl sulfide, ethyl disulfide, methyl disulfide, hydrogen sulfide, 2,5-dimethylthiophene, methyl sulfide, methyl ethyl sulfide, and negligible responses to thiophene, 2-methylthiophene, carbonyl sulfide, and carbon disulfide.

With a research-grade instrument, Bollinger et al. (1983) reported that $NO_2$, $HNO_3$, n-propyl nitrate, and $N_2O_5$ are reduced to NO by a gold-catalyzed reaction with CO. Fahey et al. (1985a), using a similar instrument with 3000 ppm CO over a gold converter at 300 °C, reported conversion efficiencies exceeding 90% for $NO_2$, $HNO_3$, $N_2O_5$, and PAN. Although at a converter temperature of 300 °C negligible response to HCN and $NH_3$ was found in the presence of water vapor, complete conversion was noted at 700 °C.

A room temperature $NO_2$ to NO converter using $FeSO_4$ has been suggested by Winfield (1977) and adopted in research grade instruments by Helas et al. (1981), Kondo et al. (1983), and Dickerson et al. (1984). A reduction in conversion efficiency has been reported under dry conditions, and conversion of PAN, $HNO_2$, and other nitrogen-containing species to NO has been noted (Fehsenfeld et al., 1987). Nonspecificity of the $FeSO_4$ converter has been observed by Fehsenfeld et al. (1987) in measuring NO and $NO_2$ in a remote environment. At $NO_x$ levels below 1 ppb, results from the $FeSO_4$ converter were biased high in the measurement of $NO_2$ (i.e., a factor of 2 at 0.1 ppb). Airborne measurements of $NO_2$ at concentrations below 0.2 ppb showed high biases of factors of 2–3 for a CLM instrument with a $FeSO_4$ converter (Gregory et al., 1990c).

In another research-grade instrument, Kley and McFarland (1980) used an Xe arc lamp to photolyze a portion of the $NO_2$ in sampled air to NO and determine the $NO_2$ concentration from the increase in NO. A fractional conversion was established using a calibration source. Interferences with the photolytic converter approach (CLM-PC) are expected from $HNO_2$, $NO_3$, $HO_2NO_2$, and $N_2O_5$, but not from $HNO_3$, n-propyl nitrate, and PAN. A detailed description of the operation (including minimization of interferent decomposition and both homogeneous and heterogeneous oxidation of NO) and performance of a CLM-PC instrument is given by Ridley et al. (1988). An artifact identified with this method is caused by nitrate-containing aerosols deposited on the surface of the photolysis tube which release NO and $NO_2$

upon irradiation (Bollinger et al., 1984). This interference is eliminated by filtering sampled air and periodic cleaning of tube surfaces.

The methods discussed above employ CLM detection of NO and are continuous. Other researchers have employed various methods of integrated sampling followed by a CLM instrument for measuring NO and $NO_2$ in the desorbed sample. Gallagher et al. (1985) have used cryosampling of stratospheric whole air samples, and Braman et al. (1986) have used copper(I) iodide coated denuder tubes to sample $NO_2$ in ambient air.

## 3.2. CLM (Luminol)

A method for the direct CLM determination of $NO_2$ was reported by Maeda et al. (1980). This method is based on the CLM reaction of gaseous $NO_2$ with a surface wetted with an alkaline solution of luminol (5-amino-2,3-dihydro-1,4-phthalazinedione). The emission is strong at wavelengths between 380 and 520 nm. The intensity of the measured light is proportional to the $NO_2$ concentration in the sampled air, and the concentration can be determined by calibration with atmospheres of known concentration.

Since the introduction of the CLM (luminol) method by Maeda et al. (1980), improvements have been made to develop an instrument suitable for use in the field (Wendel et al., 1983), and additional modifications have been made recently to produce a continuous commercial instrument (Schiff et al., 1986). Detection limits of 5–30 ppt and a response time of seconds have been claimed based on laboratory tests (Wendel et al., 1983; Schiff et al., 1986). Recent laboratory evaluation of two instruments has revealed a detection limit (i.e., twice the standard deviation of the clean air response) of 5 ppt and 95% rise and fall times of 110 and 15 s (Rickman et al., 1989). Field tests of the same instruments have shown an operating precision of $\pm 0.6$ ppb.

The original method showed no interferences from NO, $N_2O$, $NH_3$, CO, 1,2-dichloroethylene, and propylene, but positive interferences from $O_3$ and $SO_2$ and a negative interference from $CO_2$ (Maeda et al., 1980). More recently, the luminol solution has been formulated containing water, luminol, NaOH, $Na_2SO_3$, and alcohol in proportions chosen to enhance the sensitivity and minimize interferences from $O_3$, $SO_2$, and $CO_2$ (Wendel et al., 1983; Schiff et al., 1986). No interferences were reported for NO, $HNO_3$, $NH_3$, HCN, $H_2O_2$, CO, $CO_2$, and $SO_2$. The instrument has shown sensitivity to PAN (Wendel et al., 1983; Sickles, 1987) and $HNO_2$ (Rickman et al., 1989), and nonlinear response to $NO_2$ at concentrations below 10 ppb (Schiff et al., 1986; Kelly et al., 1990). The method has shown appreciable sensitivity to operating temperature which can be resolved by controlling the temperature of the reaction cell or by signal processing (Schiff et al., 1986; Bubacz et al., 1987).

Recent tests of the CLM (luminol) instrument have demonstrated the need to correct results for pressure (as might be seen in airborne applications), nonlinearity of response below 3 ppb $NO_2$, interferences from $O_3$ and PAN, and the age-dependent sensitivity of the luminol solution (Kelly et al., 1990). A manufacturer-supplied $O_3$ scrubber designed to eliminate the $O_3$ interference was also found to remove appreciable amounts of $NO_2$.

## 3.3.  Photofragmentation, Two-Photon LIF (PF/TP-LIF)

Two $NO_2$ sensors based on the measurement of the fluorescence of excited $NO_2$ have been reported by Fincher et al. (1978). One device employs a small high-pressure Xe arc flash lamp to excite $NO_2$. This device has a sensitivity of 10 ppb for an 80-s (i.e., 1024 flash lamp pulses) integration time. The other device uses a 442-nm LIF and a PM tube with photon counting of light above 600 nm. The sensitivity of this device is 1 ppb for a comparable integration time. A major drawback of these devices for broad ambient applications is limited sensitivity associated with background signals. This has been overcome with PF/TP-LIF instrumentation.

A $NO_2$ sensor incorporating PF/TP-LIF has recently been developed and deployed in the field (Rodgers et al., 1980; Davis, 1988). With this method, NO is measured in one cell using TP-LIF (see Section 2.2). An XeF excimer laser with output at 353 nm is used in a second cell to photolyze $NO_2$. The total NO signal in the second cell resulting from ambient and photofragment NO is measured as NO using TP-LIF. The $NO_2$ concentration is determined from the difference in signals of the two reaction cells and the fractional photolysis of $NO_2$. The $NO_2$ fluorescence cell is calibrated using calibration sources of NO and $NO_2$.

Since the PF/TP-LIF instrument is signal limited, the sensitivity is defined by integration time. The detection limit for $NO_2$ for a 2-min integration time is 12 ppt (Davis, 1988). The accuracy of PF/TP-LIF $NO_2$ determinations is likely to be similar to the $\pm 16\%$ reported for TP-LIF NO measurements (Davis et al., 1987). At 15 ppt NO and 50 ppt $NO_2$, the precision of $NO_2$ determinations is given as $\pm 17\%$ (Gregory et al., 1990c).

The PF/TP-LIF technique is expected to be highly specific for $NO_2$. In addition to those potential NO interferents with TP-LIF that are discussed in Section 2.2, other species that could photolyze or otherwise decompose to produce NO or $NO_2$ have been considered (Davis, 1988). Arguments were given to dismiss $HNO_3$, $HO_2NO_2$, $N_2O_5$, $CH_3ONO$, and $CH_3ONO_2$ as possible interfering species.

## 3.4.  Absorption Spectroscopy

Absorption spectroscopy is discussed in a previous section for NO (Section 2.3). Absorption methods may measure the absorption of light in the ultraviolet (UV), visible, or infrared (IR) regions of the electromagnetic spectrum. They may employ closed cells for in situ measurements (e.g., TDLAS) or open paths for remote sensing. Absorption methods require a source of radiation. Active methods utilize artificial light from a source such as an incandescent lamp or a laser, whereas passive methods use natural light from the sun or moon. Laser sources offer advantages of scanning, narrow spectral width, and high intensity, and as a result usually provide better sensitivity than nonlaser sources. In this section, several absorption techniques are addressed, including the in situ methods, TDLAS, photometry, and TTFMS, as well as remote sensors employing long-path absorption, differential optical absorption spectroscopy (DOAS), and differential absorption lidar (DIAL).

TDLAS is a well-developed technique that has been used to measure $NO_2$ as well as other species in the atmosphere. Descriptive information about the operating principle is given in Section 2.3 and the references therein. With a 150-m path length near 1600 cm$^{-1}$ the MDL is 0.1 ppb and the accuracy is ±15% (Mackay and Schiff, 1987). For a 40-m path length, the MDL for $NO_2$ is 0.5 ppb (Schiff et al., 1983). At a sampling rate giving a 4-s residence time in the White cell, stable $NO_2$ signals are achieved in approximately 1 min. Linearity has been demonstrated between the signal and $NO_2$ concentration at levels between 35 and 175 ppb. Surface losses are compensated for automatically since the $NO_2$ calibration gas, typically from a permeation tube, is introduced directly into the sampling line. TDLAS is a spectroscopic technique; as a result, the measurement of $NO_2$ using this method is highly specific.

A prototype instrument using an in situ absorption technique to measure $NO_2$ was recently reported (Jung and Kowalski, 1986). This technique employs a modified commercial $O_3$ photometric analyzer to measure the absorption of visible light by $NO_2$ at wavelengths longer than 400 nm. The signal obtained in a 1.12-m absorption cell from unscrubbed ambient air is compared with that from ambient air scrubbed of $NO_2$. A microcomputer uses Beer's law and an absorption coefficient derived from a $NO_2$ calibration source to determine the $NO_2$ concentration. Interferences from $NH_3$, NO, $O_3$, $SO_2$, and PAN have been shown to be negligible. Comparisons with commercially available CLM analyzers monitoring smog chamber experiments and ambient air have shown good agreement when $NO_2$ was expected to be present. The CLM signal was found to exceed that of the photometer when photochemical reaction products such as PAN were believed to be present. Although noise of less than 3 ppb and linear response have been demonstrated between 100 and 700 ppb, additional development and evaluation are needed to permit routine use of this technique in ambient monitoring applications.

Cryogenic sampling at 77 K combining the matrix-isolation technique (i.e., solid $CO_2$) with FTIR spectroscopy has shown promise for the sensitive determination of $NO_2$, PAN, $HNO_3$, $HNO_2$, and $N_2O_5$ (Griffith and Schuster, 1987). A theoretical MDL of 5 ppt was claimed for $NO_2$ in 15-L integrated samples of ambient air.

A laboratory prototype method, TTFMS, has been developed recently and is described in Section 2.3 (Hansen, 1989). This method has a projected MDL for $NO_2$ of 0.3 ppt.

Long-path absorption is a remote sensing technique which is typically operated over open terrain with optical path lengths of up to tens of kilometers. The absorption cross section for each species of interest is determined in the laboratory. This information is used to convert optical densities measured in the field to concentration data. For 1-h observation periods a MDL of 20 ppt has been reported using an artificial light source and a 9.2-km path length (Johnston and McKenzie, 1984). Noxon (1978) used the sun as a light source and the structure in the $NO_2$ absorption spectrum near 440 nm to obtain measures of the abundance of $NO_2$ in both the troposphere and the stratosphere. Platt and Perner (1983) have reported the application of differential optical absorption spectroscopy (DOAS) to the determination of several nitrogen-containing species. A Xe high pressure or an incandescent lamp was used with a 1- to 10-km path length. Selected applicable compounds, detection limits, and target wavelengths are NO, 400 ppt, and 226 nm; $NO_2$, 100 ppt, and 363 nm; $HNO_2$, 20 ppt, and 354 nm; and $NO_3$, 0.5 ppt, and 662 nm. The DOAS technique has been recently adapted to employ a 25-m multipass open reflection system with a path length of up to 2 km (Biermann et al., 1988). Using a 0.8-km path length and 12-min averaging times, MDLs and accuracies for $NO_2$, $HNO_2$, and $NO_3$ of 4 ppb ($\pm 10\%$), 0.6 ppb ($\pm 30\%$), and 20 ppt ($\pm 15\%$) have been reported.

Common remote sensing techniques employ light detection and ranging (LIDAR). Differential absorption lidar (DIAL) is a long-path absorption technique. This method employs light of two wavelengths propagated over a given distance at a given intensity. The concentration of the gas species of interest is related to the difference in intensities of the two wavelengths at the receiver. DIAL techniques have been applied to the ambient measurement of both NO (Alden et al., 1982) and $NO_2$ (Fredriksson and Hertz, 1984; Edner et al., 1987). Baumgartner et al. (1979) report a 5-ppb sensitivity for $NO_2$, and Staehr et al. (1985) report 10-ppb sensitivity for $NO_2$ using a laser source and a 6-km path length. DIAL methods are in the development stages for monitoring $NO_2$ in ambient atmospheres.

### 3.5.  Wet Chemical Methods

Most wet chemical methods for measuring $NO_2$ involve the collection of $NO_2$ in solution followed by a colorimetric finish using an azo dye. Many variations of this method exist including both manual and automated versions. Szonntagh (1979) traced the history of azo dye methods for $NO_2$ sampling and analysis. Nitrogen dioxide, first collected in aqueous solution, is thought to form $HNO_2$. An aromatic amine is used in the presence of an acid to react with $HNO_2$ and form a diazonium salt. The salt then rearranges and couples with another organic amine that has been added to form a red azo dye. The intensity of the color is proportional to the $NO_2$ collected and is measured using a spectrophotometer. A good overview of wet chemical methods for sampling and analysis of $NO_2$ is given by Purdue and Hauser (1980).

### 3.5.1.  Griess–Saltzman Method

In this method, air is sampled for no longer than 30 min through a fritted bubbler that contains the Griess–Saltzman reagent (Saltzman, 1954). This reagent is a solution of sulfanilic acid, NEDA, and acetic acid. Color development is complete within 15 min and is measured at 550 nm within an hour. Interferences from $SO_2$ and PAN have been noted but are usually too low in ambient air to cause significant error. Concentrations of $NO_2$ ranging from 20 to 800 ppb for 30-min sampling periods may be determined using this method. A MDL of 2 ppb and a precision of ±11% have been reported as well as a positive bias of 18% for spiked ambient air (Saltzman, 1980; Purdue and Hauser, 1980).

Calibration is usually performed statically using dilute solutions of sodium nitrite. Saltzman (1954) reported that 0.72 mol of nitrite was formed for each mole of $NO_2$ absorbed. Values of this "stoichiometric factor" ranging from 0.5 to 1.0 have been reported (Crecelius and Forwerg, 1970). The method can, however, be calibrated dynamically using calibrated $NO_2$ gas standards.

### 3.5.2.  Continuous Saltzman Method

The measurement principle is based on the Griess–Saltzman reaction. Ambient air is sampled continuously through a gas–liquid contactor, and $NO_2$ is collected on contact with an absorbing solution containing diazotizing–coupling reagents. After the color has developed, the absorbance of the solution is measured continuously with a spectrophotometer. Ozone was found to act as a negative interferent and eliminated this method as a candidate for designation as an Equivalent Method by EPA. The MDL is 10 ppb, the bias ranges from +3 to +15% at $NO_2$ concentrations between 30 and 150 ppb, and the precision is approximately ±12% (Purdue and Hauser, 1980). Although calibration can be performed using nitrite solutions or calibrated $NO_2$ gas standards, the latter approach is recommended.

### 3.5.3.  Alkaline Guaiacol Method

Various alternatives to the Griess–Saltzman method have been proposed. Recently, Baveja et al. (1984) proposed a method using alkaline o-methoxyphenol (guaiacol) as both the absorbing medium and coupling reagent. Samples are collected using fritted bubblers containing an alkaline guaiacol solution. After sampling, p-nitroaniline is added, the pH is adjusted with HCl and later with NaOH, the resulting dye is extracted in amyl alcohol, and the absorbance is read at 545 nm. Collection efficiency of 98% and a stoichiometric factor of 0.74 were reported.

### 3.5.4.  Jacobs–Hochheiser Method

This method is a modified version of the Griess–Saltzman method to permit 24-h sampling and delay in analysis times (Jacobs and Hochheiser, 1958).

Samples are collected by bubbling ambient air through a 0.1N aqueous NaOH solution using a fritted bubbler. The collected nitrite is then reacted with sulfanilamide and NEDA in acid media to form an azo dye which is measured with a spectrophotometer at 540 nm. As with the Griess–Saltzman method, dilute nitrite solutions are used for calibration.

This method was the original Reference Method designated by the EPA for $NO_2$ (Purdue and Hauser, 1980). Testing of the method showed that the originally specified $NO_2$ collection efficiency of 35% was not constant and that it varied nonlinearly with $NO_2$ concentration. In addition, interferences from NO and combinations of NO and $NO_2$ were found. As a result, in 1973 this method was withdrawn and considered unacceptable for air pollution work.

### 3.5.5.  Sodium Arsenite Method (Manual and Continuous)

This method has been designated by the EPA as an Equivalent Method in both the manual and continuous forms (Federal Register, 1986). The manual method is a 24-h integrated method similar to the Jacobs–Hochheiser method, except that sodium arsenite is added to the aqueous NaOH absorbing solution, and an orifice bubbler is used. The nitrite is reacted with sulfanilamide and NEDA in acid media to form an azo dye which is determined with a spectrophotometer. The continuous method employs the same measurement principle but uses hardware to permit continuous determination of $NO_2$ in a manner similar to that used with the continuous Saltzman method.

The overall $NO_2$ recovery is 82%. Although NO and $CO_2$ may act as interferents, their impact is minimal at typical ambient levels. Sulfur dioxide has not been tested as an interferent. The MDL is 5 ppb, the bias is $-3\%$ independent of concentration, and the precision at $NO_2$ concentrations between 30 and 150 ppb is $\pm 6$ ppb (Purdue and Hauser, 1980). Recently $HNO_2$ was found to respond equivalently to $NO_2$ (Braman et al., 1986). This interference is likely to be appreciable in urban environments during nighttime hours where concentrations above 5 ppb have been observed (Rodgers and Davis, 1989; Appel et al., 1990).

### 3.5.6.  Triethanolamine–Guaiacol–Sulfite (TGS) Method

This method has been designated by the EPA as an Equivalent Method (Federal Register, 1986). It is a manual 24-h integrated method. Samples are collected using orifice bubblers and a solution of triethanolamine (TEA), guaiacol, and sodium metabisulfite. The resulting nitrite is reacted with sulfanilamide and 8-amino-1-naphthalene sulfonic acid ammonium salt (ANSA) and the resulting azo dye is determined at 550 nm with a spectrophotometer. No interferences were found in tests with $NH_3$, CO, HCHO, NO, phenol, $O_3$, and $SO_2$. The overall $NO_2$ recovery is 93%, the MDL is 8 ppb, the bias is $-5\%$ independent of concentration, and the precision at $NO_2$ concentrations between 30 and 150 ppb is $\pm 6$ ppb (Purdue and Hauser, 1980).

### 3.5.7. Triethanolamine (TEA) Method

This method is a manual 24-h integrated method (Ellis and Margeson, 1974). Samples are collected using an aqueous solution of TEA and fritted bubblers. As with the Equivalent Methods, the resulting nitrite may be determined with a spectrophotometer after reaction with sulfanilamide and either NEDA or ANSA. Although recoveries of 80–90% were found at $NO_2$ concentrations between 20 and 350 ppb using fritted bubblers, only 50% recovery was found using orifice bubblers. Since the requirement of a fritted rather than an orifice bubbler was considered to be a major disadvantage by EPA, the development of this method as an Equivalent Method was terminated (Purdue and Hauser, 1980).

TEA has been used as the collection medium for many active and passive techniques to sample $NO_2$. Although colorimetry may be used as the analytical finish, recently IC appears to be the method of choice (Miller, 1984). Vinjamoori and Ling (1981) have used an aqueous solution of TEA, ethylene glycol, and acetone on $13 \times$ molecular sieves to sample air in the workplace for $NO_2$. Passive devices employing TEA have been used for industrial hygiene and indoor air quality sampling (Palmes et al., 1976; Wallace and Ott, 1982) as well as for ambient applications (Sickles and Michie, 1987; Mulik and Williams, 1987). Recently, a method using Whatman GF/B filters coated with an aqueous solution of TEA, ethylene glycol, and acetone has been developed for extended sampling of both $NO_2$ and $SO_2$ from ambient air (Sickles et al., 1990a). This method using IC to determine the collected $NO_2$ as nitrite and nitrate showed no interferences from NO, $NH_3$, $O_3$, $H_2S$, $CH_3SH$, and $CS_2$ but major interferences from PAN and $HNO_2$. Recovery of $NO_2$ averaged 87% in laboratory tests at concentrations between 5 and 400 ppb. Using two coated 47-mm-diam filters in series, filter temperatures above 5 °C and flow rates below 2 L min$^{-1}$ are required to ensure good $NO_2$ collection efficiency. This method has been incorporated into the design of a prototype sampler known as a transition flow reactor (TFR) system for measuring acidic deposition components (Knapp et al., 1986).

### 3.6. Other Active Methods

Several other methods for the determination of $NO_2$ have been reported. These methods include ionization spectroscopy, mass spectrometry, photothermal detection, denuders, and solid sorbents. Since they are in the early stages of development or are not being used widely, they are mentioned only briefly.

Ionization spectroscopy is a new and sensitive in situ laser technique that is currently under investigation for tropospheric measurements of NO and $NO_2$ (NASA, 1983). This method is in the early stages of development.

Atmospheric pressure ionization mass spectrometry (API/MS) has been investigated for the continuous measurement of $NO_2$ and $SO_2$ in ambient air (Benoit, 1983). A MDL of approximately 0.5 ppb was reported.

Methods employing photothermal detection of $NO_2$ have been reported (Poizat and Atkinson, 1982; Higashi et al., 1983; Adams et al., 1986). Detection is accomplished by selectively exciting transitions of $NO_2$ with a chopped continuous wave or pulsed laser source. At pressures near atmospheric, collisional de-excitation converts the absorbed energy into translational energy, leading to a temperature rise along the beam and expansion of the gas. The resulting change in refractive index of the thermal lens can be detected by spectroscopic means (Higashi et al., 1983), or the resulting pressure change can be detected with a microphone (Poizat and Atkinson, 1982). MDLs of 2–5 ppb have been reported. Development of photoacoustic methods for continuous application may be limited by acoustic noise associated with vibration and flow.

A tube or channel with its walls coated with a chemical chosen to remove a gaseous species of interest from a sample drawn through the tube under laminar conditions is known as a denuder. The concentration of the species of interest is determined by measuring the amount of the species collected on the walls or by comparing the signal strengths in the presence and absence of the denuder. Possanzini et al. (1984) recommended using a coating containing KI to collect $NO_2$. Subsequent tests showed the collection efficiency of this material to be dependent on the humidity of the sample (Sickles, 1987). An alkaline guaiacol coating on annular denuders has shown high collection efficiency for $NO_2$ (Buttini et al., 1987). After extraction using DI water, IC analysis showed quantitative recovery as $NO_2^-$. The reported MDL was 0.13 ppb for a 1 $m^3$ sample (i.e., 16 L $min^{-1}$ for 1 h), and the median precision for 14 paired samples was 3.9% expressed as relative standard deviation (RSD). A direct interference was noted with $HNO_2$, none was found with NO or PAN, and no humidity effects were observed between 20 and 80% RH. No interference tests were performed with $O_3$; however, comparison of 4-h results with those of a commercial CLM (NO + $O_3$) instrument sampling air near Rome showed good correlation (i.e., $r = .92$). Although alkaline guaiacol solutions degrade with time, no degradation effects were found for 72-h presampling or 24-h postsampling denuder storage. Longer duration storage tests (e.g., 2 weeks pre- and postsampling) and additional field evaluations are needed before this method is ready for routine application.

Recently, Adams et al. (1986) found activated $MnO_2$ to be effective under unspecified RH for removing $NO_2$ in denuder applications. Additional tests are needed under conditions representative of the ambient environment before this approach is ready for routine application.

Lipari (1984) has used a commercially available cartridge containing the sorbent Florisil (magnesium silicate) coated with diphenylamine (DPA) to sample $NO_2$ in ambient and indoor environments. The $NO_2$ reacts with the diphenylamine to form nitro and nitroso derivatives, which are subsequently detected by HPLC–UV. Although no interference from NO, $O_3$, $SO_2$, $HNO_3$, and water vapor were found, PAN produced a 50% positive interference, and $HNO_2$ was expected to interfere quantitatively. Sorbent temperature must be held below 32 °C to prevent volatilization of the diphenylamine and non-quantitative sampling. A MDL of 0.1 ppb for a 2.0-m³ air sample was claimed. The method shows promise for the sensitive determination of $NO_2$ under conditions where the noted interferents and temperature sensitivity do not pose problems.

## 3.7. Passive Samplers

Passive samplers are frequently used in industrial hygiene, indoor air, and personal exposure studies and are less frequently used in ambient air sampling. Passive NO samplers are described in Section 2.4. Namiesnik et al. (1984) have provided a good overview of passive samplers. The basis for all passive samplers is the same. The gaseous analyte molecule is transported from the bulk air to a reactive surface where the molecule impinges and is captured by chemical reaction. After exposure periods ranging from hours to days, the reactive surface is analyzed and the integrated loading of the reaction product is used to infer the average gas concentration present during the sampling period. When the transport of analyte molecules to the reactive surface is diffusion controlled, the average ambient concentration may be related directly to the ratio of the product loading to the sampling duration. This proportionality is defined as a sampler calibration factor or alternatively as the reciprocal of the sampler sampling rate.

One type of passive $NO_2$ sampler for ambient applications is the nitration plate. It is essentially an open petri dish containing TEA-impregnated filter paper. Thus, there is no diffusion barrier between the ambient air and the $NO_2$ collection surface. Nitrogen dioxide reacts with the TEA and is retained primarily as $NO_2^-$, which can be extracted and determined with a spectrophotometer or by IC. A single calibration factor is provided by the manufacturer. A recent study indicated that the calibration factors determined experimentally for the device are extremely sensitive to wind speed, $NO_2$ concentration, and temperature (Sickles and Michie, 1987). TEA is expected to collect not only $NO_2$ but also $HNO_2$, $HNO_3$, and PAN (Sickles, 1987). These results suggest that nitration plates may be useful only as qualitative indicators of ambient levels of oxides of nitrogen.

Another open surface device has been proposed by Kosmus (1985) for ambient applications. This device uses chromatographic paper in the shape of a candle that is coated with diphenylamine and is continually impregnated with a KSCN-glycerin catalyst. Nitrogen oxides, presumably $NO_2$ and NO to some extent, are collected by a catalyzed reaction with diphenylamine to form the nitrosamine. After extraction, the nitrosamine may be determined with a spectrophotometer or by differential pulse polarigraphy. Interference by iron oxide particles was noted, and interference from both PAN and $HNO_2$ is expected (Lipari, 1984). Sensitivity to wind speed as noted previously for the nitration plate is also expected. Collocated sampling was performed using four candles and a CLM NO/NO₂ analyzer at each of 14 stations (Kosmus,

1985). The nitrosamine loadings were highly correlated but in a nonlinear manner with the sum of 100% $NO_2$ plus 90% NO from the CLM instruments. These results also suggest that open surface passive samplers may be useful as qualitative indicators of ambient levels of oxides of nitrogen.

Mulik and Williams (1986) have adapted the nitration plate concept by adding diffusion barriers in their design of a passive sampling device (PSD) for $NO_2$ in ambient and personal exposure applications. The physical configuration employs a TEA-coated cellulose filter that uses two 200-mesh stainless steel diffusion screens and two stainless steel perforated plates on each side of the coated filter to act as diffusion barriers and permit $NO_2$ collection on both faces of the filter. After sampling, the filter is removed from the PSD, extracted in water, and analyzed for $NO_2^-$ by IC. A sensitivity of 0.03 ppm-h and a sampling rate of 2.6 $cm^3$ $s^{-1}$ were claimed. Comparison of PSD results with CLM determinations of $NO_2$ in laboratory tests at concentrations between 10 and 250 ppb were linearly related and highly correlated (i.e., $r$ = .996). The device exhibited increased sampling rates of approximately 50% as the wind speed increased from 20 to 45 cm $s^{-1}$, but displayed a relatively constant sampling rate at wind speeds between 45 and 300 cm $s^{-1}$ (Mulik and Williams, 1987). Interferences from PAN and $HNO_2$ are expected (Sickles, 1987). Results of TDLAS and triplicate daily PSD $NO_2$ measurements in a recent 13-day field study showed good agreement between the study average values but a correlation coefficient for daily results of only 0.47 (Mulik and Williams, 1987; Sickles et al., 1990b). Further development and testing of the PSD appears warranted.

The Palmes tube is a passive device that has been used to sample air in the workplace and indoor environments to assess personal exposure to $NO_2$ (Palmes et al., 1976; Wallace and Ott, 1982). This device and its operation are described in detail in Section 2.4. It consists of a 0.95-cm i.d. by 7.1-cm-long acrylic tube, open at one end with a TEA-coated interior surface on the closed end. Nitrogen dioxide diffuses through the 7.1-cm diffusion length where it is captured by TEA. Analysis is accomplished by extracting the TEA-coated surface and analyzing the extract for $NO_2^-$. This may be done directly by adding an aqueous solution of sulfanilamide and NEDA to the tube and determining the $NO_2^-$ concentration with a spectrophotometer at 540 nm (Palmes et al., 1976). A stoichiometric factor of unity, a linear response for dosages between 1 and 30 ppm-h, and a sampling rate of 0.02 $cm^3$ $s^{-1}$ are reported. An improvement in sensitivity from 0.3 to 0.03 ppm-h is claimed if the aqueous TEA extract is analyzed by IC using a concentrator column (Mulik and Williams, 1986). Absorption and desorption of $NO_2$ by the internal walls of the acrylic tube have been recently reported to limit applications to exposures of 0.1 ppm-h (Miller, 1988). This sensitivity can be improved to 0.03 ppm-h by using stainless steel rather than acrylic tubes. The device exhibited sampling rates increased by up to 15% as wind speed was increased from 50 to 250 cm $s^{-1}$ (Palmes et al., 1976) and decreased by 15% as temperature was reduced from 27 to 15 °C (Girman et al., 1984). Interferences from PAN and $HNO_2$ are expected (Sickles, 1987). The calibration factor for the Palmes tube is theoretically derived, and the analytical finish is calibrated by dilution of gravimetrically prepared nitrite solutions.

The performance of the Palmes tube was compared with that of two commercially available passive personal samplers, the DuPont Pro-Tek and the MDA Chronotox System (Woebkenberg, 1982). The Palmes tube displayed greater sensitivity than either of the commercial samplers and displayed adequate precision and accuracy at loadings between 1 and 80 ppm-h. Since the commercial devices may be used at only moderate to high loadings (i.e., above 5 ppm-h), they are not sufficiently sensitive for most ambient or personal exposure applications. As a result, they are not discussed further in this document.

A badge-type $NO_2$ personal sampler has been devised by Yanagisawa and Nishimura (1982). Their device uses a series of five layers of hydrophobic Teflon-type filter material as a diffusion barrier between ambient air and a TEA-coated cellulose fiber filter. Nitrogen dioxide diffuses through the hydrophobic flters to the TEA-coated surface, where it is collected. Following extraction of the TEA-coated filter in a solution of sulfanilic acid, phosphoric acid, and NEDA, a colorimetric finish at 540 nm is employed. A sensitivity of 0.07 ppm-h, a sampling rate of 1.4 $cm^3$ $s^{-1}$, and an accuracy of $\pm 20\%$ are claimed. The device exhibited increased sampling rates of up to 30% as the wind speed was increased from 15 to 400 cm $s^{-1}$. Interferences from PAN and $HNO_2$ are expected (Sickles, 1987). The calibration factor for the sampler is provided by the supplier, and the analytical finish is calibrated by dilution of gravimetrically prepared nitrite solutions.

A variation on the above approach has been proposed by Cadoff and Hodgeson (1983). The sampler is composed of a Nuclepore 47-mm filter holder with a capped base containing a TEA-coated glass fiber filter and a 0.8-$\mu$m pore size polycarbonate filter. The polycarbonate filter and the air space between this filter and the TEA-coated filter act as a diffusion barrier. A colorimetric analytical finish is employed. The performance was tested at $NO_2$ loadings between 0.06 and 1 ppm-h, and a sampling rate of 1.9 $cm^3$ $s^{-1}$ was claimed.

West and Reiszner (1978) proposed a passive $NO_2$ sampler using a silicone membrane as a diffusion barrier between ambient air and an alkaline thymol blue $NO_2$ collection solution. Collected $NO_2$ is converted to $NO_2^-$ and determined with a spectrophotometer. Results of a field comparison with the EPA-designated TGS method were not favorable and showed the proposed device to yield results approximately a factor of three higher than the TGS method. As a result, this method is not recommended.

### 3.8. Calibration

Two methods have been designated by the EPA (Code of Federal Regulations, 1987a) as alternative calibration procedures for the measurement of

$NO_2$ in the atmosphere. These methods use permeation tubes or gas phase titration (GPT) to generate known amounts of $NO_2$. Calibrations are performed using dynamic dilution with air.

A permeation tube is a porous, inert tube, usually made of Teflon, that has been partially filled with liquid $NO_2$ and sealed. Permeation of $NO_2$ through the porous tube will occur at a constant rate if the temperature of the tube and the $NO_2$ concentration gradient across the tube are held constant. The tube is maintained at a constant temperature ($\pm 0.1$ °C), and a measured flow of a dry carrier, usually nitrogen, is passed over it. The $NO_2$ permeating through the porous wall and entering the carrier stream is diluted with zero air to produce calibration $NO_2$ atmospheres of known concentrations. The permeation tube is calibrated gravimetrically by measuring the weight loss of the tube over time. The NBS provides SRM permeation tubes that emit $NO_2$ at a nominal rate of 1 $\mu g$ $min^{-1}$ (NBS, 1988). Additional information on the performance of $NO_2$ permeation tubes is given by Hughes et al. (1977). A recent report by Braman and de la Cantera (1986) indicates that permeation sources of $NO_2$ can produce atmospheres that are contaminated with other oxides of nitrogen including $HNO_3$, $HNO_2$, and NO. Further work appears warranted to define the conditions where permeation devices may be used to provide an unambiguous source of $NO_2$.

GPT employs the rapid, quantitative gas phase reaction between NO, usually from a standard gas cylinder, and $O_3$, from a stable $O_3$ generator, to produce one $NO_2$ molecule for each NO molecule consumed by reaction. When $O_3$ is added to excess NO in a titration system, the decrease in NO (and $O_3$) is equivalent to the $NO_2$ produced. Different amounts of $NO_2$ may be produced by adding different amounts of $O_3$. When the NO concentration and the flow rates entering the dynamic titration system are known accurately, the $NO_2$ concentration leaving the system can be determined accurately. The accuracy and stability of NO standard gas cylinders are described in Section 2.5.

A third source of $NO_2$ sometimes used for calibration is the cylinder of compressed gas containing $NO_2$ usually in $N_2$ (Fehsenfeld et al., 1987; Davis, 1988). Calibrations are subsequently performed using dynamic dilution with zero air. These cylinders are commercially available, and the $NO_2$ concentration should be referenced to an accepted standard. Bennett (1979) has shown that of 26 aluminum cylinders initially containing supplier-certified concentrations of $NO_2$ in $N_2$ between 100 and 300 ppb, 10 showed modest declines in $NO_2$ concentration during the first 3 months after preparation. The $NO_2$ levels in all 26 cylinders declined substantially over the 10-month study period. Schiff et al. (1983) have noted problems handling trace concentrations of $NO_2$ from a cylinder. A cylinder containing 9 ppm $NO_2$ in $N_2$ gave 15% higher readings for $NO_2$ when analysis was performed by CLM than by TDLAS (Walega et al., 1984). This discrepancy may have been due to an impurity (e.g., $HNO_3$) in the cylinder that could act as an interference with the CLM but not the TDLAS determination of $NO_2$. Davis (1988) examined a cylinder containing 44 ppm $NO_2$ in air at regular intervals over 3 years and observed a 16% change in concentration. In view of these findings, caution should be exercised if a cylinder containing $NO_2$ in $N_2$ or air is to be employed as a calibration source of $NO_2$.

## 3.9. Intercomparisons

Several intercomparisons of research grade $NO_2$ instrumentation have been conducted (Helas et al., 1981; Walega et al., 1984; Sickles et al., 1990b; Fehsenfeld et al., 1987, 1990; Gregory et al., 1990c) and are described in this section. The performance of EPA-Designated Methods based on intercomparisons and other studies is discussed in Section 3.10.

Helas et al. (1981) report the results of a field intercomparison of several $NO_2$ methods conducted in April 1979 at Deuselbach, Germany. Good agreement between a highly sensitive CLM instrument and long-path absorption was found over the 1- to 8-ppb range of observed $NO_2$ concentrations.

Walega et al. (1984) report comparisons of $NO_2$ measurements from a highly sensitive CLM instrument using a thermal $NO_2$ to NO converter with $NO_2$ measurements from a TDLAS system. Measurements of $NO_2$-spiked synthetic air conducted both in the laboratory and in the field showed good agreement. Measurements were made of ambient and captive air in downtown Los Angeles and showed maximum respective concentrations of 100 and 600 ppb. CLM results were appreciably higher than those of the TDLAS: This difference averaged 18% in the ambient air studies and 15% in the captive air studies. In the latter studies the agreement was generally within 10% in the morning but by the end of the day could be as large as 80%. This behavior was attributed to the reaction of $NO_2$ and the accumulation of photochemically produced CLM interferents such as PAN that occurred during the day.

Daily $NO_2$ concentrations determined by TDLAS, CLM (luminol), and PSDs were reported recently from a 13-day study conducted at Research Triangle Park, North Carolina, in the Fall of 1986 (Bubacz et al., 1987; Mulik and Williams, 1987; Sickles et al., 1990b). Collocated sampling was performed using a TDLAS system, two CLM (luminol) instruments, and triplicate daily PSDs. Daily average results were computed for the TDLAS, each CLM (luminol) instrument, and the PSDs. The 13-day average values from the CLM (luminol) instruments and the TDLAS system agreed to within 2 ppb; the average daily ratios of $NO_2$ by CLM (luminol) to TDLAS were 1.01 ± 0.11 (SD) and 1.19 ± 0.17; the respective correlation coefficients were high, 0.94 and 0.91; and while the results of one CLM (luminol) instrument showed no bias, results of the other were biased higher than those of the TDLAS system. The 13-day average values from the PSDs and TDLAS system agreed to within 1 ppb; the average daily ratio of $NO_2$ by PSD to TDLAS was 1.08 ± 0.32; and while there was no apparent bias, the correlation coefficient was only 0.47.

A field intercomparison of instruments designed to measure NO, $NO_x$, and $NO_y$ was conducted near Boulder, Colorado (Fehsenfeld et al., 1987). In addition, an intercomparison of $NO_2$ measurements was performed using two different $NO_2$ to NO converters prior to NO detection by CLM. The two CLM detection systems were tested and found to be equivalent. One instrument used a photolytic $NO_2$ to NO converter (CLM-PC), while the other used a $FeSO_4 \cdot 7H_2O$ surface converter. In spiking tests, the instrument with the $FeSO_4$ converter responded to $NO_2$, PAN, and $n$-propyl nitrate but not to $HNO_3$ or $NH_3$. The CLM-PC instrument responded to $NO_2$ but not significantly to $HNO_3$, $NH_3$, $n$-propyl nitrate, or PAN. For measurements of $NO_2$ + NO in ambient air, results from the two instruments agreed at concentrations above 1 ppb. However, results from the instrument with the surface converter were biased higher than those from the photolytic converter at lower NO + $NO_2$ concentrations. This discrepancy was a factor of 2 at 0.1 ppb. These results suggest that surface converters sufficiently active to convert $NO_2$ to NO can convert other nitrogen oxide species such as PAN to NO. While the use of CLM with surface $NO_2$ to NO converters may not pose a problem in many urban and suburban areas where NO and $NO_2$ are expected to be the dominant oxides of nitrogen, results cited here and elsewhere in this section suggest that surface converters are unsuitable for the interference-free measurement of $NO_2$ in ambient air containing PAN and similar compounds.

A ground-based intercomparison of $NO_2$ measurements using CLM-PC, CLM (luminol), and TDLAS research-grade instruments was performed near Boulder, Colorado (Fehsenfeld et al., 1990). Ambient concentrations ranged from 0.02 to 4 ppb. The potential interferences of $H_2O_2$, $HNO_3$, $n$-propyl nitrate, PAN, and $O_3$ were examined in spiking tests. Only the CLM (luminol) instrument displayed appreciable interferences, and they were with $O_3$ (0.6%) and PAN (24%). At ambient $NO_2$ concentrations above 2 ppb all three instruments gave similar results. Below 2 ppb, interferences from $O_3$ and PAN provided high biases to the CLM (luminol) results, but they could be corrected with measured $O_3$ and PAN results at $NO_2$ levels above 0.3 ppb. An $O_3$ scrubber added to a second CLM (luminol) instrument removed the $O_3$ interference but failed to remove PAN and appeared to remove substantial amounts (i.e., 50%) of $NO_2$. Removal of $NO_2$ in the manufacturer-supplied $O_3$ scrubber has also been reported by Kelly et al. (1990). TDLAS results compared favorably with CLM-PC at relatively high $NO_2$ levels (i.e., $>0.4$ ppb) but displayed a high bias (i.e., factor of 5) at lower $NO_2$ concentrations. No interferences or artifacts were found for the CLM-PC results.

An airborne intercomparison (i.e., CITE 2) of $NO_2$ measurements was conducted by NASA using TDLAS, PF/TP-LIF, CLM-PC, and CLM (with $FeSO_4$ converter) research-grade instruments (Gregory et al., 1990c). Sampling flights were performed primarily in the free troposphere, and $NO_2$ concentrations were below 200 ppt and generally below 100 ppt. High biases (i.e., factors of 2–3) apparently resulting from PAN interferences were present in results from the CLM instrument with the $FeSO_4$ converter, and results from this instrument were not considered in subsequent analyses. At concentrations below 200 ppt, results from the remaining three instruments were highly correlated (i.e., correlation coefficients ranged from .84 to .95) and displayed a general level of agreement to within 30 to 40%. The PF/TP-LIF results were higher than those of the CLM-PC, and the TDLAS results were the lowest. At concentrations below 50 ppt the results were poorly correlated, although the PF/TP-LIF and CLM-PC results agreed to within 20 ppt. Below 50 ppt, TDLAS results were much higher than those of the other two instruments. This bias, similar to that observed in the ground-based intercomparison of Fehsenfeld et al. (1990), was enhanced at low NO$_2$ concentrations by an error in the data reduction protocol employed in both studies.

## 3.10. Designated Methods

Sampling and analysis methodologies for NO$_2$ have been specified by the EPA (Code of Federal Regulations, 1987b). These designated methods are termed "Reference" or "Equivalent." In 1973, the original Federal Reference Method for NO$_2$, the Jacobs–Hochheiser technique, was withdrawn because of technical deficiencies (Purdue and Hauser, 1980). In 1976, the measurement principle and the associated calibration procedure on which Reference Methods for NO$_2$ must be based were specified. The measurement principle is gas phase chemiluminescence and the calibration procedure may employ either GPT of an NO standard with O$_3$ or an NO$_2$ permeation device (Code of Federal Regulations, 1987a). Since only the measurement principle and calibration procedures applicable to NO$_2$ Reference Methods were specified, different analyzers can be built and designated as Reference Methods provided they meet the performance specifications shown in Table 1 (Code of Federal Regulations, 1987b).

To be designated as an Equivalent Method, the candidate method must be based on measurement principles different from the Reference Method and meet certain performance specifications (Code of Federal Regulations, 1987b). An Equivalent Method may be either manual or automated. To be designated as Equivalent a candidate manual method must demonstrate comparability as shown in Table 2, with the Reference Method when applied simultaneously to a real atmosphere. A candidate automated method must meet the performance specifications shown in Table 1 and demonstrate comparability as shown in Table 2 with the Reference Method when applied simultaneously to a real atmosphere.

Methods designated by the EPA as Reference and Equivalent are identified in Table 3 (Federal Register, 1986). Detailed descriptions of these as well as other methods for NO$_2$ are given in previous subsections. Studies were conducted to provide a basis for the designation of methods by the EPA. Tests were performed to compare the performance of CLM, continuous colorimetric, manual sodium arsenite, and manual TGS methods (Purdue and Hauser, 1980). The methods were compared by measuring NO$_2$ in spiked and

**Table 1 Performance Specifications for Nitrogen Dioxide Automated Methods**

| Performance Parameter | Units | Nitrogen Dioxide |
|---|---|---|
| Range | ppm | 0–0.5 |
| Noise | | |
|   0% upper range limit | ppm | 0.005 |
|   80% upper range limit | ppm | 0.005 |
| Lower detectable limit | ppm | 0.01 |
| Interference equivalent | | |
|   Each interferent ($SO_2$, NO, $NH_3$, $H_2O$) | ppm | ±0.02 |
|   Total interferent | ppm | ≤0.04 |
| Zero drift, 12 and 24 h | ppm | ±0.02 |
| Span drift, 24 h | | |
|   20% of upper range limit | % | ±20.0 |
|   80% of upper range limit | % | ±5.0 |
| Lag time | min | 20 |
| Rise time | min | 15 |
| Fall time | min | 15 |
| Precision | | |
|   20% of upper range limit | ppm | 0.02 |
|   80% of upper range limit | ppm | 0.03 |

*Source.* Code of Federal Regulations (1987b).

unspiked ambient air simultaneously. Quadruplicate samples were taken for the two manual methods and duplicate analyzers were used for the two continuous methods. The $NO_2$ spikes were varied randomly from day to day over the sampling schedule and ranged from 0 to 430 ppb. Agreement both within and between methods was good: The average difference was never greater than 4 ppb. Correlation coefficients for between-method comparisons exceeded 0.985 in all cases. No between-method differences could be attributed to concentrations of NO, $CO_2$, $O_3$, total sulfur, or total suspended particulate matter in the ambient air. Significant negative interference in the continuous colorimetric method was found at concentrations of 40 and 53 ppb $NO_2$ in the presence of $O_3$ at 180 and 340 ppb. At $O_3$ concentrations of 50 ppb no

**Table 2 Comparability Test Specifications for Nitrogen Dioxide**

| Concentration Range (ppm $NO_2$) | Maximum Discrepancy Specification (ppm) |
|---|---|
| Low, 0.02–0.08 | 0.02 |
| Medium, 0.10–0.20 | 0.02 |
| High, 0.25–0.35 | 0.03 |

*Source.* Code of Federal Regulations (1987b).

**Table 3 Reference and Equivalent Methods for Nitrogen Dioxide Designated by the EPA**

| Method | Designation Number | Method Code |
|---|---|---|
| *NO$_2$ Manual Methods* (Equivalent Methods) | | |
| Sodium aresnite | EQN-1277-026 | 026 |
| Sodium aresnite/Technicon II | EQN-1277-027 | 027 |
| TGS-ANSA | EQM-1277-028 | 028 |
| *NO$_2$ Analyzers* (Reference Methods) | | |
| Beckman 952A | RFNA-0179-034 | 034 |
| Bendix 8101-B | RFNA-0479-038 | 038 |
| Bendix 8101-C | FRNA-0777-022 | 022 |
| CSI 1600 | FRNA-0977-025 | 025 |
| Meloy NA530R | RFNA-1078-031 | 031 |
| Monitor Labs 8440E | RFNA-0677-021 | 021 |
| Monitor Labs 8840 | RFNA-0280-042 | 042 |
| Philips PW9762/02 | RFNA-0879-040 | 040 |
| Thermo Electron 14B/E | RFNA-0179-035 | 035 |
| Thermo Electron 14D/E | RFNA-0279-037 | 037 |

*Source.* Federal Register (1986).

interference was detected. The performance of the CLM analyzers was judged to be superior to that of the continuous colorimetric analyzers with respect to zero drift, span drift, response times, and overall operation. Of the two manual methods, the performance of the sodium arsenite method was judged superior to the TGS method.

Eight of the Reference Methods have undergone extensive postdesignation testing in the laboratory and field (Michie et al., 1983). Performance test results have been reported and were found to meet the specifications shown in Table 1. Based on the field test results, minimum detection limits were defined as three times the precision. These MDL results ranged from 5 to 13 ppb, with an average of 9 ppb. An independent analysis of these data by Holland and McElroy (1986) also showed similar results.

Interrogation of the National Aerometric Data Bank (NADB) records for 1985 revealed that NO$_2$ data were archived from 40 states (C. Hustvedt, personal communication, 1987). Of the 335 data sets, CLM was being employed in 291 cases, and manual methods in the remainder. Of the manual methods, 42 employed the sodium arsenite method with either orifice or fritted bubblers. Interrogation of the Precision Accuracy Reporting System (PARS) data base for the State and Local Air Monitoring Stations (SLAMS) network for fourth quarter 1986 and first quarter 1987 records revealed that data were archived from tests of 114 CLM analyzers (R. C. Rhodes, personal communication, 1987). Of these, 43% were Bendix 8101C, 40% were CSI 1600, 15% were Monitor Labs 8840 and 8440E, 2% were Meloy NA 530R, and 1% was Beckman 952A.

To illustrate the precision and accuracy of the designated methods in field applications, PARS data were examined for 1983 through 1986 (Rhodes and Evans, 1988). The results, shown in Table 4 as 95% probability limits, suggest that the precision of continuous $NO_2$ analyzers falls in the range of $\pm 10$ to 15%, while the manual methods are much worse at $\pm 20$ to 50%. It should be noted that the manual precision results show a recent worsening. This trend may reflect the phasing out of manual methods in the network which was completed by the end of 1986.

The tabulated probability limits for accuracy of continuous $NO_2$ analyzers are $\pm 20\%$, while for the manual methods they are at $\pm 3$ to 7% (Rhodes and Evans, 1988). The accuracy results reflect audits of the analysis portion of the manual methods and audits of both sampling and analysis for the continuous methods. Thus, the apparent difference in accuracy may be reflecting differences in the auditing procedures employed.

## 4.   OXIDES OF NITROGEN ($NO_x$)

For the purposes of this chapter, oxides of nitrogen ($NO_x$) are considered to be the sum of NO and $NO_2$. No widely accepted methods are available for determining $NO_x$ except by determining NO and $NO_2$ individually and summing, or by converting $NO_2$ to NO and determining $NO_x$ as the total NO. Sections 2 and 3 describe methods for the determination of NO and $NO_2$.

**Table 4 National Precision and Accuracy Probability Limit Values Expressed as Percentages for Continuous and Manual Methods for $NO_2$**

| $NO_2$ Method | 1983[a] | | 1984 | | 1985 | | 1986 | |
|---|---|---|---|---|---|---|---|---|
| Continuous $NO_2$ Method | | | | | | | | |
| Precision | $-13$ | $+12$ | $-14$ | $+13$ | $-12$ | $+12$ | $-11$ | $+11$ |
| | (9299)[b] | | (8653)[b] | | (7695)[b] | | (6686)[b] | |
| Accuracy | $-19$ | $+15$ | $-21$ | $+20$ | $-20$ | $+21$ | $-21$ | $+20$ |
| | (680)[c] | | (613)[c] | | (573)[c] | | (529)[c] | |
| Manual $NO_2$ Method | | | | | | | | |
| Precision | $-19$ | $+21$ | $-21$ | $+27$ | $-27$ | $+29$ | $-48$ | $+45$ |
| | (1324)[d] | | (691)[d] | | (469)[d] | | (174)[d] | |
| Accuracy | $-5$ | $+6$ | $-6$ | $+7$ | $-3$ | $+5$ | $-4$ | $+5$ |
| | (348)[c] | | (175)[c] | | (161)[c] | | (92)[c] | |

[a]Calculated differently for 1983 than for 1984 through 1986.
[b]Number of precision checks.
[c]Number of audits; manual at 0.074 to 0.083 ppm; continuous at 0.03 to 0.08 ppm.
[d]Number of colocated samples.
*Source.* Rhodes and Evans (1988).

Commercial CLM $NO_x$ analyzers catalytically convert $NO_2$ to NO and measure $NO_x$ as the sum of the originally present NO and the converted NO. As noted in Section 3.1, $NO_2$ to NO converters used may not be specific for $NO_2$. Heated molybdenum converters, typically used in commercial analyzers, have been shown to convert PAN, $HNO_3$, and other nitroxy compounds to NO, giving rise to artifically high values for $NO_2$ and $NO_x$. In research grade CLM $NO_x$ analyzers, $FeSO_4$ converters have been shown to overestimate $NO_x$ by factors of 2–3 at concentrations below 0.2 ppb (Fehsenfeld et al., 1987; Gregory et al., 1990c).

The catalytic conversion approach will permit an accurate measure of $NO_x$ as long as the nitroxy compounds present in the sampled atmosphere are limited to NO and $NO_2$. Atmospheric concentrations of potential interferences are generally low relative to $NO_2$ (Code of Federal Regulations, 1987a). There are cases, however, where compounds other than NO and $NO_2$ contribute substantially to the atmospheric nitroxy burden. Examples include urban atmospheres such as that of Los Angeles where both PAN and $HNO_3$ levels may reach appreciable levels (Tuazon et al., 1981) and remote environments where PAN may comprise a significant fraction of the airborne nitroxy reservoir (Fehsenfeld et al., 1987; Gregory et al., 1990a).

A prototype method employing CLM has been suggested to measure $NO_x$ (Fontijn et al., 1980). This method uses the reaction between atomic hydrogen and $NO_2$ to give NO along with the subsequent CLM reaction between atomic hydrogen and NO. The emission occurs between 628 and 800 nm, and the intensity is measured by a PM tube at 640–740 nm. At a constant atomic hydrogen concentration, the light intensity is proportional to the $NO_x$ concentration. The instrument was developed for application to automotive exhaust gas. Significant interferences were noted for $O_2$ and ethene but not for $H_2O$, toluene, isopentane, CO, $CO_2$, $NH_3$, and HCN. Response was linear within 2% from 6 to 3000 ppm. Significant development is expected if the limit of detection for this technique is to be extended from 6 ppm to the ppt to ppb range appropriate for ambient air monitoring applications.

## 5. TOTAL REACTIVE ODD NITROGEN ($NO_y$)

In this chapter total reactive odd nitrogen is represented by $NO_y$. Individual components comprising $NO_y$ are NO, $NO_2$, $NO_3$, $N_2O_5$, $HNO_2$, $HNO_3$, $HO_2NO_2$, PAN, other organic nitrates, and particulate $NO_3^-$.

Although no single instrument has been devised to measure $NO_y$, researchers have combined highly sensitive research-grade CLM NO detectors with catalytic converters that are sufficiently active to reduce most of the important gas phase $NO_y$ species to NO for subsequent detection (Helas et al., 1981; Dickerson, 1984; Fahey et al., 1986). Calibrations are performed using dynamic dilution with air. Two standards are usually employed: a cylinder of compressed gas containing NO in nitrogen at an NBS-traceable concentration, and an $NO_2$ permeation tube. The NO cylinder is used to calibrate the instrument for NO, and $NO_2$ from the permeation tube is used as a surrogate to calibrate the instrument for $NO_y$.

Two types of heated converters have been employed: molybdenum and gold. As noted in Section 3.1, heated molybdenum has been shown to convert $NO_2$, $HNO_3$, PAN, methyl nitrate, ethyl nitrate and nitrite, $n$-propyl nitrate, and $n$-butyl nitrate to NO with high efficiency. Dickerson (1984) also reports that $NO_3$ and $N_2O_5$ are converted to NO on heated molybdenum, while $CH_3CN$, HCN, and $NH_3$ are not. Dickerson (1984) has coupled a 400°C molybdenum screen converter with a sensitive CLM NO detector and reported a detection limit for $NO_y$ of 25 ppt for a 20-s integration time and an accuracy of ±40% at levels well above the detection limit (Fehsenfeld et al., 1987).

A gold catalyst operated at 300 °C in the presence of 3000 ppm CO has been reported to reduce $NO_y$ to NO (Bollinger et al., 1983; Fahey et al., 1985a). Converter efficiencies near 100% were found for $NO_2$, $HNO_3$, $N_2O_5$, and PAN. Interferences in the presence of water vapor were found to be negligible for $O_3$, $NH_3$, $N_2O$, HCN, $CH_4$, and various chlorine- and sulfur-containing compounds. Fahey et al. (1986) coupled this converter with a sensitive CLM NO detector and reported a detection limit of 10 ppt for a 10-s integration time and an accuracy of ±15%.

A field intercomparison of the two instruments described above was conducted near Boulder, Colorado (Fehsenfeld et al., 1987). In this study, ambient $NO_y$ concentrations ranged from 400 ppt to over 100 ppb. Both instruments gave similar estimates of $NO_y$ concentrations under conditions that varied from representing urban to continental background air.

Using the instrument described above with a gold converter, Fahey et al. (1986) compared $NO_y$ measurements with the sum of the component species measured individually. The $NO_y$ levels systematically exceeded the sum. The difference was attributed to the presence of one or more unmeasured organic nitrate species that are similar to PAN and may be of photochemical origin.

## 6.  PEROXYACETYL NITRATE (PAN)

Several methods have been used to measure the concentration of PAN in ambient air. Stephens (1969) and Roberts (1990) have provided good overviews of many of these methods. PAN was first measured by using long-path infrared spectrometry; however, insufficient sensitivity by this technique prompted the development of other methods (Darley et al., 1963). A ground-based FTIR system with a 1-km cell has reported detection limits of 4 ppb for PAN near 790 and 1160 $cm^{-1}$ (Tuazon et al., 1978). The limited sensitivity and the complexity of FTIR systems have generally limited ambient applications of the FTIR to the relatively high concentrations associated with the Los Angeles basin. More recently, cryogenic sampling and matrix-isolation FTIR has been used to measure PAN in 15-L integrated samples of ambient air with a theoretical MDL of 50 ppt (Griffith and Schuster, 1987). As described in Section 2.3, a laboratory prototype method, TTFMS, has a projected MDL for PAN of 2 ppt (Hansen, 1989). Gas chromatography with flame ionization detection (GC-FID) may be employed to measure PAN, but this method is practical only for concentrated mixtures, above 10 ppm using a 1-mL sample loop (Meyrahn et al., 1987). The most common method is gas chromatography using electron capture detection (GC-ECD) (Darley et al., 1963; Intersociety Committee, 1972b; Stephens and Price, 1973; Singh and Salas, 1983).

## 6.1.  GC-ECD

Both manual and automated integrated sampling methods using GC-ECD have been employed (Stephens and Price, 1973; Lonneman et al., 1976). Relatively low column and detector temperatures (below 50 and 100 °C, respectively) have been used to minimize thermal decomposition of PAN. Short packed columns coated with polyethylene glycol-type stationary phases (e.g., Carbowax 400) have normally been used. Recently, improved precision and sensitivity have been achieved using fused silica capillary columns (Helmig et al., 1989; Roberts et al., 1989). Although sampling intervals are limited by the elution times of the chromatographic system, intervals of 10–15 min have been employed (Helmig et al., 1989; Nieboer and van Ham, 1976). Using packed columns, detection limits of 10 ppt have been reported using direct sampling with a 20-mL sample loop (Vierkorn-Rudolph et al., 1985) and 1–5 ppt using cryogenic enrichment of samples (Vierkorn-Rudolph et al., 1985; Singh and Salas, 1983). Capillary columns offer the potential for considerable (i.e., factor of 20) enhancement in sensitivity (Roberts et al., 1989). Accuracy estimates of ±20 to 30% have been claimed.

A comparison of two similar GC-ECD methods for airborne PAN measurements was performed (Gregory et al., 1990a). Both methods employed cryogenic enrichment of samples, used packed GC columns, and claimed detection limits below 5 ppt. Results of this study showed that at PAN concentrations below 100 ppt, agreement was approximately 17 ppt, and at higher concentrations (i.e., 100–300 ppt) the measurements agreed to within 25% (expressed as a percent difference). These findings are generally consistent with accuracy claims noted earlier.

## 6.2.  Alkaline Hydrolysis

Alkaline hydrolysis in 5% NaOH has been shown by Nicksic et al. (1967) to convert PAN quantitatively to nitrite and acetate. This permits sampling with a bubbler containing 5% NaOH and subsequent analysis for nitrite or acetate. Nitrogen dioxide is usually present in ambient air with PAN. It can interfere with PAN determination as nitrite since $NO_2$ may be collected as nitrite in alkaline solution. Acetate particles or acetic acid can interfere with PAN determination as acetate. A method involving alkaline hydrolysis followed by IC determination of acetate has been used to measure PAN in photochemical systems (Grosjean and Harrison, 1985a). The results compare favorably with those of a CLM method employing difference in $NO_x$ signals measured upstream and downstream of an alkaline bubbler. In addition to NaOH, other alkaline salts (e.g., KOH and $Na_2CO_3$) have been used to coat filters, cartridges, and annular denuders (Grosjean and Parmar, 1990; Williams and Grosjean, 1990). PAN collection efficiencies ranged from 10 to 100%, depending on the type and amount of the alkaline salt, the flow rate, and the collection device employed.

## 6.3.    GC-Alternate Detectors

As noted in Section 3.1, PAN is readily reduced to NO. Meyrahn et al. (1987) have coupled a GC to separate PAN, NO, and $NO_2$; a molybdenum converter; and a CLM NO analyzer to measure PAN as NO. With a 10-mL sample loop, a detection limit of 10 ppb was reported.

The luminol-based detector has shown sensitivity to PAN (see Section 3.2). Burkhardt et al. (1988) have used gas chromatography and a commercially available luminol-based instrument (i.e., Scintrex LMA-3 Luminox) to detect both $NO_2$ and PAN. Using a sampling interval of 40 s, linear response was claimed from 0.2 to 170 ppb $NO_2$ and from 1 to 65 ppb PAN. Although the PAN calibration was nonlinear below 1 ppb, an MDL of 0.12 ppb was reported. Drummond et al. (1989) have slightly modified the above approach by converting the PAN from the GC column to $NO_2$ and measuring the resulting $NO_2$ with a CLM (luminol) instrument.

## 6.4.    PAN Stability

PAN is an unstable gas and is subject to surface-related decomposition as well as thermal instability. PAN exists in a temperature-sensitive equilibrium with the peroxyacetyl radical and $NO_2$ (Cox and Roffey, 1977). Increased temperature favors the peroxyacetyl radical and $NO_2$ at the expense of PAN. Added $NO_2$ should force the equilibrium toward PAN and enhance its stability. In the presence of NO, peroxyacetyl radicals react rapidly to form $NO_2$ and acetoxy radicals, which decompose in $O_2$ to radicals that also convert NO to $NO_2$. As a result, the presence of NO acts to reduce PAN stability and enhance its decay rate (Lonneman et al., 1982). Stephens (1969) reported that appreciable PAN loss in a metal sampling valve was traced to decomposition on a silver-soldered joint. Meyrahn et al. (1987) reported that PAN decayed according to first-order kinetics at a rate of $2-4\%$ $h^{-1}$ in glass vessels that had been previously conditioned with PAN. They employed 200 ppm PAN in glass vessels and the noted first-order decay as the basis for one proposed method of in-field PAN calibration. In contrast, Holdren and Spicer (1984) found that without added $NO_2$, PAN at 20 ppb decayed in Tedlar bags according to first-order kinetics at a rate of 40% $h^{-1}$. The addition of 100 ppb $NO_2$ acted to stabilize PAN at 20 ppb in Tedlar bags.

A humidity-related difference in GC-ECD response has been reported (Holdren and Rasmussen, 1976). Low responses observed at humidities below 30% and PAN concentrations of 10 and 100 ppb but not 1000 ppb were attributed to sample-column interactions. This effect was not observed by Lonneman (1977). Watanabe and Stephens (1978) conducted experiments at 140 ppb and did not conclude that the reduced response was from faults in the detector or the instrument. They concluded that there was no column-related effect, and they observed surface-related sorption by PAN at 140 ppb in dry acid-washed glass flasks. They recommended that moist air be used to prepare PAN calibration mixtures to avoid potential surface-mediated effects.

Another surface-related effect has been reported for PAN analyses of remote marine air (Singh and Viezee, 1988). PAN concentrations were found to increase by 20–170 ppt, an average factor of 3.2, when the sample was stored in a glass vessel for 1–2 min prior to analysis. This effect remains to be explained.

## 6.5. Calibration

Since PAN is unstable, the preparation of reliable calibration standards is difficult. Several methods have been employed. The original method used the photolysis of ethyl nitrite in pure oxygen (Stephens, 1969). When pure PAN is desired, the reaction mixture must be purified, usually by chromatography, to remove the major by-products, acetaldehyde and methyl and ethyl nitrate (Stephens et al., 1965). For GC calibration, purification is unnecessary; the PAN concentration in the reactant matrix is established from the IR absorption spectrum and subsequently diluted to the ppb working range needed for calibration purposes (Stephens and Price, 1973).

Static mixtures of molecular chlorine, acetaldehyde, and $NO_2$ in the ratio of 2:4:4 can be photolyzed in the presence of a slight excess $NO_2$ to give a near stoichiometric yield of PAN (Gay et al., 1976). This method was adapted by Singh and Salas (1983) and later by Grosjean et al. (1984) using photolytic reactors to provide continuous PAN calibration units at concentrations between 2 and 400 ppb. In the former approach, the PAN concentration is established by measuring the change in acetaldehyde concentration across the reactor. In the latter, the PAN concentration is established by measuring the acetate in an alkaline bubbler where PAN is hydrolyzed.

A static technique involving the photolysis of acetone in the presence of $NO_2$ and air at 250 nm has been reported to produce a constant concentration of PAN (Meyrahn et al., 1987). A Penray Hg lamp is inserted into a mixture of 10 ppm $NO_2$ and 1% acetone and irradiated for 3 min to yield 8.9 ± 0.3 ppm PAN.

PAN can be synthesized in the condensed phase by the nitration of peracetic acid in hexane (Helmig et al., 1989), heptane (Nielsen et al., 1982), octane

(Holdren and Spicer, 1984), or *n*-tridecane (Gaffney et al., 1984). Purification of PAN in the liquid phase is needed using the first two methods. The resulting PAN-organic solution can be stored at $-20$ to $-80\,°C$ with losses of less than 3.6% month$^{-1}$ and can be injected directly into a vessel containing air to produce a calibration mixture. The PAN concentration is normally established by FTIR analysis of the solution or the resulting PAN-air mixture.

As noted in Section 3.1, PAN is readily reduced to NO, and CLM $NO_x$ analyzers have near quantitative responses to PAN. Thus, under some circumstances, CLM $NO_x$ response can be used for PAN calibration. One method uses the difference in $NO_x$ signal measured upstream and downstream of an alkaline bubbler (Grosjean and Harrison, 1985a). Joos et al. (1986) have coupled a CLM $NO_x$ analyzer with a GC system to permit calibration of the ECD response by reference to the CLM $NO_x$ analyzer that has been calibrated by traditional methods.

As noted previously, NO in the presence of PAN is converted to $NO_2$. Approximately four molecules of NO can react per molecule of PAN. Lonneman et al. (1982) have devised a PAN calibration procedure based on the reaction of PAN with NO in the presence of benzaldehyde, which is added to control unwanted radical chemistry and improve precision. Using this approach and an initial NO to PAN ratio of between 10 and 20 to one, the change in NO concentration is monitored with a CLM NO analyzer, the change in PAN GC-ECD response is monitored, and the resulting ratio (i.e., $\Delta NO/\Delta PAN$) is divided by the stoichiometric factor of 4.7 to arrive at a calibration factor for the ECD.

PAN and *n*-propyl nitrate have similar ECD responses. Serial dilution of the more stable compound, *n*-propyl nitrate, has been used for field operations (Vierkorn-Rudolph et al., 1985). This approach is not recommended for primary calibration, however, since it does not permit verification of quantitative delivery of PAN to the detector (Stephens and Price, 1973).

### 6.6. Other Organic Nitrates

Other organic nitrates (e.g., alkyl nitrates, peroxypropionyl nitrate (PPN), and peroxybenzoyl nitrate (PBzN)) are also present in the atmosphere but usually at lower concentrations than PAN (Fahey et al., 1986). In general, similar methods for sampling, analysis, and calibration may be used for other organic nitrates as are used for PAN (Stephens, 1969). Both FTIR and GC-ECD may be used to measure these compounds.

With MDLs of 0.1–0.4 ppb, inspection of 3000 GC-ECD chromatograms recorded at five to nine sites during the 1987 Southern California Air Quality Study yielded only seven possible (but nonprobable) observations of methyl nitrate (Grosjean et al., 1990). Roberts et al. (1989) have reported separation of PAN, PPN, and $C_1$ to $C_4$ alkyl nitrates and the potential increase in sensitivity by a factor of 20 using fused silica coated capillary columns rather than the more conventional coated packed columns. Atlas (1988) has used two 5-mg charcoal traps in series to collect $C_3$ to $C_7$ alkyl nitrates from 12- to 300-L samples at 200–400 mL min$^{-1}$ in remote atmospheres. The traps are extracted in small volumes of benzene and analyzed using capillary GC-ECD. Concentrations as low as 1 ppt were reported. PBzN may be collected as methyl benzoate using bubblers containing methanol-NaOH solutions (Appel, 1973). The resulting methyl benzoate is solvent extracted and analyzed by packed column GC-FID with an MDL of 70 ppt. Recently, a collection method using aqueous alkaline hydrolysis of PBzN to the benzoate ion followed by IC-UV analysis was reported to have a detection limit of 30 ppt in a 60-L sample (Fung and Grosjean, 1985). Using this method a median PBzN level of 0.32 ppb was reported for Los Angeles air samples.

## 7. NITRIC ACID (HNO₃)

Several methods are available for the determination of airborne concentrations of HNO₃. Among them are filtration (Okita et al., 1976; Spicer et al., 1978b), denuder tubes (Forrest et al., 1982; DeSantis et al., 1985; Ferm, 1986), CLM (Joseph and Spicer, 1978), absorption spectroscopy (Tuazon et al., 1978; Schiff et al., 1983; Biermann et al., 1988), and microcoulometry (Spicer et al., 1978b). As a result of its 2-ppb detection limit and long response time, microcoulometry has been largely replaced by other methods. Consequently, only the first four methods listed above are described in this section.

### 7.1. Filtration

Filtration techniques generally employ dual filters that rely on the collection of particulate $NO_3^-$ on the first filter and gaseous HNO₃ as $NO_3^-$ on the second filter. This method is sometimes called the filter pack (FP) method. Typically, filtration is used in conjunction with instrumental detection or subsequent chemical analysis of the material collected on the filter media. Filter extracts are usually analyzed for $NO_3^-$ using IC. Efficient HNO₃ collection has been found with nylon filters (Spicer et al., 1978b) and with filters impregnated with NaCl or NaF (Okita et al., 1976; Forrest et al., 1980; Fuglsang, 1986). The HNO₃ capacity of 47-mm-diameter NaCl-coated filters (500 $\mu$g cm$^{-2}$) far exceeds that of nylon (30 $\mu$g cm$^{-2}$) (Anlauf et al., 1986). This advantage may be offset, since the presence of the chloride ion in the NaCl-coated filter extract may hamper IC determination of $NO_3^-$. With a 47-mm-diameter nylon filter sampling at 1 m$^3$ h$^{-1}$ at a nominal HNO₃ level of 5 $\mu$g m$^{-3}$ (2 ppb), the capacity is sufficient for just over 4 days of sampling. The sensitivity of filtration and other integrative methods depends on the detection limit of the analytical finish, the variability and magnitude of the blank level, collection and extraction efficiencies, and the volume of air sampled. As an example, under an assumed IC detection limit for $NO_3^-$ of 0.05 $\mu$g mL$^{-1}$, a filter extraction volume of 10 mL, negligible blank, quantitative collection and extraction, and a sampled air volume of 24 $m^3$ (i.e., flow rate 1 $m^3$ $h^{-1}$ for 1 da), the minimum sensitivity is 0.02 $\mu g$ $m^{-3}$ (8 ppt). Precision of $\pm 10\%$ and accuracy of $+20$ to $-40\%$ are claimed for FP containing Teflon and nylon filters (Fahey et al., 1986).

Although the $HNO_3$ determination by FP methods is desirable due to low cost, simplicity, and high sensitivity, there is great difficulty in distinguishing between gaseous and particulate forms of nitrate. Errors in the measurement of gaseous $HNO_3$ may be in the form of positive artifacts due to volatilization of collected aerosol nitrates on the prefilter to form gaseous $HNO_3$ (i.e., $NH_4NO_3 \longleftrightarrow NH_3 + HNO_3$) (Appel et al., 1980); reaction of collected particulate nitrates on the prefilter with strong acids, resulting in the release of $HNO_3$ (i.e., $H_2SO_4 + 2NH_4NO_3 \rightarrow (NH_4)_2SO_4 + 2HNO_3$) (Appel and Tokiwa, 1981); or formation of $HNO_3$ on the collection medium by interaction with other $NO_x$ species (e.g., $HNO_2$ or $NO_2$) (Eatough et al., 1988; Spicer and Schumacher, 1979). Negative $HNO_3$ artifacts may result from retention of $HNO_3$ by the prefilter collection medium (Appel et al., 1984); retention of $HNO_3$ by collected particles on the prefilter (Appel et al., 1980); a low capacity for $HNO_3$ on the collection medium; or losses of $HNO_3$ by volatilization or by displacement by other acids.

Inert prefilter materials, such as Teflon, should not collect appreciable amounts of $HNO_3$ (Appel et al., 1979); this, however, does not preclude the possibility of $HNO_3$ reaction with aerosol particles collected on Teflon prefilters. In addition, some types of Teflon may sorb $HNO_3$ to a larger extent than others (Appel et al., 1988). This underscores the importance of using "inert" materials for all surfaces coming into contact with $HNO_3$ to ensure representative sampling.

## 7.2.    Denuders

To avoid some of the artifact problems associated with the use of filters, denuder tube samplers were introduced. In general, a denuder is a tube or channel that has its walls coated with or fabricated from a substance that removes the gaseous species of interest, in this case $HNO_3$ (also see Section 3.6 and Slanina et al., Chapter 3). The $HNO_3$ molecules diffuse to and impact the surface while the sample is drawn through the channel. The flow conditions are usually laminar (Re $<$ 2000) and by taking advantage of differences in diffusivities permit particles to pass through the denuder relatively undisturbed. Using the sampled air volume, the concentration of $HNO_3$ is calculated from the measured amount of $NO_3^-$ collected on the denuder walls (Ferm, 1986) or from the difference of $NO_3^-$ collected downstream in the presence and absence of the denuder (Shaw et al., 1982; Forrest et al., 1982).

Denuder tubes have employed MgO (Shaw et al., 1982), $Na_2CO_3$ (Ferm, 1986), nylon (Mulawa and Cadle, 1985), $Al_2(SO_4)_3$ (Lindqvist, 1985), $MgSO_4$, $BaSO_4$ (Klockow et al., 1989), and tungstic acid ($H_2WO_4$) (McClenny et al., 1982) to retain $HNO_3$. A coating of MgO is frequently used with the denuder difference (DD) approach where one coated denuder and one uncoated denuder are followed by nylon or NaCl-coated filters. The difference in $NO_3^-$ on the two downstream filters is attributed to $HNO_3$ (Shaw et al., 1982). The $Na_2CO_3$ coating is used in methods employing the DD or direct analysis approach. In the latter, the denuder is extracted, and the extract is analyzed for $NO_3^-$ which is attributed to $HNO_3$ (Ferm, 1986).

To maintain laminar flow, the flow rate through many conventional open channel denuders of reasonable dimensions is limited to approximately 1–2 L min$^{-1}$. As a result, long-duration samples may be required to provide sufficient analyte for quantitation. A new type of denuder, the annular denuder (AD), has been developed where the same equivalent diameter permits the flow rate to be increased by a factor of 12 (Possanzini et al., 1983). In the AD, ambient air is passed through the annular space of two concentric tubes. The outside of the inner tube and the inside of the outer tube are coated with a specific gas-absorbing substance. For collecting $HNO_3$, $Na_2CO_3$ has been used (DeSantis et al., 1985). In cases where appreciable $HNO_2$ is present, it may be cocollected on $Na_2CO_3$ and the resulting $NO_2^-$ oxidized to $NO_3^-$ over extended sampling periods by atmospheric oxidants (e.g., $O_3$), and two or more denuders are used to permit resolution of $HNO_3$ and $HNO_2$ (Febo et al., 1986; Perrino et al., 1990). The first denuder is coated with NaCl or NaF to collect $HNO_3$ as $NO_3^-$ and the downstream denuder(s) are coated with an aqueous solution of $Na_2CO_3$ and glycerol to collect $HNO_2$ as the sum of $NO_2^-$ and $NO_3^-$. For a 1-m$^3$ h$^{-1}$, 1-day sample under the same assumptions given earlier for filtration, the MDL for the AD is 0.02 μg m$^{-3}$ (8 ppt). Median precision estimates of 8 and 5% RSD have been reported for thirteen 22-h and twelve 1-wk duration samples (Sickles et al., 1989; Sickles, 1987).

Partial denuders have been fabricated of nylon filter material (Mulawa and Cadle, 1985). These denuders operated under laminar flow conditions have relied on a mathematical description of molecular diffusion to a perfect wall sink along with $HNO_3$ deposition measured along the length of the denuder to infer the sampled, ambient $HNO_3$ concentration. A refinement in the data treatment has been offered recently that considers interferent nitrate on nylon partial denuders (Febo et al., 1988). While these denuders are operated under laminar flow conditions, a recently introduced technique employs a nylon partial denuder that is operated under transition flow conditions (Re $\simeq$ 2600) (Knapp et al., 1986). This approach, called a transition flow reactor (TFR), uses a piece of nylon filter material rolled into a cylindrical shape and placed in a Teflon tube. The sample is drawn through the tube and denuder under transition flow conditions where a constant fraction of the $HNO_3$ is claimed to be collected. The tube is followed by a Teflon and a nylon filter. The $HNO_3$ is calculated by analyzing the denuder extract for $NO_3^-$ and applying the constant collection fraction. The particulate $NO_3^-$ is determined algebraically using the $NO_3^-$ measured in the extracts of the Teflon and nylon filters. For 8.5% collection efficiency and 1 m$^3$ h$^{-1}$, 1-day sample under the same assumptions given earlier for filtration, the MDL for the TFR is 0.2 μg m$^{-3}$ (0.1 ppb). A median precision of 7% RSD has been determined for seven 1-wk duration samples (Knapp et al., 1986).

Automated systems using coated denuders with thermal desorption employ $Al_2(SO_4)_3$, $MgSO_4$, $BaSO_4$, or tungstic acid (TA) to preconcentrate $HNO_3$ for subsequent delivery to instrumental detection systems. In the first case, $HNO_3$ from a 30-L sample is collected on an $Al_2(SO_4)_3$-coated denuder, thermally desorbed, thermally converted to NO, and analyzed by gas chromatography with a photoionization detector (Lindqvist, 1985). A nominal MDL of 5 ppt and precision estimate of $\pm 10\%$ were claimed.

Klockow et al. (1989) have used $MgSO_4$- and $BaSO_4$-coated denuders to collect $HNO_3$. The sample is thermally desorbed and measured with a CLM $NO_x$ analyzer. For a 30-min sample at 5 L min$^{-1}$, a nominal MDL of 0.1 $\mu g/$ m$^3$ (40 ppt) and precision estimate of $\pm 5\%$ were claimed.

A TA-coated denuder has been used to collect $HNO_3$ for analysis on an automated basis with a 40-min cycle time (McClenny et al., 1982). The collected $HNO_3$ is thermally desorbed as $NO_2$, thermally converted to NO, and measured with a commercial CLM $NO_x$ analyzer. A nominal MDL of 70 ppt and a precision estimate of $\pm 10\%$ were reported. Recent claims for a similar device with a 20-min cycle time have included MDL of 20 ppt, accuracy of 15–20%, and precision of 8% (Gregory et al., 1990b). TA-coated denuders have drawbacks: They are difficult to prepare, have low capacities, and are subject to unknown atmospheric interferences (Fellin et al., 1984; Eatough et al., 1985; Roberts et al., 1987).

## 7.3.  CLM

As noted in Section 3.1, $HNO_3$ is readily reduced to NO in the $NO_2$ to NO converters used in commercial and many research-grade CLM $NO_x$ analyzers (Joseph and Spicer, 1978; Bollinger et al., 1983; Fahey et al., 1985a; Grosjean and Harrison, 1985b; Rickman and Wright, 1986). $HNO_3$ measurements that employ CLM generally use a $NO_x$ analyzer to measure the $NO_x$ in a sampled air stream in the presence and absence of a $HNO_3$ scrubber. The difference in these $NO_x$ signals is attributed to $HNO_3$. Nylon filters have been used as a $HNO_3$ scrubber with both commercial and research-grade CLM $NO_x$ analyzers. The instrumental performance for $HNO_3$ is similar to that for $NO_2$ with the same instruments (Joseph and Spicer, 1978; Kelly et al., 1979).

In other methods employing CLM and described in Section 7.2, TA, $MgSO_4$, and $BaSO_4$ are used as regenerable $HNO_3$ scrubbers (McClenny et al., 1982; Klockow et al., 1989). With these methods, $HNO_3$ is collected on a coated denuder, the collected $HNO_3$ is thermally desorbed as $NO_x$ (regenerating the scrubber), the desorbed $NO_2$ is thermally converted to NO, and the resulting NO is measured with a commercial CLM analyzer.

## 7.4.  Absorption Spectroscopy

Absorption spectroscopy is discussed in Sections 2.3 and 3.4. Although FTIR, TDLAS, and potentially TTFMS techniques may be used to measure ambient levels of $HNO_3$, poor sensitivity limits ambient applications of FTIR. A ground-based 23-m multipass FTIR system with a 1-km path length has reported detection limits of 4 ppb near 900 cm$^{-1}$ (Tuazon et al., 1978; Biermann et al., 1988). A theoretical MDL for $HNO_3$ of 10 ppt has been claimed for 15-L integrated samples of ambient air using cryogenic sampling and matrix-isolation FTIR (Griffith and Schuster, 1987).

Cassidy and Reid (1982) report an expected MDL of 0.4 ppb using TDLAS near 1330 cm$^{-1}$. For a 40-m path length near 1720 cm$^{-1}$, the MDL is 0.4 ppb (Schiff et al., 1983). With a 150-m path length Mackay and Schiff (1987) report a MDL of 0.1 ppb and an accuracy of $\pm 20\%$. Although the volumetric residence time in the White cell of the TDLAS is 4 s, sample–surface interactions limit the response time to changes in $HNO_3$ concentration to about 5 min.

As described in Section 2.3, a laboratory prototype method, TTFMS, has been developed (Hansen, 1989). The projected MDL for $HNO_3$ is 0.3 ppt.

## 7.5. Calibration

$HNO_3$ is a highly polar material and consequently interacts readily with many surfaces (Goldan et al., 1983; Appel et al., 1988). This prevents the preparation of stable calibration mixtures in cylinders of compressed gases. Two methods, permeation devices and diffusion tubes, are generally employed to generate calibration atmospheres of $HNO_3$ (Schiff et al., 1983; Goldan et al., 1983). Permeation tubes are described for $NO_2$ in Section 3.8. Permeation tubes for $HNO_3$ are available with various emission rates from commercial suppliers. An alternative permeation device may be fabricated in the laboratory by passing carrier gas through a length of Teflon tubing that is immersed in a reservoir of $HNO_3$ and $H_2SO_4$ (Mackay and Schiff, 1987).

Diffusion tubes are generally fabricated in the laboratory (Schiff et al., 1983). A liquid mixture of $HNO_3$ and $H_2SO_4$ is held in a reservoir that is connected to a clean air dilution manifold by a capillary tube. The $HNO_3$ diffusion rate depends on the length and area of the capillary as well as the temperature of the reservoir.

Nominal $HNO_3$ emission rates for permeation tubes are provided by the supplier and for diffusion tubes may be calculated (Nelson, 1972). Although it is common to calibrate permeation tubes gravimetrically, it has been reported that non-$HNO_3$ species (i.e., $NO_2$) are also released and may account for 10–15% of the observed weight loss (Goldan et al., 1983). Since the emission rate estimate for diffusion tubes is also an approximation, the independent measurement of $HNO_3$ emission rates from permeation and diffusion tubes is recommended.

This measurement may be accomplished by pH titration or by using nylon filters, NaCl-coated filters, or caustic bubblers to collect and quantify the $HNO_3$ as $NO_3^-$. Since caustic bubblers may also collect $NO_2$ to some extent, their use could overestimate the $HNO_3$ emission rate, and a filtration technique is preferred. Alternative but more elaborate methods of confirming the $HNO_3$ emission rate employ FTIR, TDLAS, and the CLM $NO_x$ analyzer, which uses photolysis to convert $NO_2$ to NO.

## 7.6. Intercomparisons

Several field studies have been conducted that have permitted the comparison of different techniques for the measurement of $HNO_3$ (Spicer et al., 1982; Walega et al., 1984; Anlauf et al., 1985; Roberts et al., 1987; Hering et al., 1988; Solomon et al., 1988; Benner et al., 1987; Tanner et al., 1989; Sickles et al., 1990b; Gregory et al., 1990b; Dasch et al., 1989). Results from these studies suggest that the FP overestimates the $HNO_3$ concentrations and that coated denuder thermal desorption techniques in various tested configurations may not provide reliable measurements of $HNO_3$.

An intercomparison of $HNO_3$ measurement methods was conducted in Claremont, California, in August and September of 1979 (Spicer et al., 1982). Ten methods were compared: 5 FP, 2 DD, 2 CLM, and 1 FTIR. The results of five methods (2 FP, 1 DD, 1 CLM, and 1 FTIR) were in excellent agreement with median results.

Walega et al. (1984) report comparisons of CLM and TDLAS $HNO_3$ measurements of ambient and captive air performed during October and November of 1981 in Los Angeles, California. The CLM gave erratic $HNO_3$ results for ambient air. Although CLM and TDLAS measurements of $HNO_3$ in captive air samples were highly correlated, linear regression analysis indicated significant biases.

Measurements of $HNO_3$ were made during June 1982 at a rural site in Ontario using FP, TDLAS, and TA techniques (Anlauf et al., 1985). For daytime measurements, the FP and TA measurements were 16% lower than the TDLAS results. Nighttime TA results exceeded those from the FP by a factor of 2. Roberts et al. (1987) compared FP and TA measurements of $HNO_3$ made at a rural site in the Colorado mountains. The TA results were a factor of 3 higher than those of the FP. It was concluded that there are unknown atmospheric species that interfere with TA measurements of $HNO_3$.

Another $HNO_3$ intercomparison study was conducted in Claremont, California, in September 1985 (Hering et al., 1988). The methods compared include FP, DD, AD, TFR, TA, FTIR, and TDLAS. For the whole study, comparison of method means against mean of methods showed the FP to be 36% high, the DD to be 1% low, the AD to be 21% low, and the TDLAS to be 13% low. Comparison of TFR means against DD means showed the TFR to be 9% high. TDLAS gave lower daytime and higher nighttime readings than the DD. In those few cases where the $HNO_3$ concentrations were sufficiently high to be detected by the FTIR, agreement within reported uncertainties was observed between the FTIR and the FP, DD, AD, and TDLAS. Results from the TA technique were high at night and low during the day, and in view of large systematic differences, they were not included in many of the reported analyses.

During 1986, $HNO_3$ data were collected using DD and FP techniques for 24-h periods every 6 days at 8 sites in the Los Angeles basin and at one background site (Solomon et al., 1988). The annual average DD basin-wide estimate of $HNO_3$ was 4.6 $\mu g\ m^{-3}$. The corresponding FP estimate exceeded that of the DD by 3.4 $\mu g\ m^{-3}$ or by approximately 80%.

A study was conducted in January and February 1986 near Page, Arizona (Benner et al., 1987). Twelve-hour gaseous $HNO_3$ concentrations were measured with FP and AD. The mean $HNO_3$ concentration measured with the FP, 1.1 $\mu g\ m^{-3}$, exceeded that measured with the AD by 10%.

A study was conducted in July 1986 on Long Island, New York, to compare $HNO_3$ measurements resolved to a 6-h basis using high-volume FP, DD, real-time two-channel (i.e., nylon filter versus no nylon filter) CLM, and $Al_2(SO_4)_3$-coated denuder thermal desorption-to-CLM (Tanner et al., 1989). The FP results were highly correlated with those of the DD. The daytime real-time CLM results were correlated with those of the DD, but nighttime real-time CLM results exceeded DD results. This may have been caused by the retention of nighttime $HNO_2$ on the nylon filter. Results with the $Al_2(SO_4)_3$-coated denuder were scattered, mostly lower, and poorly correlated with the other methods.

Daily measurements of $HNO_3$ were made in Research Triangle Park, North Carolina, during 13 days in September and October 1986 (Sickles et al., 1990b). Comparisons of the TDLAS results with those of the AD, FP, and TFR revealed significant differences at the 0.05 level for the comparison between the TDLAS and TFR results. Significant differences were not apparent in the other two cases. Comparisons of the studywide means of daily ratios of AD, FP, and TFR to TDLAS results showed the AD to be 5% low, the FP to be 8% high, and the TFR to be 36–76% high.

An airborne intercomparison of three $HNO_3$ methods was conducted by NASA during the summer of 1986 (Gregory et al., 1990b). The methods include FP, TA-coated denuder with thermal desorption-to-CLM, and TDLAS. Sampling flights were performed primarily in the free troposphere, and the majority of the $HNO_3$ results were below 300 ppt. At concentrations below 150 ppt (i.e., typical of the free troposphere) the results were poorly correlated and none of the methods provided unambiguous $HNO_3$ results (i.e., 20% accuracy). At concentrations below 300 ppt the FP and TA denuder results were correlated ($r = .64$), with the FP high by 16%. At concentrations above 150 ppt the TDLAS results were higher than those of the other methods.

Daily measurements of $HNO_3$ were made during approximately 60 days in winter 1987/1988 at a site in Warren, Michigan, using AD and FP (Dasch et al., 1989). The relatively cold wintertime temperatures are expected to minimize positive $HNO_3$ artifacts in FP results from volatilization of collected aerosol nitrates. The average concentrations for the AD and FP agreed to within 4%.

## 8.   NITROUS ACID (HNO₂)

The measurement of $HNO_2$ in ambient atmospheres is receiving increased attention. Currently available techniques employ denuders (Ferm and Sjödin, 1985), AD (DeSantis et al., 1985; Perrino et al., 1990), CLM (Braman et al., 1986), PF/LIF (Rodgers and Davis, 1989), and absorption spectroscopy (Biermann et al., 1988).

### 8.1.   Denuders

See Sections 3.6 and 7.2 for additional discussions of denuders. As noted in Section 2.1, Braman et al. (1986) have employed a series of open channel denuders coated with materials that act to preconcentrate $HNO_3$, $HNO_2$, $NO_2$, and NO from sampled ambient air. $HNO_2$ is collected using a potassium–iron oxide-coated denuder located downstream of a TA-coated denuder that removes $HNO_3$. The $HNO_2$ is thermally desorbed from the potassium–iron oxide-coated denuder and detected as NO with a CLM NO analyzer. Although sub-ppb sensitivity is claimed, field testing is needed to demonstrate the adequacy of this method.

Nylon filter material has also been used as an open channel denuder to collect $HNO_2$ (Benner et al., 1988). Recent studies, however, have indicated that $HNO_2$ may not be retained quantitatively by nylon filters (Perrino et al., 1988; Sickles and Hodson, 1989).

Ferm and Sjödin (1985) have used two conventional open-channel $Na_2CO_3$-coated denuders in series for the determination of $HNO_2$ in ambient air. $HNO_2$ is collected quantitatively on the first denuder, while interferent artifacts from PAN and other $NO_x$ species (i.e., $NO_2$) are collected in approximately equal amounts on both denuders. Each denuder is extracted and the extract is analyzed for $NO_2^-$ using spectrophometry or IC. To correct for interferent artifacts, the difference in $NO_2^-$ found on the two denuders is attributed to $HNO_2$.

Annular denuders have also been used to measure $HNO_2$ using a similar approach (DeSantis et al., 1985; Sickles, 1987; Sickles et al., 1988; Eatough et al., 1988; Vossler et al., 1988; Koutrakis et al., 1988; Dasch et al., 1989; Appel et al., 1990; Perrino et al., 1990). The MDL for a 1-day AD sample operating at $1 m^3 h^{-1}$ assuming an extract volume of 10 mL, negligible blank, and an IC detection limit of $0.05 \mu g NO_2^- mL^{-1}$ is $0.02 \mu g m^{-3}$ (10 ppt). Estimates of precision for 1-day AD samples range from 5 to 15% (Sickles et al., 1989; Vossler et al., 1988). In those cases where denuder sampling is performed over extended periods in the presence of oxidants (i.e., $O_3$), the collected $NO_2^-$ may be oxidized to $NO_3^-$ (Febo et al., 1986; Perrino et al., 1988; Sickles and Hodson, 1989; Sickles et al., 1989). To avoid this potential for sampling artifacts, an initial denuder coated with NaCl or NaF is added to collect $HNO_3$ as $NO_3^-$ and pass $HNO_2$. The difference in the sums of $NO_2^-$ and $NO_3^-$ on the two downstream $Na_2CO_3$-glycerol-coated denuders is attributed to $HNO_2$ (Febo et al., 1986; Perrino et al., 1990).

## 8.2. CLM

It has been shown that $HNO_2$ may be measured nonspecifically as $NO_x$ with a CLM $NO_x$ analyzer (Cox, 1974; Sickles and Hodson, 1989). As noted in the previous section, Braman et al. (1986) have used a system of selective denuders to collect $HNO_2$ as well as $HNO_3$, $NO_2$, and NO for subsequent thermal desorption and detection as NO with a CLM NO analyzer.

## 8.3. PF/LIF

PF/LIF is discussed in Section 3.3 for the measurement of $NO_2$. In its present application, $HNO_2$ is photofragmented to NO and OH using radiation at 355 nm from a Nd:YAG laser (Rodgers and Davis, 1989). Appreciable amounts of NO are also produced by the photolysis of $NO_2$, which is generally present along with $HNO_2$ in ambient air. As a result, the current method is based on the detection of OH using single photon LIF. With this technique the resulting OH is excited to the $A^2\Sigma+$ state using laser radiation at 282 nm, and the fluorescence at 310 nm that accompanies the A to X transition of the excited OH is monitored. Detection limits in the low tens of ppt for 15-min integration times are claimed.

## 8.4. Absorption Spectroscopy

Although $HNO_2$ is potentially detectable (i.e., MDL of 4 ppb) using a 23-m multipass FTIR system with a 1-km path length, FTIR has not been used to measure the concentration of $HNO_2$ in ambient air (Tuazon et al., 1978). A theoretical MDL for $HNO_2$ of 10 ppt has been claimed for 15-L integrated samples of ambient air using cryogenic sampling and matrix-isolation FTIR (Griffith and Schuster, 1987). Long-path UV/visible DOAS has been used to determine $HNO_2$ as well as other trace atmospheric constituents (see Section 3.4). For a 25-m multipass, open DOAS system with a 0.8-km path length, at wavelengths near 354 nm, a MDL of 0.6 ppb is claimed (Biermann et al., 1988). For a single-pass, open DOAS system with a 10-km path length a MDL of 20 ppt has been reported (Platt and Perner, 1983).

## 8.5. Calibration

The preparation of reliable calibration mixtures containing known concentrations of $HNO_2$ is difficult. Atmospheres containing $HNO_2$ as well as $NO_2$ and NO may be produced by acidifying solutions of $NaNO_2$ with $H_2SO_4$ (Cox, 1974). Another more convenient method uses a sublimation source where $HNO_2$ is produced by subliming oxalic acid onto solid $NaNO_2$ at 30–60% RH (Braman and de la Cantera, 1986). Small concentrations of $HNO_3$, $NO_2$, and NO may also be generated using the latter technique. Both methods require independent and periodic determination of the $HNO_2$ concentration, since the source strengths are not necessarily constant.

## 8.6. Intercomparisons

In November and December 1987 a study was conducted in Long Beach, California, where simultaneous $HNO_2$ measurements were made using AD and DOAS on 6 days (Appel et al., 1990). The AD samples were integrated over 4 and 6 h, while the 15-min DOAS results were averaged to permit comparison with the AD results. The $HNO_2$ concentrations ranged from less than 1 to approximately 15 ppb, and the AD results were highly correlated with those of the DOAS. Except at the low $HNO_2$ levels that occurred during the mid-day periods where the AD exceeded the DOAS results, the AD results were 7% lower than the DOAS results. This difference is within the $\pm 30\%$ uncertainty of the DOAS results in the study.

## 9. DINITROGEN PENTOXIDE ($N_2O_5$) AND NITRATE RADICALS ($NO_3$)

The $NO_3$ radical photolyzes rapidly, and as a result ambient concentrations are low during daylight hours. $N_2O_5$ exists in a thermally sensitive equilibrium with $NO_2$ and the $NO_3$ radical and can also react heterogeneously with water vapor to produce $HNO_3$. In addition, the $NO_3$ radical reacts rapidly with NO to produce $NO_2$. In spite of their low ambient concentrations, $N_2O_5$ and the $NO_3$ radical may have important roles in both tropospheric and stratospheric $NO_x$ chemistry.

Although $N_2O_5$ has not been measured in the troposphere, it has been observed in the stratosphere using spectroscopic methods (Roscoe, 1982). In the troposphere nighttime $N_2O_5$ concentrations of up to 15 ppb have been inferred under the assumption of equilibrium using measured $NO_2$ and $NO_3$ radical concentrations (Atkinson et al., 1986). At concentrations above 5 ppb, measurement of $N_2O_5$ with FTIR spectrometry appears feasible with a 1-km path length near 1250 cm$^{-1}$. A theoretical MDL for $N_2O_5$ of 20 ppt has been claimed for 15-L integrated samples of ambient air using cryogenic sampling and matrix-isolation FTIR (Griffith and Schuster, 1987).

$N_2O_5$ is readily reduced to NO at temperatures above 200 °C and, as noted in Section 3.1, may be measured nonspecifically as $NO_x$ with CLM $NO_x$ analyzers (Bollinger et al., 1983; Fahey et al., 1985a). A $N_2O_5$ calibration system has been devised using a crystalline sample at $-80$ °C, thermal dissociation of gaseous $N_2O_5$, scavenging of the dissociation product (i.e., the $NO_3$ radical) with added NO to produce $NO_2$, and a CLM NO detector (Fahey et al., 1985b). This calibration technique focuses on the loss of NO, and an accuracy of $\pm 15\%$ is claimed.

Ambient concentrations of the $NO_3$ radical have been measured using DOAS, and concentrations between 1 and 430 ppt have been observed (Atkinson et al., 1986). Additional information on absorption spectroscopy is given in Section 3.4. For a 25-m multipass open system with 0.8 km path length, a MDL of 20 ppt is claimed (Biermann et al., 1988). For an optical path length of 17 km and a wavelength of 662 nm, the reported detection limit for the $NO_3$ radical is 1 ppt (Platt et al., 1984). Noxon (1983), using a passive absorption spectroscopic method with the moon as the light source, reports a $NO_3$ concentration of 0.25 ppt measured at 3 km altitude from Mauna Loa, Hawaii.

## 10.    PARTICULATE NITRATE (PN)

Atmospheric aerosols are chemically heterogeneous and occur in sizes ranging nominally from <0.01 to 100 $\mu$m. Many methods are available for sampling ambient aerosols, including impactors, filtration, and filtration coupled with devices to remove particles larger than a specified size (e.g., elutriators, impactors, and cyclones). The method of choice usually depends on the particle size range and the chemical composition of the aerosol of interest. As an example, ambient concentrations of particles are subject to National Ambient Air Quality Standards (Code of Federal Regulations, 1987a). This standard focuses on the concentration of particulate mass for all particles less than 10 $\mu$m in equivalent aerodynamic diameter, rather than the individual chemical species (e.g., nitrates) comprising the collected particles.

The particle size distribution of ambient PN is bimodal (Kadowaki, 1977; Wolff, 1984; Yoshizumi, 1986; Wall et al., 1988). PN is concentrated in the coarse size (i.e., greater than 2.5 $\mu$m) in marine environments, where ambient $HNO_3$ reacts with the coarse suspended sea salt (i.e., NaCl) to form $NaNO_3$. Under other circumstances, the size distribution of PN will be determined by environmental conditions and the relative presence of precursors, including $HNO_3$, $NH_3$, and acidic aerosols. For example, in the eastern United States, during the summer, when the concentration of acidic sulfates is high, the temperature is high, and the $NH_3$ emissions are low, the $NH_4NO_3 \longleftrightarrow NH_3 + HNO_3$ equilibrium is shifted to the right. This and metathetical reactions with acidic aerosols and gases make gaseous $HNO_3$ available for reaction with and retention by coarse soil-derived particles, giving rise to high concentrations of PN in the coarse size range. In contrast, some western urban areas (e.g., Los Angeles and Denver) have low $SO_2$ emissions and adequate $NH_3$ emissions to neutralize acidic aerosols. These conditions favor the concentration of PN in the fine size range, presumably as $NH_4NO_3$.

### 10.1.    Filtration

PN are generally collected by filtration techniques for subsequent analysis. Using ambient dust, John and Reischl (1978) found the filtration efficiencies of Nuclepore (polycarbonate, 0.8-$\mu$m pore) and Whatman 41 (cellulose) filters to be less than 90%. Efficiencies exceeded 99% for Gelman GA-1 (cellulose acetate), Gelman Spectrograde (glass fiber), Gelman A (glass fiber), MSA (glass fiber), and EPA grade (glass fiber) filters. For polytetrafluoroethylene

(Teflon) membrane filters, the efficiencies exceeded 99% for Fluoropore (1-$\mu$m pore), Ghia (1- to 3-$\mu$m pore), and Ghia (2- to 4-$\mu$m pore) filters but did not exceed 99% for some tests with the Fluoropore (3-$\mu$m pore) and Ghia (3- to 5-$\mu$m pore) filters.

The integrity of PN collected on filters may depend on storage conditions and other factors. Highsmith et al. (1986) have attributed weight loss observed on quartz high-volume and Teflon dichotomous filters to particle loss and volatilization during handling and shipment. Smith et al. (1978) report a 73% loss of $NO_3^-$ from Gelman AE (glass fiber) high-volume filters stored for 15 months in the open at room temperature. In contrast, filters stored for 2 years in containers at $-28$ °C showed no loss of $NO_3^-$. Witz et al. (1990) have observed 19% loss of nitrates on PM-10 samples collected on Whatman QM-A (quartz and glass fiber) filters after 1 week of room temperature storage. Nitrate losses ranged from 28 to 50% after 1 month. Dunwoody (1986), using acid-treated Whatman QM-A high volume filters, found $NO_3^-$ losses of 86% after 6–8 months of dry room temperature storage, while refrigerated filter extracts were stable over this period. Witz et al. (1990) found nitrate losses to increase with decreasing filter alkalinity, increasing acidity of the aerosol deposit, and increasing storage temperature. Dunwoody (1986) found that filters spiked with solutions containing $KNO_3$ and other salts showed no $NO_3^-$ losses over 60 days of storage. In contrast, filters spiked with $HNO_3$ lost 70–90% of the $NO_3^-$ over a period of 3 days.

Using filters to collect PN can also result in both positive and negative biases that occur during the sampling process. Some of the difficulties encountered with filtration techniques for distinguishing between particulate and gaseous nitrate are also discussed in Section 7.1. Gas–filter interactions may lead to one type of positive bias. Glass fiber filters have been employed to collect particles including PN from ambient air, and, at one point, glass fiber filters were specified by the EPA for sampling Total Suspended Particulate matter (Code of Federal Regulations, 1987a). Glass fiber filters can retain gaseous $HNO_3$ and to a lesser extent promote the oxidation of gaseous $NO_2$ leading to the formation of artifact nitrates and the resulting positive biases (Appel et al., 1979; Spicer and Schumacher, 1979). Substantial positive biases from $HNO_3$ have been reported by Appel et al. (1979) for Gelman A, GA-1, and Spectrograde; Whatman 41; MSA 1106 BH (glass fiber); and "EPA grade" filters, by Spicer and Schumacher (1979) for Millipore Nylon; Gelman E, A, AE, AA, and Spectrograde (glass fiber); MSA 1106 BH; Millipore (cellulose acetate); Gelman Microquartz (quartz fiber); and Pallflex E 70-2075W (quartz fiber) filters, and by Appel et al. (1984) for Gelman "EPA grade"; Schleicher and Schuell (S & S) (glass fiber); S & S 1 HV (glass fiber); Whatman EPM 2600 (glass fiber); Whatman EPM 1000 (glass fiber); Whatman QM-A; Pallflex 2500 QAST (quartz fiber); Gelman Microquartz; and Gelman ADL (quartz fiber) filters. Witz and Wendt (1981) have reported that the magnitudes of artifact nitrates on high-volume sampler filters were ordered as follows: Whatman EPM 1000 > Gelman AE (acid washed glass fiber) > Gelman Microquartz > Pallflex TX40H120 (Teflon-coated glass fiber) > Pallflex 2500 QAO (quartz fiber). Artifact nitrates on Gelman A and Pallflex TX40H120 filters based on laboratory tests exceeded the amount found on Pallflex QAST filters by factors of 8.6 and 3.4 (Mueller and Hidy, 1983). Substantial amounts of artifact nitrates have been reported based on field studies using Gelman A and Pallflex TX40H120 filters (Pierson et al., 1980). Higher ambient PN measurements were reported using S & S and Whatman QM-A flters than with Gelman Microquartz, Pallflex 2500 QAST or Membrana/Ghia Zefluor (Teflon) filters (Rehme et al., 1984). Small biases were also reported for EPA/ADL (quartz fiber) and Pallflex QAST filters (Spicer and Schumacher, 1979). Negligible artifact nitrates were reported for Fluoropore (Teflon) (Appel et al., 1979; Mueller and Hidy, 1983) and Ghia Zefluor filters (Appel et al., 1984), and no artifact nitrates were reported for Nuclepore (0.8-$\mu$m pore) and Millipore Mitex (Teflon) filters (Spicer and Schumacher, 1979). Good agreement was reported for PN collected on acid-treated Pallflex 2500 QAO and Fluoropore filters (Forrest et al., 1982) in contrast to tests where PN on Pallflex QAST exceeded that on Ghia Zefluor filters by 33% (Appel et al., 1984).

A second source of positive bias in using filtration for the collection of PN is the retention of gaseous $HNO_3$ by particulate matter collected on the filter. Appel et al. (1980) have reported increased retention of $HNO_3$ with mass loading of particulate matter on Ghia Zefluor filters.

Negative biases may arise from at least two sources. PN may react with cocollected acidic aerosols or gases to release $HNO_3$ from the particulate catch, leading to one type of negative bias. In the laboratory, the separate introduction of $H_2SO_4$ aerosols and gaseous HCl each resulted in appreciable losses and downstream recovery as nitrate of preloaded $NH_4NO_3$ from Ghia Zefluor filters (Appel and Tokiwa, 1981). Harker et al. (1977) reported that nitrates collected during chamber experiments on Gelman Spectrograde filters were displaced by sulfate-containing, and presumably acidic, aerosols according to a metathetical reaction. The introduction of $H_2SO_4$ aerosols to ambient particles preloaded on acid-treated Pallflex 2500 QAO filters resulted in appreciable PN losses (Forrest et al., 1980). Pierson et al. (1980) reported similar observations for PN on both glass and quartz filters. Negative correlations have been reported between the fraction of PN measured on acid gas-denuded Teflon filters and both ammonia-denuded measurements of strong acid on Teflon filters (Appel and Tokiwa, 1981) and measurements of strong acid on acid-treated Pallflex 2500 QAO filters (Forrest et al., 1982).

A second source of a negative bias with filtration for the collection of PN is the volatilization of $NH_4NO_3$. Nominal 40–50% losses of nitrate due to volatilization have been reported where laboratory air free of $HNO_3$ and $NH_3$ was drawn through Ghia Zefluor filters preloaded with $NH_4NO_3$. Ambient air drawn through acid-treated Pallflex 2500 QAO filters preloaded with $NH_4NO_3$ has shown nitrate volatilization losses ranging between 0 and 72% (Forrest et al., 1980).

Artifact nitrate formation depends to a large extent on the composition of the filter material (e.g., glass versus Teflon). Artifact nitrate formation also increases with relative humidity and decreases with temperature (Appel et al., 1979; Forrest et al., 1980). To ensure efficient particle collection and minimize artifact nitrate formation, Teflon membrane or selected quartz fiber filters are preferred over glass, Teflon-coated glass, cellulose, cellulose acetate, or polycarbonate filters. Quartz fiber filters permit sampling at high flow rates with modest pressure drops; however, they are fragile, require care in handling (Rehme et al., 1984), are subject to some positive bias from artifact nitrate formation (Appel et al., 1984), and may require pretreatment to ensure low blank levels (Leahy et al., 1980). Teflon membrane filters are the most nearly inert, but in contrast are subject to clogging with increased mass loadings (Rehme et al., 1984).

## 10.2. Denuders/Filtration

Many of the previously mentioned biases may be eliminated by deploying a combination of denuders and filters (Appel et al., 1981). For additional information on denuders see Section 7.2. Biases involving interaction of $HNO_3$ with the filter or collected sample have been circumvented to some degree through the use of denuders. During sampling, gaseous $HNO_3$ diffuses to the surface of the denuder and is collected by reaction with the denuder surface, while particles pass through the device uncollected. An inert filter material (e.g., Teflon membrane or quartz fiber) is then used to collect the particulate matter. Since the inert particulate filter is still subject to the negative biases of liberated $HNO_3$ and volatilized $NH_4NO_3$, the resulting $NO_3^-$ must be collected on a backup filter. Nylon, $Na_2CO_3$-coated, NaCl-coated, or NaF-coated filters may be used as backup filters. Although pressure drop and capacity considerations favor coated filters for high flow rate applications, due to the presence of the $Cl^-$ or $F^-$ ions, this collection method may not be compatible with an IC finish for determining $NO_3^-$.

As noted in Section 7.1, a minimum sensitivity of 0.02 $\mu g \ m^{-3}$ may be calculated under the assumptions of 0.05 $\mu g \ mL^{-1}$ analytical detection limit, 10-mL extraction volume, negligible blank, quantitative collection and extraction, and a sampled air volume of 24 $m^3$ (i.e., 1 $m^3 \ h^{-1}$ for 1 day). Thus, for a combination of a denuder and two filters, the PN sensitivity should be approximately 0.04 $\mu g \ m^{-3}$. Median precision estimates of 4–16% RSD have been reported for 22-h-duration samples of fine PN (Vossler et al., 1988; Sickles et al., 1989), and a median precision of 4% RSD was reported for twelve 1-wk-duration samples of fine PN (Sickles, 1987).

## 10.3. Analysis

After the collection of PN by the techniques discussed in the previous sections, samples are analyzed directly or indirectly for nitrate. Several methods have been used, including IC (Mulik et al., 1976), colorimetry (Mullin and Riley, 1955), derivatization/GC (Tesch et al., 1976), HPLC (Kamiura and Tanaka, 1979), voltametry (Bondini and Sawyer, 1977), ion-specific electrode (ISE) (Driscoll et al., 1972), FTIR (Bogard et al., 1982), and CLM (Yoshizumi et al., 1985). IC and colorimetry are the methods most commonly used as analytical finishes for the determination of PN.

Many of the methods for determining PN require the extraction of nitrates prior to analysis. Extraction of nitrate spiked onto nylon filters showed quantitative recovery using IC eluent solution or basic (i.e., 0.003N NaOH) solution but not using water (Hering et al., 1988). Similar tests with spiked Teflon membrane filters showed essentially quantitative recovery in each extraction medium. Other tests have shown good $NO_3^-$ recoveries from spiked and ambient nitrates on cellulose, glass fiber, and Teflon filters using IC eluent and ultrasonication, boiling deionized water, and sequential extraction in warm IC eluent and deionized water (Jenke, 1983).

In recent years IC has become a method of choice for the determination of many anions and cations in solution. IC uses conductimetric detection and a combination of resin columns to separate the ions of interest and strip or suppress the eluent from the background (Small et al., 1975; Mulik et al., 1976). The bromide and phosphate interferences noted by Mulik et al. (1976) generally do not present problems with environmental samples of PN. In some cases where filter or denuder extracts are analyzed by IC, to permit good resolution of various peaks, care must be taken to prevent excessive concentrations of chloride or $H_2O_2$ must be added to oxidize sulfite to sulfate. One recent study reported a precision estimate of 1% for replicate $NO_3^-$ measurements in extracts of ambient samples where the concentration was above 0.15 $\mu g\ mL^{-1}$ (Sickles et al., 1988). Detection limits for $NO_3^-$ of 0.025–0.1 $\mu g\ mL^{-1}$ have been reported using IC with a 0.5-mL sample loop (Anlauf et al., 1988; Mulik et al., 1976). As noted in Section 7.1, a 0.05-$\mu g$ $mL^{-1}$ analytical detection limit for $NO_3^-$ corresponds to an ambient concentration of 0.02 $\mu g\ m^{-3}$ on a single filter sampling at 1 $m^3\ h^{-1}$ for 1 day. Although nonsuppressed IC has poorer detection limits than the previously described suppressed approach, successful application to the analysis of nitrates in ambient aerosols has also been reported for nonsuppressed IC (Willison and Clark, 1984).

Various colorimetric methods for $NO_3^-$ have been used. In one widely automated method, $NO_3^-$ in the extract is reduced to $NO_2^-$, which is diazotized and determined at 550 nm using the Griess–Saltzman method (see Section 3.5) (Saltzman, 1954). The reduction may be accomplished using a copper–cadmium (Cu–Cd) reductor column (Technicon, 1972) or using hydrazine sulfate with copper as a catalyst under slightly basic conditions (Mullin and Riley, 1955; Kamphake et al., 1967). The detection limits of 0.001–0.006 $\mu g\ mL^{-1}$ claimed with these methods are somewhat more sensitive than those previously cited for the IC method. A comparison of the performance of IC with this colorimetric method for PN collected on Fluoropore filters showed excellent agreement (Fung et al., 1979).

Another colorimetric technique, the brucine method, involves the reaction of $NO_3^-$ with brucine under acidic conditions (Intersociety Committee, 1972c). The color is measured at 410 nm, and a detection limit of 0.4 $\mu$g mL$^{-1}$ is reported. Other colorimetric methods involve the nitration of 2,4-xylenol in the presence of sulfuric acid, followed by steam distillation and absorbance measurement at 435 nm (Intersociety Committee, 1972a), or the nitration of toluene in the presence of sulfuric acid, followed by extraction into toluene and absorbance measurement of the nitrotoluene—toluene complex at 284 nm (Bhatty and Townshend, 1971). The sensitivity of both methods is marginal (i.e., 1 $\mu$g mL$^{-1}$) and they are subject to interferences (Intersociety Committee, 1972a; Appel et al., 1977; Norwitz and Keliher, 1978; Kamiura and Tanaka, 1979; Bhatty and Townshend, 1971).

Using reactions similar to those described above, Tesch et al. (1976) have reacted $NO_3^-$ with benzene or other aromatic compounds in the presence of sulfuric acid and measured the resulting nitroaromatic compound using GC-EC. A sensitivity of 0.1 $\mu$g mL$^{-1}$ and applicability to determining $NO_3^-$ in saliva, blood, drinking water, and airborne particles were claimed. The above method has been modified by Tanner et al. (1979) using electron capture-sensitive fluoroaromatic derivatizing agents and a more effective catalyst (i.e., trifluoromethanesulfonic acid). With a sensitivity of 0.01 $\mu$g mL$^{-1}$, this method has been applied to microliter-sized samples and the analysis of PN.

HPLC coupled with UV detection at 210 nm has been used to measure $NO_3^-$ in the aqueous extracts of PN from glass fiber filters (Kamiura and Tanaka, 1979). No interferences were reported, and a detection limit of 0.1 $\mu$g mL$^{-1}$ was claimed.

A voltammetric technique for the measurement of $NO_3^-$ in solution has been reported by Bondini and Sawyer (1977). The technique is based on the reduction of $NO_3^-$ at a Cu–Cd catalyst that is formed on the surface of a pyrolytic graphite electrode. The detection limit is 0.06 $\mu$g mL$^{-1}$, but $NO_2^-$ is a direct interferent. Favorable comparison was reported between the results of this method and those of the Technicon (1972) colorimetric method for determining $NO_3^-$ in extracts of PN.

Ion-specific electrodes (ISE) have been used to measure $NO_3^-$ in extracts of PN with a detection limit of 1 $\mu$g mL$^{-1}$ (Simeonov and Puxbaum, 1977). ISEs suffer from poor sensitivity, potential drifts caused by variable agitation speed, frequent need for restandardization, and interferences by other ions (Driscoll et al., 1972). Spicer et al. (1978a) evaluated a nitrogen oxide gas-sensing electrode for the indirect measurement of $NO_3^-$ in solution. Since the electrode responds to $NO_2^-$ in solution, the approach was to measure $NO_2^-$ in solution before and after reducing the $NO_3^-$ to $NO_2^-$ and to attribute the difference to $NO_3^-$. Although the gas-sensing electrode was both sensitive (i.e., 0.1 $\mu$g mL$^{-1}$) and specific for $NO_2^-$, difficulties in the reduction step prevented further development.

A dry technique using FTIR for measuring $NO_3^-$ incorporated in a KBr matrix from samples of ambient PN has been reported (Bogard et al., 1982). Absorbance bands for $NO_3^-$ occur at 2430, 1384, and 840 cm$^{-1}$. Using the 1384-cm$^{-1}$ band, a detection limit of 0.1 μg $NO_3^-$ per sample was reported, although the nearby $NH_4^+$ band under same circumstances may not permit distinct resolution of $NO_3^-$. Techniques have been developed recently that permit FTIR detection of nitrates and other species in samples of ambient aerosols collected by filtration on thin Teflon membrane filters using direct transmission or by impaction using attenuated total internal reflection (Johnson and Kumar, 1987). This method is currently in the research prototype stage of its development.

The decomposition of $NO_3^-$ followed by the CLM detection of the resulting $NO_x$ (see Sections 2.1 and 3.1) has been used to determine $NO_3^-$ in PN samples. Thermal decomposition can be applied to $NO_3^-$ either on filters or in liquid extracts (Spicer et al., 1985). With this technique, $NO_3^-$ is decomposed by rapid heating to 425 °C in a $N_2$ atmosphere, and the resulting $NO_x$ (i.e., NO and $NO_2$) is measured (i.e., integrated) using a conventional CLM analyzer. Particulate nitrite is a direct interferent. A detection limit of 0.7 μg mL$^{-1}$ is claimed. Comparison of CLM and IC analyses of spiked and ambient samples showed good agreement, although the IC was more precise especially at low concentration levels. Yoshizumi et al. (1985) have modified a method developed by Cox (1980) to reduce $NO_3^-$ and $NO_2^-$ in solution and measure the evolving NO using a commercially supplied CLM instrument. The method of Yoshizumi et al. (1985) uses a flow system, does not distinguish $NO_2^-$ from $NO_3^-$ (i.e., $NO_2^-$ is a direct interferent), but has a $NO_3^-$ detection limit of 0.001 μg mL$^{-1}$. The method of Cox (1980), although using a batch approach, does distinguish $NO_2^-$ from $NO_3^-$ and has respective detection limits of 0.00005 and 0.05 μg mL$^{-1}$.

It has been suggested that volatile and nonvolatile nitrates may be distinguished by taking advantage of their different temperatures of volatilization (Yoshizumi and Hoshi, 1985). Samples of atmospheric particles collected by filtration or impaction are heated in a furnace to the optimum volatilization temperature of ammonium nitrate (i.e., 160°C). The volatilized nitrate is then collected in water for subsequent determination (e.g., by IC). A similar principle has been used in thermal denuders (Klockow et al., 1989). In this case, $HNO_3$ is collected at ambient temperature on a $MgSO_4$-coated annular denuder, and $NH_4NO_3$ is collected at 150 °C on a similar downstream denuder. After sampling, denuders are heated in turn to 700 °C to liberate $NO_x$ for determination by a CLM $NO_x$ analyzer. Sturges and Harrison (1988) have reported several potential interferences with the volatilization approach. Nitric acid from volatilized $NH_4NO_3$ in the presence of NaCl, for example, will cause displacement of the chloride as HCl and formation of nonvolatile $NaNO_3$. Differences in the thermal stabilities of ammonium sulfate/nitrate double salts were also demonstrated. These observations cast doubt on the feasibility of thermal speciation of PN.

## 11.  NITROUS OXIDE (N$_2$O)

Ambient N$_2$O levels have been measured by several methods, including infrared spectroscopy (both absorption and emissions spectra), mass spectrometry, manometry, and gas chromatography coupled with thermal conductivity,
flame ionization, ultrasonic phase shift, helium ionization, and electron capture detectors (Pierotti and Rasmussen, 1977). The most commonly used
method employs GC-ECD with a detection limit of 20 ppb (Thijsse, 1978)
and a precision of $\pm 3\%$ at the background level of 330 ppb (Cicerone et al.,
1978). Cassidy and Reid (1982) also report an expected MDL of 20 ppb for
N$_2$O using TDLAS near 1150 cm$^{-1}$ (see Section 2.3 for more on TDLAS).
As described in Section 2.3, a laboratory prototype method, TTFMS, has
been developed with a projected MDL for N$_2$O of 3 ppt (Hansen, 1989).
    Calibration can be performed using commercially supplied cylinders of
compressed gas (Thijsse, 1978), dilution of pure N$_2$O, N$_2$O permeation tubes
(Cicerone et al., 1978) or gravimetric preparation of calibration mixtures
(Komhyr et al., 1988). SRM mixtures of N$_2$O and CO$_2$ in air are also available
at nominal N$_2$O concentrations of 300 and 330 ppb (NBS, 1988).


## 12.  SUMMARY

Since the publication in 1971 of the original version of *Air Quality Criteria
for Nitrogen Oxides* (EPA, 1971), changes have occurred in the technology
associated with the sampling and analysis for ambient oxides of nitrogen and
related species. During the 1970s, roughly the period between publication of
the original Criteria Document and its first update and revision, several events
occurred that focused on the determination of NO$_2$ in ambient air. In 1973,
the original Reference Method was withdrawn because of unresolvable technical difficulties. Major methods development efforts over the next 3–4 years
yielded both automated and manual methods that were suitable for the determination of NO$_2$ in ambient air. As a result, EPA designated a new Reference Method and Equivalent Methods for NO$_2$. The Reference Method
specifies a measurement principle and calibration procedures, namely gas
phase chemiluminescence (GP CLM) with calibration using either gas phase
titration of NO with O$_3$ or a NO$_2$ permeation device. The sodium arsenite
method in both the manual and continuous forms and the TGS method were
also designated as Equivalent Methods. Subsequently, commercial GP CLM
instruments were designated as Reference Methods. The sensitivity of these
devices was in the low ppb range and while the GP CLM instruments were
recognized as being susceptible to interferences by other nitroxy species, it
was believed that the atmospheric concentrations of these compounds were
generally low relative to NO$_2$.
    In the 1980s, additional developments have occurred. Information from

air quality monitoring networks is now readily available and has shown the GP CLM instruments to have nominal precision and accuracy of $\pm 10$ to 15% and 20%, respectively, and to have replaced manual methods to a large extent in network applications. Heightened interest in the research community on the speciation of atmospheric trace gases and specifically nitrogen-containing species has prompted a new wave of methods development. While the basic design and performance of the commercial instruments have remained essentially unchanged, researchers have improved GP CLM measurement technology and refined other instrumental methods to permit the determination of NO, $NO_2$, and $NO_y$ in the low ppt range. Although GP CLM NO detectors coupled with catalytic $NO_2$ to NO converters are still not specific for $NO_2$, they have proven useful for measuring $NO_y$, and GP CLM NO detectors coupled with photolytic $NO_2$ to NO converters have shown improved specificity for $NO_2$.

A continuous liquid phase CLM device for sensitively detecting $NO_2$ has been developed and may be suitable to measure $NO_2$ if interference problems can be overcome. Passive samplers for $NO_2$ have been used, primarily for workplace and indoor applications, but hold promise for ambient measurements as well. Gas chromatography with electron capture detection is useful in the determination of PAN, other organic nitrates, and $N_2O$.

Laser-induced fluorescence has been introduced to detect NO, $NO_2$, and $HNO_2$ with high sensitivity and specificity. Tunable diode laser spectroscopy has been used to detect NO, $NO_2$, and $HNO_3$. Long-path spectroscopy has also been used to detect NO, $NO_2$, $HNO_2$, and $NO_3$. Two-tone frequency-modulated spectroscopy holds promise for the sensitive measurement of NO, $NO_2$, PAN, $HNO_3$, and $N_2O$. These spectroscopic methods are research tools and are not yet easily or economically suited for routine monitoring.

Interest in acidification of the environment has resulted in the development of methods for $HNO_2$ and $HNO_3$. Integrative methods using denuders have been introduced to permit sensitive determination of these and other species. In recent years, the potential for artifacts in using filters for sampling particulate matter and specifically particulate nitrate has been recognized. This has given rise to careful characterization of filter media for potential artifacts and the use of combinations of denuders and filters to permit more specific determination of nitrogen-containing gases and particulate nitrates in ambient air.

The characteristics and status of measurement methodologies for most of the ambient oxides of nitrogen addressed in this chapter are summarized in Table 5. The pertinent features include method type, development stage, sample duration, and performance parameters. NOx is not included in Table 5 because no specific method is available except as the sum of NO and NO2. N2O5 is not included because it has not been measured in the troposphere. For more details on specific methods, the text and cited references should be consulted.

**Table 5  Selected Instruments and Methods for Determining Ambient Oxides of Nitrogen**

| Species | Methods[a] | Type[b] | Development Stage[c] | Sample Duration | Performance Precision | Performance Accuracy | Performance MDL | Comments |
|---|---|---|---|---|---|---|---|---|
| NO | CLM (NO+O$_3$) | I, A, C | C | 5 min | ≤10% | ≤20% | ≤9 ppb | — |
|  | CLM (NO+O$_3$) | I, A, C | R | 2–60 s | 20 ppt | 20 ppt | 5 ppt | — |
|  | TP-LIF | I, A, C | R | 30 s | — | 16% | 10 ppt | — |
|  | TDLAS | I, A, C | R, C | 60 s | — | — | 0.5 ppb | 40m path length |
|  | TTFMS | I, A, C | L | 60 s | — | — | 4 ppt | 100-m path length |
|  | PSD | I, P, IN | C | 24 h | — | — | 70 ppb-h[d] | — |
| NO$_2$ | CLM (NO+O$_3$) | I, A, C | C | 5 min | 10% | 20% | 9 ppb | EPA reference method; many interferences |
|  | CLM (NO+O$_3$) | I, A, C | R | <100 s | 20 ppt | 30% | 10–25 ppt | Uses thermal or photolytic converters |
|  | CLM (Luminol) | I, A, C | C | 100 s | 0.6 ppb | — | 10 ppt | Interferences: PAN, HNO$_2$, O$_3$; |
|  | TP-LIF | I, A, C | R | 2 min | 20 ppt | 16% | 12 ppt | — |
|  | TDLAS | I, A, C | R, C | 60 s | — | 15% | 100 ppt | 150-m path length |
|  | TTFMS | I, A, C | L | 60 s | — | — | 0.3 ppt | 100-m path length |
|  | DOAS | R, A, C | R, C | 12 min | — | 10% | 4 ppb | 800-m path length |
|  | DIAL | R, A, C | R | — | — | — | 10 ppb | 6-km path length |
|  | Bubbler | I, A, IN | RM | 24 h | 6 ppb | 20% | 8 ppb[d] | EPA equivalent methods[e] |
|  | TEA Filter | I, A, IN | L | 24 h | 15% | 10% | 0.2 ppb[d] | Interferences: PAN and HNO$_2$[e] |
|  | Guaiacol Denuder | I, A, IN | L | 1 h | 4% | — | 0.1 ppb[d] | Stability of extract uncertain |

| Species | Method | | | | | | | Comments |
|---|---|---|---|---|---|---|---|---|
| | DPA Cartridge | I, A, IN | L | 8 h | 8% | — | 0.1 ppb[d] | DPA may volatize; interferences: $HNO_2$ and PAN |
| | TEA PSD | I, P, IN | L | 24 h | 30% | — | 30 ppb-h[d] | Similar to Palmes Tube; interferences as above[e] |
| $NO_y$ | CLM (NO+$O_3$) | I, A, C | R | 10 s | — | 15% | 10 ppt | CO with AU reducing catalyst |
| PAN | GC-ECD | I, A, IN | R, RM | 15 min | — | 30% | 10 ppt[d] | Sensitivity can be enhanced by using cryogenic sampling and capillary columns |
| | GC-CLM | I, A, IN | L | — | — | — | — | CLM (NO+$O_3$) and (Luminol) reported |
| | Alkaline Hydrolysis | I, A, IN | L | — | — | — | — | Detected as $NO_2$ or Acetate[e] |
| | TTFMS | I, A, C | L | 60 s | — | — | 2 ppt | 100-m path length |
| Other Nitrates | GC-ECD | I, A, IN | R | 24 h | — | — | 1 ppt[d] | Sample collected on charcoal |
| $HNO_3$ | Filter | I, A, IN | R, RM | 24 h | 10% | 20% | 8 ppt[d] | May be nylon or CaCl-impregnated filter; subject to artifacts[e] |
| | Denuder | I, A, IN | R, RM | 24 h | 8% | — | 8 ppd[d] | Not subject to above artifacts[e] |
| | TDLAS | I, A, C | R, C | 5 min | — | 20% | 100 ppt | 150-m path length |
| | TTFMS | I, A, C | L | 60 s | — | — | 0.3 ppt | 100-path length |
| $HNO_2$ | Denuder | I, A, IN | R, RM | 24 h | 15% | — | 10 ppt[d] | Annular denuder preferred[e] |
| | LIF | I, A, C | R | 15 min | — | — | 20 ppt | OH detected following photofragmentation |

**Table 5** (*Continued*)

| Species | Methods[a] | Type[b] | Development Stage[c] | Sample Duration | Performance Precision | Accuracy | MDL | Comments |
|---|---|---|---|---|---|---|---|---|
| | DOAS | R, A, C | R, C | 12 min | — | 30% | 600 ppt | 800-m path length |
| NO$_3$ | DOAS | R, A, C | R, C | 12 min | — | 15% | 20 ppt | 800-m path length |
| Particulate NO$_3$ | Filter | I, A, IN | R, RM | 24 h | 10% | — | 40 ng/m$^{3d}$ | Teflon or quartz fiber filters followed by nylon or coated filter; open filters subject to artifacts[e] |
| | Denuder/Filters(s) | I, A, IN | R, RM | 24 h | 10% | — | 40 ng/m$^{3d}$ | Use of denuders avoids artifacts; denuders collect $HNO_3$ and $NH_3$; teflon and nylon filters used |
| N$_2$O | GC-ECD | I, A, IN | R, RM | 15 min | 3% | — | 20 ppb$^d$ | — |
| | TTFMS | I, A, C L | | 60 s | — | — | 3 ppt | 100-m path length |

[a] CLM (NO+O$_3$) = Chemiluminescent using NO+O$_3$ reaction;
CLM (Luminol) = Chemiluminescent using reaction with Luminol;
TP-LIP = Two photon laser induced fluorescence;
LIF = Laser induced fluorescence;
TDLAS = Tuneable diode laser spectroscopy;
TTFMS = Two-tone frequency modulated spectroscopy;
DOAS = Differential optical absorption spectroscopy;
DIAL = Differential absorption lidar;
TEA = Triethanolamine;
DPA = Diphenylamine;

PSD     = Passive sampling device;

GC-ECD = Gas chromatography with electron capture detector;

CG-CLM = Gas chromatography with CLM detector.

[b] I   = In sit;

R   = Remote;

A   = Active;

P   = Passive;

C   = Continuous;

IN = Integrative.

[c] L   = Laboratory prototype;

R   = Research tool;

C   = Commercially available;

RM = Routine method.

[d] Depends on the sampled air volume (i.e., flow rate *and* sampling duration).

[e] Uses IC or colorimetric analytical finish.

*Acknowledgment*

Support for preparation of this manuscript was provided in part by the United States Environmental Protection Agency under purchase orders 8D1418NALX and 9D2062NALX. The manuscript has not been subjected to Agency review and, therefore, does not necessarily reflect the views of the Agency. As a result, no official endorsement should be inferred.

# REFERENCES

Adams, K. M., Japar, S. M., and Pierson, W. R. (1986). "Development of a $MnO_2$-coated, cylindrical denuder for removing $NO_2$ from atmospheric samples." *Atmos. Environ.* **20**, 1211–1215.

Alden, M., Edner, H., and Svanberg, S. (1982). "Laser monitoring of atmospheric NO using ultraviolet differential-absorption techniques." *Opt. Lett.* **7**, 543–545.

Anlauf, K. G., Fellin, P., Wiebe, H. A., Schiff, H. I., Mackay, G. I., Braman, R. S., and Gilbert, R. (1985). "A comparison of three methods for measurement of atmospheric nitric acid and aerosol nitrate and ammonium." *Atmos. Environ.* **19**, 325–333.

Anlauf, K. G., Wiebe, H. A., and Fellin, P. (1986). "Characterization of several integrative sampling methods for nitric acid, sulphur dioxide and atmospheric particles." *J. Air Pollut. Control Assoc.* **36**, 715–723.

Anlauf, K. G., MacTavish, D. C., Wiebe, H. A., Schiff, H. I., and Mackay, G. I. (1988). "Measurement of atmospheric nitric acid by the filter method and comparisons with the tuneable diode laser and other methods." *Atmos. Environ.* **22**, 1579–1586.

Appel, B. R. (1973). "A new and more sensitive procedure for analysis of peroxybenzoyl nitrate." *J. Air Pollut. Control Assoc.* **23**, 1042–1044.

Appel, B. R., and Tokiwa, Y. (1981). "Atmospheric particulate nitrate sampling errors due to reactions with particulate and gaseous strong acids." *Atmos. Environ.* **15**, 1087–1089.

Appel, B. R., Hofter, E. M., Kothny, E. L., and Wall, S. M. (1977). "Interference in 2,4-xylenol procedure for nitrate determination in atmospheric aerosols." *Environ. Sci. Technol.* **11**, 189–190.

Appel, B. R., Wall, S. M., Tokiwa, Y., and Haik, M. (1979). "Interference effects in sampling particulate nitrate in ambient air." *Atmos. Environ.* **13**, 319–325.

Appel, B. R., Wall, S. M., Tokiwa, Y., and Haik, M. (1980). "Simultaneous nitric acid, particulate nitrate and acidity measurements in ambient air." *Atmos. Environ.* **14**, 549–554.

Appel, B. R., Tokiwa, Y., and Haik, M. (1981). "Sampling of nitrates in ambient air." *Atmos. Environ.* **15**, 283–289.

Appel, B. R., Tokiwa, Y., Haik, M., and Kothny, E. L. (1984). "Artifact particulate sulfate and nitrate formation on filter media." *Atmos. Environ.* **18**, 409–416.

Appel, B. R., Povard, V., and Kothny, E. L. (1988). "Loss of nitric acid within inlet devices intended to exclude coarse particles during atmospheric sampling." *Atmos. Environ.* **22**, 2535–2540.

Appel, B. R., Winer, A. M., Tokiwa, Y., and Biermann, H. W. (1990). "Comparison of atmospheric nitrous acid measurements by annular denuder and differential optical absorption systems." *Atmos. Environ.* **24A**, 611–616.

Atkinson, R., Winer, A. M., and Pitts, J. N., Jr. (1986). "Estimation of nighttime $N_2O_5$ concentrations from ambient $NO_2$ and $NO_3$ radical concentrations and the role of $N_2O_5$ in nighttime chemistry." *Atmos. Environ.* **20**, 331–339.

Atlas, E. (1988). "Evidence for $\geq C_3$ alkyl nitrates in rural and remote atmospheres." *Nature (London)* **331**, 426–428.

Baumgartner, R. A., Fletcher, L. D., and Hawley, J. G. (1979). "A comparison of lidar and air quality station $NO_2$ measurements." *J. Air Pollut. Control Assoc.* **29**, 1162–1165.

Baveja, A. K., Chaube, A., and Gupta, V. K. (1984). "Extractive spectrophotometric method for the determination of atmospheric nitrogen dioxide." *Atmos. Environ.* **18**, 989–993.

Benner, C. L., Eatough, D. J., Eatough, N. L., and Bhardwaja, P. (1987). *Evaluation of an Annular Denuder Method for the Collection of Atmospheric Nitrogeneous Species in the Southwest Desert.* 80th Annual Air Pollution Control Association Meeting, New York, Paper 63.6.

Benner, C. L., Eatough, N. L., Lewis, E. A., Eatough, D. J., Huang, A. A., and Ellis, E. C. (1988). "Diffusion coefficients for ambient nitric and nitrous acids from denuder experiments in the 1985 nitrogen species methods comparison study." *Atmos. Environ.* **22**, 1669–1672.

Bennett, B. I. (1979). *Stability Evaluation of Ambient Concentrations of Sulfur Dioxide, Nitric Oxide, and Nitrogen Dioxide Contained in Compressed Gas Cylinders.* U.S. Environmental Protection Agency, Washington, DC, Publ. No. EPA 600/4-79-006.

Benoit, F. M. (1983). "Detection of nitrogen and sulfur dioxides in the atmosphere by atmospheric pressure ionization mass spectrometry." *Anal. Chem.* **55**, 2097–2099.

Bhatty, M. K., and Townshend, A. (1971). "Spectrophotometric determination of small amounts of nitrate and nitrite by conversion to nitrotoluene and extraction into toluene." *Anal. Chim. Acta* **56**, 55–60.

Biermann, H. W., Tuazon, E. C., Winer, A. M., Wallington, T. J., and Pitts, J. N., Jr. (1988). "Simultaneous absolute measurements of gaseous nitrogen species in urban ambient air by long pathlength infrared and ultraviolet-visible spectroscopy." *Atmos. Environ.* **22**, 1545–1554.

Bogard, J. S., Johnson, S. A., Kumar, R., and Cunningham, P. T. (1982). "Quantitative analysis of nitrate ion in ambient aerosols by Fourier-Transform Infrared Spectroscopy." *Environ. Sci. Technol.* **16**, 136–140.

Bollinger, M. J., Sievers, R. E., Fahey, D. W., and Fehsenfeld, F. C. (1983). "Conversion of nitrogen dioxide, nitric acid, and n-propyl nitrate to nitric oxide by gold-catalyzed reduction with carbon monoxide." *Anal. Chem.* **55**, 1980–1986.

Bollinger, M. J., Hahn, C. J., Parrish, D. D., Murphy, P. C., Albritton, D. L., and Fehsenfeld, F. C. (1984). "$NO_x$ measurements in clean continental air and analysis of the contributing meterology." *J. Geophys. Res.* **89**, 9623–9631.

Bondini, M. E., and Sawyer, D. T. (1977). "Voltammetric determination of nitrate ion at parts-per-billion levels." *Anal. Chem.* **49**, 485–489.

Bradshaw, J. D., Rodgers, M. O., and Davis, D. D. (1982). "Single photon laser-induced fluorescence detection of NO and $SO_2$ for atmospheric conditions of composition and pressure." *Appl. Opt.* **21**, 2493–2500.

Bradshaw, J. D., Rodgers, M. O., Sandholm, S. T., KeSheng, S., and Davis, D. D. (1985). "A two-photon laser-induced fluorescence field instrument for ground-based and airborne measurements of atmospheric NO." *J. Geophys. Res.* **90**, 12861–12873.

Braman, R. S., and de la Cantera, M. A. (1986). "Sublimaton sources for nitrous acid and other nitrogen compounds in air." *Anal. Chem.* **58**, 1533–1537.

Braman, R. S., de la Cantera, M. A., and Han, Q. X. (1986). "Sequential, selective hollow tube preconcentration and chemiluminescence analysis system for nitrogen oxide compounds in air." *Anal. Chem.* **58**, 1537–1541.

Bubacz, D. K., Daughtrey, E. H., Pleil, J. D., and Kronmiller, K. G. (1987). Luminox measurements of ambient $NO_2$. In *Proceedings of the 1987 EPA/APCA Symposium on Measurement of Toxic and Related Air Pollutants*. Environmental Protection Agency, Washington, DC, Publ. No. EPA 600/9-87-010, pp. 398–403.

Burkhardt, M. R., Maniga, N. I., and Stedman, D. H. (1988). "Gas chromatographic method for measuring nitrogen dioxide and peroxyacetyl nitrate in air without compressed gas cylinders." *Anal. Chem.* **60**, 816–819.

Butcher, S. S., and Ruff, R. E. (1971). "Effect of inlet residence time on analysis of atmospheric nitrogen oxides and ozone." *Anal. Chem.* **43**, 1890–1892.

Buttini, P., DiPalo, V., and Possanzini, M. (1987). "Coupling of denuder and ion chromatographic techniques for $NO_2$ trace level determination air." *Sci. Total Environ.* **61**, 59–72.

Cadoff, B. C., and Hodgeson, J. (1983). "Passive sampler for ambient levels of nitrogen dioxide." *Anal. Chem.* **55**, 2083–2085.

Carroll, M. A., McFarland, M., Ridley, B. A., and Albritton, D. L. (1985). "Ground-based nitric oxide measurements at Wallops Island, Virginia." *J. Geophys. Res.* **90**, 12853–12860.

Cassidy, D. T., and Reid, J. (1982). "Atmospheric pressure monitoring of trace gases using tunable diode lasers." *Appl. Opt.* **21**, 1185–1190.

Cicerone, R. J., Shetter, J. D., Stedman, D. H., Kelly, T. J., and Liu, S. C. (1978). "Atmospheric $N_2O$: Measurements to determine its sources, sinks, and variations." *J. Geophys. Res.* **83**, 3042–3050.

Code of Federal Regulations, Title 40, Part 50 (1987a). *National Primary and Secondary Ambient Air Quality Standards*, Revised as of July 1, 1987. U.S. Government Printing Office, Washington, DC (40 CFR 50).

Code of Federal Regulations, Title 40, Part 53. (1987b). *Ambient Air Monitoring Reference and Equivalent Methods*, Revised as of July 1, 1987. U.S. Government Printing Office, Washington, DC (40 CFR 53).

Cox, R. A. (1974). "The photolysis of gaseous nitrous acid." *J. Photochem.* **3**, 175–188.

Cox, R. A., and Roffey, M. J. (1977). "Thermal decomposition of peroxyacetylnitrate in the presence of nitric oxide." *Environ. Sci. Technol.* **11**, 900–906.

Cox, R. D. (1980). "Determination of nitrate and nitrite at the parts per billion level by chemiluminescence." *Anal. Chem.* **52**, 332–335.

Crecelius, H.-J., and Forwerg, W. (1970). "Investigations of the 'Saltzman Factor.' " *Staub—Reinhalt, Luft* **30**, 23–25.

Darley, E. F., Kettner, K. A., and Stephens, E. R. (1963). "Analysis of peroxyacyl nitrates by gas chromatography with electron capture detection." *Anal. Chem.* **35**, 589–591.

Dasch, J. M., Cadle, S. H., Kennedy, K. G., and Mulawa, P. A. (1989). "Comparison of annular denuders and filter packs for atmospheric sampling." *Atmos. Environ.* **23**, 2775–2782.

Davis, D. D. (1988). *Atmospheric Nitrogen Oxides, Their Detection and Chemistry. Third Year Report to Coordinating Research Council.* Georgia Institute of Technology, Atlanta, pp. 1–13.

Davis, D. D., Bradshaw, J. D., Rodgers, M. O., Sandholm, S. T., and KeSheng, S. (1987). "Free tropospheric and boundary layer measurements of NO over the central and eastern North Pacific Ocean." *J. Geophys. Res.* **92**, 2049–2070.

Delany, A. C., Dickerson, R. R., Melchior, F. L., Jr., and Wartburg, A. F. (1982). "Modification of a commercial $NO_x$ detector for high sensitivity." *Rev. Sci. Instrum.* **53**, 1899–1902.

DeSantis, F., Febo, A., Perrino, C., Possanzini, M., and Liberti, A. (1985). "Simultaneous measurements of nitric acid, nitrous acid, hydrogen chloride and sulfur dioxide in air by means of high-efficiency annular denuders." In *Proceedings of the Workshop on Advancements in Air Pollution Monitoring Equipment and Procedures, 2–6 June 1985, Freiburg (FRG)*, pp. 68–75.

Dickerson, R. R. (1984). "Measurements of reactive nitrogen compounds in the free troposphere." *Atmos. Environ.* **18**, 2585–2593.

Dickerson, R. R., Delany, A. C., and Wartburg, A. F. (1984). "Further modification of a commercial $NO_x$ detector for high sensitivity." *Rev. Sci. Instrum.* **55**, 1995–1998.

Driscoll, J. N., Berger, A. W., Becker, J. H., Funkhouser, J. T., and Valentine, J. R. (1972). "Determination of oxides of nitrogen in combination effluents with a nitrate ion selective electrode." *J. Air Pollut. Control Assoc.* **22**, 119–122.

Drummond, J. W., Volz, A., and Ehhalt, D. H. (1985). "An optimized chemiluminescence detector for tropospheric NO measurements." *J. Atmos. Chem.* **2**, 287–306.

Drummond, J. W., Schiff, H. I., Karecki, D. R., and Mackay, G. I. (1989). *Measurements of $NO_2$, $NO_x$, $O_3$, PAN, $HNO_3$, $H_2O_2$, and $H_2CO$ during the Southern California Air Quality Study.* Paper presented at the 82nd Annual Meeting of the Air and Waste Management Association, Anaheim, CA, Paper 139.4.

Dunwoody, C. L. (1986). "Rapid nitrate loss from $PM_{10}$ filters." *J. Air Pollut. Control Assoc.* **36**, 817–818.

Eatough, D. J., White, V. F., Hansen, L. D., Eatough, N. L., and Ellis, E. C. (1985). "Hydration of nitric acid and its collection in the atmosphere by diffusion denuders." *Anal. Chem.* **57**, 743–748.

Eatough, N. L., McGregor, S., Lewis, E. A., Eatough, D. J., Huang, A. A., and Ellis, E. C. (1988). "Comparison of six denuder methods and a filter pack for the collection of ambient $HNO_3(g)$, $HNO_2(g)$, and $SO_2(g)$ in the 1985 NSMC study." *Atmos. Environ.* **22**, 1601–1618.

Edner, H., Fredriksson, K., Sunesson, A., Svanberg, S., Uneus, L., and Wendt, W. (1987). "Mobile remote sensing system for atmospheric monitoring." *Appl. Opt.* **26**, 4330–4338.

Ellis, E. C., and Margeson, J. H. (1974). *Evaluation of Triethanolamine Procedure for Determination of Nitrogen Dioxide in Ambient Air.* U.S. Environmental Protection Agency, Washington, DC, Publ. No. EPA-650/4-74-031.

Environmental Protection Agency (EPA) (1971). *Air Quality Criteria for Nitrogen Oxides.* U.S. Environmental Protection Agency, Washington, DC, Publ. No. AP–84.

Fahey, D. W., Eubank, C. S., Hubler, G., and Fehsenfeld, F. C. (1985a). "Evaluation of a catalytic reduction technique for the measurement of total reactive odd-nitrogen $NO_y$ in the atmosphere." *J. Atmos. Chem.* **3**, 434–468.

Fahey, D. W., Eubank, C. S., Hubler, G., and Fehsenfeld, F. C. (1985b). "A calibrated source of $N_2O_5$." *Atmos. Environ.* **19**, 1883–1890.

Fahey, D. W., Hubler, G., Parrish, D. D., Williams, E. J., Norton, R. B., Ridley, B. A., Singh, H. B., Lin, S. C., and Fehsenfeld, F. C. (1986). "Reactive nitrogen species in the troposphere: Measurements of NO, $NO_2$, $HNO_3$, particulate nitrate, peroxyacetyl nitrate (PAN), $O_3$, and total reactive odd nitrogen ($NO_y$) at Niwot Ridge, Colorado." *J. Geophys. Res.* **91**, 9781–9793.

Febo, A., DeSantis, F., and Perrino, C. (1986). "Measurement of atmospheric nitrous and nitric acid by means of annular denuders." In *Physico-Chemical Behavior of Atmospheric Pollutants, Air Pollution Research Report 2.* Reidel, Dordrecht, pp. 121–125.

Febo, A., DeSantis, F., Liberti, A., and Perrino, C. (1988). "Nitric acid-nitrate aerosol measurements by a diffusion denuder: A performance evaluation." *Atmos. Environ.* **22**, 2062–2064.

*Federal Register* (1986). **51**(53), March 19, p. 9595.

Fehsenfeld, F. C., Dickerson, R. R., Hubler, G., Luke, W. T., Nunnermacker, L. J., Williams, E. J., Roberts, J. M., Calvert, J. G., Curran, C. M., Delany, A. C., Eubank, C. S., Fahey, D. W., Fried, A., Gandrud, B. W., Langford, A. O., Murphy, P. C., Norton, R. B., Pickering, K. E., and Ridley, B. A. (1987). "A ground-based intercomparison of NO, $NO_x$, and $NO_y$ measurement techniques." *J. Geophys. Res.* **92**, 14710–14722.

Fehsenfeld, F. C., Drummond, J. W., Roychowdhury, U. K., Galvin, P. J., Williams, E. J., Buhr, M. P., Parrish, D. D., Hubler, G., Langford, A. O., Calvert, J. G., Ridley, B. A., Grahek, F., Heikes, B., Kok, G., Shetter, J., Walega, J., Elsworth, C. M., Norton, R. B., Fahey, D. W., Murphy, P. C., Havermale, C., Mohnen, V. A., Demerjian, K. L., Mackay, G. I., and Schiff, H. I. (1990). "Intercomparison of $NO_2$ measurement techniques." *J. Geophys. Res.* **95**(D4), 3579–3597.

Fellin, P., Brown, P. M., and Caton, R. B. (1984). *Development of a Nitric Acid Sampler.* Concord Scientific Corporation, Downsview, Ontario.

Ferm, M. (1986). "A $Na_2CO_3$-coated denuder and filter for determination of gaseous $HNO_3$ and particulate $NO_3^-$ in the atmosphere." *Atmos. Environ.* **20**, 1193–1201.

Ferm, M., and Sjödin, A. (1985). "A sodium carbonate coated denuder for determination of nitrous acid in the atmosphere." *Atmos. Environ.* **19**, 979–983.

Fincher, C. L., Tucker, A. W., and Birnbaum, M. (1978). "Laser and flash lamp fluorescence $NO_2$ monitors: A comparison." *Laser Spectrosc.* **158**, 137–140.

Folsom, B. A., and Courtney, C. W. (1979). "Accuracy of chemiluminescent analyzers measuring nitric oxide in stack gases." *J. Air Pollut. Control Assoc.* **29**, 1166–1169.

Fontijn, A., Volltrauer, H. N., and Frenchu, W. R. (1980). "$NO_x$ monitor based on an H-atom direct chemiluminescence method." *Environ. Sci. Technol.* **14**, 324–328.

Forrest, J., Tanner, R. L., Spandau, D., D'Ottavio, T., and Newman, L. (1980). "Determination of total inorganic nitrate utilizing collection of nitric acid on NaCl-impregnated filters." *Atmos. Environ.* **14**, 137–144.

Forrest, J., Spandau, D. J., Tanner, R. L., and Newman, L. (1982). "Determination of atmospheric nitrate and nitric acid employing a diffusion denuder with a filter pack." *Atmos. Environ.* **16**, 1473–1485.

Fredriksson, K. A., and Hertz, H. M. (1984). "Evaluation of the DIAL technique for studies on $NO_2$ using a mobile lidar system." *Appl. Opt.* **23**, 1403–1411.

Fuglsang, K. (1986). *A Filter Pack for Determination of Total Ammonia, Total Nitrate, Sulfur Dioxide, and Sulfate in the Atmosphere*. Risø National Laboratory, Roskilde, Denmark, Report MST-LUFT-A-103.

Fung, K. K., and Grosjean, D. (1985). "Peroxybenzoylnitrate: Measurements in smog chambers and urban air." *Sci. Total Environ.* **46**, 29–40.

Fung, K. K., Heisler, S. L., Price, A., Nuesca, B. V., and Mueller, P. K. (1979). "Comparison of ion chromatography and automated wet chemical methods for analysis of sulfate and nitrate in ambient particulate filter samples." In E. Sawicki and J. D. Mulick, eds., *Ion Chromatographic Analysis of Environmental Pollutants*. Ann Arbor Sci. Publ., Ann Arbor, MI, Vol. 2, pp. 203–209.

Gaffney, J. S., Fajer, R., and Senum, G. I. (1984). "An improved procedure for high purity gaseous peroxyacyl nitrate production: Use of heavy lipid solvents." *Atmos. Environ.* **18**, 215–218.

Gallagher, C. C., Forsberg, C. A., Pierl, R. V., Faucher, G. A., and Calo, J. M. (1985). "Nitric oxide and nitrogen dioxide content of whole air samples obtained at altitudes from 12 to 30 km." *J. Geophys. Res.* **90**, 7899–7912.

Gay, B. W., Jr., Noonan, R. C., Bufalini, J. J., and Hanst, P. L. (1976). "Photochemical synthesis of peroxyacyl nitrates in gas phase via chlorine-aldehyde reaction." *Environ. Sci. Technol.* **10**, 82–85.

Girman, J. R., Hodgson, A. T., Robinson, B. K., and Traynor, G. W. (1984). "Laboratory studies of the temperature dependence of the Palmes $NO_2$ passive sampler." In *Proceedings of the National Symposium on Recent Advances in Pollutant Monitoring of Ambient Air and Stationary Sources*. U. S. Enviromental Protection Agency, Washington, DC, Publ. No. EPA-600/9-84-001, pp. 152–166.

Goldan, P. D., Kuster, W. C., Albrittion, D. L., Fehsenfeld, F. C., Connell, P. S., Norton, R. B., and Huebert, B. J. (1983). "Calibration and tests of the filter-collection method for measuring clean-air, ambient levels of nitric acid." *Atmos. Environ.* **17**, 1355–1364.

Gregory, G. L., Hoell, J. M., Jr., Ridley, B. A., Singh, H. B., Gandrud, B., Salas, L. J., and Shetter, J. (1990a). "An intercomparison of airborne PAN measurements." *J. Geophys. Res.* **95**(D7), 10077–10087.

Gregory, G. L., Hoell, J. M., Jr., Huebert, B. J., Van Bramer, S. E., LeBel, P. J., Vay, S. A., Marinaro, R. M., Schiff, H. I., Hastie, D. K., Mackay, G. I., and Karecki, D. R. (1990b). "An intercomparison of airborne nitric acid measurements." *J. Geophys. Res.* **95**(D7), 10089–10102.

Gregory, G. L., Hoell, J. M., Jr., Carroll, M. A., Ridley, B. A., Davis, D. D., Bradshaw, J., Rodgers, M. O., Sandholm, S. T., Schiff, H. I., Hastie, D. R., Karecki, D. R., Mackay, G. I., Harris, G. W., Torres, A. L., and Fried, A. (1990c). "An intercomparison of airborne nitrogen dioxide instruments." *J. Geophys. Res.* **95**(D7), 10103–10127.

Gregory, G. L., Hoell, J. M., Jr., Torres, A. L., Carroll, M. A., Ridley, B. A., Rogers, M. O., Bradshaw, J., Sandholm, S., and Davis, D. D. (1990d). "An intercomparison of airborne nitric oxide measurements: A second opportunity." *J. Geophys. Res.* **95**(D7), 10129–10138.

Griffith, D. W., and Schuster, G. (1987). "Atmospheric trace gas analysis using matrix isolation-Fourier transform infrared spectroscopy." *J. Atmos. Chem.* **5**, 59–81.

Grosjean, D., and Harrison, J. (1985a). "Peroxyacetyl nitrate: Comparison of alkaline hydrolysis and chemiluminescence methods." *Environ. Sci. Technol.* **19**, 749–752.

Grosjean, D., and Harrison, J. (1985b). "Response of chemiluminescence $NO_x$ analyzers and ultraviolet ozone analyzers to organic air pollutants." *Environ. Sci. Technol.* **19**, 862–865.

Grosjean, D., and Parmar, S. S. (1990). "Interferences from aldehydes and peroxyacetyl nitrate when sampling urban air organic acids on alkaline traps." *Environ. Sci. Technol.* **24**, 1021–1026.

Grosjean, D., Fung, K., Collins, J., Harrison, J., and Breitung, E. (1984). "Portable generator for on-site calibration of peroxyacetyl nitrate analyzers." *Anal. Chem.* **56**, 569–573.

Grosjean, D., Parmar, S. S., and Williams, E. L., III. (1990). "Southern California air quality study: A search for methyl nitrate." *Atmos. Environ.* **24A**, 1207–1210.

Hansen, D. A. (1989). "Measuring trace gases with FM spectroscopy." *EPRI J.*, June, 42–43.

Harker, A. B., Richards, L. W., and Clark, W. E. (1977). "The effect of atmospheric $SO_2$ photochemistry upon observed nitrate concentrations in aerosols." *Atmos. Environ.* **11**, 87–91.

Helas, G., Flanz, M., and Warneck, P. (1981). "Improved $NO_x$ monitor for measurements in tropospheric clean air regions." *Int. J. Environ. Anal. Chem.* **10**, 155–166.

Helas, G., Broll, A., Rumpel, K.-J., and Warneck, P. (1987). "On the origins of night-time NO at a rural measurement site." *Atmos. Environ.* **21**, 2285–2295.

Helmig, D., Muller, J., and Klein, W. (1989). "Improvements in analysis of atmospheric peroxyacetyl nitrate (PAN)." *Atmos. Environ.* **23**, 2187–2192.

Hering, S. V., Lawson, D. R., Allegrini, I., Febo, A., Perrino, C., Possanzini, M., Sickles, J. E., II, Anlauf, K. G., Wiebe, A., Appel, B. R., John, W., Ondo, J., Wall, S., Braman, R. S., Sutton, R., Cass, G. R., Solomon, P. A., Eatough, D. J., Eatough, N. L., Ellis, E. C., Grosjean, D., Hicks, B. B., Womack, J. D., Horrocks, J., Knapp, K. T., Ellestad, T. G., Paur, R. J., Mitchell, W. J., Pleasant, M., Peake, E., MacLean, A., Pierson, W. R., Brachaczek, W., Schiff, H. I.,

Mackay, G. I., Spicer, C. W., Stedman, D. H., Winer, A. M., Biermann, H. W., and Tuazon, E. C. (1988). "The nitric acid shootout: Field comparison of measurement methods." *Atmos. Environ.* **22**, 1519–1539.

Higashi, T., Imasaka, T., and Ishibashi, N. (1983). "Thermal lens spectrophotometry with argon laser excitation source for nitrogen dioxide determination." *Anal. Chem.* **55**, 1907–1910.

Highsmith, V. R., Bond, A. E., and Howes, J. E., Jr. (1986). "Particle and substrate losses from Teflon and quartz filters." *Atmos. Environ.* **20**, 1413–1417.

Hoell, J. M., Jr., Gregory, G. L., McDougal, D. S., Carroll, M. A., McFarland, M., Ridley, B. A., Davis, D. D., Bradshaw, J., Rodgers, M. O., and Torres, A. L. (1985). "An intercomparison of nitric oxide measurement techniques." *J. Geophys. Res.* **90**, 12843–12851.

Hoell, J. M., Jr., Gregory, G. L., McDougal, D. S., Torres, A. L., Davis, D. D., Bradshaw, J., Rodgers, M. O., Ridley, B. A., and Carroll, M. A. (1987). "Airborne intercomparison of nitric oxide measurement techniques." *J. Geophys. Res.* **92**, 1995–2008.

Holdren, M. W., and Rasmussen, R. A. (1976). "Moisture anomaly in analysis of peroxacetyl nitrate (PAN)." *Environ. Sci. Technol.* **10**, 185–187.

Holdren, M. W., and Spicer, C. W. (1984). "Field calibration procedure for peroxyacetyl nitrate." *Environ. Sci. Technol.* **18**, 113–116.

Holland, D. M., and McElroy, F. F. (1986). "Analytical method comparisons by estimates of precision and lower detection limit." *Environ. Sci. Technol.* **20**, 1157–1161.

Hughes, E. E., Rook, H. L., Deardorff, E. R., Margeson, J. H., and Fuerst, R. G. (1977). "Performance of a nitrogen dioxide permeation device." *Anal. Chem.* **49**, 1823–1829.

Intersociety Committee (1972a). "Tentative method of analysis for nitrate in atmospheric particulate matter (2,4-Xylenol Method) 12306-01-70T." In *Methods of Air Sampling and Analysis*. Am. Public Health Assoc., Washington, DC.

Intersociety Committee (1972b). "Tentative method of analysis for peroxyacetyl nitrate (PAN) in the atmosphere (gas chromatographic method) 44301-01-70T." In *Methods of Air Sampling and Analysis*. Am. Public Health Assoc., Washington, DC.

Intersociety Committee (1972c). "Tentative method of analysis for nitrate in atmospheric particulate matter (Brucine Method) 12306-02-72T." *Health Lab. Sci.* **9**, 324–326.

Jacobs, M. B., and Hochheiser, S. (1958). "Continuous sampling and ultramicrodetermination of nitrogen dioxide in air." *Anal. Chem.* **30**, 426–428.

Jenke, D. R. (1983). "Comparison of three methods for the extraction of selected anions from media used for the collection of airborne particulates." *J. Air Pollut. Control Assoc.* **33**, 765–767.

John, W., and Reischl, G. (1978). "Measurements of the filtration efficiencies of selected filter types." *Atmos. Environ.* **12**, 2015–2019.

Johnson, S. A., and Kumar, R. (1987). "Comparison of three aerosol sampling techniques and the differences in the nitrate determined by each." In *Proceedings of the 1987 EPA/APCA Symposium on Measurement of Toxic and Related Air Pol-*

*lutants.* U.S. Environmental Protection Agency, Washington, DC, Publ. No. EPA 600/9-87-010, pp. 183–188.

Johnston, P. V., and McKenzie, R. L. (1984). "Long-path absorption measurements of tropospheric $NO_2$ in rural New Zealand." *Geophys. Res. Lett.* **11**, 69–72.

Joos, L. F., Landolt, W. F., and Leuenberger, H. (1986). "Calibration of peroxyacetyl nitrate measurements with an $NO_x$ analyzer." *Environ. Sci. Technol.* **20**, 1269–1273.

Joseph, D. W., and Spicer, C. W. (1978). "Chemiluminescence method for atmospheric monitoring of $HNO_3$ and nitrogen oxides." *Anal. Chem.* **50**, 1400–1403.

Joshi, S. B., and Bufalini, J. J. (1978). "Halocarbon interferences in chemiluminescent measurements of $NO_x$." *Environ. Sci. Technol.* **12**, 597–599.

Jung, J., and Kowalski, J. (1986). "Direct ambient nitrogen dioxide measurement by visible light absorption." In *Proceedings of the 1986 EPA/APCA Symposium on Measurement of Toxic Air Pollutants*. U.S. Environmental Protection Agency, Washington, DC, Publ. No. EPA 600/9-86-013, pp. 845–855.

Kadowaki, S. (1977). "Size distribution and chemical composition of atmospheric particulate nitrate in the Nagoya area." *Atmos. Environ.* **11**, 671–675.

Kamiura, T., and Tanaka, M. (1979). "Determination of nitrate in suspended particulate matter by high-performance liquid chromatography with U.V. detection." *Anal. Chim. Acta* **110**, 117–122.

Kamphake, L. J., Hannah, S. A., and Cohen, J. M. (1967). "Automated analysis for nitrate by hydrazine reduction." *Water Res.* **1**, 205–216.

Kelly, T. J. (1986). *Modifications of Commercial Oxides of Nitrogen Detectors for Improved Response*. Brookhaven National Laboratory, Upton, NY, Report No. BNL 38000.

Kelly, T. J., Stedman, D. H., and Kok, G. L. (1979). "Measurements of $H_2O_2$ and $HNO_3$ in rural air." *Geophys. Res. Lett.* **6**, 375–378.

Kelly, T. J., Spicer, C. W., and Ward, G. F. (1990). "An assessment of the luminol chemiluminescence technique for measurement of $NO_2$ in ambient air." *Atmos. Environ.* **24A**, 2397–2403.

Klapheck, K., and Winkler, P. (1985). "Sensitivity loss of a $NO_x$-chemiluminescence analyzer due to deposit formation." *Atmos. Environ.* **19**, 1545–1548.

Kley, D., and McFarland, M. (1980). "Chemiluminescence detector for NO and $NO_2$." *Atmos. Technol.* **12**, 63–68.

Klockow, D., Niessner, R., Malejczyk, M., Kiendl, H., von Berg, B., Keuken, M. P., Wayers-Ypelaan, A., and Slanina, J. (1989). "Determination of nitric acid and ammonium nitrate by means of a computer-controlled thermodenuder system." *Atmos. Environ.* **23**, 1131–1138.

Knapp, K. T., Durham, J. L., and Ellestad, T. G. (1986). "Pollutant sampler for measurements of atmospheric acidic dry deposition." *Environ. Sci. Technol.* **20**, 633–637.

Komhyr, W. D., Dutton, E. D., and Thompson, T. M. (1988). "A general gravimetric dilution technique for preparing trace calibration gases: $N_2O$ calibration gas preparation." *Environ. Sci. Technol.* **22**, 845–848.

Kondo, Y., Iwata, A., and Takagi, M. (1983). "A chemiluminescent $NO_x$-detector for the aircraft measurement." *J. Meteorol. Soc. Jpn.* **61**, 756–762.

Kondo, Y., Matthews, W. A., Iwata, A., Morita, Y., and Takagi, M. (1987). "Aircraft measurements of oxides of nitrogen along the eastern rim of the Asian continent: Winter observations." *J. Atmos. Chem.* **5**, 37–58.

Kosmus, W. (1985). "Summation method for monitoring nitrogen oxides." *Int. J. Environ. Anal. Chem.* **22**, 269–279.

Koutrakis, P., Wolfson, J. M., Slater, J. L., Brauer, M., Spengler, J. D., Stevens, R. K., and Stone, C. L. (1988). "Evaluation of an annular denuder/filter pack system to collect acidic aerosols and gases." *Environ. Sci. Technol.* **22**, 1463–1468.

Leahy, D. F., Phillips, M. F., Garber, R. W., and Tanner, R. L. (1980). "Filter material for sampling of ambient aerosols." *Anal. Chem.* **52**, 1779–1780.

Lindqvist, F. (1985). "Determination of nitric acid in ambient air by gas chromatography/photoionization detection after collection in a denuder." *J. Air Pollut. Control Assoc.* **35**, 19–23.

Lipari, F. (1984). "New solid-sorbent method for ambient nitrogen dioxide monitoring." *Anal. Chem.* **56**, 1820–1826.

Lonneman, W. A. (1977). "PAN measurement in dry and humid atmospheres." *Environ. Sci. Technol.* **11**, 194–195.

Lonneman, W. A., Bufalini, J. J., and Seila, R. L. (1976). "PAN and oxidant measurement in ambient atmospheres." *Environ. Sci. Technol.* **4**, 374–380.

Lonneman, W. A., Bufalini, J. J., and Namie, G. R. (1982). "Calibration procedure for PAN based on its thermal decomposition in the presence of nitric oxide." *Environ. Sci. Technol.* **16**, 655–660.

Mackay, G. I., and Schiff, H. I. (1987). "Reference measurements of $HNO_3$ and $NO_2$ by tunable diode laser absorption spectroscopy." In *Proceedings of the 1987 EPA/ APCA Symposium on Measurement of Toxic and Related Air Pollutants*, U.S. Environmental Protection Agency, Washington, DC, Publ. No. EPA 600/9-87-010, pp. 366–372.

MacPhee, R. D., Higuchi, J. E., and Leh, F. K. V. (1976). "Comparison of oxidant measurement methods, ultraviolet photometry, and moisture effects." In *Proceedings of Ozone/Oxidants Interactions with the Total Environment Specialty Conference*. Air Pollut. Control Assoc., Washington, DC, pp. 166–175.

Maeda, Y., Aoki, K., and Munemori, M. (1980). "Chemiluminescence method for the determination of nitrogen oxide." *Anal. Chem.* **52**, 307–311.

Matthews, R. D., Sawyer, R. F., and Schefer, R. W. (1977). "Interferences in chemiluminescent measurement of NO and $NO_2$ emissions from combustion systems." *Environ. Sci. Technol.* **11**, 1092–1096.

McClenny, W. A., Galley, P. C., Braman, R. S., and Shelley, T. J. (1982). "Tungstic acid technique for monitoring nitric acid and ammonia in ambient air." *Anal. Chem.* **54**, 365–369.

Meyrahn, H., Helas, G., and Warneck, P. (1987). "Gas chromatographic determination of peroxyacetyl nitrate: Two convenient calibration techniques." *J. Atmos. Chem.* **5**, 405–415.

Michie, R. M., Jr., McElroy, F. F., Sokash, J. A., Thompson, V. L., and Fritschel, B. P. (1983). *Performance Test Results and Comparative Data for Designated Reference and Equivalent Methods for Nitrogen Dioxide*. U.S. Environmental Protection Agency, Washington, DC, Publ. No. EPA 600/4-83-019.

Miller, D. P. (1984). "Ion chromatographic analysis of Palmes tubes for nitrite." *Atmos. Environ.* **18**, 891–892.

Miller, D. P. (1988). "Low-level determination of nitrogen dioxide in ambient air using the Palmes tube." *Atmos. Environ.* **22**, 945–947.

Mueller, P. K., and Hidy, G. M. (1983). *The Sulfate Regional Experiment: Report of Findings*. Electric Power Research Institute, Palo Alto, CA, Publ. No. EPRI EA-1901, Vol. 1.

Mulawa, P. A., and Cadle, S. H. (1985). "A comparison of nitric acid and particulate nitrate measurements by the penetration and denuder difference methods." *Atmos. Environ.* **19**, 1317–1324.

Mulik, J. D., and Williams, D. E. (1986). "Passive sampling devices for $NO_2$." In *Proceedings of the 1986 EPA/APCA Symposium on Measurement of Toxic Air Pollutants*. U.S. Enviromental Protection Agency, Washington, DC, Publ. No. EPA 600/9-86-013, pp. 61–70.

Mulik, J. D., and Williams, D. E. (1987). "Passive sampling device measurements of $NO_2$ in ambient air." In *Proceedings of the 1987 EPA/APCA Symposium on Measurement of Toxic and Related Air Pollutants*. U.S. Environmental Protection Agency, Washington, DC, Publ. No. EPA 600/9-87-010, pp. 387–397.

Mulik, J., Puckett, R., Williams, D., and Sawicki, E. (1976). "Ion chromatographic analysis of sulfate and nitrate in ambient aerosols." *Anal. Lett.* **9**, 653–663.

Mullin, J. B., and Riley, J. P. (1955). "The spectrophotometric determination of nitrate in natural waters, with particular reference to sea-water." *Anal. Chim. Acta* **12**, 464–480.

Namiesnik, J., Gorecki, T., Kozlowski, E., Torres, L. and Mathieu, J. (1984). "Passive dosimeters—an approach to atmospheric pollutants analysis." *Sci. Total Environ.* **38**, 225–258.

National Aeronautics and Space Administration (NASA) (1983). *Assessment of Techniques for Measuring Tropospheric $N_xO_y$*. NASA, Washington, DC, NASA Conf. Publ. 2292.

National Bureau of Standards (NBS) (1988). *NBS Standard Reference Material Catalog 1988–89*. U.S. Government Printing Office, Washington, DC, NBS Spec. Publ. 260.

Nelson, G. O. (1972). *Controlled test atmospheres principles and techniques*. Ann Arbor Sci. Publ., Ann Arbor, MI, pp. 126–134.

Nicksic, S. W., Harkins, J., and Mueller, P. K. (1967). "Some analyses for PAN and studies of its structure." *Atmos. Environ.* **1**, 11–18.

Nieboer, H., and van Ham, J. (1976). "Peroxyacetyl nitrate (PAN) in relation to ozone and some meteorological parameters at Delft in the Netherlands." *Atmos. Environ.* **10**, 115–120.

Nielsen, T., Hansen, A. M., and Thomsen, E. L. (1982). "A convenient method for preparation of pure standards of peroxyacetyl nitrate for atmospheric analysis." *Atmos. Environ.* **16**, 2447–2450.

Norwitz, G., and Keliher, P. N. (1978). "Inorganic interferences in the 2,4-xylenol spectrophotometric method for nitrate and their elimination." *Anal. Chim. Acta* **98**, 323–333.

Noxon, J. F. (1978). "Tropospheric $NO_2$." *J. Geophys. Res.* **83**, 3051–3057.

Noxon, J. F. (1983). "$NO_3$ and $NO_2$ in the mid-Pacific troposphere." *J. Geophys. Res.* **88**, 11017–11021.

Okita, T., Morimoto, S., Izawa, M., and Konno, S. (1976). "Measurement of gaseous and particulate nitrates in the atmosphere." *Atmos. Environ.* **10**, 1085–1089.

Palmes, E. D., and Tomczyk, C. (1979). "Personal sampler for $NO_x$." *Am. Ind. Hyg. Assoc. J.* **40**, 588–591.

Palmes, E. D., Gunnison, A. F., DiMattio, J., and Tomczyk, C. (1976). "Personal sampler for nitrogen dioxide." *Am. Ind. Hyg. Assoc. J.* **37**, 570–577.

Perrino, C., DeSantis, F., and Febo, A. (1988). "Uptake of nitrous acid and nitrogen oxides by nylon surfaces: Implications for nitric acid measurement." *Atmos. Environ.* **22**, 1925–1930.

Perrino, C., DeSantis, F., and Febo, A. (1990). "Criteria for the choice of a denuder sampling technique devoted to the measurement of atmospheric nitrous and nitric acids." *Atmos. Environ.* **24A**, 617–626.

Pierotti, D., and Rasmussen, R. A. (1977). "The atmospheric distribution of nitrous oxide." *J. Geophys. Res.* **82**, 5823–5832.

Pierson, W. R., Brachaczek, W. W., Korniski, T. J., Truex, T., and Butler, J. W. (1980). "Artifact formation of sulfate, nitrate, and hydrogen ion on backup filters: Allegheny Mountain experiment." *J. Air Pollut. Control Assoc.* **30**, 30–34.

Platt, U. F., and Perner, D. (1983). "Measurements of atmospheric trace gases by long path differential UV/visible absorption spectroscopy." Chapter 2.6. In D. K. Killinger and A. Mooradian, eds., *Optical and Laser Remote Sensing*. Springer-Verlag, New York, pp. 97–105.

Platt, U. F., Winer, A. M., Biermann, H. W., Atkinson, R., and Pitts, J. N., Jr. (1984). "Measurement of nitrate radical concentrations in continental air." *Environ. Sci. Technol.* **18**, 365–369.

Poizat, O., and Atkinson, G. H. (1982). "Determination of nitrogen dioxide by visible photoacoustic spectroscopy." *Anal. Chem.* **54**, 1485–1489.

Possanzini, M., Febo, A., and Liberti, A. (1983). "New design of a high-performance denuder for the sampling of atmospheric pollutants." *Atmos. Environ.* **17**, 2605–2610.

Possanzini, M., Febo, A., and Cecchini, F. (1984). "Development of a KI annular denuder for $NO_2$ collection." *Anal. Lett.* **17**, 887–896.

Purdue, L. J., and Hauser, T. R. (1980). "Review of U. S. Environmental Protection Agency $NO_2$ monitoring methodology requirements." In S. D. Lee, ed., *Nitrogen Oxides and Their Effects on Health*. Ann Arbor Sci. Publ., Ann Arbor, MI, Chapter 4, pp. 51–76.

Rehme, K. A., Smith, C. F., Beard, M. E., and Fitz-Simmons, T. (1984). *Investigation of Filter Media for Use in the Determination of Ambient Particulate Matter*. Environmental Protection Agency, Washington, DC, Publ. No. EPA-600/S4-84-048.

Rhodes, R. C., and Evans, E. G. (1988). *Precision and Accuracy Assessments for State and Local Air Monitoring Networks*. U.S. Environmental Protection Agency, Washington, DC, Publ. EPA-600/S4-88-007.

Rickman, E. E., Jr., and Wright, R. S. (1986). *Interference of Nitrogeneous Compounds on Chemiluminescent Measurement of Nitrogen Dioxide*. Research Triangle Institute, Research Triangle Park, NC, Report No. RTI/3180/24-01F.

Rickman, E. E., Jr., Green, A. H., Wright, R. S., and Sickles, J. E., II. (1989). *Laboratory and Field Evaluations of Extrasensitive Sulfur Dioxide and Nitrogen Dioxide Analyzers for Acid Deposition Monitoring.* Research Triangle Institute, Research Triangle Park, NC, Report No. RTI/3999/18-04F.

Ridley, B. A., and Howlett, L. C. (1974). "An instrument for nitric oxide measurements in the stratosphere." *Rev. Sci. Instrum.* **45**, 742–746.

Ridley, B. A., Carroll, M. A., Gregory, G. L., and Sachse, G. W. (1988). "NO and $NO_2$ in the troposphere: Technique and measurements in regions of a folded tropopause." *J. Geophys. Res.* **93**, 15813–15830.

Roberts, J. M. (1990). "The atmospheric chemistry of organic nitrates." *Atmos. Environ.* **24A**, 243–287.

Roberts, J. M., Norton, R. B., Goldan, P. D., and Fehsenfeld, F. C. (1987). "Evaluation of the tungsten oxide denuder tube technique as a method for the measurement of low concentrations of nitric acid in the troposphere." *J. Atmos. Chem.* **5**, 217–238.

Roberts, J. M., Fajer, R. W., and Springston, S. R. (1989). "Capillary gas chromatographic separation of alkyl nitrates and peroxycarboxylic nitric anhydrides." *Anal. Chem.* **61**, 771–772.

Rodgers, M. O., and Davis, D. D. (1989). "A UV-photofragmentation/laser-induced fluorescence sensor for the atmospheric detection of HONO." *Environ. Sci. Technol.* **23**, 1106–1112.

Rodgers, M. O., Asai, K., and Davis, D. D. (1980). "Photofragmentation-laser induced fluorescence: A new method for detecting atmospheric trace gases." *Appl. Opt.* **19**, 3597–3605.

Roscoe, H. K. (1982). "Tentative observation of stratospheric $N_2O_5$." *Geophys. Res. Lett.* **9**, 901–902.

Saltzman, B. E. (1954). "Colorimetric microdetermination of nitrogen dioxide in the atmosphere." *Anal. Chem.* **26**, 1949–1955.

Saltzman, B. E. (1980). "Critique of measurement techniques for ambient nitrogen oxides." In S. D. Lee, ed., *Nitrogen Oxides and Their Effects on Health.* Ann Arbor Sci. Publ. Ann Arbor, MI, Chapter 3, pp. 31–50.

Schiff, H. I., Hastie, D. R., Mackay, G. I., Iguchi, T., and Ridley, B. A. (1983). "Tunable diode laser systems for measuring trace gases in tropospheric air." *Environ. Sci. Technol.* **17**, 352–364A.

Schiff, H. I., Mackay, G. I., Castledine, C., Harris, G. W., and Tran, Q. (1986). "A sensitive direct measurement $NO_2$ instrument." In *Proceedings of the 1986 EPA/APCA Symposium on Measurement of Toxic Air Pollutants.* U.S. Enviromental Protection Agency, Washington, DC, Publ. No. EPA 600/9-86-013, pp. 834–844.

Schwartz, S. E. (1983). *Mass Transport Considerations Pertinent to Aqueous-Phase Reactions of Gases in Liquid-Water Clouds.* Brookhaven National Laboratory, Upton, NY, Report No. BNL 34174, p. 9.

Shaw, R. W., Jr., Stevens, R. K., Bowermaster, J., Tesch, J. W., and Tew, E. (1982). "Measurements of atmospheric nitrate and nitric acid: The denuder difference experiment." *Atmos. Environ.* **16**, 845–853.

Shores, R. C., and Smith, F. (1984). *Stability Evaluation of Sulfur Dioxide, Nitric Oxide and Carbon Monoxide Gases in Cylinders.* U.S. Environmental Protection Agency, Washington, DC, Publ. No. EPA-600/4-84-086.

Sickles, J. E., II (1987). *Sampling and Analytical Methods Development for Dry Deposition Monitoring.* Research Triangle Institute, Research Triangle Park, NC, Report RTI/2823/00-15F.

Sickles, J. E., II, and Hodson, L. L. (1989). "Fate of nitrous acid on selected collection surfaces." *Atmos. Environ.* **23**, 2321–2324.

Sickles, J. E., II, and Michie, R. M. (1987). "Evaluation of the performance of sulfation and nitration plates." *Atmos. Environ.* **21**, 1385–1391.

Sickles, J. E., II, and Wright, R. S. (1979). *Atmospheric Chemistry of Selected Sulfur Containing Compounds: Outdoor Smog Chamber Study Phase I.* U.S. Environmental Protection Agency, Washington, DC, Publ. No. EPA-600/7-79-227, pp. 45–49.

Sickles, J. E., II, Perrino, C., Allegrini, I., Febo, A., Possanzini, M., and Paur, R. J. (1988). "Sampling and analysis of ambient air near Los Angeles using an annular denuder system." *Atmos. Environ.* **22**, 1619–1625.

Sickles, J. E., II, Hodson, L. L., Rickman, E. E., Jr., Saeger, M. L., Hardison, D. L., Turner, A. R., Sokol, C. K., Estes, E. D., and Paur, R. J. (1989). "Comparison of the annular denuder system and the transition flow reactor for measurements of selected dry deposition species." *J. Air Pollut. Control Assoc.* **39**, 1218–1224.

Sickles, J. E., II, Groshe, P. M., Hodson, L. L., Salmons, C. A., Cox, K. W., Turner, A. R., and Estes, E. D. (1990a). "Development of a method for the sampling and analysis of sulfur dioxide and nitrogen dioxide from ambient air." *Anal. Chem.* **62**, 338–346.

Sickles, J. E., II, Hodson, L. L., McClenny, W. A., Paur, R. J., Ellestad, T. G., Mulik, J. D., Anlauf, K. G., Wiebe, H. A., Mackay, G. I., Schiff, H. I., and Bubacz, D. K. (1990b). "Field comparison of methods for the measurement of contributors to acidic dry deposition." *Atmos. Environ.* **24A**, 155–165.

Simeonov, V., and Puxbaum, H. (1977). "A comparative study on the nitrate determination in airborne dust." *Mikrochim. Acta* **2**, 397–403.

Singh, H. B., and Salas, L. J. (1983). "Methodology for the analysis of peroxyacetyl nitrate (PAN) in the unpolluted atmosphere." *Atmos. Environ.* **17**, 1507–1516.

Singh, H. B., and Viezee, W. (1988). "Enhancement of PAN abundance in the Pacific marine air upon contact with selected surfaces." *Atmos. Environ.* **22**, 419–422.

Small, H., Stevens, T. S., and Bauman, W. C. (1975). "Novel ion exchange chromatographic method using conductimetric detection." *Anal. Chem.* **47**, 1801–1809.

Smith, J. P., Grosjean, D., and Pitts, J. N., Jr. (1978). "Observation of significant losses of particulate nitrate and ammonium from high volume glass fiber filter samples stored at room temperature." *J. Air Pollut. Control Assoc.* **28**, 930–933.

Solomon, P. A., Fall, T., Salmon, L., Lin, P., Vasquez, F., and Cass, G. R. (1988). *Acquisition of Acid Vapor and Aerosol Concentration Data for Use in Dry Deposition Studies in the South Coast Air Basin.* California Institute of Technology, Pasadena, EQL Report 25 to CAARB, Vol. 1.

Spicer, C. W., and Schumacher, P. M. (1979). "Particulate nitrate: Laboratory and field studies of major sampling interferences." *Atmos. Environ.* **13**, 543–552.

Spicer, C. W., Schumacher, P. M., Kouyoumjian, J. A., and Joseph, D. A. (1978a). *Sampling and Analytical Methodology for Atmospheric Particulate Nitrates.* U.S. Environmental Protection Agency, Washington, DC, Publ. No. EPA-600/2-78-067.

Spicer, C. W., Ward, G. F., and Gay, B. W., Jr. (1978b). "A further evaluation of microcoulometry for atmospheric nitric acid monitoring." *Anal. Lett.* **A11**(1), 85–95.

Spicer, C. W., Howes, J. E., Jr., Bishop, T. A., Arnold, L. H., and Stevens, R. K. (1982). "Nitric acid measurement methods: An intercomparison." *Atmos. Environ.* **16**, 1487–1500.

Spicer, C. W., Joseph, D. W., and Schumacher, P. M. (1985). "Determination of nitrate in atmospheric particulate matter by thermal decomposition and chemiluminescence." *Anal. Chem.* **57**, 2338–2341.

Staehr, W., Lahmann, W., and Weitkamp, C. (1985). "Range-resolved differential absorption lidar: Optimization of range and sensitivity." *Appl. Opt.* **24**, 1950–1956.

Stephens, E. R. (1969). The formation, reactions, and properties of peroxyacyl nitrates (PANs) in photochemical air pollution. *Adv. Environ. Sci.* **1**, 119–146.

Stephens, E. R., and Price, M. A. (1973). "Analysis of an important air pollutant: Peroxyacetyl nitrate." *J. Chem. Educ.* **60**, 351–354.

Stephens, E. R., Burleson, F. R., and Cardiff, E. A. (1965). "The production of pure peroxyacyl nitrates." *J. Air Pollut. Control Assoc.* **15**, 87–89.

Sturges, W. T., and Harrison, R. M. (1988). "Thermal speciation of atmospheric nitrate and chloride: A critical evaluation." *Environ. Sci. Technol.* **22**, 1305–1311.

Szonntagh, E. L. (1979). "Colorimetric azo dye methods for the atmospheric analysis of nitrogen dioxide: Historical development." *Period. Polytech., Chem. Eng.* **23**, 207–215.

Tanner, R. L., Fajer, R., and Gaffney, J. (1979). "Determination of parts-per-billion concentrations of aqueous nitrate by derivatization gas chromatography with electron capture detection." *Anal. Chem.* **51**, 865–870.

Tanner, R. L., Daum, P. H., and Kelly, T. J. (1983). "New instrumentation for airborne acid rain research." *Int. J. Environ. Anal. Chem.* **13**, 323–335.

Tanner, R. L., Kelly, T. J., Dezaro, D. A., and Forrest, J. (1989). "A comparison of filter, denuder, and real-time chemilumescence techniques for nitric acid determination in ambient air." *Atmos. Environ.* **23**, 2213–2222.

Technicon (1972). *Industrial Method No. 158-71W/Tentative, Nitrate and Nitrite in Water and Seawater.* Technicon Industrial Systems, Tarrytown, NY.

Tesch, J. W., Rehg, W. R., and Sievers, R. E. (1976). "Microdetermination of nitrates and nitrites in saliva, blood, water, and suspended particulates in air by gas chromatography." *J. Chromatogr.* **126**, 743–755.

Thijsse, T. R. (1978). "Gas chromatographic measurement of nitrous oxide and carbon dioxide in air using electron capture detection." *Atmos. Environ.* **12**, 2001–2003.

Torres, A. L. (1985). "Nitric oxide measurements at a nonurban eastern United States site: Wallops instrument results from July 1983 GTE/CITE mission." *J. Geophys. Res.* **90**, 12875–12880.

Tuazon, E. C., Graham, R. A., Winer, A. M., Easton, R. R., Pitts, J. N., Jr., and Hanst, P. L. (1978). "A kilometer pathlength Fourier-transform infrared system for the study of trace pollutants in ambient and synthetic atmospheres." *Atmos. Environ.* **12**, 865–875.

Tuazon, E. C., Winer, A. M., Graham, R. A., and Pitts, J. N., Jr. (1981). *Atmospheric Measurements of Trace Pollutants: Long Path Fourier Transform Infrared Spec-*

*troscopy.* U.S. Environmental Protection Agency, Washington, DC, Publ. No. EPA-600/S3-81-026.

Vierkorn-Rudolph, B., Rudolph, J., and Diederich, S. (1985). "Determination of peroxyacetyl nitrate (PAN) in unpolluted areas." *Int. J. Environ. Anal. Chem.* **20**, 131–140.

Vinjamoori, D. V., and Ling, C.-S. (1981). "Personal monitoring method for nitrogen dioxide and sulfur dioxide with solid sorbent sampling and ion chromatographic deterination." *Anal. Chem.* **53**, 1689–1691.

Vossler, T. L., Stevens, R. K., Paur, R. J., Baumgardner, R. E., and Bell, J. P. (1988). "Evaluation of improved inlets and annular denuder systems to measure inorganic air pollutants." *Atmos. Environ.* **22**, 1729–1736.

Walega, J. G., Stedman, D. H., Shetter, R. E., Mackay, G. I., Iguchi, T., and Schiff, H. I. (1984). "Comparison of a chemiluminescent and a tunable diode laser absorption technique for the measurement of nitrogen oxide, nitrogen dioxide, and nitric acid." *Environ. Sci. Technol.* **18**, 823–826.

Wall, S. M., John, W., and Ondo, J. L. (1988). "Measurement of aerosol size distributions for nitrate and major ionic species." *Atmos. Environ.* **22**, 1649–1656.

Wallace, L. A., and Ott, W. R. (1982). "Personal monitors: A state-of-the art survey." *J. Air Pollut. Control Assoc.* **32**, 601–610.

Watanabe, I., and Stephens, E. R. (1978). "Reexamination of moisture anomly in analysis of peroxyacetyl nitrate." *Environ. Sci. Technol.* **12**, 222–223.

Wendel, G. J., Stedman, D. H., and Cantrell, C. A. (1983). "Luminol-based nitrogen dioxide detector." *Anal. Chem.* **55**, 937–940.

West, P. W., and Reiszner, K. D. (1978). *Personal Monitor for Nitrogen Dioxide.* U.S. Environmental Protection Agency, Washington, DC, Publ. No. EPA-600/2-78-001.

Williams, E. L., III, and Grosjean, D. (1990). "Removal of atmospheric oxidants with annular denuders." *Environ. Sci. Technol.* **24**, 811–824.

Willison, M. J., and Clark, A. G. (1984). "Analysis of atmospheric aerosols by non-suppressed ion chromatography." *Anal. Chem.* **56**, 1037–1039.

Winer, A. M., Peters, J. W., Smith, J. P., and Pitts, J. N., Jr. (1974). "Response of commercial chemiluminescent NO-NO$_2$ analyzers to other nitrogen-containing compounds." *Environ. Sci. Technol.* **8**, 1118–1121.

Winfield, T. W. (1977). *A Method for Converting NO$_2$ to NO by Ferrous Sulfate Prior to Chemiluminescent Measurements.* Presented at American Chemical Society Meeting, New Orleans, LA.

Witz, S., and Wendt, J. G. (1981). "Artifact sulfate and nitrate formation at two sites in the South Coast Air Basin." *Environ. Sci. Technol.* **15**, 79–83.

Witz, S., Eden, R. W., Wadley, M. W., Dunwoody, C., Papa, R., and Torre, K. J. (1990). "Rapid loss of particulate nitrate, chloride and ammonium on quartz fiber filters during storage." *J. Air Waste Manage. Assoc.* **40**, 53–61.

Woebkenberg, M. L. (1982). "A comparison of three passive personal sampling methods for NO$_2$." *Am. Ind. Hyg. Assoc. J.* **43**, 553–561.

Wolff, G. T. (1984). "On the nature of nitrate in coarse continental aerosols." *Atmos. Environ.* **18**, 977–981.

Wright, R. S., Tew, E. L., Decker, C. E., von Lemden, D. J., and Barnard, W. F.

(1987). "Performance audits of EPA protocol gases and inspection and maintenance calibration gases." *J. Air Pollut. Control Asssoc.* **37**, 384–385.

Yanagisawa, Y., and Nishimura, H. (1982). "A badge-type personal sampler for measurement of personal exposure to $NO_2$ and NO in air." *Environ. Int.* **8**, 235–242.

Yoshizumi, K. (1986). "Regional size distributions of sulfate and nitrate in the Tokyo metropolitan area in summer." *Atmos. Environ.* **20**, 763–766.

Yoshizumi, K., and Hoshi, A. (1985). "Size distributions of ammonium nitrate and sodium nitrate in atmospheric aerosols." *Environ. Sci. Technol.* **19**, 258–261.

Yoshizumi, K., Aoki, K., Matsuoka, T., and Asakura, S. (1985). "Determination of nitrate by a flow system with a chemiluminescent $NO_x$ analyzer." *Anal. Chem.* **57**, 737–740.

Zafiriou, O. C., and True, M. B. (1986). "Interferences in environmental analysis of NO by NO plus $O_3$ detectors: A rapid screening technique." *Environ. Sci. Technol.* **20**, 594–596.

# 3

# THE APPLICATION OF DENUDER SYSTEMS TO THE ANALYSIS OF ATMOSPHERIC COMPONENTS

*J. Slanina, P. J. de Wild, and G. P. Wyers*

*Netherlands Energy Research Foundation ECN, The Netherlands*

## 1. INTRODUCTION

Artifacts observed in sampling atmospheric trace components employing filter techniques or bubbler systems were originally the main reason to develop methods based on the separation of gases and particulate matter by denuder systems (Niessner and Klockow, 1980). In fact, denuders were used in much

*Gaseous Pollutants: Characterization and Cycling,* Edited by Jerome O. Nriagu.
ISBN 0-471-54898-7  © 1992 John Wiley & Sons, Inc.

of the earlier work (Forrest et al., 1982) to trap gaseous components that could interfere in filter measurements. In a late stage, denuders were used as sampling systems to measure either gaseous components or particulate matter, converted to the gaseous phase by, for example, heating the air sample. In the early stages of development, 1979–1987, research was generally directed to demonstrating the selectivity of denuder-based methods. In this period it was clearly demonstrated that denuder systems were generally superior to the former sampling methods, especially if reactive gases were to be measured. Many procedures for the application of denuders to measure a large number of components in the atmosphere have recently been summarized in an excellent review by Ali et al. (1989), in which generally accepted theoretical description of the absorption processes of the different denuder types, cylindrical and annular, is presented. The conventional denuder methods, however, are not readily adaptable for application in monitoring networks. The evolution of denuder systems, from an intrinsic laboratory instrument to a device applicable to different aspects of monitoring, has taken place in the last few years. Several aspects tied in with this development, such as mechanization, automation, quality control, and intercomparison with other methods, are highlighted in this chapter.

The first denuders were based on the cylindrical design, with the disadvantage that only very small sampling rates could be used. Therefore, unless long sampling times are employed, very low amounts of the components of interest have to be analyzed and as a consequence ultra-trace methodology must be applied to achieve sufficiently low detection limits and to avoid contamination problems in sample preparation. This situation has prohibited a widespread application of denuders. The introduction of annual devices, allowing much larger sampling rates, eliminated this problem at least partly (Vossler et al., 1988). Unfortunately, coating and extraction of the trapped species are labor intensive and limit is use. Thermodenuder systems (e.g., Braman et al., 1982; Klockow et al., 1989; Slanina et al., 1985) alleviate this problem because the species is trapped on a coating and liberated by heating in an easily detectable form. These devices can operate automatically and be used as monitors. The main drawback of these systems is that the selectivity can be problematic in some cases, due to either the limited selectivity of the coating or unwanted conversions of interfering species in the heating step. Wet denuder systems in different forms (Dasgupta, 1984; Keuken et al., 1988) offer an alternative for easy operation. In these systems the species of interest is taken up by a solution, present as an aqueous layer either on the walls of an annular denuder or inside semipermeable tubes. The solution is analyzed on-line or transferred to a fraction collector for later analysis.

The development of these mechanized and automated systems opens up the possibilities of incorporating denuder systems in monitoing activities. Assessment of the concentrations of reactive species is not only important for acquiring a general knowledge on atmospheric chemistry, but also for establishing pollution/effect relations. Species like nitric acid, nitrous acid, and ammonia play an important role in acidification and nitrogen input in ecosystems. On many locations, the nitrogen load is determined by these compounds and not by deposition of $NO_x$. Measurement of deposition velocities, or surface resistances, of ammonia and nitric acid has recently been performed by means of denuder systems and is also directly effect related. Measurement of deposition velocities by means of the gradient method calls for a precision in the order of 0.5–5%, depending on species and surface characteristics. Here, a good precision is, next to selectivity, important as well.

## 2. PRINCIPLES OF THE DENUDER TECHNIQUE

The denuder in its most elementary form is a cylindrical tube coated with a reagent, which selectively samples a specific gaseous component from an aerosol. This concept is schematically shown in Fig. 1. Air is sucked through the denuder for a certain period and the sample volume is measured by a gas meter or estimated from sample time and flow, the latter being controlled by a critical orifice. After sampling, the coating of the denuder is extracted and analyzed for the compound of interest. The coating of the denuder should be selective for the trace gases of interest, stable under ambient temperatures and relative humidities, and compatible with the analytical chemical techniques used for determining the concentrations in the extract. The theoretical fundamentals underlying the absorption of gaseous components for both cylindrical and annular denuders have recently been discussed by several authors (e.g., Febo et al., 1989; Ali et al., 1989) and therefore only the relevant equations for the determination of collection efficiencies are presented here. The equation for diffusional loss of a gaseous species from an aerosol passing

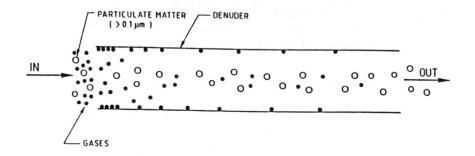

**Figure 1.** Diffusional separation of gases and particulate matter in a denuder tube.

a denuder under laminar flow conditions was developed by Gormley and Kennedy (1949):

$$\frac{C}{C_0} = 0.819 \exp(-14.63\delta) + 0.097 \exp(-89.2\delta)$$

$$+ 0.019 \exp(-212\delta) + \cdots$$

$$\text{with } \delta = \frac{\pi DL}{4F}$$

where $C_0$ and $C$ are the concentrations at the entrance and exit of the denuder respectively, $D$ is the diffusion coefficient of the collected species in air ($cm^2/s$), $L$ is the length of the denuder (cm), and $F$ is the flow rate ($cm^3/s$). The equation is valid for a laminar flow regime, that is, with a Reynolds number below 2100. The Reynolds number (Re) for flow through a circular tube can be calculated from

$$\text{Re} = \frac{4F}{\eta v d}$$

where $v$ is the kinematic viscosity of air (0.152 $cm^2/s$ stp) and $d$ the internal diameter of the tube (cm). The laminar flow regime in a cylindrical tube is established after the gas has passed a certain distance $l$, the so-called laminar flow subduction zone. The entrance part of the denuder is therefore generally left uncoated and the length of the laminar flow subduction zone can be calculated from

$$l = 0.07\text{Re } d$$

With an annular denuder much higher flow rates can be employed. the equation for the collection efficiency in an annular denuder was determined empirically by Possanzini et al. (1983):

$$\frac{C}{C_0} = 0.819 \exp(-22.53\delta)$$

$$\text{with } \delta = \left(\frac{\eta DL}{4F}\right)\left(\frac{d_1 + d_2}{d_2 - d_1}\right)$$

where $d_1$ and $d_2$ are the outer diameter of the inner tube and the inner diameter of the outer tube, respectively. Winiwarter (1989) has presented an expression for the collection efficiency in annular denuders, starting from Fick's law. His conclusions agree with earlier work, based on the Gormley and Kennedy equations (Possanzini et al., 1983). McMurry and Stolzenburg (1987) have presented a theory that describes the penetration of gaseous components in denuders as a function of the mass accommodation coefficient. Corsi et al. (1988) developed a numerical solution for computing the collection efficiency of membrane-based diffusion scrubbers, a version of the recently developed solution-based denuders. A number of studies on the comparison of theoretical and empirical behavior have been published (e.g., Brauer et al., 1989; Murphy and Fahey, 1987), which show that a reasonable agreement exists between theoretical performance and experimental results. In general, it can be concluded that adequate theoretical descriptions of absorption of different species in both cylindrical and annular denuder systems are available.

Deposition of particulate matter in a denuder must be avoided. Assuming that the flow inside the denuder is strictly laminar, particle losses may arise from diffusion or from gravitational settling. Diffusional loss is important for small particle diameters (<0.1 μm), whereas loss by sedimentation becomes significant for coarse (>1 μm) particles (Fig. 2) (e.g., Ferm, 1979; Niessner et al., 1988). A theoretical approach to estimate the deposition rate of particles is presented by Febo et al. (1989). Sedimentation of coarse particles can be prevented by operating the denuders in a vertical position. Alternatively, a preseparator such as a cyclone or an impactor can be used to remove large particles from the sample stream. However, since denuder systems are generally applied to measure reactive, polar compounds ($HNO_3$, $HNO_2$, $NH_3$, $H_2O_2$, HCl, HF), losses in the inlet system can occur easily, especially if a

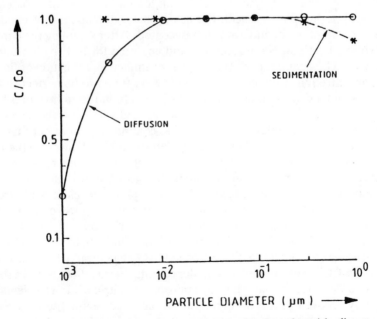

**Figure 2.** Loss of particulate matter in a denuder as a function of particle diameter: flow, 2 L/min; denuder length, 90 cm; i.d. 0.6 cm. (After Niessner et al. (1988).)

preseparator is used, since these compounds absorb readily on almost any surface. These losses will depend on the sample flow. Cylindrical denuder systems, which sample at flows ranging from 0.4 to 2 L/min are more prone to inlet losses than, for example, annular devices with sampling rates in the order of 5–30 L/min. To account for these losses the inlet can be leached and analyzed as well (Ferm, 1986), although some particulate matter will also be deposited at the inlet due to the turbulent flow regime. Losses in the inlet of annular systems have been evaluated by Vossler et al. (1988) and Brauer et al. (1989). Both studies show that the losses of gaseous components on glass or glass coated with PTFE are at an acceptable level at sample flow rates of 10–20 L/min and sampling periods of 1 to 24 h.

Other sampling artifacts arising from, for example, the decomposition of volatile particulate matter or from collection of compounds interfering in the analytical technique are discussed elsewhere in this manuscript.

## 3.  CYLINDRICAL DENUDERS

Cylindrical denuders have been applied to the measurement of many atmospheric constituents such as $HNO_3$, $HNO_2$, $NH_3$, $NO_2$, $SO_2$, HCl, and organic compounds. An overview is given in Table 1. The first application of a denuder for sampling of atmospheric trace gases, rather than for removal of gaseous impurities from aerosol, was described in 1979 by Ferm. For sampling of ammonia he employed Pyrex tubes, 50 cm long and 3 mm i.d., coated with oxalic acid or phosphorous acid. For achievement of laminar flow conditions the first 15 cm of the denuder was left uncoated. After sampling the denuders are extracted with an NaOH solution and analyzed for $NH_4^+$ by ion-selective electrodes. The denuder performance was examined as a function of flow and $NH_3$ concentration. Both coating reagents were found to be perfect sinks for ammonia and a collection efficiency of 90% was found for a flow of 3 L/min. The determined diffusion coefficient in air agreed with the value assessed in the literature. Ferm preferred oxalic acid as coating reagent rather than phosphorous acid, which has a low point of deliquescence. However, the disadvantage of an oxalic acid coating is that, especially at higher ambient temperature, a fraction may be lost by sublimation, together with the chemisorbed ammonia. Other acidic coatings, such as phosphoric or citric acid, are also used as denuder coatings for ammonia. However, extreme care should be taken to avoid contamination of the denuders with ambient $NH_3$ before or after exposure (Ferm, 1979; Dimmock and Marshall, 1986). For sampling of acidic gases such as $HNO_3$, HCl, $SO_2$, and $HNO_2$, a neutral or alkaline reagent ($NaCl$, $Na_2CO_3$, $NaF$) is used as denuder coating. Ferm (1986) tested a sodium carbonate denuder for sampling of atmospheric nitric acid. The denuder is heated by an oven at a few degrees above ambient temperature to prevent dissolution of the coating, for example, during fog conditions. Eatough et al. (1985) noticed a slight dependency of the diffusion coefficient for $HNO_3$ on

**Table 1 Cylindrical Denuders: An Overview**

| Sampled Species | Quantitative Aspects | | System Characteristics | | References |
|---|---|---|---|---|---|
| | Detection Limit | Precision (RSD)[a] | Coating | (1) Dimensions ($l \times$ i.d.)<br>(2) Sample Flow<br>(3) Sample Time | |
| $NH_3$ | 8.5 ng/m³ | 15% | Oxalic acid | (1) 50 × 0.3 cm<br>(2) 10 L/min<br>(3) 24 h | Ferm (1979) |
| $NH_3$ | Not reported | 5% for 43 µg/m³ cal. gas | Citric acid | (1) 60 × 0.3 cm<br>(2) 2 L/min<br>(3) 3.5 h | Bos (1980) |
| $HNO_3$<br>$NH_4NO_3$ | Not reported | Not reported | $Na_2CO_3$ | (1) 46 × 0.4 cm<br>(2) 10–30 L/min<br>(3) 1 h | Forrest et al. (1982) |
| $NH_3$<br>$HNO_3$ | Not reported<br>Not reported | Not reported<br>Not reported | Oxalic acid<br>KOH | (1) 30 × 0.4 cm<br>(2) 12 L/min<br>(3) 24 h | Lewin and Hansen (1984) |
| $HNO_2$ | 0.5 nmol/m³ | 10–20% | $Na_2CO_3$ | (1) 75 × 0.4 cm<br>(2) 2 L/min<br>(3) 24 h | Ferm and Sjödin (1985) |
| $HNO_3$<br>$NO_3^-$<br>$SO_2$<br>$SO_4^{2-}$<br>$NO_2$ | 0.5 nmol/m³<br>3 nmol/m³<br>2 nmol/m³<br>2 nmol/m³<br>Not reported | 7.5%<br>Not reported<br>Not reported<br>Not reported<br>Not reported | $Na_2Co_3$<br><br><br><br>Activated $MnO_2$ | (1) 75 × 0.4 cm<br>(2) 2 L/min<br>(3) 24 h<br><br>(1) 40 × 0.4 cm, 5 cm coated<br>(2) 0.2–0.7 L/min | Ferm (1986)<br><br><br><br>Adams et al. (1986) |
| $HCl$ | (50/t) µg/m³<br>$t$ = sampling time (min) | 0.02 µg/m³ for 0.16–0.55 µg/m³ | NaF | (1) 95 × 0.3 cm<br>(2) 2 L/min | Dimmock and Marshall (1987) |
| $SO_2$ | Not reported | Not reported | $Na_2CO_3$/glycerol | (1) 95 × 1 cm, coiled<br>(2) 10 L/min | Pui et al. (1990) |

[a]RSD,

135

relative humidity due to formation of the hydrate $HNO_3 \cdot xH_2O$. This implies that for some trace gases ($HNO_3$, $NH_3$) the collection efficiency will be a function of the atmospheric water content.

Particulate nitrates and ammonium salts can interfere with sampling of ammonia and nitric acid by denuders. Artifacts may arise from deposition of particles on the coating, but also from dissociation of particles through shifts in solid–gas equilibria induced by rapid removal of the gaseous components in the denuder. Several studies have shown that the interference from particular matter is not significant when denuders are operated in a vertical position. Ferm (1979) found that interference of particulate $NH_4^+$ in ambient aerosol (0.1–4 μm) is of the same order of magnitude as the detection limit for the $NH_3$ measurement. Particle interference in the oxalic acid denuder was also investigated by Dimmock and Marshall (1986), using artificial ammonium sulfate aerosol. Sampling of gaseous ammonia in ambient air was compared to sampling in the presence of ammonium sulfate particles, but no significant effect was determined on the determination of ammonia. These observations are supported by an earlier study of Eatough et al. (1985) who performed laboratory and field measurements with a nylon denuder for $HNO_3$ and found that sampling of $HNO_3$ does not cause a detectable shift in the equilibrium between gaseous and particulate nitrate. However, Pratsinis et al. (1989) have demonstrated that problems due to dissociation of $NH_4NO_3$ could occur if large amounts of small particles (<0.1 μm) are present at temperatures above 35 °C. These conclusions agree with earlier work by Slanina (1982).

Coatings such as $Na_2CO_3$ or NaF will also quantitatively retain over acidic compounds such as $HNO_2$, HCl, and $SO_2$. An NaF-coated denuder for sampling of HCl was shown to perform satisfactorily and free from interference by particulate chlorides (Dimmock and Marshall, 1987). Sodium carbonate denuders are also applied for sampling of $HNO_2$ and $SO_2$ (e.g., Ferm and Sjödin, 1985; Ferm, 1986). However, for nitrous acid interferences were found from NO and $NO_2$ (artifact formation of $HNO_2$ at the denuder surface) and from hydrolysis of PAN (Ferm and Sjödin, 1985). It is possible to correct for these interferences by sampling with several denuders in series, on the condition that the interfering substance is collected at a lesser efficiency than the compound of interest. It was shown by Febo et al. (1989) that interferences by different compounds can be deduced from the deposition pattern within a denuder.

Recently a coiled denuder has been suggested for atmospheric sampling. Pui et al. (1990) describe a three-turn coil with a diameter of 10 cm, made from a 1-cm-i.d. glass tube. The advantages of this design are its compactness (only 6 cm long) and the high mass-transfer efficiency of helical flow (relative to straight flow). The collection efficiency for $SO_2$, using a $Na_2CO_3$/glycerine coating, was better than 99% at a flow of 10 L/min. Particle loss due to diffusion, sedimentation, and inertial impaction was less than 2% for neutral particles between 0.1 and 2 μm. A high loss was observed for particles >2 μm and therefore the coiled denuder should be used with a suitable separator (cyclone, impactor) for coarse particles. Sampling problems were encountered at low humidities due to an increased loss of electrostatically charged particles.

Sampling of atmospheric trace gases by denuders was a significant improvement over filter and impinger techniques, which are prone to sampling artifacts. Impingers suffer from lack of selectivity for gaseous and particulate matter, whereas filters can yield erroneous results due to reactions between gases and particles on the filter or to volatilization of particles. Cylindrical denuders are selective, inexpensive, and easy to apply in the field, but the coating and extraction procedures are rather laborious (risk of contamination) and the sample solutions have to be analyzed off-line. An attempt was made by Bos (1980) to automate a cylindrical denuder for $NH_3$. His ammonia monitor featured automatic coating, extraction, and colorimetric analysis. The main disadvantage of using simple tubular denuders, however, is the low flow rate of 0.5–3 L/min, which necessitates long sampling periods (at least several hours for most trace gases) and may lead to absorption losses in the inlet system. A time resolution of several hours may be sufficient for some monitoring applications, but for studies of dry deposition using micrometeorological methods a time resolution better than 20 min–1 h is needed. Attempts have been made to reduce this problem using multiple denuders which are sampled simultaneously and extracted with the same aliquot of wash solution (Forrest et al., 1982; Lewin and Hansen, 1984).

## 4. ANNULAR DENUDER SYSTEMS

As stated before, the problem connected with the low sample flows of cylindrical denuders have prompted several research groups to investigate devices capable of being used with higher flow rates. A general discussion on the optimum application of annular denuders, regarding coating and setup (such as placing devices in series to characterize interferences) has been given by Perrino et al. (1990), with emphasis on the measurement of nitric and nitrous acid. An overview of the applications of annular denuders is given in Table 2. A simple way to increase the flow rate in denuders is to insert a rod in the middle of a cylindrical device, allowing sampling rates up to 5 L/min (Keuken, 1989). The annular denuder developed by researchers of CNR (Possanzini et al., 1983), however, allowed sampling rates of 20–50 L/min, depending on the design of the annular system. The annular devices offer, in principle, all possibilities of the cylindrical systems plus extra options as a function of the larger air sample taken. Compounds such as organic substances and $H_2O_2$ are present in such low concentrations that they can be measured only by high-volume denuder systems.

Possanzini et al. (1987) describe a method for C1–C3 aldehydes, based on annular denuders coated with 2,4-dinitrophenylhydrazine. The aldehydes are analyzed as hydrazone derivatives by HPLC in combination with UV and

**Table 2 Annular Denuder Systems: An Overview**

| Sampled Species | Quantitative Aspects | | | System Characteristics | | |
| | Detection Limit | Precision (RSD) | | Coating | (1) Dimensions[a] (2) Sample Flow (3) Sample Time | References |
| --- | --- | --- | --- | --- | --- | --- |
| $NO_2$ | Not reported | Not reported | | KI | (1) 20 × 1.32 cm<br>20 × 2.32 cm<br>23 × 3.32 cm<br>0.16 cm<br>(2) 1–10 L/min | Possanzini et al. (1984) |
| $NH_3$<br>$NH_4^+$ | Not reported<br>Not reported | Not reported<br>Not reported | | Oxalic acid | (1) 25 × ? cm<br>0.16 cm<br>(2) 20 L/min<br>(3) 0.5 h | Allegrini et al. (1984) |
| $HNO_3$<br>$SO_2$<br>$NH_3$ | ≈0.1 $\mu g/m^3$<br>≈0.1 $\mu g/m^3$<br>Not reproted | ±4% (0.2–1.6 $\mu g/m^3$)<br>±5% (1–6 $\mu g/m^3$)<br>Not reported | | $Na_2CO_3$/glycerol | (1) 38.1 × ? cm<br>0.15 cm<br>(2) 15 L/min<br>(3) 6 h<br>16 h | Stevens et al. (1985) |
| Tetra-alkyl-lead | Not reported | 5% | | IC1–PEG–Carbowax 600 matrix[b] | (1) 12 × 1.3 cm<br>0.15 cm<br>(2) 1 L/min<br>(3) 24 h | Febo et al. (1986) |
| HCB[c]<br>α-HCH[d]<br>γ-HCH | 0.04 $ng/m^3$<br>0.03 $ng/m^3$<br>0.05 $ng/m^3$ | Not reported<br>Not reported<br>Not reported | | Tenax | (1) ?<br>(2) 16.7 L/min<br>(3) 24 h | Johnson et al. (1986) |

| Species | Detection limit | Precision/accuracy | Coating | Sampling conditions | Reference |
|---|---|---|---|---|---|
| $CH_2O$ | $\approx 0.4$ ppbv | Not reported | 2,4-DNP[e] | (1) 20 × 1.3 cm, 0.15 cm; (2) 2 L/min; (3) 4 h | Possanzini et al. (1987) |
| $CH_3CHO$ | $\approx 0.2$ ppbv | Not reported | | | |
| $C_2H_5CHO$ | $\approx 0.2$ ppbv (100 L air) | Not reported | | | |
| $HNO_3$ | $\geq 0.1$ µg/m³ (3 m³ air) | <7% (0.2–1.1 µg/m³) | $Na_2CO_3$/glycerol | (1) 22 × 3.5 cm, 0.15 cm; (2) 15 L/min; (3) 3 h | Allegrini et al. (1987) |
| $HNO_2$ | $\geq 0.1$ µg/m³ (3 m³ air) | 10% (0.8–3.4 µg/m³) | | | |
| $SO_2$ | $\geq 0.1$ µg/m³ (3 m³ air) | 5% (2.8–9.9 µg/m³) | | | |
| $NO_2$ | $\geq 0.1$ µg/m³ (3 m³ air) | 6% (8.1–20. µg/m³) | | | |
| $NH_3$ | 0.14 µg/m³ (3 m³ air) | 4% (1.9–7.5 µg/m³) | | | |
| $H_2O_2$ | 0.05 µg/m³ | Not reported | $Ti(IV)-H_2SO_4$ | (1) 10 × 1.3 cm, 0.15 cm; (2) 10–20 L/min; (3) 1 h | Possanzini et al. (1988) |
| HCOOH | 2 nmol/m³ | 3% | KOH | (1) 30 × 0.9 cm, 0.15 cm; 90 × 0.9 cm, 0.15 cm; 30 × 0.9 cm, 0.15 cm; (2) 9 L/min; (3) 3 h | Rosenberg et al. (1988) |
| $CH_3COOH$ | 14 nmol/m³ | 8% | | | |
| $NHO_3$ | 1.5 nmol/m³ ($NO_3^-$) | 14% | NaF | | |
| $H_2SO_4$ | 1.5 nmol/m³ ($SO_4^{2-}$) | 23% | NaF (heated) | | |
| $NH_4^+$ | 2 nmol/m³ | 16% | | | |
| $NO_3^-$ | 1.5 nmol/m³ | 12% | | | |
| $Cl^-$ | 3 nmol/m³ | Not reported | | | |
| HCl | Not reported | Not reported | $H_3PO_4$ | (1) 26.5 × 2.4 cm, 0.1 cm; 24.2 × 2.4 cm, 0.1 cm; (2) 10 L/min | Koutrakis et al. (1988) |
| $NH_3$ | Not reported | 9% | $Na_2CO_3$/glycerol | | |
| $HNO_3$ | Not reported | 2.6% | | | |
| $HNO_2$ | Not reported | 7.6% | | | |
| $SO_2$ | Not reported | 2.6% | | | |
| $CH_2O$ | 0.3 µg/m³ | Not reported | $NaHSO_4$/triethanolamine | (1) #; (2) 2.5 L/min; (3) 24 h | Ali et al. (1989, p. 763), Cecchini et al. (1985) |

139

**Table 2** (*Continued*)

| | Quantitative Aspects | | System Characteristics | | |
|---|---|---|---|---|---|
| Sampled Species | Detection Limit | Precision (RSD) | Coating | (1) Dimensions[a]<br>(2) Sample Flow<br>(3) Sample Time | References |
| PAH[f] | Not reported | Not reported | Silicon grease | (1) 20.3 × 0.8 cm<br>0.32 cm<br>(2) 100–200 L/min<br>(3) 24 h | Coutant et al. (1989b) |
| HNO₃<br>HNO₂ | Not reported<br>Not reported | Not reported<br>Not reported | NaCl + 2<br>Na₂CO₃/glycerol in<br>series | (1) 20 × 3.3 cm<br>0.15 cm<br>10 × 3.3 cm<br>0.15 cm<br>(2) 15 L/min<br>(3) 12 h (day)<br>24 h | Perrino et al. (1990) |

[a]Length × inner diameter of the outer tube.
[b]Polyethyleneglycol.
[c]Hexachlorobenzene.
[d]Hexachlorohexane.
[e]Dinitrophenylhydrazine.
[f]Poly aromatic hydrocarbon(s).
?, Unmentioned.

voltammetric detection. Johnson et al. (1986) applied a tenax-coated annular denuder to measure chlorinated hydrocarbons such as PCBs. Rosenberg et al. (1988) have developed a system consisting of a number of annular denuders in series, which samples organic acids as well as $HNO_3$, HCl, and $NH_3$. Coutant et al. (1989a) have devised a high-volume system based on 12 concentric aluminum tubes coated with silicone grease, capable of sampling organic vapors at a rate of 100 L/min. A technique for the measurement of $H_2O_2$ by means of a series of annular denuders coated with a $Ti(VI)/H_2SO_4$ mixture has been reported by Possanzini et al. (1988).

Due to their high sampling rates, annular denuders can, more easily than cylindrical devices, be adapted for application in monitoring networks and in studies dedicated, for example, to the determination of dry deposition. The potential of annular denuders as monitoring devices has led to extensive intercomparison measurements to establish quality control. Most of these experiments were performed in the United States, and some were performed in Europe. The "nitric acid shootout" described by Hering et al. (1988) revealed that $HNO_3$ measurements from three annular denuder systems differed by about 30%. However, the results from two denuders agreed reasonably well with measurements by denuder difference systems and optical methods. A comparison between annular denuders and filterpacks under winter conditions, during which dissociation of ammonium nitrate is not very important, indicated good agreement for $HNO_3$, HCl, and, to a lesser extent, $SO_2$ (Dasch et al., 1989). An intercomparison of different denuder and filterpack systems for sampling of $HNO_3$, $HNO_2$, and $SO_2$ (Eatough et al., 1988) led to the conclusion that the annular systems performed quite well.

Intercomparison with other methods showed that sampling of organic acids by means of annular denuders compares favorably with other methods (Keene et al., 1989). Sickles et al. (1990) describe an intercomparison experiment for sampling of a number of acidifying compounds with annular denuders, denuder difference systems, and a tunable diode laser absorption spectrometer (TDLAS) and arrived at the conclusion that the results of the annular denuder systems are consistent with the other investigated methods. An intercomparison experiment between annular denuders and differential optical absorption spectrometry, described by Appel et al. (1990), showed good agreement on the results for nitrous acid. Wiebe et al. (1990) report the results of an intercomparison experiment on ammonia, with the conclusion that fair agreement was found between annular and cylindrical denuders and between the annular denuder and Fourier transform infrared spectroscopy. Two intercomparison experiments, coordinated by the C.E.C., have been organized in Europe. Demuynck (1985) reports on the results obtained at the Schauinsland, Germany, indicating fairly good results for annular denuder systems. A large intercomparison experiment in Rome (Allegrini, 1990) on measurement of nitric acid indicated good agreement between the different annular denuder techniques.

The conclusion from the results of all these intercomparison experiments is that the accuracy and precision of annular denuder measurements are sufficient for many applications. A number of unexplained results makes it clear that further work on intercomparison is necessary.

## 5. THERMODENUDER SYSTEMS

The first generation of thermodenuders concerned heated denuder tubes in which specific aerosol particles are evaporated or decomposed and the released gases/vapors are collected on the coating. These denuders were washed in the same manner as conventional cylindrical and annular devices. In more recent designs the collected species is thermally desorbed and detected by an on-line monitor. In some applications the desorption step is accompanied by (or followed by) conversion of the species into an easily detectable compound. The coating of these thermodenuders is continuously regenerated at each heating step and therefore laborious coating and extraction procedures are avoided. The thermodenuder can easily be automated and is, in principle, highly suited for application in computerized monitoring networks. An overview of thermodenuder systems is given in Table 3.

Several thermodenuder systems have been described for the analysis of atmospheric sulfates. Thermal treatment of aerosol containing sulfates allows separate sampling of sulfuric acid and neutral sulfate particles (Cobourn et al., 1978; Niessner and Klockow, 1980). The thermodenuder designed and tested by Niessner and Klockow (1980) is a cylindrical (glass) denuder, part of which is heated, in series with a PTFE filter. Sulfuric acid droplets evaporate in the heated part of the denuder and the vapor is collected on the uncoated denuder wall. The temperature of the thermodenuder (137 °C) was chosen such that 90% of the sulfuric acid was collected, whereas $(NH_4)_2SO_4$ and $(NH_4)HSO_4$ were not decomposed significantly, but collected on the filter. After sampling, the denuder is rinsed with water, and sulfate is determined in the extract. Experiments with the same denuder coated with NaCl showed that 99% of sulfuric acid was collected and that NaCl can be considered an irreversible sink for $H_2SO_4$ for sampling periods up to 24 h. Interferences from $SO_2$, ammonium and sodium sulfates were investigated. The only potential interferents found were $(NH_4)H_3(SO_4)_2$ and $NaH_3(SO_4)_2$.

The method described above was extended by Slanina et al. (1981) to accommodate for simultaneous sampling of all $SO_4^{2-}$ and $NO_3^-$ compounds. This apparatus consisted of several denuders, at both ambient and elevated temperatures, connected in series. NaF-, $H_3PO_4$-, and NaOH-coated denuders at ambient temperature were used to collect (free) $HNO_3$, $NH_3$, and $SO_2$, respectively. A NaF-coated thermodenuder heated at 115–135 °C served to retain $H_2SO_4$ and to decompose $NH_4NO_3$ and collect the resulting $HNO_3$. Ammonia from dissociation of $NH_4NO_3$ was collected in a $H_3PO_4$-coated denuder. In a second thermodenuder coated with NaF and heated at 215–235 °C ammonium (bi)sulfates were decomposed and the resulting $H_2SO_4$

collected. The last denuder in series, coated with $H_3PO_4$, sampled $NH_3$ released from dissociation of the ammonium (bi)sulfates. After sampling, the denuders were rinsed with water. This denuder system was tested in the laboratory and in the field (Slanina et al., 1981; Slanina, 1982). It became apparent that the potential of this method was good, but that sampling of ammonium nitrate and ammonium sulfates needed further improvement.

Lindqvist (1985b) used a heated denuder tube coated with manganese and palladium oxides to sample sulfuric acid aerosol. Interferences were selectively being removed from the sample by a series of predenuders. After sampling, the trapped sulfur species was thermally desorbed at 800 °C and converted to hydrogen sulfide which could be determined by gas chromatography. Slanina et al. (1985) describe a thermodenuder system consisting of two copper(II) oxide-coated denuder tubes in series at temperatures of 120 and 240 °C, respectively, for the selective collection and determination of airborne sulfuric acid and ammonium sulfates. This method is based on the adsorption of sulfuric acid on copper/copper(II) oxide and the decomposition of ammonium sulfate at temperatures above 220 °C. The resulting copper sulfate can be decomposed to sulfur dioxide at 800 °C, which is analyzed on-line by flame photometric detection. Interferences from $SO_2$ and reduced sulfur compounds are removed by predenuders coated with potassium carbonate and active charcoal. The copper/copper(II) oxide denuder tube is also applied for the measurement of sulfur dioxide at very low concentration levels (Slanina et al., 1987). $H_2S$ is removed by a $NaHSO_4/Ag_2SO_4$ predenuder. The collection efficiency of this system is adequate at relative humidities above 10%.

Similar thermodenuder systems have been developed for the measurement of nitrogen compounds. Lindqvist (1985a) used a thermodenuder coated with $Al_2(SO_4)_3$ for preconcentration of $HNO_3$. After sampling, $HNO_3$ is thermally desorbed as $NO_x$, which, after conversion to NO, is analyzed by gas chromatography with photo-ionization detection. A thermodenuder coated with tungstic acid has been described to measure ambient $NH_3$ and $HNO_3$ concentrations (McClenny et al., 1982; Braman et al., 1982). After sampling, the denuder is heated and the $NH_3$ and $HNO_3$ chemisorbed on the coating are thermally desorbed at a temperature of 350 °C. $HNO_3$ is desorbed as $NO_2$, whereas $NH_3$ is desorbed as a mixture of $NH_3$ and NO (Roberts et al., 1988). After conversion to NO by a gold converter at 600 °C the purge gas is analyzed by a $NO_x$ monitor. Braman et al. (1982) used a second $WO_3$-coated denuder at ambient temperature to temporarily trap desorbed $NH_3$ while $HNO_3$ is being analyzed. Alternatively, signals from $NH_3$ and $HNO_3$ can be separated using their differences in desorption temperature (McClenny et al., 1982; Roberts et al., 1988). In an analogous way, interferences from alkylamines can be separated from signals from $NH_3$ and $HNO_3$ using temperature programmed desorption. However, at high humidities alkylamines may hydrolize to $NH_3$ on the denuder coating (Roberts et al., 1988). No significant interferences were detected from $NO_2$, PAN, or VOCs. The coa-

**Table 3 Thermodenuder Systems: An Overview**

| Sampled Species | Quantitative Aspects | | | System Characteristics | | |
|---|---|---|---|---|---|---|
| | Sampling (s) or Desorbing (d) temperature (°C) | Detection Limit | Precision (RSD) | Coating | (1) Dimensions[a] (2) Sample Flow (3) Sample Time | References |
| $H_2SO_4$ | 140 (s) | 0.7 µg | 2% (2 µg) | NaCl | (1) 90 × 0.6 cm (2) 40 L/h (3) 3–10 min | Niessner and Klockow (1980) |
| $HNO_3$ | Ambient (s) | Not reported | Not reported | NaCl | | |
| $NH_4NO_3$ | Ambient (s) | Not reported | Not reported | NaCl | | |
| $HNO_3$ | Ambient (s) | 0.2 µg/m³ | 9% | NaF | (1) 340 × 0.6 cm (total): 60 cm + | Slanina et al. (1981), Slanina (1982) |
| $NH_3$ | Ambient (s) | Not reported | Not reported | $H_3PO_4$ | 40 cm + | |
| $SO_2$ | Ambient (s) | Not reported | Not reported | NaOH | 40 cm + | |
| $H_2SO_4$ | 140 (s) | 0.4 µg/m³ | 11% | NaF | 60 cm + | |
| $NH_3NO_3^-$ ($NH_4NO_3$) | Ambient (s) | Not reported | 19% | $H_3PO_4$ | 40 cm + | |
| $H_2SO_4$ ↰ Ammonium sulfate | 220 (s) | 0.4 µg/m³ | ±6% | NaF | 60 cm + / 40 cm + | |
| $NH_3$ ↰ Ammonium sulfate | Ambient (s) | Not reported | Not reported | $H_3PO_4$ | (2) 1–2 L/min (3) 4 h | Braman et al. (1982) |
| $HNO_3$ | 350 (d) | 0.01 ppb (20 L) | 5.3% | $WO_3$ | (1) 45 × 0.6 cm (o.d.) (2) 1–2 L/min (3) 30 min | |
| $NH_3$ | 350 (d) | 0.01 ppb (20 L) | 8.2% | $WO_3$ | | |
| $HNO_3$ | 540 (d) | 0.014 µg/m³ (30 L) | 6–15% | $Al_2(SO_4)_3$ | (1) 45 × 0.5 cm (2) ? (3) ±30 min | Lindqvist (1985b) |

144

| Compound | | Detection limit | Accuracy | Coating | Dimensions[a] | Reference |
|---|---|---|---|---|---|---|
| $H_2SO_4$ ←<br>Ammonium sulfate ←<br>($SO_2$) ← | 120 (s)<br>220 (s)<br>(800) (d) | 0.02 µg/m$^3$<br>0.02 µg/m$^3$<br>(both 0.1 µg/m$^3$ at 5 min sampling) | 13%<br>6%<br>(both at 30 min. sampling) | Cu/CuO | (1) 50 × 0.6 cm<br>(2) 0.5 L/min<br>(3) 60 min | Slanina et al. (1985) |
| $SO_2$ | 800 (d) | 0.015–0.1 µg/m$^3$ | <3% | Cu/CuO | (1) 50 × 0.6 cm<br>(2) 0.5 L/min<br>(3) 30 min | Slanina et al. (1987) |
| $H_2SO_4$<br>Ammonium sulfate | 130 (s)<br>700 (d) | 0.1 µg/m$^3$<br>1.0 µg/m$^3$ | Not reported<br>Not reported | Uncoated FPD[b] | (1) ? stainless steel tube<br>(2) 18 L/min<br>(3) 30 min | Böhm and Israël (1987) |
| $HNO_3$<br>$NH_4NO_3$ | Ambient (s)<br>150 (s) | 0.1 µg/m$^3$<br>0.1 µg/m$^3$ | <5%<br><5% | $MgSO_4$ | (1) 25 × 2.8 cm<br>0.15 cm (annulus)<br>(2) 5 L/min<br>(3) 0.5 h | Klockow et al. (1989) |
| $HNO_3$<br>$NH_4NO_3$ | Ambient (s)<br>140 (s) | 0.06 µg/m$^3$<br>0.06 µg/m$^3$ | <5%<br><5% | $BaSO_4$ + active carbon + NaF | (1) 50 × 0.6 cm<br>(2) 0.66 L/min<br>(3) 24 h | Langford et al. (1989) |
| $NH_3$ | 350–380 (d) | ≤0.05 ppbv | ≈20% >0.2 ppbv<br>≈30% <0.2 ppbv | $MoO_3$ rod inside quartz tube | (1) 50 × 0.4 cm<br>0.04 cm (annulus)<br>(2) 6–120 L/h | Keuken et al. (1989) |
| $NH_3$ | 700 (d) | 0.1 µg/m$^3$ | <5% (1–10 µg/m$^3$) | $V_2O_5$ | (1) 25 × 2.8 cm<br>0.15 cm (annulus)<br>(2) 10 L/min<br>(3) 10–20 min | Keuken et al. (1989) |
| Cd | 200 (s) | 0.2 µg/m$^3$ | Not reported | Ag | (1) 50 × 0.6 cm<br>(2) 1 L/min<br>(3) 30 min | Larjava et al. (1990) |

[a]Length × inner diameter of the outer tube.
[b]Flame photometric detector.
?, Unmentioned.

ting of this thermodenuder is difficult to apply, however (vacuum deposition from a tungsten filament), and tends to thermally degrade at the high temperatures required for the desorption of ammonia (Roberts et al., 1988).

A number of problems associated with the tungstic acid thermodenuder were circumvented by Langford et al. (1989) who developed a molybdenum(VI) oxide annular denuder system for ammonia measurements. The denuder consists of a molybdenum rod suspended inside a quartz tube. The surface of the rod is oxidized in situ to $MoO_3$, whereas the outer wall of the annulus is left uncoated. An important advantage of this denuder system is that on heating the denuder a reproducible fraction of $NH_3$ is catalytically oxidized to NO on the $MoO_3$ surface, thereby obviating the need of a secondary converter. The purge gas is analyzed by a $NO_x$ monitor. This surface is not selective but also adsorbs $HNO_3$ and $NO_2$. Separation between signals from these gases and $NH_3$ was obtained by temperature-programmed desorption. Alkylamines will potentially interfere with the ammonia measurements.

Keuken et al. (1989) have developed an annular thermodenuder coated with vanadium pentoxide for measurement of $NH_3$. After sampling, the denuder is rapidly heated to 700 °C by a preheated, moveable furnace and ammonia is desorbed and (catalytically) converted to NO and $NO_2$ (Fig. 3). Nitric acid is to some extent also collected by the $V_2O_5$ coating, but the collection efficiency of the coating is a factor 10 higher for ammonia. PAN and alkylamines will interfere with the ammonia measurement but are generally present in low concentrations relative to $NH_3$. Interference from another, so far unidentified, nitrogen compound was sometimes detected. However, this interference is homogeneously collected throughout the entire denuder,

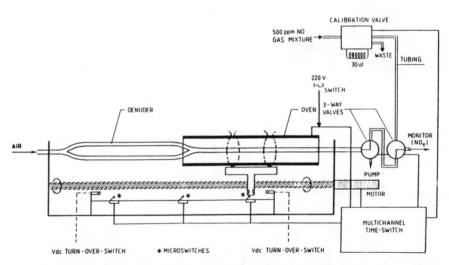

**Figure 3.** The ammonia thermodenuder system of Keuken et al. (1989), based on adsorption of $NH_3$ on a $V_2O_5$-coated denuder.

whereas ammonia is trapped in the first segment. By separately heating and analyzing both parts of the denuder the $NH_3$ response can be corrected for this interference. Intercomparison of measurements by this thermodenuder system, a phosphoric acid-coated annular denuder, a wet annular denuder system (see below), and a tunable diode laser absorption spectrometer showed good agreement.

Two thermodenuder systems for the simultaneous measurement of $HNO_3$ and $NH_4NO_3$ are described by Klockow et al. (1989). Both systems contain three denuders in series: Two denuders are at ambient temperature during sampling, while the last one is heated at 150 °C. One system uses quartz annular denuders coated with $MgSO_4$, and the other system consists of, respectively, a $BaSO_4$-coated, an active charcoal-coated, and an NaF-coated glassy carbon denuder. Nitric acid is sampled in the first denuder, a nitrogen-containing, unknown interference is sampled in all three denuders of the $MgSO_4$- coated system, but only in the first two denuders of the other system, and ammonium nitrate is decomposed and the resulting $HNO_3$ collected in the third denuder. In the $MgSO_4$-coated system the three denuders are heated consecutively at 700 °C and the released $NO_x$ is analyzed by a chemiluminescence monitor. The $NO_x$ response for the middle denuder is used to correct the signals of $HNO_3$ and $NH_4NO_3$ for the interfering substance. The other thermodenuder system uses an active charcoal denuder to trap the interfering compound such that $NH_4NO_3$ is sampled free of interferences.

A new application of the denuder technique was presented by Larjava et al. (1990) who developed a thermodenuder for the analysis of volatile heavy metals in flue gas. The behavior of heavy metal species under different physical and chemical conditions was investigated and denuders coated by elemental silver were found to be efficient for sampling of elemental cadmium at 200 °C. From the obtained experimental data the diffusion coefficient for Cd could be calculated.

Intercomparison experiments for a number of thermodenuders have been described. In particular, the tungstic acid denuder has been frequently compared with other measurement techniques for ammonia and nitric acid, such as the denuder difference method (Appel et al., 1985, 1988), tunable diode laser absorption, and filterpack techniques (Anlauf et al., 1985), conventional cylindrical and annular denuders, and filterpacks (Appel et al., 1988). Several experiments have indicated that the tungstic acid denuder is not a very reliable method for measurement of nitric acid or ammonia. One reason is the poor stability of the $WO_3$ coating at high temperatures (e.g., Roberts et al., 1988). Generally, $HNO_3$ results were higher than those measured by the denuder difference and filterpack methods. A slightly modified version of the $MgSO_4$- coated thermodenuder for $HNO_3$ and $NH_4NO_3$ (Klockow et al., 1989) was used in an intercomparison exercise in Rome in 1988 (Allegrini et al., 1990) which included measurements by cylindrical and annular denuders and filterpacks. The nitric acid measurements of the thermodenuder were in agreement with other techniques, but the collection efficiency of ammonium nitrate was too low.

## 6.  SOLUTION-BASED DENUDER SYSTEMS

The first description of a solution-based denuder system is presented by Das-gupta (1984). The device, a diffusion scrubber, consists of a tubular micro-porous membrane surrounded by an outer tube. The air sample is sucked either through the membrane tube or through the annulus between membrane and outer jacket. The gaseous compound of interest diffuses toward the membrane and is collected in a suitable scrubber solution which flows on the other side of the membrane (in counterflow). For collection of ammonia a

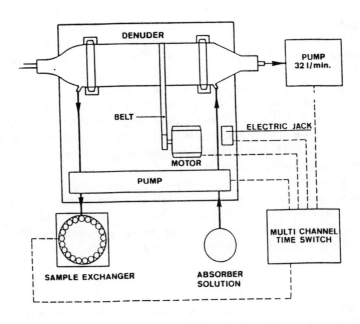

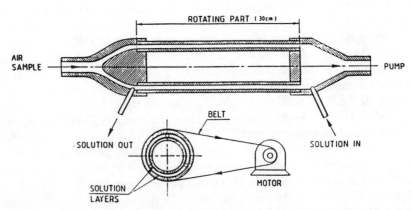

**Figure 4.** Schematic illustration of the rotating wet annular denuder for simultaneous sampling of several gaseous pollutants. (Keuken et al. (1988).)

formic acid solution was chosen. In subsequent work of Dasgupta et al. (1986) a diffusion scrubber with a dilute buffered formaldehyde solution was investigated for sampling of $SO_2$. Other applications include the measurement of $HNO_3$, HCHO, and $H_2O_2$ (Dasguta and Philips, 1987; Dasgupta et al., 1988). The diffusion scrubber has a very simple design and is well suited for application in a continuous-flow analysis system with on-line detection of the collected species, as illustrated by Lindgren and Dasgupta (1989) who used a diffusion scrubber coupled to an ion chromatograph for measurement of $SO_2$. In actual sampling situations, however, varying relative humidities may lead to loss of water from the scrubber solution. A possible solution to this problem is given by Dasgupta et al. (1986).

A so-called wet annular denuder system for simultaneous sampling of $NH_3$, $HNO_3$, $HNO_2$, HCL, $SO_2$, and $H_2O_2$ in ambient air has been developed by Keuken et al. (1988). The gases are collected by absorption in solutions present in the annuli of the denuder tubes (Fig. 4). For sampling of $NH_3$, $HNO_3$, $HNO_2$, and HCl a formic acid solution at pH $= 4$ is used. $SO_2$ and $H_2O_2$ are sampled in a $NaH_2PO_4$ solution with additions of formaldehyde and PHOPA to stabilize these compounds. During sampling the denuders are rotated along their axes such that the entire wall of the annular gap is wetted by the absorption solution. After sampling, the solutions are extracted from the denuders and stored in a fraction collector. Sampling, extraction, and refilling of the denuder are performed automatically. The wet annular denuder system has been extended with an on-line detection system for ammonia, based on application of a porous membrane and conductivity detection. In this way a fully automated monitor for ammonia was obtained with retention of the compact design (Wyers et al., 1991). Initial results indicate a very good precision and accuracy, in the order of 1% or better. A comparison of the wet denuder technique with conventional denuders and filterpacks (Allegrini, 1990) gave satisfactory results. A reproducibility on the order of 5–10% was obtained for ammonia and nitric acid.

## 7.  CONCLUSIONS

Denuder systems are gaining an important role in the analysis of gaseous and particulate components in the atmosphere. They offer possibilities to measure reactive components and discriminate between gaseous and particulate phases which are difficult to match by other methods. The application of denuder systems is now expanding into the areas of monitoring, deposition studies, and effect-related investigations. The range of available denuder systems enables this extension. The various denuder systems have not yet been developed to the point that they can be used with ease and confidence by every scientist or technician engaged in environmental research. However, a number of recent developments indicates that a better state of affairs may be reached in the future for the routine application of denuders to the measurement of a number of important atmospheric constituents.

## REFERENCES

Adams, K. M., Japar, S. M., and Pierson, W. R. (1986). "Development of a MnO₂-coated, cylindrical denuder for removing $NO_2$ from atmospheric samples." *Atmos. Environ.* **20**(6), 1211–1215.

Ali, Z., Thomas, C. L. P., and Alder, J. F. (1989). "Denuder tubes for sampling of gaseous species." *Analyst* **114**, 759–769.

Allegrini, I., De Santis, F., di Palo, V., and Liberti, A. (1984). "Measurement of particulate and gaseous ammonia at a suburban area by means of diffusion tubes (denuders)." *J. Aerosol Sci.* **15**(4), 465–471.

Allegrini, I., De Santis, F., di Palo, V., Febo, A., Perrino, C., and Possanzini, M. (1987). "Annular denuder method for sampling reactive gases and aerosols in the atmosphere." *Sci. Total Environ.* **67**, 1–16.

Anlauf, K. G., Fellin, P., Wiebe, H. A., Schiff, H. I., Mackay, G. I., Braman, R. S., and Gilbert, R. (1985). "A Comparison of three methods for measurement of atmospheric nitric acid and aerosol nitrate and ammonium." *Atmos. Environ.* **19**(2), 325–333.

Appel, B. R., Tokiwa, Y., and Kothny, E. L. (1985). "Evaluation of an automated tungstic acid technique for nitric acid and ammonia." In *Proceedings of the Fifth Annual National Symposium on Recent Advances in the Measurement of Air Pollutants.* Raleigh, NC.

Appel, B. R., Tokiwa, Y., Kothny, E. L., Wu, R., and Povard, V. (1988). "Evaluation of procedures for measuring atmospheric nitric acid and ammonia." *Atmos. Environ.* **22**(8), 1565–1573.

Appel, B. R., Winer, A. M., Tokiwa, Y., and Biermann, H. W. (1990). "Comparison of atmospheric nitrous acid measurements by annular denuder and differential optical absorption systems." *Atmos. Environ.* **24A**(3), 611–616.

Böhm, R., and Israël, G. W. (1987). "A continuous sulfuric acid and sulfate monitor for ambient air." *J. Aerosol Sci.* **18**(6), 857–860.

Bos, R. (1980). "Automatic measurement of atmospheric ammonia." *J. Air Pollut. Control Assoc.* **30**(11), 1222–1224.

Braman, R. S., and Shelly, T. J. (1988). "Tungsten acid for preconcentration and determination of gaseous and particulate ammonia and nitric acid in ambient air." *Anal. Chem.* **54**, 358–364.

Brauer, M., Koutrakis, P., Wolfson, J. M., and Spengler, J. D. (1989). "Evaluation of the gas collection of an annular denuder system under simulated atmospheric conditions." *Atmos. Environ.* **23**, 1981–1986.

Cecchini, F., Febo, A., and Possanzini, M. (1985). *Anal. Lett.* **18**, 681.

Cobourn, W. G., Husar, R. B., and Husar, J. D. (1978). "Continuous in situ monitoring of ambient particulate sulfur using flame photometry and thermal analysis." *Atmos. Environ.* **12**, 89–98.

Corsi, R. L., Chang, D. P. Y., and Larock, B. R. (1988). "A numerical solution for mass transport in membrane-based diffusion scrubbers." *Environ. Sci. Technol.* **22**(5), 561–565.

Coutant, R. W., Callahan, P. J., and Chuang, J. C. (1989a). *Annular Denuder Sampler for Phase-Distributed Semivolatile Organic Chemicals.* Environmental Protection Agency, Research Triangle Park, NC, Report EPA/600/3-89/029.

Coutant, R. W., Callahan, P. J., Kuhlman, M. R. and Lewis, R. G. (1989b). "Design and performance of a high-volume compound annular denuder." *Atmos. Environ.* **23**(10), 2205–2211.

Dasch, J. M., Cadle, S. H., Kennedy, K. G., and Mulawa, P. A. (1989). "Comparison of annular denuders and filter packs for atmospheric sampling." *Atmos. Environ.* **23**(12), 2775–2782.

Dasgupta, P. K. (1984). "A diffusion scrubber for the collection of atmospheric gases." *Atmos. Environ.* **18**(8), 1593–1599.

Dasgupta, P. K. and Lindgren, P. F. (1989). "Inlet pressure effects on the collection efficiency of diffusion scrubbers." *Environ. Sci. Technol.* **23**, 895–897.

Dasgupta, P. K., McDowell, W. L., and Rhee, J. S. (1986). "Porous membrane-based diffusion scrubber for the sampling of atmospheric gases." *Analyst* **111**.

Dasgupta, P. K., Dong, S., Hwang, H., Yang, H. C., and Genfa, Z. (1988). "Continuous liquid-phase fluorometry coupled to a diffusion scrubber for the real-time determination of atmospheric formaldehyde, hydrogen peroxide and sulfur dioxide." *Atmos. Environ.* **22**(5), 949–963.

Demuynck, M. (1985). *Field Experiments on the Use of Denuders to Measure Acid Deposition*, Experimental Workshop Schauinsland 21-26/10/85. Black Forest, Federal Republic of Germany, 55.

Dimmock, N. A., and Marshall, G. B. (1986). "The determination of free ammonia in ambient air with diffusion/denuder tubes." *Anal. Chim. Acta* **185**, 159–169.

Dimmock, N. A., and Marshall, G. B. (1987). "The determination of hydrogen chloride in ambient air with diffusion/denuder tubes." *Anal. Chim. Acta* **202**, 49–59.

Durham, J. L., Spiller, L. L., and Ellestad, T. G. (1987). "Nitric acid-nitrate aerosol measurements by a diffusion denuder: A performance evaluation." *Atmos. Environ.* **21**(3), 589–598.

Eatough, D. J., White, V. F., Hansen, L. D., Eatough, N. L., and Ellis, E. C. (1985). "Hydration of nitric acid and its collection in the atmosphere by diffusion denuders." *Anal. Chem.* **57**, 743–748.

Eatough, N. L., McGregor, S., Lewis, E. A., Eatough, D. J., Huang, A. A., and Ellis, E. C. (1988). Comparison of six denuder methods and a filter pack for the collection of ambient $HNO_3(g)$, $HNO_2(g)$ and $SO_2(g)$ in the 1985 NSMC study." *Atmos. Environ.* **22**(8), 1601–1618.

Febo, A., di Palo, V., and Possanzini, M. (1986). "The determination of tetraalkyl lead in air by a denuder diffusion technique." *Sci. Total Environ.* **48**, 187–194.

Febo, A., De Santis, F., Perrino, C., and Giusto, M. (1989). "Evaluation of laboratory and field performance of denuder tubes: A theoretical approach." *Atmos. Environ.* (7), 1517–1530.

Ferm, M. (1979). "Method for determination of atmospheric ammonia." *Atmos. Environ.* **13**, 1385–1393.

Ferm, M. (1986). "A $Na_2CO_2$-coated denuder and filter for determination of gaseous $HNO_3$ and particulate $NO_3^-$ in the atmosphere." *Atmos. Environ.* **20**(6), 1193–1201.

Ferm, M., and Sjödin, A. (1985). "A sodium carbonate coated denuder for determination of nitrous acid in the atmosphere." *Atmos. Environ.* **19**(6), 979–983.

Forrest, J., Spandau, D. J., Tanner, R. L., and Newman, L. (1982). "Determination of atmospheric nitrate and nitric acid employing a diffusion denuder with a filter pack." *Atmos. Environ.* **16**, 1473–1485.

Gormley, P. G., and Kennedy, M. (1949). "Diffusion from a stream flowing through a cylindrical tube." *Proc. R. Ir. Acad.* **12**, 163–169.

Hering, S. V., Lawson, D. R., Allegrini, I., Febo, A., Perrino, C., Possanzini, M., Sickles, J. E., II, Anlauf, K. G., Wiebe, A., Appel, B. R., John, W., Ondo, J., Wall, S., Braman, R. S., Sutton, R., Cass, G. R., Solomon, P. A., Eatough, D. J., Eatough, N. L., Ellis, E. C., Grosjean, D., Hicks, B. B., Womack, J. D., Horrocks, J., Knapp, K. T., Ellestad, T. G., Paur, R. J., Mitchell, W. J., Pleasant, M., Peake, E., MacLean, A., Pierson, W. R., Brachaczek, W., Schiff, H. I., Mackay, G. I., Spicer, C. W., Stedman, D. H., Winer, A. M., Biermann, H. W., and Tuazon, E. C. (1988). "The nitric acid shootout: Field comparison of measurement methods." *Atmos. Environ.* **22**(88), 1519–1539.

Johnson, N. D., Barton, S. C., Thomas, G. H. S., Lane, D. A., and Schroeder, W. H. (1986). *Field Evaluation of a Diffusion Denuder Based Gas/Particle Sampler for Chlorinated Organic Compounds.* 79th Annual Meeting of the Air Pollution Control Association, Minneapolis, MN, June 22–27, 1986.

Kenne, W. C., Talbot, R. W., Andreae, M. O., Beecher, K., Berresheim, H., Castro, M., Farmer, C., Galloway, J. N., Hoffmann, M. R., Li, S.-M., Maben, J. R., Munger, J. W., Norton, R. B., Pszenny, A. A. P., Puxbaum, H., Westberg, H., and Winiwarter, W. (1989). "An intercomparison of measurement systems for vapor and particulate phase concentrations of formic and acetic acids." *J. Geophys. Res.* **94**(D5), 6457–6471.

Keuken, M. P. (1989). "The Determination of Acid-Deposition Related Compounds in the Lower Atmosphere." Netherlands Energy Research Foundation, Report ECN- 215.

Keuken, M. P., Schoonebeek, C. A. M., van Wensveen-Louter, and Slanina, J. (1988). "Simultaneous sampling of $NH_3$, $HNO_3$, HCl, $SO_2$ and $H_2O_2$ in ambient air by a wet annular denuder system." *Atmos. Environ.* **22**(11), 2541–2548.

Keuken, M. P., Wayers-IJpelaan, A., Möls, J. J., Otjes, R. P., and Slanina, J. (1989). "The determination of ammonia in ambient air by an automated thermodenuder system." *Atmos. Environ.* **23**(10), 2177–2185.

Klockow, D., Niessner, R., Malejczyk, M., Kiendl, H., and von Berg, B. (1989). "Determination of nitric acid and ammonium nitrate by means of a computer-controlled thermodenuder system." *Atmos. Environ.* **23**(5), 1131–1138.

Koutrakis, P., Wolfson, J. M., Slater, J. L., Brauer, M., Spengler, J. D., Stevens, R. K., and Stone, C. L. (1988). "Evaluation of an annular denuder/filter pack system to collect acidic aerosols and gases." *Environ. Sci. Technol.* **22**(12), 1463–1468.

Langford, A. O., Goldan, P. D., and Fehsenfeld, F. C. (1989). "A molybdenum oxide annular denuder system for gas phase ambient ammonia measurements." *J. Atmos. Chem.* **8**, 359–376.

Larjava, K., Reith, J., and Klockow, D. (1990). "Development and laboratory investigations of a denuder sampling system for the determination of heavy metal species in flue gases at elevated temperatures." *Int. J. Environ. Anal. Chem.* **38**, 31–45.

Lewin, E. E., and Hansen, K. A. (1984). "Diffusion denuder assembly for collection and determination of gases in air." *Anal. Chem.* **56**(4), 842–845.

Lindgren, P. F., and Dasgupta, P. K. (1989). "Measurement of atmospheric sulfur dioxide by diffusion scrubber coupled ion chromatography." *Anal. Chem.* **61**, 19–24.

Lindqvist, F. (1985a). "Determination of nitric acid in ambient air by gas chromatography/photo ionization detection after collection in a denuder." *J. Air Pollut. Control Assoc.* (JAPCA) **35**, 19–23.

Lindqvist, F. (1985b). "Determination of ambient sulfuric acid aerosol by gas chromatography/photo ionization detection after pre-concentration in a denuder." *Atmos. Environ.* **19**(10), 1671–1680.

McMurry, P. H., and Stolzenburg, M. R. (1987). "Mass accommodation coefficients from penetration measurements in laminar tube flow." *Atmos. Environ.* **21**(5), 1231–1234.

Murphy, D. M., and Fahey, D. W. (1987). "Mathematical treatment of the wall loss of a trace species in denuder and catalytic converter tubes." *Anal. Chem.* **59**, 2753–2759.

Niessner, R., and Klockow, D. (1980). "A thermoanalytical approach to speciation of atmospheric strong acids." *Int. J. Environ. Anal. Chem.* **8**(3), 163–175.

Niessner, R., Klockow, D., Plomp, A., and Slanina, J. (1988). "Development of a preseparator for denuder sampling." *Int. J. Environ. Anal. Chem.* **32**(3–4), 243–254.

Perrino, C., De Santis, F., and Febo, A. (1990). "Criteria for the choice of a denuder sampling technique devoted to the measurement of atmospheric nitrous and nitric acids." *Atmos. Environ.* **24A**(3), 617–626.

Possanzini, M., Febo, A., and Liberti, A. (1983). "New design of a high-performance denuder for the sampling of atmospheric pollutants." *Atmos. Environ.* **17**(12), 2605–2610.

Possanzini, M., Febo, A., and Cecchini, F. (1984). "Development of a KI annular denuder for $NO_2$ collection". *Anal. Lett.* **17**(A10), 887–896.

Possanzini, M., Ciccioli, P., di Palo, V., and Draisci, R. (1987). "Determination of low boiling aldehydes in air and exhaust gases by using annular denuders combined with HPLC techniques." *Chromatographia* **23**(11), 829–834.

Possanzini, M., di Palo, V., and Liberti, A. (1988). "Annular denuder method for determination of $H_2O_2$ in the ambient atmosphere." *Sci. Total Environ.* **77**, 203–214.

Pratsinis, S. E., Xu, M., Biswas, P., and Willeke, K. (1989). "Theory for aerosol sampling through annular diffusion denuders." *J. Aerosol Sci.* **20**(8), 1597–1600.

Pui, D. Y. H., Lewis, C. W., Tsai, C. -J., and Liu, B. Y. H. (1990). "A compact coiled denuder for atmospheric sampling." *Environ. Sci. Technol.* **24**(3), 307–312.

Roberts, J. M., Langford, A. O., Goldan, P. D., and Fehsenfeld, F. C. (1988). "Ammonia measurements at Niwot Ridge, Colorado and Point Arena, California, using the tungsten oxide denuder tube technique," *J. Atmos. Chem.* **7**, 137–152.

Rosenberg, C., Winiwarter, W., Gregori, M., Pech, G., Casensky, V., and Puxbaum, H. (1988). "Determination of inorganic and organic volatile acids, $NH_3$, particulate $SO_4^{2-}$, $NO_3^-$ and $Cl^-$ in ambient air with an annular diffusion denuder system." *Fresenius Z. Anal. Chem.* **331**, 1–7.

Sickles, J. E., II, Hodson, L. L., McClenny, W. A., Paur, R. J., Ellestad, T. G., Mulik, J. D., Anlauf, K. G., Wiebe, H. A., Mackay, G. I., Schiff, H. I., and

Bubacz, D. K. (1990). "Field comparison of methods for the measurement of gaseous and particulate contributors to acidic dry deposition." *Atmos. Environ.* **24A**(1), 155–165.

Slanina, J. (1982). "Measurement of strong acids and the corresponding ammonium salts in the Netherlands." *VDI-Ber.* **429**.

Slanina, J., van Lamoen-Doornenbal, Lingerak, W. A., and Meilof, W., Klockow, D., and Niessner, R. (1981). "Application of a thermo-denuder analyser to the determination of $H_2SO_4$, $HNO_3$ and $NH_3$ in air." *Int. J. Environ. Anal. Chem.* **9**, 59–70.

Slanina, J., Schoonebeek, C. A. M., Klockow, D., and Niesser, R. (1985). "Determination of sulfuric acid and ammonium sulfates by means of a computer-controlled thermodenuder system." *Anal. Chem.* **57**(9), 1955–1960.

Slanina, J., Keuken, M. P., and Schoonebeek, C. A. M. (1987). "Determination of sulfur dioxide in ambient air by a computer-controlled thermodenuder system." *Anal. Chem.* **59**, 2764–2766.

Stevens, R. K., Paur, R. J., Allegrini, I., De Santis, F., Febo, A., Perrino, C., Possanzini, M., Cox, K. W., Estes, E. E., Turner, A. R., and Sickles, J. E., II. (1985). "Measurement of $HNO_3$, $SO_2$, $NH_3$ and particulate nitrate with an annular denuder system." In *Proceedings of the Fifth Annual National Symposium on Recent Advances in Pollutant Monitoring of Ambient Air and Stationary Sources*. Environmental Protection Agency, Research Triangle Park, NC, EPA Report 55–71.

Vossler, T. L., Stevens, R. K., Paul, R. J., Baumgardner, R. E., and Bell, J. P. (1988). "Evaluation of improved inlets and annular denuder systems to measure inorganic air pollutants." *Atmos. Environ.* **22**, 1729–1736.

Wiebe, H. A., Anlauf, K. G., Tuazon, E. C., Winer, A. M., Biermann, H. W., Appel, B. R., Solomon, P. A., Cass, G. R., Ellestad, T. G., Knapp, K. T., Peake, E., Spicer, C. W., and Lawson, D. R. (1990). "A comparison of measurements of atmospheric ammonia by filter packs, transition-flow reactors, simple and annular denuders and fourier transform infrared spectroscopy." *Atmos. Environ.* **24A**(5), 1019–1028.

Winiwarter, W. (1989). "A calculation procedure for the determination of the collection efficiency in annular denuders." *Atmos. Environ.* **23**(9), 1997–2002.

Wyers, G. P. et al. (1991). In preparation.

# 4

# GAS PHASE ORGANICS IN THE MARINE ATMOSPHERE

## *J. C. Marty and E. Jalliffier-Merlon*

*Observatoire Océanologique de Villefranche-sur-Mer, Laboratoire de Physique et Chimie Marines, UA CNRS 353, Quai de la Darse, BP 08, F-06230 Villefranche-sur-Mer, France*

*Gaseous Pollutants: Characterization and Cycling,* Edited by Jerome O. Nriagu.
ISBN 0-471-54898-7 © 1992 John Wiley & Sons, Inc.

## 1.  INTRODUCTION

In spite of considerable work done on gas phase compounds in the urban environment, only a few studies have been undertaken on their cycle in the marine environment. The origin of many of the organic trace pollutants found in marine air is related to combustion. Initially generated in the gas phase, the least volatile of them undergo a gas–particle conversion step shortly after emission. But a large part of the more volatile components will escape particle incorporation and persist in the atmospheric gas phase over the oceans. In addition to anthropogenic compounds, high quantities of organics principally in the gas phase are produced by natural vegetation, and, in particular, terpenoids compounds, which are photochemically highly reactive. The products of the photochemical reaction and the unreacted part of these compounds are transported over the oceans. In marine air, compounds of other sources are present in the gas phase, such as the biologically produced organics transported in the atmosphere by evaporation from seawater or from aerosols emitted by bursting air bubbles and spray.

So the determination of organic constituents in the gas phase of the marine atmosphere is rather complex because there are many compounds originating from multiple sources and because of the low levels of concentration, which make the collection of uncontaminated samples from ships difficult.

In this paper we review the works done on gas phase organics in the remote marine atmosphere ranging from $C_{10}$ to $C_{30}$ (expressed in $n$-alkanes chain length) by focusing on sampling methods and source-related compounds.

## 2.  SAMPLING AND ANALYSIS

### 2.1.  Definition

Similar to dissolved material from seawater, gas phase compounds are defined by the mode of sampling, that is, they are passed through a glass fiber filter (e.g., Whatman GF/F) and are trapped on an adsorbent.

Commercially available glass fiber filters are reported to be more than 99.7% efficient at 5 cm s$^{-1}$ face velocity for particles greater than 0.3 μm diameter, and more than 98% efficient for particles with diameters greater than 0.03 μm (Butcher and Charlson, 1972), and they are also said to collect submicrometer particles down to 0.0025 μm (John and Reischl, 1978). Thus, the gas phase material is constituted by truly gaseous compounds and very small particles <0.03–0.003 μm in diameter. Because of the low concentrations of organics in marine air on both particulate and gaseous phase, it is necessary to process large quantities of air. Gas phase compounds have been trapped on different adsorbents. The most commonly used are polyurethane foam, florisil, and Tenax. Some studies have used charcoal traps as absorbents. But there is a great difference between polyurethane foam (PUF) and florisil on the one hand and Tenax and charcoal trap on the other. For the first group, large volumes of air are collected and compounds are extracted in solvent then separated in various classes and analyzed. This analytical scheme, including solvent evaporation steps, limits the range of volatility of compounds analyzed. For the second group (tenax, charcoal), only low volumes of air are processed and compounds can be directly analyzed without sample preseparation after thermal or solvent desorption leading to a range of hydrocarbons $>C_6$.

## 2.2. Polyurethane Foams

PUF was first adapted for large-volume air sampling of organochlorine compounds (Bidleman and Olney, 1974). Data have been obtained using PUF by Eichmann et al. (1979, 1980) (alkanes), Marty (1981) (alkanes, fatty acids), Marty et al. (1984) (polycyclic aromatic hydrocarbons, PAH), and Zafiriou et al. (1985) (alkanes). Flow rate through PUF has been used in a range of $2–25$ m$^3$ h$^{-1}$ leading to sample volumes of $100–1000$ m$^3$ in 1 or 2 days.

This is important because compound volatility and ambient temperature influence the efficiency of sample collection for each compound. Sampling efficiencies for selected compounds have been estimated by Simon and Bidleman (1979) for polychlorinated biphenyls and by Atlas and Giam (1986) from the data of Burdick and Bidleman (1981) and You and Bidleman (1984). Breakthrough volumes (m$^3$) at 20 °C on PUF column (7.8 cm diam, 7.5 cm thick) with flow rates $0.3–0.5$ m$^3$ h$^{-1}$ were estimated between 120 for fluorene and 10 000 for pyrene. Van Vaeck et al. (1984) have determined quantitatively the volatilization losses of aerosol organic constituents upon prolonged high-volume sampling. This volatilization artifact has been shown to be important for compounds of intermediate vapor pressure (e.g., n- alkanes from eicosane to tetracosane, PAH from phenanthrene/anthracene to benz[a]anthracene/chrysene).

A difficulty in the utilization of PUF for sampling of organics is the problem of blanks. Multiple solvent-cleaning procedures were developed (Marty, 1981; Peltzer et al., 1984), but despite exhaustive extractions, a "polymeric" contaminating material is obtained; fortunately, this does not interfer in the analytic procedure.

The range of compounds that can be analyzed on PUF is $>C_{15}$.

## 2.3. Florisil

Florisil is a synthetic magnesium silicate which has been used to trap organic compounds from the marine atmosphere by Atlas and Giam (1981), Giam et al. (1975) because it is more efficient than PUF for trapping more volatile compounds. Lipari and Swarin (1985) have used florisil for analysis of formaldehyde in air. This adsorbent has been used in the same conditions (flow rate, high volume of air trapped) as PUF and is easier to clean by combustion at 320–400 °C for 12–24 h. However, as noticed by Atlas and Giam (1986), chemosorption or irreversible adsorption may occur on florisil for some compounds and some deactivation is possible by moisture.

## 2.4.  Tenax

Tenax is probably the most extensively employed adsorbent for sampling gas phase compounds from the atmosphere (Class and Ballschmiter, 1986; Termonia and Alaerts, 1985; Yokouchi et al., 1983). In the marine environment, Eichmann et al. (1979) have used it for collecting *n*- alkanes in the range $C_9$–$C_{17}$ by drawing 20–60 L of air through adsorption tubes (50 mm long, 4 mm inner diameter). This corresponds to sampling times of 2–6 h. The adsorption tube was directly connected to the gas chromatograph.

A problem with using Tenax, a porous polymer resin, is its susceptibility to degradation (Josephson, 1981; Hauser et al., 1983; Middleditch et al., 1988; Pellizzari et al., 1984). Tenax is easily contaminated by improper handling in the laboratory or during field experiments (Middleditch et al., 1988; Pellizzari et al., 1976; Pellizzari and Krost, 1985). Special sampling cartridges have been developed by Russwurm et al. (1981).

Tenax GC has been used for PAH concentration by Barns et al. (1981), Jones et al. (1976), with subsequent desorption by solvents.

Due to low volumes of air sampled, and to preserve the range of volatility of compounds, analysis from Tenax is performed without pretreatment, by direct injection on GC, by thermal desorption, or by extraction with a minimum quantity of solvent. Tenax has also been used with silicagel because its capacity of adsorption for the highly volatile compounds is limited (Spingarn and Northington, 1982).

## 2.5.  Charcoal Traps

A few studies have used active charcoal traps to collect gas phase organics from the atmosphere by Grob and Grob (1971), Hov et al. (1983), and Jalliffier-Merlon and Marty (1988; this last reference applies only to marine air). A considerable advantage of charcoal is a high specific surface that permits the condensation of high quantities of compounds. Therefore, only low quantities of adsorbent and very low quantities of solvent for elution are necessary. This increases the sensitivity of the method to the nanogram per liter level. Moreover, the efficiency of active charcoal is not changed or is only slightly changed by moisture and $CO_2$ concentration (Andersson et al., 1984). Thermodesorption from charcoal traps has also been used (Cobb et al., 1986; Parkes et al., 1975). The range of compounds that can be analyzed by this system is in the range of $C_6$–$C_{20}$ alkanes, as for Tenax.

## 2.6.  Other Techniques

Other solid adsorbents have been used for the collection of atmospheric samples but not for marine samples: Amberlite XAD-2 has been used for

PAH by Andersson et al. (1983), and for other compounds by Langhorst (1980). Chromosorb 102 has been used by Barns et al. (1981). A list of supports used in collection of atmospheric gas phase samples is given in Table 1.

Direct collection of atmosphere (followed by concentration in GC with cryogenic trap) has also been used in 0.5- to 25-L containers: glass containers (Nelson and Quigley, 1982; Westberg and Rasmussen, 1972), stainless steel containers (Bonsang and Lambert, 1985; Greenberg and Zimmerman, 1984; Rudolph and Ehhalt, 1981; Singh and Salas, 1982; Tille et al., 1985), Teflon bags (Lilian, 1972; Seila, 1979), Tedlar bags (Arnts et al., 1982; Lonneman et al., 1978). This approach has mainly been applied, using cryogenic trap and GC, in the $C_1$–$C_{12}$ volatility range.

## 2.7.  Analysis

All types of analyses used in atmospheric gas phase organic compounds are based on high-resolution capillary gas chromatography. Depending on the range of volatility, different phases are used, essentially nonpolar with film thicknesses from 0.12 to 5.0 μm (Table 2). All types of injection systems have been used: on column, split splitless, solid (Grob) injection, headspace thermodesorption (with backflushing), and so on.

Thermodesorption is used on compounds trapped on a solid adsorbent in stainless steel or Pyrex tubing. This trap is placed in a micro oven and flushed with helium or nitrogen. The temperature is raised very rapidly (1000 °C min$^{-1}$) at a preprogrammed temperature (180–280 °C). Adsorbed substances eluted by gas vector are condensed on a cryogenic trap ($-30$ to $-120$ °C) or directly in the head of a chromatographic column.

Thermodesorption offers the advantage of highly sensitive automatization. But only one analysis is possible and some chemical modifications for the more thermolabile species are likely.

Desorption of solid adsorbents by solvents, usually $CS_2$, and subsequent injection in GC is commonly employed in spite of difficulties of manipulation.

Detection is performed by FID or ECD for most of the injections after extraction from the adsorbent and chemical treatment. Mass spectrometry is used for direct injection of the material eluted from the adsorbent.

## 3.  CHARACTERIZATION OF COMPOUNDS

### 3.1.  General

The first report of gas phase organic substances in the marine atmosphere is from Wade and Quinn (1974) who used a tower in Bermuda and sampling aboard the R.V. *Trident*. Analysis of the collected air samples indicated that 95% of the heavier hydrocarbons (range $C_{14}$–$C_{32}$) near Bermuda were found in the porous polyurethane, and 5% retained by glass fiber filters. During

**Table 1 Sorbents Used in the Collection of Gas Phase Samples**

| Sorbent | Composition | Trademark | Specific Surface Area ($m^2 \, g^{-1}$) | Mean Pore Diameter (Å) | Temperature Limit (°C) |
|---|---|---|---|---|---|
| Chromosorb 102 | Styrene–divinylbenzene | Johns Manville Corp. | 300–400 | 80 | 250 |
| Chromosorb 106 | Polystyrene cross-linked resin | Johns Manville Corp. | 600–800 | 50 | 225 |
| Tenax GC Tenax TA | Poly-2,6-diphenyl-*p*-phenylene oxide | Enka Research Institute Arhen | 19–30 | 720 | 375 |
| Porapak P | Styrene–ethylvinyl–benzene–divinylbenzene | Waters Associates Inc. | 50–100 | 75 | 250 |
| Porapak Q | Ethylvinylbenzene–divinylbenzene | Waters Associates Inc. | 550–650 | 75 | 250 |
| Amberlite XAD2 | Styrene–divinylbenzene | Rohm and Haas Co. | 300–370 | 90 | 150 |
| Amberlite XAD4 | Styrene–divinylbenzene | Rohm and Haas Co. | 750–850 | 50 | 150 |
| Carbopack | Black carbon graphitized | Supelco Inc. | 100 | — | >400 |
| Carbopack C | Black carbon graphitized | Supelco Inc. | 12 | — | >400 |
| Activated charcoal | — | Bender-Hobein AG | 1000–1300 | — | 700 |
| Carbosieve SIII | Molecular sieve of carbon | Supelco Inc. | 550 | 15–40 | >400 |
| Spherocarb | Molecular sieve of carbon | Perkin-Elmer Corp. | 1200 | 15 | 300 |

**Table 2 Phase Thckness in High-Resolution GC,**
**Corresponding Range of *n*-Alkanes, and**
**Corresponding Range of Ebullition Points**

| ex: CP-Sil 5 CB Film Thickness ($\mu$m) | *n*-Alkane Range | Ebullition Point Range (°C) |
|---|---|---|
| 0.12 | $C_{10}-C_{60}$ | 170–650 |
| 0.4 | $C_6 -C_{35}$ | 70–500 |
| 1.2 | $C_3 -C_{25}$ | −42–400 |
| 5 | $C_1 -C_{15}$ | −164–270 |

the cruise between Bermuda and the east coast of the United States they found a constant total concentration of hydrocarbons (gaseous + particulate). But, there was a change in the ratio 95% gaseous/5% particulate in Bermuda to 60% gaseous/40% particulate near land. Based on the distribution of hydrocarbons, gas phase hydrocarbons were of petroleum origin. The percentage of petroleum character was greater near shore (Garrett and Smagin, 1976). This situation was thought to reflect a longer residence time for gas phase hydrocarbons and a possibility of particle to gas conversion with aging of particles. Duce (1978) has made estimates of the organic carbon concentrations over marine areas, from 8 ng m$^{-3}$ for the $C_3-C_{12}$ range (from Rasmussen, 1974) to 0.18 $\mu$g m$^{-3}$ for the $C_{14}-C_{32}$ range (from Quinn and Wade, 1974). This is to be compared with POC estimations of 0.2–2.4 $\mu$g m$^{-3}$.

Since these works, all attempts to separate particulate and gas phase organics in the marine atmosphere have reported, from a global point of view, a concentration of one order of magnitude higher in the gas phase. Atlas and Giam (1981) reported 90% of organic pollutants in the gas phase from Enewetak Atoll in the north Pacific ocean (polychlorobiphenyls, total DDT, dieldrin, chlordane, phtalate esters), Duce and Gagosian (1982), Eichmann et al. (1979, 1980), Marty (1981), for *n*-alkanes and Marty et al. (1984) for PAH report total concentrations in the gas phase about 10 times higher than in particulates. Attempts have been made to characterize organic components from air (see the review from Simoneit and Mazurek, 1981). Below, we discuss the three major sources of gas phase compounds from the atmosphere, namely natural (terrestrial and marine) and anthropogenic sources and their evolution through the marine environment.

## 3.2. Natural Continental Sources

Organic substances emitted from natural terrestrial sources are principally terpenes. Terpene production by forests has been extensively studied in coniferous forests (Arnts et al., 1982; Hov et al., 1983; Lamb et al., 1985; Roberts et al., 1983; Tingey et al., 1980; Whitby and Coffey, 1977; Yokouchi et al., 1983), in rural environments (Greenberg and Zimmerman, 1984; Holdren et al., 1979; Lonneman et al., 1978), and in equatorial forests (Crutzen et al., 1985; Greenberg and Zimmerman, 1984; Greenberg et al., 1984; Gregory et al., 1986; Zimmerman et al., 1988).

The compounds principally emitted are α-pinene, β-pinene, Δ-3-carene, camphene, and *d*-limonene. These compounds are highly reactive, and atmospheric photochemical reactions of terpenoid compounds have been studied by Atkinson (1986), Grimsrud et al. (1975), Jacob and Wofsy (1988), Lilian (1972), Westberg and Rasmussen (1972), and Winer et al. (1976).

Terpenes have rarely been encountered in the marine atmosphere. Residence times are relatively short (from minutes to 2 days). They have been detected in appreciable amounts in coastal zones (Jalliffier-Merlon and Marty, 1988). Some biological terpenes have been reported in the marine environment, for example, halogenated monoterpenes (Fenical, 1982; Moore, 1976). Isopropyltoluene, a compound with aromatic structure, is often reported with terpenes because it is emitted by vegetation and has been firstly detected in air by Rasmussen (1970).

Much work has been done on *n*-alkanes. They are found far from sources because their residence times are long (day to year). *n*-Alkanes from terrestrial vegetation are characterized by the predominance of odd over even number of carbon atoms in the range $C_{23}-C_{33}$. Due to high vapor pressure these compounds are often reported in the particulate phase but are not present in the gas phase. A similar predominance of odd over even *n*-alcanes has been noted by Eichmann et al. (1979) in the range $C_{10}-C_{16}$ without definitive explanation.

### 3.3. Marine Sources

Organic compounds from marine sources can be emitted from seawater and particularly from the sea surface microlayer by direct evaporation at the sea surface or from particles by bubble bursting (Marty, 1981; Marty and Saliot, 1980). Among the various gas phase compounds identified (*n*-alkanes, isoprenoids, alkenes, aldehydes, ketones, and various halogenated and sulfured compounds) the sea can act as a source for a number of them, depending on their volatility.

#### 3.3.1. Alkanes

*n*-Alkanes from benthic and pelagic algae are dominated by the presence of *n*-$C_{15}$ for brown algae and *n*-$C_{17}$ for red algae (Youngblood et al., 1971). Even *n*-alkanes *n*-$C_{18}$, *n*-$C_{20}$, *n*-$C_{22}$ have been related to bacterial activity (Han and Calvin, 1969; Nishimura and Baker, 1986). Isoprenoids pristane and phytane were reported in benthic algae (Clark and Blumer, 1967). Evidence of marine sources for gas phase *n*-alkanes are seen from data of Marty (1981) who reports the abundance of *n*-$C_{17}$ in gaseous compounds from the Gulf of Guinea (equatorial Atlantic). In Fig. 1 gaseous *n*-alkanes distribution

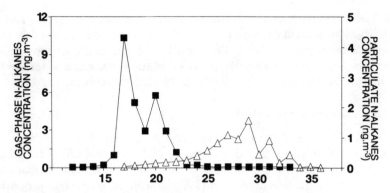

**Figure 1** Particulate ($\triangle$) and gas phase ($\blacksquare$) *n*-alkanes from the Gulf of Guinea, Equatorial Atlantic (Romancap cruise of the R.V. *Capricorne*, June 1977). (From Marty (1981).)

dominated by the *n*-$C_{17}$ with a concentration of 10 ng m$^{-3}$ is in opposition to the particulate material from the same sample, which shows a regular distribution of *n*- alkanes from $C_{20}$ to $C_{35}$. The odd over even predominance of $C_{29}$, $C_{31}$, $C_{33}$ is characteristic of vegetal terrestrial input.

The partition between particulate and gas phase is situated around the *n*-$C_{22}$ (Fig. 1). This is consistent with the observations of Junge (1977) and Marty (1981), who showed that compounds having a vapor pressure value up to 10$^{-5}$ mm Hg might be present as gases in the atmosphere. A predominance of $C_{15}$ and $C_{17}$ *n*-alkanes in various atmospheric marine gas phases has also been noticed by Duce and Gagosian (1982) and Eichmann et al. (1979, 1980) and in Zafiriou et al. (1985).

Recent work in the Mediterranean Sea (Sicre et al., 1991) shows that low molecular weight *n*-$C_{17}$ and *n*-$C_{20}$ are dominant homologs in gas phase *n*-alkanes and points out the presence of a hump of unresolved alkanes centered in the low molecular range. $C_{19}$, $C_{20}$, and $C_{22}$ homologs dominate in the vapor phase and are attributed to a marine bacterial origin. The authors suggest that a marine source for these compounds is likely, although no prevalence of *n*-$C_{15}$ or *n*-$C_{17}$ over even alkanes *n*-$C_{16}$ and *n*-$C_{18}$ has been noticed.

### 3.3.2. Alkenes

Alkenes from the atmosphere have been principally studied in the range $C_2$–$C_6$ (Bonsang et al., 1988; Kanakidou et al., 1988). There is a lack of data in the field of biogenic alkenes in the marine environment. These compounds are very reactive (Rudolph et al., 1986) and can initiate a photochemical production of ozone (Koppman and Rudolph, 1988) and carbonyl compounds.

Alkenes *n*-$C_{17}$:1 and $C_{21}$:6 are very important components from algae (Blumer et al., 1971). *n*-$C_{17}$:1 has been detected in some species of chlorophycea and $C_{19}$:5 in rhodophycea (Youngblood et al., 1971). *n*-$C_{17}$:1 and *n*-$C_{17}$:2 have been identified in bacteria (Han and Calvin, 1969; Paoletti et al.,

1976). Schwarzenbach et al. (1978) reported the occurrence of $C_{15}$ and $C_{17}$ monoenes as prominent compounds in marsh creeks. To our knowledge, these compounds have not yet been identified in the gas phase of atmospheric marine samples, but have been detected as volatiles in surface water samples from the Mediterranean Sea (E. Jalliffier-Merlon, unpublished results).

### 3.3.3. Aldehydes and Ketones

Volatile aldehydes and ketones have been measured in surface waters as produced by phytoplankton (Gschwend et al., 1980; Jalliffier-Merlon and Marty, 1991; Kikuchi et al., 1974; Whelan et al., 1982). Autooxidation of unsaturated fatty acids is one of the mechanism of aldehydes production (Frankel et al., 1982; Mead et al., 1986; Neff et al., 1983; Schauenstein et al., 1977). Photochemical reactions on humic substances in seawater can produce aldehydes (Momzikoff et al., 1983; Zafiriou, 1983). Aldehydes and ketones are highly reactive in the atmosphere. Their rapid tropospheric photolysis diminishes any contribution of long-range transport. In spite of relatively high concentrations reported in coastal seawaters during phytoplanktonic blooms (Gschwend et al., 1980; Jalliffier-Merlon and Marty, 1991; Schwarzenbach et al., 1978) only a few reports on aldehydes in marine atmosphere are available (Atlas et al., 1984; Zafiriou et al., 1980). However, these compounds could be potentially useful markers of phytoplanktonic specific productivity (Jalliffier-Merlon and Marty, 1991).

### 3.3.4. Organohalogenated Compounds

Halogenated compounds are principally man-made products. However, some of them—polybromomethanes, alkyl monohalogens, and dihalogens—are natural products in the marine environment (Gschwend et al., 1985; Helz and Hsu, 1978; Moore, 1976; Newman and Gschwend, 1987). These volatile halogens are very stable. Residence times vary from a month to a year. Some of these compounds emitted by benthic algae can be transferred to the atmosphere where they play a significant role in the global cycle of chlorine, bromine, and iodine, as pointed out by Rasmussen et al. (1982) and Singh et al. (1983) for continental atmosphere and by Andreae and Barnard (1984) and Class et al. (1986) for marine atmosphere.

### 3.3.5. Sulfured Compounds

Sulfured compounds have been principally studied in the gaseous (volatility $< C_6$) range ($H_2S-DMS-COS-CS_2$). In the range of volatility studied here, dimethyl disulfide (DMDS) and dimethyl trisulfide (DMTs) have short residence times in the atmosphere and therefore may play an important role in local atmospheric sulfur budget. Moreover, the rapid tropospheric oxidation of these gases might be expected to produce $SO_2$ and $SO_4$.

DMDS has been reported in surface seawater from coastal area by E. Jalliffier-Merlon (unpublished), Mantoura et al. (1982), and Schwarzenbach et al. (1978). Banwart and Bremmer (1974) and Rasmussen (1974) indicate that DMDS is emitted in gas phase from cultures of bacteria and algae. Steudler and Peterson (1984) indicate significant presence of DMDS over salt marshes and tidal creeks.

## 3.4. Anthropogenic Sources

Hydrocarbons and halogenated hydrocarbons are introduced in the environment by human activity. For halogenated hydrocarbons the major source is evaporation of various solvents and fuels used in industry. Aromatics are issued from petroleum products and pyrolysis of these petroleum products and from other combustibles such as wood and coal (Junk and Ford, 1980). Combustion products may originate from forest fires as well. Recent works have been realized on anthropogenic emissions in urban environments (e.g., Bouchertall, 1986; Foster et al., 1988; Grosjean and Fung, 1984; Nelson and Quigley, 1982).

It is widely accepted that most of the compounds related to combustion processes are initially generated in the gas phase and undergo homogenous and/or heterogeneous condensation from further cooling of the emission (Van Vaeck et al., 1984). Therefore, the PAH appear in the gas phase and condense on submicrometer particles (which present the largest specific area) and halogenated hydrocarbons are principally emitted by evaporation in the gas phase.

### 3.4.1. Aromatic Hydrocarbons

Some authors have reported the presence of volatile PAH in seawater (McDonald et al., 1984; Sauer, 1981; Schwarzenbach et al., 1978). Alkylbenzenes have been detected in surface waters from Peru upwelling by Gschwend et al. (1980).

However, PAH have rarely been identified in the marine remote atmosphere. They have been analyzed in aerosols by Grimalt et al. (1988) and Sicre et al. (1987a,b) in the Mediterranean aerosol, by Lunde (1976) and Lunde and Björseth (1977) in Norwegian marine air, by Ketseridis et al. (1976) and Marty (1981) in Atlantic air. They are below the detection limit in Enewetak Atoll equatorial Atlantic samples (Gagosian et al., 1982).

Marty et al. (1984) report gas phase hydrocarbons in the remote marine atmosphere from tropical and equatorial Atlantic air. Concentrations of gas phase PAH in the atmosphere ranging from 7 to 18 ng m$^{-3}$ are about one order of magnitude higher than those of particulate PAH. This fractionation between the gaseous and the particulate phase is identical to that of $n$-alkanes from the same samples (Marty, 1981). In Fig. 2 the quasi-exclusive dominance of phenanthrene over other less volatile major PAH in the gas phase is strongly different from the particulate phase where phenanthrene is accompanied by other PAH (fluoranthene, pyrene, chrysene, etc.). The predominance of the parent compound in the distribution of alkylated components in the phen-

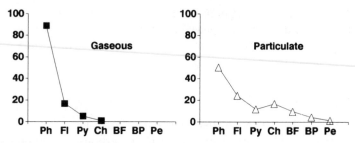

**Figure 2.** Relative distribution (%) of major PAH from atmospheric sample from the gulf of Guinea, Equatorial Atlantic (Romancap cruise of the R.V. *Capricorne*, June 1977). (From Marty et al. (1984).) Ph, phenanthrene; Fl, fluoranthene; Py, pyrene; Ch, chrysene + tryphenylene + benzo[a]anthracene; BF, benzo[j] + benzo[k]fluoranthene; BP, benzo[a]pyrene + benzo[e]pyrene; Pe, perylene.

anthrene series (Fig. 3) suggests a fossil fuel combustion origin. The predominance of the gas phase for PAH has been noted by Cautreels and Van Cauwenberghe (1978) for continental atmosphere. As pointed out by Marty et al. (1984), gaseous PAH must be less subject to photooxidation than the rate predicted by laboratory models in order to reach sampling sites far from sources. Neff (1979) discusses this aspect and shows that photooxidation may be slow for some PAH in the presence of oxygenated derivatives such as phenols present in the atmosphere, as well as through singlet oxygen capture by complex metallic ions.

Greenberg and Zimmerman (1984), analyzing aromatics from benzene to naphthalene, found the same discrepancy between the time for transport and calculated residence times. They suggest that the transport is through water masses. Since the residence times of aromatics in marine waters are estimated to about a week (Whittle et al., 1982), transport by water masses and subsequent volatilization is a possible mechanism for the introduction of PAH in the atmosphere.

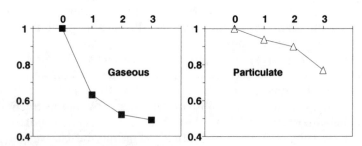

**Figure 3.** Distribution of alkylated homologues ($C_1$–$C_4$) of phenanthrene (normalized to phenanthrene) from atmospheric sample, for dissolved and particulate species, from the Gulf of Guinea, Equatorial Atlantic (Romancap cruise of the R.V. *Capricorne*, June 1977). (From Marty et al. (1984).)

Partition between the gas phase and the particulate phase has been reviewed by Pankow (1987). The lifetime in the atmosphere of submicrometer particles is high because they are too small to fall rapidly and they are not efficiently scavenged by rain.

It is suggested by Bidleman (1988) that there are two types of binding between PAH or pesticides and anthropogenic particles: a nonexchangeable fraction highly adsorbed on active sites and an exchangeable fraction slightly bound to the particle. A permanent exchange between the exchangeable fraction and the gas phase might explain a persistence of PAH in the gas phase far from original sources.

### 3.4.2. Halogenated Hydrocarbons

These compounds are very stable in the environment. Therefore, their concentrations in the gas phase of the marine remote atmosphere must be significant. Some data have been obtained in seawater from the coastal zone by Kennicut et al. (1988) and McDonald et al. (1984). Halogenated hydrocarbons have been analyzed in the marine atmosphere (Class and Ballschmiter, 1986; Cronn et al., 1977; Kirschmer and Ballschmiter, 1983; Rasmussen and Khalil, 1983). Major organochlorinated compounds encountered are tetrachloromethane ($CCl_4$), chloroform ($CHCl_3$), 1,1,1- trichloroethane ($CH_3-CCl_3$), 1,2-dichloroethane ($CH_2-CH_2Cl$), trichloroethene ($HClC=CCl_2$), and tetrachloroethene ($Cl_2C=CCl_3$). These compounds reflect an interhemispheric variation of 50 to 100%, which exhibits the predominance of industrial sources in the Northern Hemisphere. There is a constant increase in concentration of these compounds in the troposphere linked to their high atmospheric stability (from hours to some days for unsaturated compounds and from 5 to 75 years for saturated ones) (Lovelock, 1977; Singh et al., 1976).

Atlas and Giam (1981) report 90% of organic pollutants in the gas phase from Enewetak Atoll. Concentrations of 0.003–2.68 ng m$^{-3}$ of polychlorobiphenyls, total DDT, dieldrin, chlordane, and phtalate esters plasticizers were reported enhancing the long-range atmospheric transport of organic pollutants to remote marine areas.

### REFERENCES

Andersson, K., Levin, J.-O., and Nilsson, C.-A. (1983). "Sampling and analysis of particulate and gaseous aromatic hydrocarbons from coal tar sources in the working environment." *Chemosphere* **12**(2), 197–207.

Andersson, K., Levin, J.-O., Lindahl, R., and Nilsson, C.-A. (1984). "Influence of air humidity on sampling efficiency of some solid adsorbents used for sampling organics from work-room air." *Chemosphere* **13**(3), 437–444.

Andreae, M. O., and Barnard, W. R. (1984). "The marine chemistry of dimethyl sulfide." *Mar. Chem.* **14**, 267–279.

Arnts, R. R., Petersen, W. B., Seila, R. L., and Gay, B. W., Jr. (1982). "Estimates of a-pinene emissions from a loblolly pine forest using an atmospheric diffusion model." *Atmos. Environ.* **16**(9), 2127–2137.

Atkinson, R. (1986). "Kinetics and mechanisms of the gas-phase reactions of the hydroxyl radical with organic compounds under atmospheric conditions." *Chem. Rev.* **86**, 69–201.

Atlas, E., and Giam, C. S. (1981). "Global transport of organic pollutants: Ambient concentrations in the remote marine atmosphere." *Science* **211**, 163–165.

Atlas, E., and Giam, C. S. (1986). "Sampling organic compounds for marine pollution studies." *NATO ASI Ser. G* **9**, 209–230.

Atlas, E., Sullivan, K., and Giam, C. S. (1984). "Vapor-phase aldehydes in marine air." *Searex Newsl.* **1**, 10–13.

Banwart, W., and Bremmer, J. (1974). "Gas chromatographic identification of sulfur gases in soil atmosphere." *Soil Biol. Biochem.* **6**, 113–115.

Barns, R. D., Law, L. M., and McLeod, A. J. (1981). "Comparison of some porous polymers as adsorbents for collection of odor samples and the application of the technique to an environmental malodor." *Analyst (London)* **106**(1261), 412–418.

Bidleman, T. F. (1988). "Atmospheric processes." *Environ. Sci. Technol.,* **22**(4), 361–367.

Bidleman, T. F., and Olney, C. E. (1974). "High volume collection of atmosphere PCB." *Bull. Environ. Contam. Toxicol.* **11**, 442–447.

Blumer, M., Guillard, R. R. L., and Chase, T. (1971). "Hydrocarbons of marine phytoplankton." *Mar. Biol. (Berlin)* **8**, 183–189.

Bonsang, B., and Lambert, G. (1985). "Nonmethane hydrocarbons in an oceanic atmosphere." *J. Atmos. Chem.* **2**, 257–271.

Bonsang, B., Kanakidou, M., Lambert, G., and Monfray, P. (1988). "The marine source of $C_2$-$C_6$ aliphatic hydrocarbons." *J. Atmos. Chem.* **6**, 3–20.

Bouchertall, F. (1986). "Volatile hydrocarbons in the atmosphere of the Kiel bight (western Baltic)." *Mar. Chem.* **19**, 153–160.

Burdick, N. F., and Bidleman, T. F. (1981). "Frontal movements of hexachlorobenzene and polychlorinated biphenyl vapors through polyurethane foam." *Anal. Chem.* **53**, 1926–1929.

Butcher, S. S., and Charlson, R. J. (1972). *An Introduction to Air Chemistry.* Academic Press, New York, p. 49.

Cautreels, W., and Van Cauwenberghe, K. (1978). "Experiments on the distribution of organic pollutants between airborne particulate matter and the corresponding gas phase." *Atmos. Environ.* **12**, 1133–1141.

Clark, R. C., Jr., and Blumer, M. (1967). "Distribution of *n*-paraffins in marine organisms and sediment." *Limnol. Oceanogr.* **12**, 79–87.

Class, T., and Ballschmiter, K. (1986). "Chemistry of organic traces in air VI. Distribution of chlorinated $C_1$-$C_4$ hydrocarbons in air over the northern and southern Atlantic ocean." *Chemosphere* **15**(4), 413–427.

Class, T., Kohnle, R., and Ballschmiter, K. (1986). "Chemistry of organic traces in air VII: Bromo- and bromochloromethanes in air over the Atlantic ocean." *Chemosphere* **15**(4), 429–436.

Cobb, G. P., Braman, R. S., and Hua, K. M. (1986). "Carbon hollow tubes as collectors in the thermal desorption/gas chromatographic analysis of atmospheric organic compounds." *Anal. Chem.* **58**(11), 2213–2217.

Cronn, D. R., Rasmussen, R. A., Robinson, E., and Harsch, D. E. (1977). "Halogenated compound identification and measurement in the troposphere and lower stratosphere." *J. Geophys. Res.* **82**(37), 5935–5944.

Crutzen, P. J., Delany, A. C., Greenberg, J., Haagenson, P., Heidt, L., Lueb, R., Pollock, W., Seiler, W., Wartburg, A., and Zimmerman, P. (1985). "Tropospheric chemical composition measurements in Brazil during the dry season." *J. Atmos. Chem.* **2**, 233–256.

Duce, R. A. (1978). "Speculation on the budget of particulate and vapor phase nonmethane organic carbon in the global troposphere." *Pure Appl. Geophys.* **116**, 244–273.

Duce, R. A., and Gagosian, R. B. (1982). "The input of atmospheric $n$-$C_{10}$ to $n$-$C_{30}$ alkanes to the ocean." *J. Geophys. Res.* **87**(C9), 7192–7200.

Eichmann, R., Neuling, P., Ketseridis, G., Hahn, J., Jeanicke, R., and Junge, C. (1979). "N-alkanes studies in the troposphere. I. Gas and particulate concentrations in north Atlantic air." *Atmos. Environ.* **13**, 587–599.

Eichmann, R., Ketseridis, G., Schebeske, G., Jeanicke, R., Hahn, J., Warneck, P., and Junge, C. (1980). "N-alkanes studies in the troposphere. II. Gas and particulate concentration in Indian ocean air." *Atmos. Environ.* **14**, 695–703.

Fenical, W. (1982). "Natural products chemistry in the marine environment." *Science* **215**, 923–928.

Foster, P., Laffond, M., and Jacob, V. (1988). "Determination of organic compounds in the Grenoble area. Characterization and evolution." In *Field Measurements and Their Interpretation, COST 611, Physico-Chemical Behaviour of Atmospheric Pollutants*. Villefranche-sur-mer, France, Air Pollut. Res. Rep. 14, pp. 196–211.

Frankel, E. N., Neff, W. E., Selke, E., and Weisleder, D. (1982). "Photosensitized oxidation of methyl linoleate: secondary and volatile thermal decomposition products." *Lipids* **17**(1), 11–18.

Gagosian, R. B., Zafiriou, O. C., Peltzer, E. T., and Alford, J. B. (1982). "Lipids in aerosols from the tropical north Pacific: temporal variability." *J. Geophys. Res.* **87**, 11133–11144.

Garrett, W. D., and Smagin, V. M. (1976). "Determination of the atmospheric contribution of petroleum hydrocarbons to the ocean. *WMO* [*Publ.*] **440**(6), 1–27.

Giam, C. S., Chan, H. S., and Neff, G. S. (1975). "Rapid and inexpensive method for detection of polychlorinated biphenyls and phtalates in air." *Anal. Chem.* **47**, 2139–2320.

Greenberg, J. P., and Zimmerman, P. R. (1984). "Nonmethane hydrocarbons in remote tropical, continental, and marine atmospheres." *J. Geophys. Res.* **89**(D3), 4767–4778.

Greenberg, J. P., Zimmerman, P. R., Heidt, L., and Pollock, W. (1984). "Hydrocarbon and carbon emissions from biomass burning in Brazil." *J. Geophys. Res.* **89**(D1), 1350–1354.

Gregory, G. L., Hariss, R. C., Talbot, R. W., Rasmussen, R. A., Garstang, M., Andreae, M. O., Hinton, R. R., Browell, E. V., Beck, S. M., Sebacher, D. I., Khalil, M. A. K., Ferek, R. J., and Harriss, S. V. (1986). "Air chemistry over the tropical forest of Guyana." *J. Geophys. Res.* **91**(D8), 8603–8612.

Grimalt, J., Albaiges, J., Sicre, M. A., Marty, J.-C., and Saliot, A. (1988). "Aerosol transport of polynuclear aromatic hydrocarbons over the Mediterranean Sea." *Naturwissenschaften* **75**, 39–42.

Grimsrud, E. P., Westberg, H. H., and Rasmussen, R. A. (1975). "Atmospheric reactivity of monoterpene hydrocarbons, $NO_x$ photooxidation and ozonolysis." *Int. J. Chem. Kinet. Symp.* **1**, 183–195.

Grob, K., and Grob, G. (1971). "Gas-liquid chromatographic-mass spectrometric investigation of $C_6$-$C_{20}$ organic compounds in an urban atmosphere. An application of ultra trace analysis on capillary columns." *J. Chromatogr.* **62**, 1–13.

Grosjean, D., and Fung, K. (1984). "Hydrocarbons and carbonyls in Los Angeles air." *J. Air Pollut. Control Assoc.* **34**, 587–543.

Gschwend, P., Zafiriou, O. C., and Gagosian, R. B. (1980). "Volatile organic compounds in seawater from the Peru upwelling region." *Limnol. Oceanogr.* **25**(6), 1044–1053.

Gschwend, P., McFarlane, J. K., and Newman, K. A. (1985). "Volatile halogenated organic compounds released to seawater from temperate marine macroalgae." *Science* **227**, 1033–1035.

Han, J., and Calvin, M. (1969). "Hydrocarbons distribution of algae and bacteria and microbiological activity in sediments." *Proc. Natl. Acad. Sci. U.S.A.*, **64**, 436–443.

Hauser, R., Scott, Dr. R., and Midgett, M. R. (1983). "Monitoring airborne organics." *Environ. Sci. Technol.* **17**, 86A–96A.

Helz, G. R., and Hsu, R. Y. (1978). "Volatile chloro- and bromocarbons in coastal waters." *Limnol. Oceanogr.*, 23(5), 858–869.

Holdren, M. W., Westberg, H. H., and Hill, H. H. (1979). *Analytical Methodology for the Identification and Quantitation of Vapor Phase Organic Pollutants.* CRC-APRAC, Coord. Res. Counc. Air. Pollut. Res. Advis. Comm., New York, Interim Report CAPA-11-71.

Hov, Ø., Schjoldager, J., and Wathne, B. M. (1983). "Measurement and modeling of the concentrations of terpenes in coniferous forest air." *J. Geophys. Res.* **88**(C15), 10679–10688.

Jacob, D. J., and Wofsy, S. C. (1988). "Photochemistry of biogenic emissions over the Amazon forest." *J. Geophys. Res.* **93**(D2), 1477–1486.

Jalliffier-Merlon, E., and Marty, J.-C. (1988). "Behaviour of some volatile organic atmospheric pollutants in the Rhône delta system." In *Field Measurements and Their Interpretation, COST 611, Physico-Chemical Behaviour of Atmospheric Pollutants.* Villefranche-sur-mer, France, Air Pollut. Res. Rep. 14, pp. 133–139.

Jalliffier-Merlon, E., Marty, J. C., Denant, V., and Seliot, A. (1991). "Phytofanketonic sources of volatile aldehydes in the river Rhône estuary." *Est. Coast. Shelf. Sci.,* **32**, 463–482.

John, W., and Reischl, G. (1978). "Measurements of the filtration efficiencies of selected filter types." *Atmos. Environ.* **12**, 2015–2019.

Jones, P. W., Giamar, R. D., Strup, P. E., and Stanford, T. B. (1976). "Efficient collection of polycyclic organic compounds from combustion effluents." *Environ. Sci. Technol.* **10**, 806–810.

Josephson, J. (1981). "Monitoring airborne organics." *Environ. Sci. Technol.* **15**, 731–733.

Junge, C. E. (1977). "Basic considerations about trace constituents in the atmosphere as related to the fate of global pollutants." In Suffet, I. H. ed., *Fate of Pollutants in the Air and Water Environments.* Wiley, New York, Vol. 8, pp. 7–25.

Junk, G. A., and Ford, C. S. (1980). "A review of organic emissions from selected combustion processes." *Chemosphere* **9**, 187–230.

Kanakidou, M., Bonsang, B., Le Roulley, J. C., Lambert, G., Martin, D., and Sennequier, G. (1988). "Marine source of atmospheric acetylene." *Nature (London)* **333**, 51–52.

Kennicut, M. C., Brooks, J. M., Atlas, E. L., and Giam, C. S. (1988). "Organic compounds of environmental concern in the Gulf of Mexico: A review." *Aquat. Toxicol.* **11**, 191–212.

Ketseridis, G., Hahn, J., Jeanicke, R., and Junge, C. (1976). "The organic constituents of atmospheric particulate matter." *Atmos. Environ.* **10**, 603–610.

Kikuchi, T., Mimura, T., Moriwaki, Y., Ando, M., and Negoro K. (1974). "Metabolites of a diatom *Synedra rumpens* Kütz. isolated from water in Lake Biwa. Identification of odorous compounds, *n*-hexanal and *n*-heptanal, and analysis of fatty acids." *Chem. Pharm. Bull.* **22**(4), 945–949.

Kirschmer, P., and Ballschmiter, K. (1983). "Baseline studies of the global pollution. VIII The complex pattern of $C_1$-$C_4$ organohalogens in the continental and marine background air." *Int. J. Anal. Chem.* **14**, 275–284.

Koppman, R., and Rudolph, J. (1988). "The distribution of light alkenes in the maritime atmosphere and their potential impact on photochemical ozone formation." In *Field Measurements and Their Interpretation, COST 611, Physico-Chemical Behaviour of Atmospheric Pollutants.* Villefranche-sur-mer, France, Air Pollut. Res. Rep. 14, pp. 140–145.

Lamb, B., Westberg, H., Allwine, G., and Quarles, T. (1985). "Biogenic hydrocarbon emissions from deciduous and coniferous trees in the United States." *J. Geophys. Res.* **90**(D1), 2380–2390.

Langhorst, M. L. (1980). "Solid absorbent collection and gas chromatographic determination of the polypropylene glycol butyl ether ester of 2,4,5-T in air." *Am. Ind. Hyg. Assoc. J.* **41**(5), 238–233.

Lilian, D. (1972). "Formation and destruction of ozone in a simulated natural system (Nitrogen dioxide + α-pinene + h$\nu$)." *Adv. Chem. Ser.* **113**, 211–218.

Lipari, F., and Swarin, S. J. (1985). "2,4-Dinitrophenylhydrazine-coated Florisil sampling cartridges for the determination of formaldehyde in air." *Environ. Sci. Technol.* **14**(4), 44–48.

Lonneman, W. A., Seila, R. L., and Bufalini, J. J. (1978). "Ambient air hydrocarbon concentrations in Florida." *Environ. Sci. Technol.* **12**(4), 459–463.

Lovelock, J. E. (1977). "Methyl chloroform in the troposphere as an indicator of OH radical abundance." *Nature (London)* **267**, 32.

Lunde, G. (1976). "Long-range aerial transmission of organic micropollutants." *Ambio* **5**, 207.

Lunde, G., and Björseth, A. (1977). "Polycyclic aromatic hydrocarbons in long-range transported aerosols." *Nature (London)* **268**, 518–519.

Mantoura, R. F. C., Gschwend, P. M., Zafiriou, O. C., and Clarke, K. R. (1982). "Volatile organic compounds at a coastal site. 2. Short term variations." *Environ. Sci. Technol.* **16**(1), 38–45.

Marty, J. -C. (1981). "Chimie de l'interface air-mer: Accumulation des lipides dans la microcouche, leur éjection et leur évaporation dans l'atmosphère." Thèse d'état, Univesité Pierre et Marie Curie, Paris, 309 pp.

Marty, J. -C., and Saliot, A. (1980). "Ejection et évaporation: Deux mécanismes de transfert de la matière organique, acide gras, de la mer vers l'atmosphère. *Océanis* **6**, 181–191.

Marty, J. -C., Tissier, M. J., and Saliot, A. (1984). "Gaseous and particulate polycyclic aromatic hydrocarbons (PAH) from the marine atmosphere." *Atmos. Environ.* **18**(10), 2183–2190.

McDonald, T. J., Brooks, J. M., and Kennicut, M. C. (1984). "Release of volatile liquid hydrocarbons from spilled petroleum." *Bull. Environ. Contam. Toxicol.* **32**, 621–628.

Mead, J. F., Alfin- Slater, R. B., Howton, D. R., and Popjak, G. (1986). *Lipids. Chemistry, Biochemistry and Nutrition.* Plenum Press, New York, 486 pp.

Middleditch, B. S., Zlatkis, A., and Schwartz, R. D. (1988). "Trace analysis of volatile polar organics: problems and prospects."*J. Chromatogr. Sci.* **26**, 150–152.

Momzikoff, A., Santus, R., and Giraud, M. (1983). "A study of the photosensitizing properties of seawater." *Mar. Chem.* **12**, 1–14.

Moore, R. E. (1976). "Volatiles compounds from the marine algae." *Acc. Chem. Res.* **10**, 40–47.

Neff, J. M. (1979). *Polycyclic Aromatic Hydrocarbons in the Aquatic Environment. Sources, Fates and Biological Effects.* Applied Science, London.

Neff, W. E., Frankel, E. N., Selke, E., and Weisleder, D. (1983). "Photosensitized oxidation of methyl linoleate monohydroperoxides: hydroperoxy cyclic peroxides, dihydroperoxides, keto esters and volatile thermal decomposition products." *Lipids* **18**(12), 868–876.

Nelson, P. F., and Quigley, S. M. (1982). "Non-methane hydrocarbons in the atmosphere of Sydney, Australia." *Environ. Sci. Technol.* **16**(10), 650–655.

Newman, K. A., and Gschwend, P. M. (1987). "A method for quantitative determination of volatile organic compounds in marine macroalgae." *Limnol. Oceanogr.* **32**(3), 702–708.

Nishimura, M., and Baker, E. W. (1986). "Possible origin of *n*-alkanes with a remarkable odd-to-even predominance in recent marine sediments." *Geochim. Cosmochim. Acta* **50**, 299–305.

Pankow, J. F. (1987). "Review and comparative analysis of the theories on partitioning between the gas and aerosol particulate phases in the atmosphere." *Atmos. Environ.* **21**(11), 2275–2283.

Paoletti, C., Pushparaj, B., Florenzano, G., Capella, P., and Lercker, G. (1976). "Unsaponifiable matter of green and blue-green algal lipids as a factor of biochemical differentiation of their biomasses. I. Total unsaponifiable and hydrocarbon fraction." *Lipids* **11**, 258–265.

Parkes, D. G., Ganz, C. R., Polinsky, A., and Schulze, J. (1975). "A simple gas chromatographic method for the analysis of trace organics in ambient air." *Am. Ind. Hyg. Assoc. J.,* March, 165–173.

Pellizzari, E. D., and Krost, K. J. (1985). "Chemical transformations during ambient air sampling for organic vapors." *Anal. Chem.* **56**(11), 1813–1819.

Pellizzari, E. D., Brunch, J. E., Berkley, R. E., and McRae, J. (1976). "Collection and analysis of trace organic vapor pollutants in ambient atmospheres. The performance of a Tenax GC cartridge sampler for hazardous vapors." *Anal. Lett.* **9**, 45–63.

Pellizzari, E. D., Demian, B., and Krost, K. (1984). "Sampling of organic compounds in the presence of reactive inorganic gases with Tenax G. C. *Anal. Chem.* **56**(4), 793–798.

Peltzer, E. T., Alford, J. B., and Gagosian, R. B. (1984). "Methodology for sampling and analysis of lipids in aerosols from the remote atmosphere." *Tech. Rep. WHOI* **84-9**, 1–104.

Rasmussen, R. A. (1970). "Isoprene identified as a forest-type emission to the atmosphere." *Environ. Sci. Technol.* **4**(8), 667–673.

Rasmussen, R. A. (1974). "Emission of biogenic hydrogen sulfide." *Tellus* **26**(1-2), 254–260.

Rasmussen, R. A., and Khalil, M. A. K. (1983). "Natural and anthropogenic trace gases in the lower troposphere of the Arctic." *Chemosphere* **12**(3), 371–375.

Rasmussen, R. A., Khalil, M. A. K., Gunawardena, R., and Hoydt, S. D. (1982). "Atmospheric methyl iodide ($CH_3I$)." *J. Geophys. Res.* **87**, 3086–3090.

Roberts, J. M., Fehsenfeld, F. C., Albritton, D. L., and Sievers, R. E. (1983). "Measurement of monoterpene hydrocarbons at Niwot Ridge, Colorado." *J. Geophys. Res.* **88**(C15), 10667–10678.

Rudolph, J., and Ehhalt, D. H. (1981). "Measurements of $C_2$-$C_6$ hydrocarbons over the north Atlantic." *J. Geophys. Res.* **86**(C12), 11959–11964.

Rudolph, J., Johnen, F. J., and Khedim, A. (1986). Problems connected with the analysis of halocarbons and hydrocarbons in the non urban atmosphere. *3rd Workshop Chem. Anal. Hydrocarbons, Lausanne, Suisse, March, 1986.*

Russwurm, G. M., Stikeleather, J. A., Killough, P. H., and Windsor, J. G., Jr. (1981). "Design of a sample cartridge for the collection of organic vapors." *Atmos. Environ.* **15**, 929–931.

Schauenstein, E., Esterbauer, M., and Zollner, H. (1977). "Aldehydes in biological systems: their natural occurrences and biological activity." *Pion,* London, 192 pp.

Sauer, T. C., Jr. (1981). "Volatile liquid hydrocarbon characterization of underwater hydrocarbon vents and formation waters from offshore production operations." *Environ. Sci. Technol.* **15**, 917–922.

Schwarzenbach, R. P., Bromund, R. H., Gschwend, P. H., and Zafiriou, O. C. (1978). "Volatile organic compounds in coastal seawater." *Org. Geochem.* **1**, 93–107.

Seila, R. L. (1979). "Non-urban hydrocarbon concentration in the ambient air north of Houston, Texas." *U.S. Environ. Prot. Agency, Off. Res. Dev. [Rep.] EPA* **EPA-600/3-79-010.**

Sicre, M. A., Marty, J. -C., Saliot, A., Aparicio, X., Grimalt, J., and Albaiges, J. (1987a). "Aliphatic and aromatic hydrocarbons in different sized aerosols over the Mediterranean Sea: Occurrence and origin." *Atmos. Environ.* **21**, 2247–2259.

Sicre, M. A., Marty, J. -C., Saliot, A., Aparicio, X., Grimalt, J., and Albaiges, J. (1987b). "Aliphatic and aromatic hydrocarbons in the Mediterranean aerosol." *Int. J. Environ. Anal. Chem.* **29**, 73–94.

Sicre, M. A. Marty, J. C., Lorre, A., and Saliot, A. (1990). "Airborne and vapor phase hydrocarbons over the Mediterranean Sea." *Geophys. Res. lett.*, **17**(12), 2161–2164.

Simon, C. G., and Bidleman, T. F. (1979). "Sampling airborne polychlorinated biphenyls with polyurethane foam. Chromatographic approach to determining retention efficiencies." *Anal. Chem.* **51**, 1110–1113.

Simoneit, B. R. T., and Mazurek, M. A. (1981). "Air pollution: The organic components." *CRC Crit. Rev. Environ. Control* **12**, 218–276.

Singh, H. B., and Salas, L. J. (1982). "Measurements of selected light hydrocarbons over the Pacific ocean: Latitudinal and seasonal variations." *Geophys. Res. Lett.* **9**(8), 842–845.

Singh, H. B., Fowler, D. P., and Peyton, T. O. (1976). "Atmospheric carbon tetrachloride: another man-made pollutants." *Science* **192**, 1231–1234.

Singh, H. B., Salas, L. J., and Stiles, R. E. (1983). "Methyl halides in and over the eastern Pacific (40°N-32°S)." *J. Geophys. Res.* **88**, 3086–3090.

Spingarn, N. E., and Northington, D. J. (1982). "Analysis of volatile hazardous substances by GC/MS." *J. Chromatogr. Sci.* **20**, 286–288.

Steudler, P. A., and Peterson, B. J. (1984). "Contribution of gaseous sulfur from salt marshes to the global sulphur cycle." *Nature (London)* **311**, 455–457.

Termonia, M., and Alaerts, G. (1985). "Trace analysis of organic pollutants in air by polymer adsorption and dual flamme ionization detection-electron-capture detection capillary gas chromatography." *J. Chromatogr.* **328**, 367–371.

Tille, K. J. W., Savelsberg, M., and Bachmann, K. (1985). "Airborne measurements of nonmethane hydrocarbons over western Europe: vertical distributions, seasonal cycles of mixing ratios and source strengths." *Atmos. Environ.* **19**(11), 1751–1760.

Tingey, D. T., Manning, M., Grothaus, L. C., and Burns, W. F. (1980). "Influence of light and temperature on monoterpene emission rates from slash pine." *Plant Physiol.* **65**, 797–801.

Van Vaeck, L., Van Cauwenberghe, K., and Janssens, J. (1984). "The gas-particle distribution of organic aerosol constituants: Measurements of the volatilization artefact in Hi-vol. cascade impactor sampling." *Atmos. Environ.* **18**, 417–430.

Wade, T. C., and Quinn, J. G. (1974). In R. A. Duce, P. L. Parker, and C. S. Giam, eds., *Pollutant Transfer to the Marine Environment*. University of Rhode Island, Kingston, p. 25.

Westberg, H. H., and Rasmussen, R. A. (1972). "Atmospheric photochemical reactivity of monoterpene hydrocarbons." *Chemosphere* **4**, 163–168.

Whelan, J. K., Tarafa, M. E., and Hunt, J. M. (1982). "Volatile $C_1$-$C_8$ organic compounds in macroalgae." *Nature (London)* **299**, 50–52.

Whitby, R. A., and Coffey, P. E. (1977). "Measurement of terpenes and other organics in an Adirondack mountain pine forest." *J. Geophys. Res.* **82**(37), 5928–5934.

Whittle, K. J., Hardy, R., Mackie, P. R., and McGill, A. S. (1982). "A quantitative assessment of the source and fate of petroleum compounds in the marine environment." *Philos. Trans. R. Soc. London, Ser. B* **297**, 193–218.

Winer, A. M., Lloyd, A. C., Darnall, K. R., and Pitts, J. N., Jr. (1976). "Relative rate constants for the reaction of the hydroxyl radical with selected ketones, chloroethenes, and monoterpenes hydrocarbons." *J. Phys. Chem.* **80**(14), 1635–1639.

Yokouchi, Y., Okinawa, M., Ambe, Y., and Fuwa, K. (1983). "Seasonal variation of monoterpene in the atmosphere of a pine forest." *Atmos. Environ.* **17**(4), 743–750.

You, F., and Bidleman, T. F. (1984). "Influence of volatility on the collection of polycyclic aromatic hydrocarbon vapors with polyurethane foam." *Environ. Sci. Technol.* **18**, 330–333.

Youngblood, W. W., Blumer, M., Guillard, R. L., and Fiore, F. (1971). "Saturated and unsaturated hydrocarbons in marine benthic algae." *Mar. Biol. (Berlin)* **8**, 190–201.

Zafiriou, O. C. (1983). "Natural water photochemistry." In J. P. Riley and R. Chester, eds., *Chemical Oceanography.* Academic Press, London, pp. 339–376.

Zafiriou, O. C., Alford, J., Herrera, M., Peltzer, E. T., Gagosian, R. B., and Liu, S. C. (1980). "Formaldehyde in remote marine air and rain: Flux measurements and estimates." *Geophys. Res. Lett.* **7**(5), 341–344.

Zafiriou, O. C., Gagosian, R. B., Peltzer, E. T., and Alford, J. B. (1985). "Air to sea fluxes of lipids at Enewetak Atoll." *J. Geophys. Res.* **90**, 2409–2423.

Zimmerman, P. R., Greenberg, J. P., and Westberg, C. E. (1988). "Measurements of atmospheric hydrocarbons and biogenic emissions fluxes in the Amazon boundary layer." *J. Geophys. Res.* **93**(D2), 1407–1416.

# 5

# ATMOSPHERIC DISTRIBUTION AND SOURCES OF NONMETHANE HYDROCARBONS

*Hanwant B. Singh*

NASA Ames Research Center, Moffett Field, California

*and*

*Patrick B. Zimmerman*

National Center for Atmospheric Research, Boulder, Colorado

*Gaseous Pollutants: Characterization and Cycling,* Edited by Jerome O. Nriagu.
ISBN 0-471-54898-7   © 1992 John Wiley & Sons, Inc.

## 1.  INTRODUCTION AND OVERVIEW

Natural and man-made emissions of nonmethane hydrocarbon (NMHC) into the global atmosphere are estimated to be in the vicinity of 500–1000 Tg/yr (Tg $= 10^{12}$ g). This carbon flux is comparable to or larger than that of methane. The complexity of NMHC emitted into the atmosphere is enormous and more than $10^3$ different species are involved. Atmospheric lifetimes of these species can range from a few hours to several months. The first impetus for studying NMHC emissions and chemistry came in the early 1950s following the recognition by Haagen-Smit (1952) that the Los Angeles smog was created as a result of photochemical reactions involving NMHC and nitrogen oxides ($NO_x$), and that ozone ($O_3$) was a principal component of smog. In the late 1960s it was further recognized that high $O_3$ levels existed throughout the country and NMHC and $NO_x$ emissions were once again responsible. Mechanisms of smog formation were studied extensively and salient processes were isolated (Altshuller and Bufalini, 1971; Niki et al., 1972; Demerjian et al., 1974). In parallel with these postindustrial developments were ideas proposed by Went (1960a,b) who suggested that the plant biomass was a major source of terpenoid hydrocarbons and attributed the occurrence of blue haze to aerosol formation from these emissions. Early smog chamber studies led the United States Environmental Protection Agency (EPA) to conclude that the best way to control the national $O_3$ problem was to curtail the emissions of NMHC from industrial and transportation sources (U.S. Department of Health, Education and Welfare (USHEW), 1970; U.S. Environmental Protection Agency (USEPA), 1984), even though it was evident that natural NMHC emissions could also react to produce $O_3$ (Ripperton et al., 1971). To this date, the role of natural emissions in $O_3$ formation continues to be a matter of active research and controversy (Dimitriades, 1981; Bufalini and Arnts, 1981; Altshuller, 1983; Trainer et al., 1987; Chameides et al., 1988).

In the last two decades evidence also began to accumulate that many of the NMHC had either sufficiently large sources or sufficiently long lifetime to make them ubiquitous in the global troposphere (Robinson, 1978; Duce et al., 1983). It further became evident that biogenic emissions were not limited to terpenoid-type materials but isoprene and a variety of alkanes, alkenes, and oxygenated hydrocarbon were also involved (Sanadze and Kalandadze, 1966; Rasmussen, 1970; Zimmerman et al., 1978; Altshuller, 1983; Isidorov et al., 1985). The role of oceans in both light ($C_2$–$C_5$) and very heavy ($C_9$–$C_{28}$) hydrocarbon emissions was recognized (Swinnerton and Lamontagne, 1974; Eichmann et al., 1980). During this time a number of field, laboratory, and modeling studies were performed to characterize the sources, distributions, and chemistry of this complex group of chemicals. In addition to their well-known role as precursors of urban smog, NMHC have been proposed as (1) significant contributors to the global budget of carbon monoxide (Robinson and Robbins, 1969; Zimmerman et al., 1978), (2) carriers of reactive nitrogen through their oxidation products such as peroxyacetyl nitrate (Crutzen, 1979; Singh and Hanst, 1981), (3) possible sinks for Cl atoms in the troposphere and lower stratosphere (Singh, 1977; Rudolph et al., 1981; Singh and Kasting, 1988), (4) tracers of atmospheric transport (Rasmussen and Khalil, 1983b; Ehhalt et al., 1985), (5) possible indicators of tropospheric OH radical concentrations (Calvert, 1976; Singh et al., 1981, 1985; Roberts et al., 1984), (6) either directly or indirectly, contributors to the budget of tropospheric ozone, (7) a potentially important link in the global carbon cycle, and (8) biologically important compounds that can act as pheromones and allelopathic agents and affect herbivory (Harborne, 1988).

Despite significant recent interest in NMHC, our knowledge is quite incomplete, in part because of the extreme complexity of the sources, the large variety of species involved, and often the lack of suitable measurement techniques. In this study we evaluate the present state of measurement technology, and assess our knowledge of the distribution, sources, sinks, and trends of primary gaseous NMHC in various atmospheric compartments with emphasis on literature published subsequent to the review by Duce et al. (1983). No attempt is made to specifically deal with oxygenated species or aerosols.

## 2. NONMETHANE HYDROCARBONS OF INTEREST

### 2.1. Chemical Species

There are in excess of $10^3$ NMHC species that are emitted to the atmosphere from anthropogenic and biogenic sources. Over 850 different hydrocarbons in the vapor over high-test gasoline have been separated. Hampton et al. (1982, 1983) have observed nearly 400 different species from vehicle-exhaust polluted air; over 300 of these were identified. In addition, a large number of species from natural vegetation, composed of primary and oxygenated hydrocarbons, have also been identified. Table 1 provides an example of

**Table 1 Organics in Volatile Emissions of Arboreous Plants**

| Compound | Plant Species[a] | Compound | Plant Species[a] |
|---|---|---|---|
| Propylene | 8,9 | Chloroform | 16 |
| Butylene | 1–12,15,16 | Dimethyl sulfide | 15 |
| Isoprene | (1–4)**,5–17 | Santene | 8,11 |
| Hydrocarbon $C_5H_{10}$ | 1,2,5–9,13–15 | Cyclofenchene | 8–10,15 |
| 2-Methylbutane | 1–8,12 | Bornilene | 8–17 |
| 2,3-Deimethyl butadiene | 8,13,14 | Tricyclene | 7,8*,9–10 |
| Hydrocarbon $C_8H_{12}$ | 8,9 | α-Thujene | 9,10,13–17 |
| Hydrocarbon $C_9H_{10}$ | 1,6,11,16 | ga-Pinene | 3,(7,8,11,12,14)*,(9,10,13,15,17)**,16 |
| Methanol | 15 | δ-Fenchene | 9,10,12 |
| Ethanol | 3–5,10,12–15 | ε-Fenchene | 9,12 |
| 3-Hexene-l-ol | 1,2,5*,7,8 | α-Fenchene | 7–17 |
| Propanal | 6 | β-Fenchene | 7–10,12 |
| Isobutanal | 16 | Camphene | 3,(7,9,10,12)*,*,(8,11)**,13–17 |
| Crotonal | 17 | Sabinene | 12,13,15,17,(14,16)* |
| Isobutenal | 1–4,9,7 | β-Pinene | 3,7–11,15 |
| Acetone | 1–17 | Myrcene | 7–13,14*,15–17 |
| Butanone-2 | 8,13–15 | | |

| | | | | |
|---|---|---|---|---|
| Methyl vinyl ketone | 2,4 | | 3-Carene | 7*,8–10,12–15,16*,17 |
| Pentanone-2 | 6,8,15 | | α-Phellandrene | 3,8–11 |
| Pentanone-3 | 5,7–10,15,16 | | β-Phellandrene | 8–11,14 |
| Furan | 6 | | α-Terpinene | 8–10,17 |
| 2-Methyl furan | 1–5,12,13 | | β-Terpinene | 3,8–10,13–15 |
| 3-Methyl furan | 1–5,12,13 | | γ-Terpinene | 9,13–15,17 |
| Ethyl furan | 1,2,4,8 | | Limonene | 3,7–17 |
| Vinyl furan | 1,2,4,8 | | Terpinolene | 9–17 |
| Ethyl acetate | 9 | | 1,8-Cineol | 3,8,14 |
| 3-Hexene-l-ol acetate | 1,4–8 | | Fenchone | 16* |
| Methyl (α-methyl) | | | Thujone | 16** |
| butyrate | 14** | | Camphor | 8 |
| Methyl (α-methyl) | | | | |
| capronolate | 14 | | p-Cymene | 7–17 |
| Methyl chloride | 14–16 | | Menthane | 14 |

[a]1, Bay-leaved willow; 2, aspen; 3, balsam poplar; 4, European oak; 5, European birch; 6, sorb; 7, European larch; 8,European fir; 9, Scots pine; 10, Siberian pine; 11, silver fir; 12, common juniper; 13, Zeravshan juniper; 14, pencil cedar; 15, evergreen cypress; 16, northern white cedar; 17,Chinese arbor vitae.

*An important VOC component; **the dominant VOC component.

*Source*: Isidorov et al. (1985).

some of the volatile organic chemicals that have been identified as emissions from vegetation (Isidorov et al., 1985). Primary hydrocarbons most often measured from biogenic sources include ethylene, propylene, ethane, and propane from seawater, and isoprene, α-pinene, β-pinene, limonene, carene, and myrcene from vegetation. Each of these hydrocarbons is further capable of atmospheric oxidation to produce many new products. Needless to say, the numbers of species involved are very large. Table 2 provides some perspective on the important primary NMHC and typical bulk mixing ratios that are likely to be present in various compartments of the atmosphere. Based on present knowledge, it is reasonable to expect that at a minimum some 100 or so select species must be measured to quantify the burden of atmospheric NMHC.

## 2.2.   Atmospheric Lifetimes

To a large degree the atmospheric behavior of trace species is dictated by their lifetimes. Highly reactive species react near the vicinity of the sources, while slow-reacting species may be transported over long distances. The major removal process for primary gaseous NMHC appears to be reaction with OH radicals and to a lesser degree with $O_3$. As shown in Table 3, OH concentrations (also $O_3$) are strongly dependent on season and latitude. Thus, the lifetimes of NMHC are a strong function of season and geography. As a group, NMHC do not absorb UV radiation in the troposphere, and photolysis as a sink process is negligible. (This, of course, is not true of oxygenated species.) Primary hydrocarbons are also relatively insoluble and not likely to be removed by physical processes such as dry and wet deposition. The lifetime of an NMHC can then be defined as

$$\tau = \{k_{OH}[OH] + k_{O3}[O_3]\}^{-1} \qquad (1)$$

where $k_{OH}$ and $k_{O3}$ are reaction rates with OH and $O_3$, respectively. In Table 4 these lifetimes (equation 1) are calculated for a number of selected NMHC under typical atmospheric conditions. Given these lifetimes of a few hours to a few months, it is to be expected that the composition of NMHC would vary significantly both in space and time. It is important to recognize that the highly reactive hydrocarbons do not need to be highly abundant. Table 5 shows this point by considering NMHC abundances that would be needed if their reactivity was to be equal to that of methane. Thus, isoprene present at less than 0.1 ppb (ppb = $10^{-9}$ v/v) would be as effective in chemistry as 1700 ppb of methane.

## 2.3.   Measurement Techniques

Many of the most common techniques used to measure atmospheric NMHC employ a gas chromatograph (GC) equipped with a flame ionization detector

**Table 2  Key NMHC Species Expected to be Present in Various Atmospheric Compartments**

| Atmospheric Compartment | Typical Average NMHC Mixing Ratio (ppbC) | Significant NMHC Species[a] |
|---|---|---|
| Stratosphere | | Acetylene, ethane*, propane*, butanes (*i,n*). |
| Marine boundary layer and free troposphere | 0–5[b] 5–20 NH 3–5 SH | All of the above plus pentanes (*i,n*), hexanes, ethylene*, propylene*, butenes (all isomers)*, pentenes (all isomers)*, dimethyl sulfide*, benzene, toluene, xylenes (*o,m,p*), $C_9$–$C_{28}$ *n*-alkanes*. |
| Rural/remote terrestrial atmospheres | 20–100 | All of the above plus heptanes (*n*), octanes (*n*), ethyl benzene, styrene, isoprene* α-pinene*, β-pinene*, myrcene*, Δ-3-carene*, camphene, *d*-limonene*, terpinolene*, and other pinenes. |
| Urban atmospheres | 100–1,000 (Max. >10 000) | All of the above plus hexenes (all isomers), heptenes, (all isomers), cyclo pentene, cyclohexanes, methyl pentanes (2 and 3), dimethyl butane, dimethyl pentanes (2,4 and 2,3), methyl cyclopentane, methyl hexane, methyl heptane, trimethyl pentane (all isomers), ethyl hexane, ethyl cyclohexane, dodecane, propyl benzenes (*i,n*), ethyl toluenes (*o,m,p*), trimethyl benzenes (all isomers), methyl styrene, diethyl benzene (1,4 and 1,3) and many others. |

[a]In all these compartments, a complex array of oxygenated species (aldehydes, ketones, alcohols, acids) is also expected to be present due to both direct emissions and chemical oxidation.

[b]Excludes halogenated species. Mixing ratios decrease rapidly with altitude in the stratosphere.

*Significant natural biogenic sources are known to exist.

Table 3 Surface Mean OH Radical Concentrations at Noontime

| Latitude (°N) | OH Concentration ($10^6$ molecules $cm^{-3}$) | | | |
|---|---|---|---|---|
| | March 30 | June 30 | September 30 | December 30 |
| 5 | 4.18 | 3.64 | 3.48 | 5.84 |
| 25 | 3.54 | 3.77 | 4.04 | 2.94 |
| 45 | 0.61 | 3.70 | 2.09 | 0.04 |
| 65 | 0.03 | 1.09 | 0.28 | 0.00 |
| 85 | 0.00 | 0.31 | 0.00 | 0.00 |

Note. The OH field corresponds to a global mean methyl chloroform lifetime of 5.5 years.
Source: Kanakidou et al. (1991).

(FID) and/or a photoionization detector (PID). In very limited instances GC-MS and Fourier transform infrared (FTIR) spectroscopic techniques have also been employed. Both packed and capillary column GCs are widely used. All methods involve batch processing and are inherently slow. Typically light hydrocarbons ($C_2$–$C_5$) are separated by packed columns and heavier hydrocarbons ($C_5$–$C_{12}$) by capillary columns (Singh, 1980; Greenberg and Zimmerman, 1984; Bonsang and Lambert, 1985; Matuska et al., 1986). Several hundred milliliters of ambient air (typically 100–2000 mL) is preconcentrated at liquid oxygen or argon temperature prior to GC separation and detection. With a 1-L sample and a FID, a detection sensitivity of about 10 ppt (ppt = $10^{-12}$ v/v) is achievable.

In principle, it is possible to cryotrap and analyze an air sample in real time without going through a process of sample storage. However, in practice, largely due to logistical constraints, air samples are often collected in a variety of ways and stored for extended time periods (days to months) prior to analysis. Samples are generally collected in (1) Teflon or Tedlar bags, (2) solid sorbents, or (3) stainless steel cannisters (electropolished and unpolished). Singh (1980) has discussed the advantages and disadvantages of each of these techniques. Because of difficulties involved in handling, Teflon/Tedlar bags are seldom used. Solid sorbents, such as Tenax, are convenient to use and offer many advantages, including ease of handling, nearly unlimited sample volumes, and operation at ambient temperatures. There are disadvantages, of course, because Tenax, the most commonly used solid sorbent, is unable to quantitatively collect light hydrocarbons ($C_1$–$C_7$) and there may be problems of artifact formation due to $O_3$ and $NO_2$ reactions. The most widely used sampling method is electropolished stainless steel canisters in which whole air samples are collected at pressures of 30–100 psig. While a variety of hydrocarbon species are stable in these canisters for long time periods, significant departures can occur (Singh, 1980). These departures may be related to the specific electropolishing, preconditioning, and sampling procedures used by specific investigators. Holdren et al. (1979) find substantial losses (20–100%) for isoprene and terpenes over a 3-week storage period.

**Table 4 Estimated Atmospheric Lifetimes of Selected NMHC**

| Hydrocarbons | $K_{OH} \times 10^{12}$ cm$^3$ molecule$^{-1}$s$^{-1}$ | $K_{O3} \times 10^{15}$ cm$^3$ molecule$^{-1}$s$^{-1}$ | Lifetime[a] (days) Tropics | Lifetime[a] (days) Global |
|---|---|---|---|---|
| Ethane | 11.0 exp $(-1090/T)$[b] | — | 20.0 | 92 |
| Propane | 16.0 exp $(-800/T)$ | — | 5.3 | 22 |
| n-Butane | 18.0 exp $(-554/T)$ | — | 2.1 | 80 |
| n-Pentane | 3.7 | — | 1.5 | 52 |
| Acetylene | 2.0 exp $(-252/T)$ | — | 6.7 | 24 |
| Ethylene | 7.8 | 9.0 exp $(-2566/T)$ | 0.7 | 1.9 |
| Propene | 4.1 exp $(546/T)$ | 6.1 exp $(-1912/T)$ | 0.2 | 0.5 |
| Benzene | 1.2 | — | 4.8 | 16 |
| Toluene | 6.4 | — | 0.9 | 3.0 |
| o-Xylene | 14.0 | — | 0.4 | 0.4 |
| Isoprene | 24.0 ep $(409/T)$ | 16.0 exp $(-2144/T)$ | 0.07 | 0.2 |
| $\alpha$-Pinene | 14.0 exp $(446/T)$ | 0.94 exp $(-731/T)$ | 0.07 | 0.1 |
| $\beta$-Pinene | 24.0 exp $(358/T)$ | 0.021 | 0.08 | 0.2 |
| $C_{20}H_{42}$ | 20.4 exp $(358/T)$ | — | 5.1 | 21.3 |

[a]Tropics: $T = 298$ K; OH $= 2 \times 10^6$ molecules cm$^{-3}$, O$_3 = 7.4 \times 10^{11}$ molecules cm$^{-3}$ ($\approx 30$ ppb); day $= 24$ h. Global: $T = 275$ K; OH $= 6 \times 10^5$ molecules cm$^{-3}$, O$_3 = 7.4 \times 10^{11}$ molecules cm$^{-3}$.
[b]Rate costants are as compiled by Warneck (1988).

**Table 5  Mixing Ratios of NMHC Required for Reactivity Equal to That of Methane**

| Hydrocarbons | Equivalent Reactivity Mixing Ratio (ppb) |
|---|---|
| Methane | 1750 |
| Ethane | 40 |
| Propane | 9.6 |
| *n*-butane | 3.5 |
| Acetylene | 10.4 |
| Ethylene | 1.0 |
| Propylene | 0.3 |
| Benzene | 7.0 |
| Toluene | 1.3 |
| Isoprene | 0.08 |

*Note.* Based on OH reactivity alone; $T = 275$ K.

Others (Greenberg and Zimmerman, 1984) report less significant losses. Singh et al. (1988) note that ethylene concentrations at the low ppt level actually increase in such sampling canisters over a period of several days. Comparison of cryosampling followed by analysis in real time, with whole air sample collection and subsequent analysis, have shown good agreement for light alkanes and other species but significant differences in measured alkene concentrations (Greenberg et al., 1990). In their comparison study no provisions were made to remove ozone from the directly sampled air inlet stream. It is likely that $O_3$ destroys alkenes during the cryogenic sampling process. Experimental evidence indicates that $O_3$ itself is generally destroyed relatively quickly in stainless steel vessels and hence poses less of a problem when whole air samples are collected in cannisters.

Although a variety of NMHC measurements have been performed, there are several uncertainties that remain unresolved. For example, in one atmospheric intercomparison study conducted in a remote area of the Pacific, involving only light hydrocarbons, which were all analyzed using gas chromatography with flame ionization detection, significant differences in hydrocarbon quantitation were evident (Carsey, 1987). Other intercomparisons of specific hydrocarbons in urban air have generally shown reasonable agreement for selected species. The following major weaknesses exist in our present measurement capabilities.

## Sampling

- Possible destruction of species by reaction with $O_3$, $NO_2$ and/or artifact formation during sampling
- Loss (or gain) in contact with surfaces, especially over extended storage periods lasting several days to months

- Sample contamination and unrepresentative sampling

## Detection

- Improper and incomplete identification of chromatographic peaks due to lack of GC-MS confirmation
- Lack of chromatographic separation resulting in incorrectly identified and overlapping peaks
- Lack of FID selectivity and sensitivity, and a clear need for more sensitive and specific detectors
- Lack of continuous (slow and fast response) instrumentation

## Calibration

- Lack of uniformly available stable standards at low (ppb or ppt) concentrations
- Wide use of carbon response with FID leading to errors in quantitation that may approach 10% for hydrocarbons and are substantially higher for oxygenated organics

## Intercomparisons

- Lack of intercomparison studies at low concentration synthetic air mixtures of NMHCs
- Lack of intercomparison studies in ambient rural/remote environments using independent measurement techniques

Overall, the basic tools for measurements of NMHC are in hand, but many improvements are necessary before uniformly reliable data can be obtained. Hydrocarbon measurements are particularly challenging in remote environments where concentrations are extremely low and in studies of biogenic emission contributions to urban air where a huge range of compounds and concentrations are present. In this situation biogenic hydrocarbons often constitute a small but chemically important and difficult to differentiate part of the urban air mix. The FID has undergone little change in the last two decades. Sample analysis is cumbersome and usually requires several hours from sample introduction to final tabulation. It is therefore difficult to acquire the temporal and spatial resolution required to adequately define the hydrocarbon distribution so that photochemical models of their effects on the chemistry of the atmosphere can be adequately tested. In addition, since there have been no instruments with the capability for fast continuous operation, it has been difficult to take advantage of micrometeorological advances, such as eddy correlation, to measure hydrocarbon fluxes. There is a continuing need for more sensitive and specific detectors, and for fast continuous instruments that can be used to measure fluxes of specific classes of hydrocarbons.

## 3.  ATMOSPHERIC DISTRIBUTIONS

### 3.1.  Urban Environments

For the last thirty years NMHC have been measured in a variety of urban atmospheres in many countries. Recently published studies provide a significant body of data from the United States (Sexton and Westburg, 1984; Singh et al., 1985; Seila et al., 1989), the United Kingdom (Colbeck and Harrison, 1985), and Australia (Nelson et al., 1983). These studies typically show that some 100–1000 ppbC of NMHC are present in most urban locations and the individual species have a strong resemblance to gasoline and its combustion products. A quick idea of the NMHC distribution of urban atmospheres can be had from the results in Table 6 based on measurements from 39 urban atmospheres in the United States for the period of 1984 to 1986 (Seila et al., 1989). Based on the median profile of these cities, 442 ppbC of total NMHC is present and is predominantly made up of alkanes (57.7%), aromatics (26.3%), alkenes (13.1%), and acetylene (2.9%). Most biogenic hydrocarbons, while undoubtedly present in low concentrations relative to anthropogenic hydrocarbons, were not specifically differentiated in these studies.

Strong diurnal and seasonal variations in the distribution of urban/rural hydrocarbons are present and are largely dictated by their reactivity and prevailing meteorology. Figure 1 shows such a diurnal profile for a variety of aromatic hydrocarbons measured in southern California. The afternoon minimum in all cases is due to deep convective mixing and chemical loss by OH radicals. Singh et al. (1985) have analyzed the ratios of these hydrocarbons relative to benzene and estimate that the data are consistent with an OH abundance of $2.6 \times 10^6$ molecules $cm^{-3}$. Very little information on the seasonal cycle of anthropogenic NMHC in urban environments is available, but a pattern emerges suggesting higher values in winter compared to summer as a result of typical atmospheric stagnation and lower removal rates (Table 3).

It is further noted that the advent of the catalytic converter to reduce NMHC emissions has resulted in significant changes in the composition of the gasoline as well as emissions from tail pipes. Emissions data suggest that the 1982 car fleet (postcatalytic) should emit about a quarter as much NMHC as the precatalytic car fleet (Sigsby et al., 1987). Studies of Lonneman et al. (1986) inside Lincoln tunnel from 1970 and 1982 provide direct evidence. These measurements show that under comparable traffic density and air ventilation rates, the NMHC emissions from the 1980 fleet were only about a quarter of the 1970 fleet. This study and the tail-pipe emission studies confirm that significant reductions in NMHC have been achieved. Although these reductions do not directly translate into similar reductions in ambient levels because of the complexity of the urban sources (e.g., automobile exhaust, gasoline spillage and evaporation, natural gas, industrial emissions), it is indeed plausible that significant changes in emissions have occurred. Singh et al. (1985) use available measurements to suggest that dramatic reductions in the ambient concentrations of benzene and toluene in the south coast of California have already occurred (Fig. 2).

## 3.2. Rural/Nonurban Environments

### 3.2.1. Anthropogenic Hydrocarbons

Typical rural NMHC mixing ratios are in the range of 20–100 ppbC. Rural NMHC, principally of anthropogenic origin, have been measured around the world and many of the data have been summarized in a number of studies (Hov et al., 1990; Colbeck and Harrison, 1985; Sexton and Westberg, 1984; Seila, 1984; Kanakidou et al., 1989; Rudolph and Koppmann, 1989). Although many of the hydrocarbons are clearly of anthropogenic origin, a larger contribution from biogenic sources is often present. Table 7 shows the median mixing ratios of NMHC measured at three selected sites in southern Appalachia (Seila, 1984). A comparison with Table 6 shows many similar chemicals present at much lower concentrations. Table 8 further shows that total NMHC mixing ratios were 40–90 ppbC, with alkanes constituting the largest known fraction. This is largely due to their longer lifetime compared to those of alkenes and aromatics. It is further instructive to note that the largest fraction in Table 8 is made up of "unknown" species. In both urban and rural environments some 15–40% of carbon is typically made up of unidentified chemicals.

### 3.2.2. Biogenic Hydrocarbons

As discussed earlier (Tables 1 and 2) many NMHC species are emitted by plant species, isoprene and $\alpha$-pinene being the two principal ones (Rasmussen, 1972; Zimmerman, 1979; Isidorov et al., 1985). In North America, major forest species such as pines, firs, junipers, spruce cedars, and redwoods are terpene emitters, while oaks, aspens, willow, poplar's, and sycamore are mainly isoprene emitters, and several varieties (e.g., spruce, sweet gum, eucalyptus) emit both. These biogenic hydrocarbons are highly reactive and have atmospheric lifetimes of minutes to hours (Table 3). In rural and, in some cases, urban areas, the atmospheric abundance of these species can compete with the anthropogenic burden in $O_3$ formation (Trainer et al., 1987; Chameides et al., 1988). The high reactivity of these species ensures that typically they are most abundant in the surface of the boundary layer and in forest canopies.

Table 9 summarizes some of the more recent measurements of key biogenic species from a variety of locations. Isoprene is the most abundant species in this group. Figures 3 and 4 show the diurnal variation in isoprene and $\alpha$-pinene from a forested site in Brazil and Japan, respectively (Rasmussen and Khalil, 1988; Yokouchi et al., 1983). The maximum isoprene abundance is during the time of day with maximum temperatures and insolation resulting

**Table 6 Concentration Statistics (ppbC) Based on Measurements in 39 U.S. Cities from 1984–1985**

| Compound | $N^a$ | Median | Minimum | 25% | 75% | Maximum |
|---|---|---|---|---|---|---|
| Isopentane | 832 | 45.3 | 1.4 | 26.2 | 71.6 | 3393 |
| n-Butane | 833 | 40.3 | 4.5 | 23.9 | 65.5 | 5448 |
| Toluene | 836 | 33.8 | 2.7 | 20.6 | 56.6 | 1299 |
| Propane | 835 | 23.5 | 1.8 | 12.2 | 45.2 | 393 |
| Ethane | 830 | 23.3 | 0.6 | 12.4 | 41.0 | 470 |
| n-Pentane | 834 | 22.0 | 1.0 | 12.5 | 36.0 | 1450 |
| Ethylene | 707 | 21.4 | 1.2 | 13.2 | 35.8 | 1001 |
| m- and p-Xylene | 836 | 18.1 | 1.3 | 11.3 | 30.0 | 338 |
| 2-Methylpentane | 836 | 14.9 | 1.2 | 8.5 | 23.5 | 647 |
| Isobutane | 835 | 14.8 | 1.4 | 8.4 | 28.6 | 1433 |
| Acetylene | 709 | 12.9 | — | 7.3 | 23.2 | 114 |
| Benzene | 835 | 12.6 | 1.0 | 7.9 | 19.9 | 273 |
| n-Hexane, 2-ethyl-1-butene | 836 | 11.0 | 0.8 | 6.2 | 18.4 | 601 |
| 3-Methylpentane | 831 | 10.7 | 0.1 | 6.4 | 16.6 | 351 |
| 1,2,4-Trimethylbenzene | 828 | 10.6 | — | 6.7 | 17.1 | 81 |
| Propylene | 835 | 7.7 | 0.4 | 4.3 | 14.3 | 455 |
| 2-Methylhexane | 763 | 7.3 | 0.2 | 4.5 | 11.7 | 173 |
| o-Xylene | 831 | 7.2 | 0.9 | 4.7 | 11.6 | 79 |
| 2,2,4-Trimethylpentane | 835 | 6.8 | 0.4 | 3.9 | 11.6 | 106 |
| Methylcyclopentane | 834 | 6.4 | 0.5 | 3.7 | 10.3 | 293 |
| 3-Methylhexane | 828 | 5.9 | 0.3 | 3.5 | 9.7 | 168 |
| 2-Methylpropene, butene-1 | 827 | 5.9 | — | 3.8 | 9.8 | 365 |
| Ethylbenzene | 836 | 5.9 | 0.7 | 3.6 | 9.8 | 159 |
| m-Ethyltoluene | 832 | 5.3 | 0.1 | 3.3 | 8.6 | 83 |

| | | | | | | N |
|---|---|---|---|---|---|---|
| n-Heptane | 831 | 4.7 | 0.1 | 2.8 | 8.2 | 233 |
| 2,3-Dimethylbutane | 834 | 3.8 | 0.3 | 2.3 | 6.1 | 177 |
| c-2-Pentene | 750 | 3.6 | — | 1.9 | 6.0 | 339 |
| 1,2,3-Trimethylbenzene | 758 | 3.4 | 0.1 | 1.6 | 5.7 | 1701 |
| Methylcyclohexane | 836 | 3.4 | 0.3 | 2.0 | 6.0 | 184 |
| n-Decane | 835 | 3.3 | 0.2 | 1.9 | 6.0 | 138 |
| 1,3,5-Trimethylbenzene | 825 | 3.0 | 0.3 | 2.0 | 5.1 | 51 |
| $C_{11}$ Aromatic | 773 | 3.0 | 0.2 | 1.8 | 4.7 | 71 |
| tert-2-Pentene | 807 | 2.9 | 0.1 | 1.5 | 4.7 | 291 |
| o-Ethyltoluene | 836 | 2.9 | 0.2 | 1.9 | 4.6 | 54 |
| p-Ethyltoluene | 831 | 2.8 | 0.1 | 1.8 | 4.7 | 54 |
| $C_{10}$ Aromatic | 832 | 2.8 | 0.2 | 1.8 | 4.5 | 235 |
| n-Octane | 799 | 2.6 | 0.2 | 1.6 | 4.6 | 163 |
| 2-Methyl-1-butene | 822 | 2.6 | 0.1 | 1.4 | 4.4 | 242 |
| 1,2-Dimethyl-3-ethylbenzene | 756 | 2.5 | 0.2 | 1.6 | 4.3 | 149 |
| tert-2-Butene | 811 | 2.5 | 0.1 | 1.4 | 4.2 | 337 |
| 2,3,4-Trimethylpentane | 833 | 2.5 | 0.1 | 1.5 | 4.4 | 78 |
| 2-Methylheptane | 820 | 2.5 | 0.1 | 1.3 | 4.2 | 75 |
| 1,4-Diethylbenzene | 821 | 2.4 | 0.1 | 1.5 | 4.0 | 33 |
| 3-Methylheptane | 832 | 2.2 | 0.1 | 1.4 | 3.9 | 109 |
| n-Nonane | 821 | 2.2 | 0.2 | 1.3 | 4.2 | 89 |
| Cyclohexane | 817 | 2.2 | 0.2 | 1.1 | 4.8 | 409 |
| 2,4-Dimethylpentane | 827 | 2.2 | 0.2 | 1.3 | 3.8 | 72 |
| Cyclopentane | 823 | 2.1 | 0.1 | 1.2 | 3.2 | 104 |

[a]N, number of samples.

*Source.* Adapted from Seila et al. (1989).

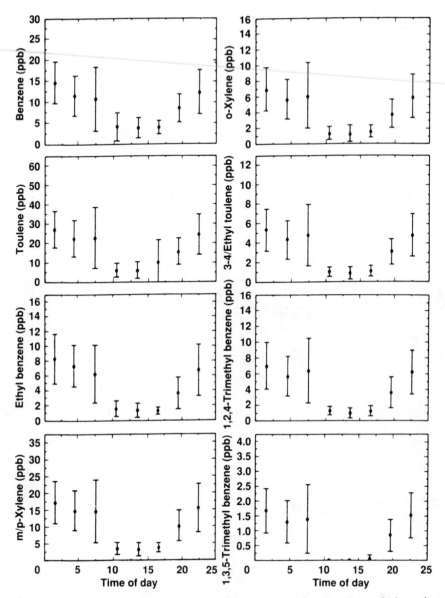

**Figure 1.** Diurnal behavior of aromatic hydrocarbons based on measurements at a site in southern California.

in peak emissions related to photosynthesis or photorespiration. At night emissions quickly cease since there is no isoprene storage reservoir within stems or leaves, and the atmospheric reservoir is depleted by reaction with $O_3$ and $NO_3$ radicals. In contrast to isoprene, large quantities of terpenes can be stored in certain plant species. These terpene reservoirs leak to the at-

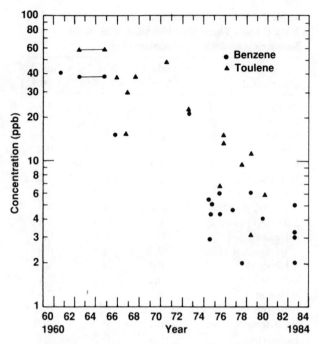

**Figure 2.** Benzene and toluene trend in the atmosphere of the south coast air basin of California.

mosphere independent of light at rates that are dependent on leaf diffusivity and on terpene temperature, volatility, and solubility (Tingey, 1981; Tingey et al., 1991). Pinene diurnal behavior can thus be quite different since emissions can continue at night. Highest values of α-pinene may be encountered at night (Fig. 4) when removal by photochemical reactions and ozonolysis, and convective mixing is at its minimum. It is commonly observed that above the boundary layer (lower free troposphere) virtually no isoprene or terpenes are present. Even within the boundary layer significant gradients can be expected. Figure 5 shows the daytime vertical distributions and variability of isoprene and α/β-pinenes at a rural site (Jachin, Alabama) in the United States.

Few long-term studies of NMHC in rural or forested areas have been performed. In one such study, Seila (1984) found evidence of a seasonal cycle with highest concentrations of isoprene and α-pinene in summer (Fig. 6). This result is also consistent with the findings of Zimmerman et al. (1984) who observed higher mixing ratios of isoprene and α-pinene at Niwot Ridge, Colorado, in July compared to November (Table 9), but observed little difference in the abundance of other monoterpenes. Seila (1984) find that the abundance of the "unknown" species is very large and they also show a seasonal cycle similar to that of isoprene and α-pinene. No systematic seasonal cycle in the anthropogenic group (alkanes, alkenes, acetylene, aromatics) was

**Table 7 Median Concentrations (ppbC) of Selected NMHC from Three Rural/Remote Sites in the Southern Appalachian Mountains**

| Compounds | Site 1 ($N = 9$) | Site 2 ($N = 6$) | Site 3 ($N = 11$) |
|---|---|---|---|
| Acetylene | 0.8 | 1.0 | 0.8 |
| Ethane | 2.6 | 9.1 | 6.7 |
| Propane | 3.4 | 6.0 | 5.6 |
| Isobutane | 0.7 | 2.3 | 0.9 |
| n-Butane | 1.1 | 2.5 | 2.0 |
| Isopentane | 0.9 | 3.8 | 1.5 |
| n-Pentane | 0.7 | 1.0 | 1.0 |
| 2-Methyl pentane | 0.2 | 0.6 | 0.4 |
| 3-Methyl pentane | 0.1 | 0.4 | 0.2 |
| n-Hexane | 0.2 | 0.5 | 0.3 |
| 2,4-Dimethyl pentane | 0.3 | 0.3 | 0.4 |
| n-Heptane | 0.1 | 0.2 | 0.2 |
| n-Octane | 0.1 | 0.1 | 0.1 |
| n-Nonane | 0.1 | 0.1 | 0.1 |
| Ethylene | 1.7 | 3.2 | 0.8 |
| Propylene | 0.1 | 0.7 | 0.3 |
| Isobutylene | 0.1 | 0.2 | 0.2 |
| Benzene | 0.9 | 1.3 | 1.1 |
| Toluene | 1.0 | 2.6 | 1.4 |
| Ethyl benzene | 0.3 | 0.5 | 0.4 |
| m/p-Xylene | 0.6 | 1.0 | 0.7 |
| o-Xylene | 0.3 | 0.4 | 0.4 |
| Isoprene | 0.1 | 0.8 | 0.5 |
| α-pinene | 0.3 | 0.3 | 4.3 |

*Note*: N, number of samples analyzed at each site. Site 1: Roan Mountain, elevation 1.8 km; Site 2: Grandfather Mountain, elevation 1.7 km; Site 3: Linville Gorge, elevation 1 km.

*Source*: Adapted from Seila (1984).

observed. This indirect evidence suggests that many of the unidentified hydrocarbons may also be of natural biogenic origin. Although there is considerable scatter, Fig. 7 shows evidence of a similar annual cycle in the emission of α-pinene in a Japanese Pine forest (Yokouchi et al., 1983). As we shall see later, it is reasonable to expect that biogenic emissions would be at their maximum under conditions that represent high temperatures, high humidity, and high insolation.

### 3.3. Remote Marine Environments

In remote atmospheres the most abundant NMHC are the longest lived group of light hydrocarbons, some aromatics, and alkenes of possibly oceanic origin.

**Table 8 NMHC Composition in the Southern Appalachian Mountains of Northwest North Carolina**

| Site | Total NMHC (ppbC)[a] | % of Total NMHC | | | | | |
| --- | --- | --- | --- | --- | --- | --- | --- |
| | | Acetylene | Alkanes | Alkenes | Aromatics | Biogenics | Unknowns |
| Roan Mountain[b] | 40.3 | 2 | 30 | 6 | 8 | 1 | 53 |
| Grandfather Mountain[b] | 78.8 | 1 | 39 | 6 | 8 | 1 | 45 |
| Linville George[b] | 85.2 | 1 | 27 | 2 | 5 | 6 | 59 |
| Rich Mt. * | 72.7 | 1 | 31 | 2 | 7 | 2 | 57 |
| Dear Park | 68.2 | 1 | 36 | 4 | 8 | 1 | 51 |

[a]Based on an analysis of 43 samples from 5 sites.
[b]Hill sites 1–2 km above sea level.
*Source*: Adapted from Seila (1984).

Table 9 Typical Mean Concentrations of Isoprene and Terpene from Selected Terrestrial Sites

| Species | Niwot Ridge[1,2] (3 km elevation, rocky mountains) | | | Georgia, USSR[3] (Forest) | Rome, Italy[4] (Forest) | Brazil, Amazon Basin[5] (Forest) | | Kenya[5] (Agricultural land) | | Nigeria[5] (Coastal tropical) | Appalachian Mountains[6] (Forested hill sites) Sep. 81 to Oct. 82 | | | | Southwest France[7] (Rural) |
|---|---|---|---|---|---|---|---|---|---|---|---|---|---|---|---|
| | S | S | F | S | S | F | S | DS | WS | S | | | | | S |
| Isoprene | | 0.63 | 0.11 | 1.4 | — | 5.45 | 2.04 | 0.04 | <0.01 | 1.21 | 0.02 | 0.1 | 1.1 | <.01 | 0.19 |
| α-Pinene | 0.05 | 0.14 | 0.07 | 0.8 | 1.5 | 0.2 | 0.10 | <0.01 | <0.01 | 0.06 | 0.03 | 0.04 | 0.04 | 0.05 | — |
| β-Pinene | 0.97 | 0.08 | 0.07 | 0.43 | 0.18 | 0.01 | 0.03 | <0.01 | <0.01 | <0.01 | — | — | — | — | — |
| Δ-3-Carene | 0.05 | | | 0.90 | 0.06 | <0.01 | 0.01 | <0.01 | <0.01 | <0.01 | — | — | — | — | — |
| Camphene | 0.04 | 0.04 | 0.05 | 0.09 | — | 0.04 | 0.03 | <0.01 | <0.01 | 0.01 | — | — | — | — | — |
| d-Limonene | 0.03 | 0.05 | 0.03 | 0.08 | 0.04 | — | — | — | — | — | — | — | — | — | — |
| Myrcene | — | | | 0.68 | — | — | — | — | — | — | — | — | — | — | — |

Note. S, summer; F, fall; DS, dry season; WS, wet season. 1, Roberts et al. (1983); 2, Greenberg and Zimmerman (1984); 3, Shaw et al. (1983); 4, Ciccioli et al. (1984); 5, Zimmerman et al. (1988); 6, Seila (1984); 7, Kanakidou et al. (1988).

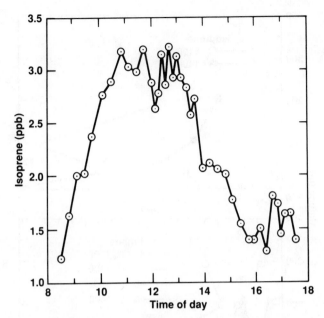

**Figure 3.** Diurnal behavior of isoprene near ground level in the Brazilian Amazon basin (From Rasmussen and Khalil (1988).)

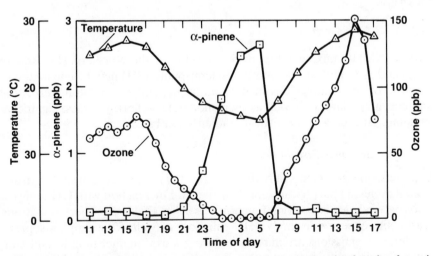

**Figure 4.** Diurnal variation in the concentration of α-pinene in the air of a pine forest in Tosukuba, Japan. (From Yokouchi et al. (1983).)

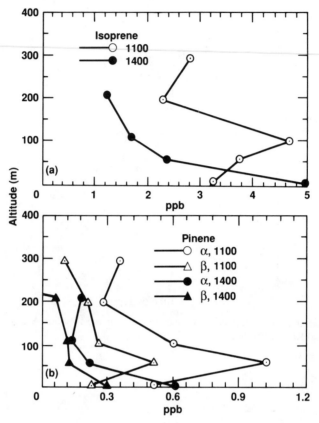

**Figure 5.** Vertical profile of isoprene and α/β pinenes at a rural site (Jachin, Alabama) in the United States.

Typical NMHC mixing ratios of 5–20 ppbC in the Northern Hemisphere (NH) and 3–5 ppbC in the Southern Hemisphere (SH) may be encountered. Because of the complexities in their anthropogenic and natural (terrestrial and oceanic) sources, as well as their moderately short lifetimes, significant seasonal, latitudinal, and vertical gradients can be expected.

### 3.3.1.   Seasonal Distributions

Long-term measurements of NMHC are few but are expected to show a strong annual cycles, in part because of their removal by reaction with OH radicals, whose abundance is a strong function of season (Table 3). There is a strong seasonality expected in the NMHC emissions as well; for example, isoprene and terpene emissions are maximum during warm summer months and minimum during cold winter months. Emissions of anthropogenic NMHC are also expected to vary with season in a more modest fashion. The available

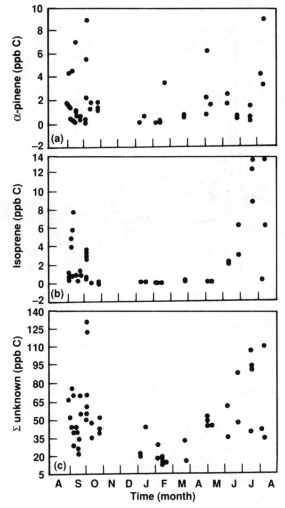

**Figure 6.** Seasonal variation of α-pinene, isoprene, and sum of unknown compounds based on measurements in southern Appalachian mountains. (From Seila (1984).)

U.S. data suggest that the anthropogenic emissions may be about 10% higher in spring and summer than in fall and winter (National Acid Precipitation Assessment Program (NAPAP), 1990).

Figure 8 shows the seasonal cycle of ethane based on measurements at marine coastal sites by Blake and Rowland (1986a) from a variety of latitudes in different seasons. Model tests at mid and high latitudes suggest that the annual cycle is predominantly derived from reaction with OH radicals (Kasting and Singh, 1986; Kanakidou et al., 1991). Figure 9 shows a similar behavior for propane from sites in the NH and SH. These annual cycles are not limited

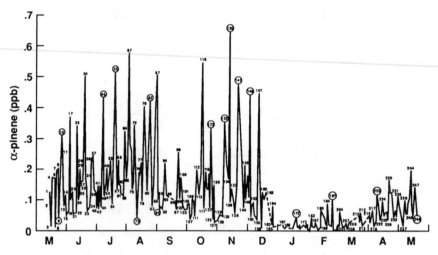

**Figure 7.** Seasonal variation in the concentration of α-pinene in a pine forest in Tosukuba, Japan. Mean temperature for July–Sep. is 24 °C and for Jan.–Feb. is 4 °C. (From Yokouchi et al. (1983).)

to the marine environment and can also be seen in the continental free troposphere. Figure 10 shows these data for ethane, propane, and benzene from a high-altitude site (1.8 km ASL) in Germany (Hahn and Matuska, 1990). All of these data are consistent with the view that remote and free tropospheric seasonal cycles are principally controlled by the abundance of the OH radical.

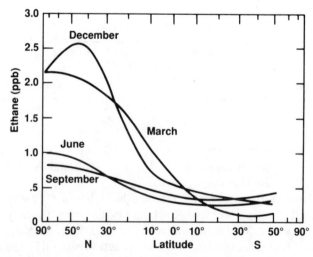

**Figure 8.** Surface level latitudinal variations of ethane in different seasons. (Adapted from Blake and Rowland (1986a).)

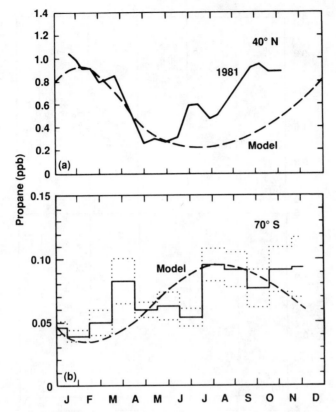

**Figure 9.** Seasonal variation of propane at 40°N and 70°S. (From Singh and Salas (1982) and Rudolph et al. (1989).)

Superimposed on the OH lead removal cycle are source and meteorological signatures. There are instances where these may dominate over the OH removal process. One region where this may happen is the Arctic, where trappings of pollutants in the dark and cold winter/spring period is well known. Table 10 shows the measurements of select NMHC in spring and summer months. A big systemic seasonal change is evident. Only part of this is attributable to OH removal. A significant component is also due to the fact that as the summer approaches dilution caused by significant advective and vertical mixing causes NMHC concentrations to decline.

### 3.3.2. Latitudinal Distributions

There is a large body of data dealing with the latitudinal variations of light NMHC ($C_2$–$C_5$) and limited information on others. A bulk of these measurements are from surface platforms, but limited free tropospheric aircraft

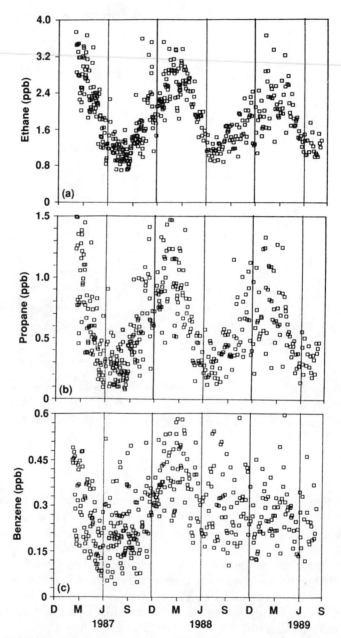

**Figure 10.** Seasonal cycles of ethane, propane and benzene at a high-altitude site (1.8 km ASL) in Germany. (From Hahn and Matuska (1990).)

**Table 10 Seasonal Variations in NMHC Abundance in the Arctic Region**

| Hydro-carbons | Spitsbergen, Norway (76–80°N)[a] | | | Barrow, Alaska (71°N)[b] | | |
|---|---|---|---|---|---|---|
| | Spring (March) | Summer (July) | Ratio (Spring/Summer) | Spring (March) | Summer (June) | Ratio (Spring/Summer) |
| $C_2H_6$ | 3950 | 1195 | 3.3 | 2800 | 1200 | 2.4 |
| $C_2H_4$ | 156 | 255 | 0.6 | 65 | 65 | 1.0 |
| $C_2H_2$ | 954 | 67 | 14.3 | 600 | 200 | 3.0 |
| $C_3H_8$ | 2156 | 87 | 24.8 | 1200 | 200 | 6.0 |
| $i$-$C_4H_{10}$ | 390 | <20 | >20 | — | — | — |
| $n$-$C_4H_{10}$ | 805 | <20 | >40 | — | — | — |
| $C_6H_6$ | 307 | 66 | 4.7 | 290 | 70 | 4.1 |
| $C_7H_8$ | 51 | 10 | 5.1 | 60 | 20 | 3.0 |

[a]From Hov et al. (1984).
[b]From Rasmussen and Khalil (1983).

data have also become available in recent years. The most common feature of these data are the following:

- Concentrations of a large number of NMHC are high in the Northern Hemisphere and decrease toward the equator with relatively constant values in the Southern Hemisphere.
- Alkenes, which have very short lifetimes, are relatively uniformly distributed in the marine boundary layer.
- Superimposed on the latitudinal cycles are the seasonal cycles of NMHC.

Systematic hemispheric gradients have been observed both in the Pacific and the Atlantic for the most stable of NMHC, namely, ethane, propane, and acetylene. All three appear to have strong sources in the NH. Figure 11 shows these latitudinal gradients as measured by a variety investigators. Despite the existence of a systematic trend, there is a large amount of scatter. This is expected because of the relatively short lifetime of these species (weeks to months), the proximity of continental sources, and seasonal effects. The latitudinal trend for higher alkanes, for example, iso/$n$-butane, iso/$n$-pentane, and hexanes have also been observed but the data are far more scattered. Table 11 summarizes the ranges of these concentrations based on a variety of studies (Singh et al., 1988; Rudolph and Johnen, 1990). Butanes for example are present in the 100- to 500-ppt range north of 15°N and are barely detectable at 10–20 ppt south of 15°N. It is probably fair to add that these latitudinal gradients are most profound near the surface only. More sensitive measurements of >$C_4$ alkane in the free troposphere are needed to define systematic trends. As noted earlier, these higher alkanes are relatively reactive and may participate in tropospheric chemistry directly and via sequestration of reactive nitrogen (Calvert and Madronich, 1987; Singh, 1987).

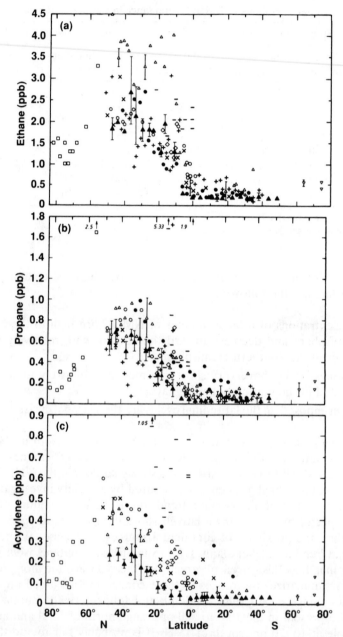

**Figure 11.** Latitudinal distributions of ethane, propane, and butane based on surface measurements. Adapted from Roemer and Hout (1988); dashes, Greenberg et al. (1984); open circles, squares, triangles, and diamonds, Rudolph et al. (1985); pluses, Bonsang and Lambert (1985); dark circles and triangles, Singh and Salas (1982) and Singh et al. (1988); crosses, Tille and Bachmann (1986).

**Table 11  Typical Surface Level Mixing Ratios of $C_4$-$C_6$
Alkanes in Marine Atmospheres**

| | Mixing Ratios[a] (ppt) | |
|---|---|---|
| Hydrocarbons | NH | SH |
| $n$-Butane | 20–150 | 10–50 |
| Isobutane | 15–250 | 10–50 |
| $n$-Pentane | 10–500 | 10–200 |
| Isopentane | 10–300 | 10–100 |
| $n$-Hexane | 10–300 | 10–100 |

[a]NH, Northern Hemisphere; SH: Southern Hemisphere.

*Source.* Rudolph and Ehhalt (1981), Singh and Salas (1982), Singh
et al. (1988), Greenberg et al. (1990), Rudolph and Johnen (1990).

It has been evident for some time that ethylene and propylene mixing
ratios over the oceans have been somewhat uniform (Rudolph and Ehhalt,
1981; Singh and Salas, 1982). Ocean water supersaturations coupled with
uniform distributions in the marine boundary layer have been used to infer
a sizeable oceanic source. Figure 12 shows the distribution of these alkenes
at several latitudes. Clearly, a large scatter in the data exists. This is in part
due to measurement problems, but is also due to the fact that the oceanic
sources of alkenes are by no means uniform. As we shall see in the next
section, atmospheric data are incompatible with direct estimates of the oceanic
source.

Aromatic hydrocarbons are also present in the global atmosphere, with
benzene being the dominant species. Table 12 summarizes these data from a
variety of sources. Often high concentrations of these species have been
encountered over the oceans. These can be attributed to ocean water contamination from oil seepage and spillage (Greenberg and Zimmerman, 1984).

### 3.3.3.  Vertical Distributions

A number of aircraft studies have attempted to describe the vertical distributions of NMHC. The salient among these are Tille et al. (1985), Rudolph
(1988), Singh et al. (1988), and Greenberg et al. (1990). A typical feature of
these profiles is relatively high concentrations near the ground and lower
concentrations aloft. A composite of the midlatitude ethane and propane
profiles compiled by Singh et al. (1988) from a variety of studies is shown in
Fig. 13. It is noted that often unusually high concentrations of even highly
reactive species are encountered in the free troposphere (Ehhalt et al., 1985;
Rudolph, 1988; Singh et al., 1988). Such phenomena have also been noted
for sulfur gases and radon, a radioactive gas with a 3.8-day half-life, and are
often attributed to rapid vertical mixing via processes such as cloud pumping
(Chatfield and Crutzen, 1987; Kritz et al., 1990). The vertical distributions

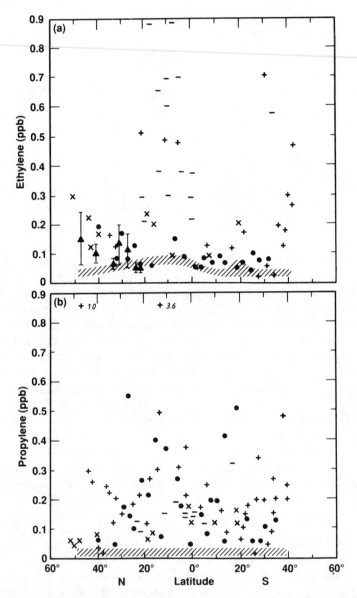

**Figure 12.** Latitudinal distributions of ethylene and propylene based on surface measurements. Shaded area from Rudolph and Johnen (1990); other symbols same as in Fig. 11.

of higher alkanes are similar but more variable. Data for $C_4$–$C_5$ alkane are shown in Table 13. Vertical structures of ethylene and propylene have also been published but the data must be considered much less reliable. An example of the measured two-dimensional vertical structure of ethylene and propylene as measured by Rudolph (1988) with grab sampling techniques is

**Table 12 Aromatic Hydrocarbons over the Pacific and Atlantic**

| Hydrocarbons | Mixing Ratios (ppt) | | | |
| --- | --- | --- | --- | --- |
| | 40–50° N | NH[a] | SH[a] | 40–50° S |
| Benzene | 100–200 | 49 ± 25 | 10 ± 2 | 10–20 |
| Toluene | 20–150 | 20 ± 12 | 6 ± 2 | 1–5 |
| Ethyl benzene | 3–5 | 8 ± 4 | 4 ± 2 | ≤3 |
| m/p-Xylene | 1–5 | 25 ± 12 | 13 ± 5 | ≤5 |
| o-Xylene | 3–5 | 14 ± 6 | 7 ± 3 | 1–4 |

[a]30° S–42° N.

*Source.* Tille and Bachmann (1986), Nutmagul and Cronn (1985), Rudolph et al. (1984b), Rasmussen and Khalil (1983b).

shown in Fig. 14. These vertical structures are inconsistent with model predictions that are based on eddy diffusion. Fast cloud pumping processes must be invoked to explain high values of alkenes in the free troposphere. Greenberg et al. (1990) have attempted to gather free tropospheric data from sampling at Mauna Loa (3.4 km ASL) during the prevailing down-slope flow conditions. Both surface and free tropospheric data from this site are summarized in Table 14. Median free tropospheric concentrations of 37 ppt ethylene and 6 ppt propylene are much lower than those reported by Rudolph (1988). There is little doubt that large inhomogeneities in the atmospheric abundance of highly reactive species such as alkenes are present and neither the measurements nor the models are able to provide a reasonable description of their distributions.

### 3.4.  Stratospheric Measurements

Very few measurements of NMHC that extend above the tropopause are available and the data are limited to ethane, propane, and acetylene. Grab sampling as well as spectroscopic techniques have been employed. Figure 15 shows the distribution of ethane and acetylene recently derived by Rinsland et al. (1987) for the 25–31°N latitude band from the ATMOS Fourier transform spectrometer during the SPACE 3 shuttle mission of April 30 to May 1. Ethane mixing ratios up to about 20 km are detectable. Data for ethane are in general agreement with the more limited measurements of Rudolph et al. (1981) and Aiken et al. (1987). Rudolph et al. (1981) report data up to 30 km, with ethane reaching about 10 ppt at 30 km. Acetylene has also been measured in the stratosphere (Cronn and Robinson, 1979; Rudolph et al., 1984a; Aiken et al., 1987; Rinsland et al., 1987). Once again concentrations above 20 km are below 20 ppt and barely detectable. The only propane data are from Rudolph et al. (1981), and show that propane mixing ratios at 15 km are 20–30 ppt and that propane is undetectable above 20 km. Overall,

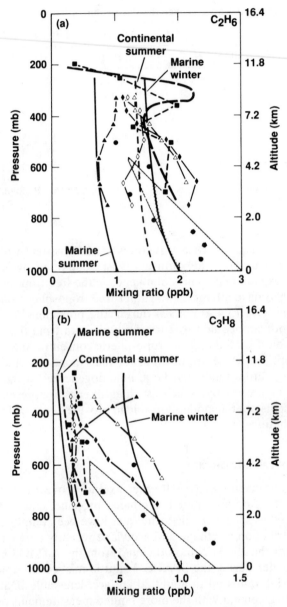

**Figure 13.** Vertical profiles of ethane and propane based on measurements at midlatitudes (38–50°N). Light dashed and solid curves are model predictions from Kasting and Singh (1986); heavy dashed curve, Ehhalt et al. (1985); solid circles and squares, Rudolph et al. (1982, 1984b); hatched area, Tille et al. (1985); open triangles, solid triangles, open diamonds, solid diamonds, Singh et al. (1988).

**Table 13 Vertical Structures of Butanes and Pentanes (ppb)**

Singh et al. (1988)

| Alt. (km) | Iso-C$_4$ C-S | Iso-C$_4$ M-S | Iso-C$_4$ M-W | n-C$_4$ C-S | n-C$_4$ M-S | n-C$_4$ M-W | Iso-C$_5$ C-S | Iso-C$_5$ M-W | n-C$_5$ C-S |
|---|---|---|---|---|---|---|---|---|---|
| 2.5 | — | 0.01 | 0.04 | — | 0.01 | 0.07 | — | 0.10 | — |
| 3.5 | 0.11 | 0.01 | 0.05 | 0.18 | 0.01 | 0.09 | 0.07 | 0.10 | 0.10 |
| 4.5 | 0.10 | 0.01 | 0.06 | 0.16 | 0.02 | 0.07 | 0.06 | 0.13 | 0.10 |
| 5.5 | 0.07 | 0.01 | 0.07 | 0.12 | 0.01 | 0.007 | 0.05 | 0.10 | 0.09 |
| 6.5 | 0.05 | 0.02 | 0.04 | 0.09 | 0.02 | 0.05 | 0.05 | 0.08 | 0.09 |
| 7.5 | 0.04 | 0.03 | 0.03 | 0.06 | 0.04 | 0.09 | 0.04 | 0.09 | 0.09 |
| 8.5 | 0.03 | 0.05 | 0.04 | 0.04 | 0.06 | 0.06 | 0.02 | 0.14 | 0.09 |

Rudolph (1988)

| Alt. (km) | Iso-C$_4$ M-S | Iso-C$_4$ M-W | n-C$_4$ M-S | n-C$_4$ M-W | Iso-C$_5$ M-S | Iso-C$_5$ M-W | n-C$_5$ M-S | n-C$_5$ M-W |
|---|---|---|---|---|---|---|---|---|
| 1–3 | 0.02 | 0.16 | 0.02 | 0.53 | — | 1.5 | — | 0.11 |
| 3–6 | 0.02 | 0.12 | 0.02 | 0.37 | — | 1.4 | — | 0.07 |

*Note.* Midlatitude profiles (40–55°N). C, continental; M, marine; S, summer; W, winter.

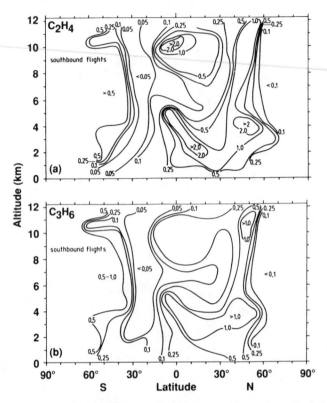

**Figure 14.** Two-dimensional structure of ethylene and propylene. (From Rudolph (1988).)

**Table 14 Mixing Ratios (ppt) Derived for Dry (Free Tropospheric) and Moist (Surface) Air Sampled at the Mauna Loa Observatory**

| | Dry Free Tropospheric Air | | Moist Surface Air | |
|---|---|---|---|---|
| Hydrocarbons | Median | Interquartile Range | Median | Interquartile Range |
| Ethane | 741 | 651–832 | 897 | 730–1009 |
| Ethylene | 37 | 20–57 | 81 | 40–127 |
| Acetylene | 75 | 58–90 | 109 | 85–145 |
| Propane | 26 | 16–44 | 53 | 38–79 |
| Propylene | 6 | 2–12 | 29 | 13–52 |
| Isobutane | 3 | 2–8 | 10 | 6–15 |
| n-Butane | 4 | 2–9 | 12 | 8–21 |
| Isopentane | 0 | 0–2 | 6 | 3–12 |
| n-Pentane | 0 | 0–0 | 7 | 4–14 |
| Isoprene | 0 | 0–2 | 12 | 5–54 |
| Benzene | 12 | 8–14 | 21 | 18–31 |
| Toluene | 3 | 1–5 | 7 | 6–18 |

*Source.* Greenberg et al. (1991).

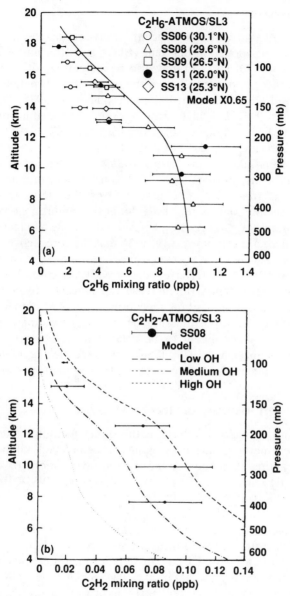

**Figure 15.** Volume mixing ratios of ethane and acetylene from the ATMOS spectra at 25–30°N. (From Rinsland et al. (1987).)

only the most stable NMHC have been detected in the stratosphere and the measurements are quite limited. Future direct emissions of NMHC by sizeable supersonic aircraft fleets may change the NMHC composition of the stratosphere, leading to significant sequestration of the oxides of nitrogen in the form of PAN via OH and Cl radial chemistry (Singh and Hanst, 1981; Singh and Kasting, 1988).

## 3.5.  Global Trends

It is now evident that methane concentrations are increasing in the troposphere at a rate of about 1% per year (Blake and Rowland, 1986b). There are indications that $O_3$ and CO may also be increasing, although these data are less conclusive, in part because of the temporal and spatial variability of these short-lived species (Logan, 1985; Rinsland et al., 1985; Volz and Kley, 1988). Ehhalt et al. (1991) have recently examined the atmospheric spectra obtained at the Jungfraujoch observatory (Switzerland) in 1951 and recently (1984 to 1988) to study trends of ethane, the most stable of the NMHC (Fig. 16). They find that atmospheric concentrations of ethane have increased by about 35% in the last 40 years (0.9 ± 0.3%/yr). It is further argued that this may be a global trend. No conclusive data on the global trends of other NMHC are available, but there is little doubt that anthropogenic sources for some of the species (e.g., acetylene, benzene) dominate and increasing trends must exist. There are estimates suggesting that the forest cover of the globe is nearly 40% less than it was several centuries ago. It is conceivable that emissions of natural NMHC have been greatly affected. Future warming scenarios have the potential to increase the natural emissions of NMHC through changes in species carbon allocation, increased photosynthetic rates, and increased terpene vapor pressure. However, since emissions are highly species specific and are likely to be related to net primary productivity (P. R. Zimmerman et al., 1978, 1988), and predicted regional changes in these are highly uncertain, future emission forecasting is perilous.

## 3.6.  NMHC, OH Radicals, and the Air Mass Age

Because of the difficulty involved in the direct measurements of the OH radical, indirect methods have been extensively employed. Globally averaged OH concentrations have been deduced from observations of select species with known OH reactivities (Singh, 1977; Volz et al., 1981). Because of the varying reactivities of NMHC, the ratios of species often can be used to provide clues on the OH radical abundance or on the age of the air mass when OH is known. In situ average concentrations of OH have been deduced from NMHC measurements following tagged urban air plumes (Calvert, 1976; Singh et al., 1981) as well as from single point measurements using reliable meteorological analysis (Roberts et al., 1984; Singh et al., 1985).

The following simple equation can be used to measure the average OH concentration:

$$[OH] = \frac{\ln[(C_A/C_B)_{t_1}/(C_A/C_B)_{t_2}]}{(k_A - k_B)(t_2 - t_1)} \tag{2}$$

where

$$(t_2 - t_1) = \text{transport time}$$

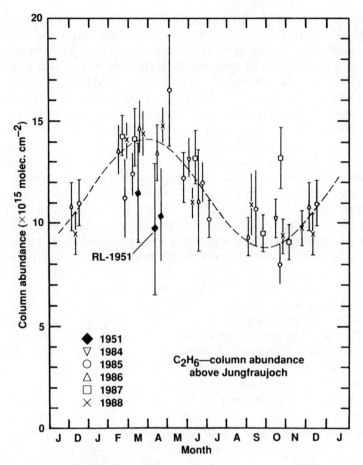

**Figure 16.** Ethane column abundance above Jungfraujoch from past (1951) and recent (1984–1988) spectra. (From Ehhalt et al. (1991).)

$C_A, C_B$ = atmospheric mixing ratios of two different hydrocarbons, A and B

$k_A, k_B$ = the second-order rate constants of OH with compounds A and B, respectively, $cm^3$ molecule$^{-1}$ s$^{-1}$

$[OH]$ = mean OH concentration, molecule cm$^{-3}$

$(C_A/C_B)_{t_1}$ = average of the ratios of the concentrations of compounds A and B at time $t_1$

$(C_A/C_B)_{t_2}$ = average of the ratios of the concentrations of compounds A and B in the same air mass at time $t_2$

The assumptions inherent in this calculation include that there are no significant sources of the compounds along the trajectory of the air mass, the sole sink of the hydrocarbons is reaction with OH, and the background tropospheric concentrations are negligible compared to those in the air mass. Calvert (1976) and Singh et al. (1981) used data from Lagrangian experiments in which sampling was done within air masses tracked by constant altitude balloons. The reactivity of a variety of NMHC was normalized against acetylene as the air masses moved over a period of several hours. Typical daytime mean OH concentrations of $2-3 \times 10^6$ molecules cm$^{-3}$ were calculated, but the estimate depended strongly on the hydrocarbon used. Roberts et al. (1984) have employed the same concept to estimate OH from single site measurements of aromatic hydrocarbons at Niwot Ridge, CO. By using trajectory analysis they determined both the source (southern California) and the transport time. Published literature provided information on the mixing ratios at the source. The change in the ratio of aromatic hydrocarbons from the source (southern California) to the receptor (Niwot Ridge) provided a direct calculation of the mean OH abundance during this time. Singh et al. (1985) used similar assumptions at the diurnal time scale to estimate morning OH concentrations in southern California. In all instances, daytime OH concentrations typically in the range of $2-3 \times 10^6$ molecules cm$^{-3}$ were obtained. Of course, if the mean values of the OH concentrations (e.g., from models) are known, then equation 2 can be used to estimate the age of the air mass. Mean toluene/benzene ratios are typically 2–3 in most urban areas and about 0.3 at remote sites. Assuming mean OH numbers densities ($\approx 10^6$ molecules cm$^{-3}$), one can calculate that such air masses are only 5–10 days away from their urban source (Nutmagul and Cronn, 1985; Singh et al., 1985).

Because these are not controlled experiments and often the airmass history is poorly known, results must be considered only approximate. Recently, McKeen et al. (1990) have discussed the errors in these methods for quantitative OH determinations. Despite shortcomings, such results have provided valuable insights into the photochemical smog system in the absence of more direct information.

## 4. SOURCES OF NMHC IN THE UNITED STATES AND THE WORLD

Virtually all anthropogenic activities related to the use or transfer of energy release NMHC into the atmosphere. In the industrialized countries fuel combustion, natural gas emissions, industrial processing of chemicals, and waste are a major source of NMHC. In less developed parts of the world, biomass burning is an added component. There are, of course, large natural emissions from terrestrial and oceanic sources. Quantitative knowledge on the size of these emissions is rudimentary. As stated earlier, we have now developed a reasonable understanding of the removal processes of many individual species. These removal estimates provide a useful constraint on the budget of many hydrocarbon species. Direct estimates of emissions are often made by using absolute or relative emission factors that are determined in controlled laboratory and field experiments. In the case of plant emissions there are further complications because these emissions are a strong function of environmental variables such as temperature, sunlight, and sometimes relative humidity. Extrapolations have been empirical in nature, because the mechanisms of NMHC synthesis and their release from plants are just beginning to be understood. Few vegetation species have been measured and the taxonomic relationships between vegetation species and emission species are not well characterized. Here we present a brief summary of NMHC emissions based on present knowledge. Many of the estimates are derived from a number of published studies that attempt to quantify both anthropogenic and natural sources of NMHC for the United States and the world (National Academy of Sciences (NAS), 1976; Zimmerman et al., 1978; Zimmerman, 1979; Eichmann et al., 1980; Duce et al., 1983; Greenberg et al., 1984; Andreae et al., 1985; Ehhalt et al., 1986; Sawada and Tatsuka, 1986; Lamb et al., 1987; Rasmussen and Khalil, 1988, USEPA, 1990; Hegg et al., 1990; Lobert et al., 1991; Middleton and Stockwell, 1990; NAPAP, 1990). While U.S. data are somewhat reliable, global estimates are often based on simple extrapolations and can have significant errors. It is likely that many of these estimates are in error by a factor of three or more for annual averages and by much more for specific time periods.

## 4.1.  Anthropogenic Sources

Detailed studies of anthropogenic emissions of organic compounds have been performed in the United States (Wagner et al., 1986; D. Zimmerman et al., 1988; USEPA, 1990; Middleton and Stockwell, 1990; NAPAP, 1990). Figure 17 shows the trends in reactive volatile organic compounds (predominantly NMHC) in the United States (USEPA, 1990). A total of 19 Tg/yr is a sig-

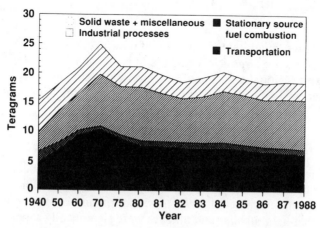

**Figure 17.** Emission trends of reactive volatile organic compounds in the United States (1940–1988). (From USEPA (1990).)

nificant reduction from the 25 Tg/yr released in 1970. This decline in emissions is a direct result of control strategies such as catalytic convertors and better fuel efficiencies (Sigsby et al., 1987). NAPAP (1990) suggests that only a modest seasonal variation in these emissions (spring, 27%; summer, 27%; fall, 23%; and winter, 23%) is present. Table 15 provides breakdown of individual categories for these emissions for the United States and the world. Global estimates take into account the fact that the rest of the world consumes nearly 2.5 times as much fuel as the United States, but much of the rest of the world has not adopted technologies such as catalytic convertors. Recognizing that 2–3% of NMHC is emitted as a fraction of fuel consumed, we estimate a global transportation source of 22 Tg/yr and stationary sources near 4 Tg/yr. Another major source of NMHC emissions is the refining and transportation of gasoline and natural gas production and seepage. Much of the latter is made up of light alkanes. USEPA (1990) estimates an 8.5 Tg/yr source for the United States, and the world estimate is taken to be twice as large.

Forest fires, open burning, and incineration are small anthropogenic sources in the United States. However, biomass burning is a major source of global NMHC emissions. Based on direct measurements of grassland and tropical forest burning, Greenberg et al. (1984) find that on the average the emissions of NMHC are about 1.1% (0.4 to 2.8% range) of those of $CO_2$. By using an estimated source of $3.1 \times 10^{15}$ gC/yr of $CO_2$ from biomass burning (Seiler and Crutzen, 1980), an NMHC source of 37 TgC/yr can be derived. Comparable results can also be deduced from the data of Hegg et al. (1990) and Lobert et al. (1991). These NMHC are made up of a variety of species including alkene, alkane, aromatics, and oxygenates, as summarized in Table 16. Overall, it is estimated that anthropogenic emissions of NMHC in the United States and the world are nearly 20 and 100 Tg/yr, respectively.

## 4.2. Natural Sources

As stated earlier, quantitative estimates of natural emissions have been made but uncertainties are large. Nevertheless, it is reasonable to believe that both on global and regional scales natural sources may dominate over anthropogenic emissions. Two types of natural sources are identified: oceanic and plant emissions.

### 4.2.1. Oceanic Sources

The existence of marine sources of light hydrocarbons has been known for over two decades from direct analysis of seawater (Swinnerton and Linnenbom, 1967; Lamontagne et al., 1974; Swinnerton and Lamontagne, 1974). It was found that light alkanes and alkenes were significantly supersaturated in seawater compared to their atmospheric burden. It was also observed that near the equator (10°N to 15°S) seawater concentrations of ethylene and

**Table 15 Estimates of NMHC Emissions from the United States and the World**

| Source Type | United States 1988 (Tg/yr) | World (Tg/yr) |
|---|---|---|
| **Anthropogenic** | | |
| Transportation | 6.1 | 22 |
| Stationary source fuel combustion | 0.9 | 4 |
| Industrial precesses including natural gas production | 8.5 | 17 |
| Forest fires/open burning/incineration | 1.4 | 45 |
| Organic solvents | 2.4 | 15 |
| *Subtotal* | 19.3 | 103 |
| **Natural** | | |
| Oceans | | |
| Light hydrocarbons | — | 5–10 |
| $C_9$–$C_{28}$ *n*-alkanes | — | 1–26 |
| Dimethyl sulfide | — | 20–40 |
| Microbial production | 1 | 6 |
| Isoprene | 7–20 | 350–450* |
| Terpenes and other biogenics | 25–50 | 480* |
| *Subtotal* | 33–71 | 862–1012 |
| **Total Emissions** | 52–90 | 965–1115 |

*Uncertanties in these numbers are of the order of a factor of three. Extrapolation from U.S. inventories would suggest that these sources may be overestimated.

*Source.* See text.

propylene were significantly elevated. Using these measurements and the model of Liss and Slater (1974), Ehhalt et al. (1986) calculate that a global oceanic source of NMHC of the order of 21 Tg/yr may exist, but 70% of the emissions would be in the tropics. This oceanic source is composed predominantly of ethylene and propylene (19 Tg/yr) with the remainder mainly ethane and propane. More recent measurements of Bonsang et al. (1988) from the intertropical Indian Ocean (15°N to 25°S; 40–56°E) show a much greater degree of supersaturation. When extrapolated to the globe, a marine NMHC source of nearly 52 TgC/yr, with about half accounted for by ethylene (15 TgC/yr) and propylene (11 TgC/yr), is calculated. These authors also report sizeable emissions of butene (4 TgC/yr), pentene (3.5 TgC/yr), and hexene (2.5 TgC/yr). The remainder ($\approx$16 TgC/yr) is due to $C_2$–$C_6$ alkanes. The possibility of a small oceanic source of acetylene has also been suggested (Kanakidou et al., 1988). More recently, Plass et al. (1989) observed both NMHC supersaturation in the Atlantic ocean (30°N to 30°S) as well as the elevated levels of ethylene and propylene near the equator. These seawater concentrations are much lower than those observed by Bonsang et al. (1988), but are very similar to those of Swinnerton and Lamontagne (1974). Table

**Table 16 Composition of NMHC Emissions from Biomass Burning**

| HC Group | Cerrado (Grasslands) | Selva (Tropical forest) |
|---|---|---|
| Alkanes | 20% (69% ethane, 21% propane) | 35% (43% ethane, 44% propane) |
| Alkenes | 44% (45% ethylene, 28% propylene) | 46% (59% ethylene, 25% propylene) |
| Alkynes | 10% | 1% |
| Aromatics | 13% (64% benzene, 25% toluene) | 13% (58% benzene, 29% toluene) |
| Oxygenated | 10% (52% furan, 36% 2-methylfuran) | 5% (72% furan, 19% 2-methylfuran) |

*Note.* Total NMHC emissions $\approx$ 1.1% of $CO_2$ emissions (on a carbon basis) $\approx$ 37 TgC/yr.

*Source.* Greenberg et al. (1984).

17 summarizes these seawater measurements and provides a corresponding emission flux from the oceanic source. Extrapolation of this flux to global waters (Table 17) suggests that oceans may contribute only about 5 Tg/yr of NMHC, corresponding to a mean flux of $7 \times 10^8$ molecules $cm^{-2} s^{-1}$. One reason for the unusually high concentrations measured by Bonsang et al. (1988) may be due to their relative proximity to coastal waters of high productivity. If one assumes that areas of high emissions represent 5–10% of oceanic surface, it is likely that the global oceanic source is in the 5- to 10-Tg/yr range. Clearly, more seawater data are needed to reliably estimate the oceanic source.

Model runs based on atmospheric observations constrain the marine flux of NMHC to between 1 and $20 \times 10^{10}$ molecules $cm^{-2} s^{-1}$ (Donahue and Prinn, 1990). This is inconsistent with the estimates of atmospheric flux from air–sea exchange models and direct seawater measurements ($\approx 10^9$ molecules $cm^{-2} s^{-1}$). The northern hemispheric fluxes reported in Table 17 provide an upper limit equilibrium boundary layer mixing ratios of 15 ppt ethylene and 5 ppt propylene. These mixing ratios are much lower than those measured (Fig. 12). A great deal of this uncertainty must be attributed to the quality and paucity of the field data. There is the further possibility that in the case of ethane, propane, and ethylene large terrestrial sources impact the marine boundary layer. As we shall see later, oceans may only contribute about 10% to the global ethylene emissions. A larger oceanic emission is that of dimethyl sulfide, which is a sulfur containing hydrocarbon (Lovelock et al., 1972). Its marine source has been estimated to be in the vicinity of 20–40 TgC/yr (Andreae et al., 1985). In many ways this organic will behave in a manner similar to alkenes, since its lifetime is estimated to be only a few hours.

Very heavy NMHC ($C_9$–$C_{28}$ $n$-alkanes) have also been measured in remote marine environments in the air as well in seawater. Based on $C_9$–$C_{28}$ $n$-alkane measurements over the Atlantic and the Indian oceans, Eichmann et al. (1980) estimate that 26 TgC/yr of these very heavy NMHC may be emitted by the oceans. Duce et al. (1983), however, point out that other similar measurements show these heavy alkanes to be present at one to two orders of magnitude lower concentrations than those used by Eichmann et al. (1980). Clearly a global extrapolation of these measurements is unwarranted, and total emissions may be anywhere between 1 and 26 Tg/yr. It is evident that estimates of NMHC oceanic fluxes of both light and heavy NMHC are based on extremely limited and often conflicting measurements, and rely far more on extrapolations than on real knowledge of distributions and production mechanisms.

### 4.2.2. *Plant Emissions*

Present quantitative knowledge of plant emissions is subject to large uncertainties. Zimmerman (1979) has measured these emissions from a variety of vegetation, and these results have been summarized by Altshuller (1983).

**Table 17 The $C_2$–$C_4$ NMHC Seawater Concentrations and the Oceanic Emission Rates**

| Hydrocarbons | Seawater Concentrations[a] (nL/L) | | Emission Flux ($10^7$ molecules cm$^{-2}$ s$^{-1}$) | | Source (Tg/yr) | | |
|---|---|---|---|---|---|---|---|
| | NH | SH | NH | SH | NH | SH | Total |
| Ethane | 3.4 | 0.36 | 31.0 | 5.5 | 0.70 | 0.19 | 0.89 |
| Propane | 0.98 | 0.16 | 8.9 | 2.3 | 0.30 | 0.12 | 0.42 |
| Isobutane | 0.08 | 0.02* | 0.7 | 0.3* | 0.03 | 0.02 | 0.05 |
| n-Butane | 0.21 | 0.3* | 1.8 | 0.5* | 0.08 | 0.03 | 0.11 |
| Ethene | 4.3 | 1.4 | 40.0 | 18.0 | 0.85 | 0.57 | 1.42 |
| Propene | 2.2 | 0.64 | 21.0 | 8.5 | 0.67 | 0.41 | 1.08 |
| 1-Butene | 1.2 | 0.38 | 12.0 | 5.0 | 0.51 | 0.32 | 0.83 |
| Acetylene | 0.09 | 0.06* | 0.9 | 0.8* | 0.02 | 0.02 | 0.04 |

[a]30°N to 30°S.

*Several measurements are below the lower limit of detection.

*Source*. Adapted from Plass et al. (1989).

Some qualitative patterns are evident: (1) emissions are highest in warmer temperatures, (2) isoprene emissions are strongly effected by the availability of sunlight, and (3) the composition of NMHC is complex and very few mechanistic details are known. It is becoming increasingly clear that in addition to isoprene and monoterpenes, a variety of other more complex NMHC are also emitted.

Some attempts have been made to develop VOC emission inventories for the United States. The first of these efforts was made by Zimmerman (1979) who divided United States into four regions and, based on the emission rates measured from a variety of plant species, estimated the total emissions of isoprene and terpenes. Lamb et al. (1987) have further refined these studies by using available biomass and land use data and have devised algorithms for the available emission rates. Figure 18 shows the emission rates from deciduous and coniferous plants as a function of temperature. It is clear from

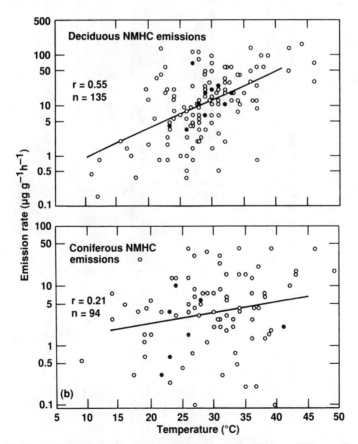

**Figure 18.** Total NMHC emissions from deciduous (*a*) and coniferous (*b*) trees. (From Lamb et al. (1987).)

these data that extremely large errors in the extrapolation of emission rates are involved. When these data are further limited to isoprene or α-pinene emitters, the uncertainties are nearly as large (Lamb et al., 1987). Overall isoprene and terpenes emission rates vary over a wide range of 1–50 μg g$^{-1}$ h$^{-1}$ and 0.1–10 μg g$^{-1}$ h$^{-1}$, respectively. These emission rates increase with temperature at a rate of 5–30% °C$^{-1}$.

A comparison of United States emissions inventories as derived by Zimmerman (1979) and Lamb et al. (1987) is shown in Fig. 19. Figure 19a clearly

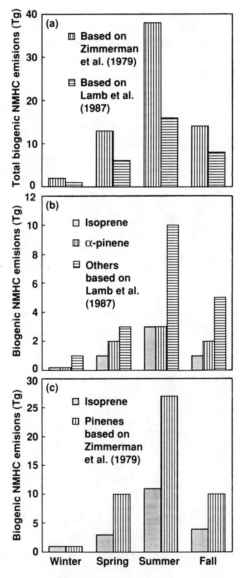

**Figure 19.** Biogenic NMHC emissions from the United States.

shows that total biogenic NMHC emissions are largest in summer but the 31 Tg/yr estimated by Lamb et al. is less than half the 70 Tg/yr rate calculated by Zimmerman. (The latter included Alaska and a soil emission factor and should be reduced by about 15% before it can be directly compared to the 31 Tg/yr estimated by Lamb et al.) Lamb et al. (1987) also find hydrocarbons other than isoprene and α-pinene to be major players (Fig. 19*b*), while Zimmerman attributes most of the emissions to isoprene and monoterpenes (Fig. 19*c*). Both inventories used similar emission rate algorithms. The differences in estimated emissions are primarily due to (1) differences in spatial and temporal resolution of the inventories, (2) methods of calculating seasonal variations in biomass, and (3) the land-use and vegetation distribution categories. The variability between these estimates illustrates the sensitivity of emissions estimates at this large scale to different techniques in vegetation mapping and seasonal biomass dynamics. Recent advances in mapping vegetation distributions and seasonal variations with satellite imagery provide a means of further increasing the accuracy of NMHC emissions estimates.

Several global NMHC emission rate estimates have been calculated by extrapolating from regional scales. Zimmerman (1979) determined a global annual emission rate of 350 Tg/yr of isoprene and 480 Tg/yr of monoterpenes by scaling up the United States emission rates to the global level using a net primary productivity ratio. Rasmussen and Khalil (1988) used previous empirical measurements of the isoprene emission rate from a few species as the basis for inventories of several ecosystem types, and extrapolated to a global estimate of 450 Tg/yr.

Part of the uncertainty with national and global source estimates is associated with emission rate factors and algorithms. Biochemical and physiological studies of NMHC release are beginning to provide a clearer understanding of the mechanisms controlling NMHC emissions (Monson et al., 1991; Sharkey et al., 1991). These studies are providing the basis of mechanistic emission rate algorithms which can increase the accuracy of NMHC emission rate estimates (Guenther et al., 1991). In addition, major uncertainties in emission estimates are due to extrapolations of specific emission factors to vegetation species that have not been previously measured. This error can be alleviated through ambient measurement programs which use hydrocarbon ratios to initialize emissions, and physiological studies to extrapolate seasonally.

## 4.3. Source of Individual NMHC Species

In this section we present our best estimates of the sources of individual hydrocarbon species, recognizing that large uncertainties currently exist. In the United States fairly detailed emissions inventories of the anthropogenic sources have been developed for each of the hydrocarbon species (Wagner et al., 1986; D. Zimmerman et al., 1988; Middleton and Stockwell, 1990; USEPA, 1990; NAPAP, 1990). Global anthropogenic sources are largely based on extrapolations of fuel usage and principally rely on data collected in the industrial world (NAS, 1976; Ehhalt et al., 1986; USEPA, 1990; Middleton and Stockwell, 1990). The global biomass burning source of NMHC is determined from field and laboratory studies of Greenberg et al. (1984), Hegg et al. (1990), and Lobert et al. (1991). Oceanic sources are calculated from a variety of field studies performed over the last two decades (Swinnerton and Lamontagne, 1974; Andreae et al., 1985; Bonsang et al., 1988; Plass et al., 1989). Biogenic plant emissions are estimated from NAS (1976), Zimmerman et al. (1978), Zimmerman (1979), Altshuller (1983), Duce et al. (1983), Lamb et al. (1987), and Rasmussen and Khalil (1988).

Table 18 uses available data on sources and sinks of NMHC to provide a knowledge of the source term and the nature of sources. The best estimate of the global ethane source is 10–15 Tg/yr and can be derived from its atmospheric distribution and the assumption that OH reaction is the principle removal mechanism. The uncertainty in this range is largely attributable to the uncertainty in global OH estimates. The biomass burning source of ethane is estimated to be 4–5 Tg/yr. Ethane source from oil combustion and natural gas emissions is estimated to be 2–3 Tg/yr. Oceanic source of nearly 1 Tg/yr can also be calculated. Direct biogenic emissions from plants and vegetation cannot be presently calculated but may be significant.

Recently detailed global emission inventory for ethylene has been devised (Sawada and Tatsuka, 1986). A 35-Tg/yr (range 18–45 Tg/yr) source of ethylene, which is 74% natural and 26% anthropogenic, is estimated (Table 19). Assuming a global mean atmospheric residence time of about 2 days, we calculate a global mean mixing ratio of 48 ppt. This is in good agreement with many studies of the atmospheric composition of ethylene (e.g., Greenberg et al., 1990). Only about 10% of this abundance (2–3 Tg/yr) is attributable to the oceanic source. Disagreements in the marine budgets of ethylene (Donahue and Prinn, 1990) probably arise from the incorrect assumption of a dominant marine source unaffected by terrestrial influences.

Acetylene is a near exclusive product of combustion, although a small oceanic source (<0.1 Tg/yr) has been identified. A biomass burning source of 2–4 Tg/yr and auto emission source of 1–2 Tg/yr is estimated. This 3- to 6-Tg/yr source is generally consistent with the measured atmospheric abundance of acetylene. There has been a great deal of work with dimethyl sulfide because of interest in the sulfur cycle, and its source is estimated to be 25–50 Tg(S)/yr or 20–40 TgC/yr. Atmospheric measurements of propane would suggest that a source somewhat larger than that of ethane is present (15–20 Tg/yr). Propane is not emitted from either the combustion or evaporation of fossil fuel (NAS, 1976; Sigsby et al., 1987). Biomass burning, natural gas emissions, and oceanic emissions are thought to be the three major sources for propane, with estimated emissions of 1.5–2.5 Tg/yr, 1–2 Tg/yr, and 0.5–1.0 Tg/yr, respectively. These sources are clearly inadequate to account for the atmospheric burden of propane and sizeable unknown (biogenic?) sources are clearly present. Combustion is a major source of propylene. Emissions from automobiles and biomass burning are estimated to be about 1.5 and 5 Tg/yr, respectively. The oceanic source of propylene is calculated to be 2–3

**Table 18 Global Sources of Individual NMHC Species**

| Hydrocarbons | World (Tg/Yr) | United States (Tg/Yr) | Major Sources |
|---|---|---|---|
| Ethane | 10–15 | 1–2 | Natural gas emissions, biomass burning, oceans, grasses and other vegetation |
| Ethylene | 20–45 | 1–3 | Fuel combustion, biomass burning, oceans, terrestrial ecosystems |
| Acetylene | 3–6 | 0.5 | Autoemissions, biomass burning |
| Dimethyl sulfide | 20–40 | 0.01–0.1 | Oceanic, terrestrial biogenic emissions |
| Propane | 15–20 | 0.3–1 | Natural gas, biomass burning, oceans, grasses and other vegetation |
| Propylene | 7–12 | 0.5 | Fuel combustion, biomass burning, oceans |
| n-Butane | 1–2 | 0.2 | Fuel combustion, natural gas, biomass burning, oceans |
| Isobutane | 1–2 | 0.3 | Fuel combustion, natural gas, biomass burning, oceans |
| Butenes | 2–3 | 0.2 | Fuel combustion, biomass burning, oceans |
| n-Pentane | 1–2 | 0.3 | Fuel combustion, natural gas |
| Isopentane | 2–3 | 0.6 | Fuel combustion, natural gas |
| Benzene | 4–5 | 0.6 | Fuel combustion, biomass burning |
| Toluene | 4–5 | 1.3 | Fuel combustion, solvents, biomass burning |
| Xylenes (o, m, p) | 2–3 | 0.7 | Fuel combustion, solvents, biomass burning |
| Isoprene | 350–450 | 7–20 | Forest/plant emissions |
| Monoterpenes and other biogenies | 400–500 | 25–50 | Forest/plant emissions |

Source. See text.

**Table 19  Global Sources of Ethylene**

| | Source | Ethylene Emission | | |
| | Material | Range | Mean | |
| Ethylene Sources | ($10^3$ Tg/yr) | (Tg/yr) | (Tg/yr) | % |
|---|---|---|---|---|
| Natural | | | | |
| Terrestrial ecosystem | | 16.6–29.0 | 23.3 | 65.8 |
| Aquatic system | | | 2.9 | 8.2 |
| *Subtotal* | | | 26.2 | 74.0 |
| Anthropogenic | | | | |
| Coal combustion | 2.67 | | 0.42 | 1.2 |
| Fuel oil combustion | 2.49 | | 1.54 | 4.3 |
| Gas combustion | 1.06 | | 0.00 | 0.0 |
| Refuse incineration | 0.50 | | 0.10 | 0.3 |
| Leakage from industry | 0.03 | | 0.03 | 0.1 |
| Biomass burning | 7.90 | 1.8–14.8 | 7.10 | 20.1 |
| *Subtotal* | 14.65 | | 9.19 | 26.0 |
| Total | | 18.4–44.6 | 35.4 | 100.0 |

*Source.* Adapted from Sawada and Tatsuka (1986).

Tg/yr. Butanes and pentanes are nearly exclusive products of combustion, natural gas emissions with perhaps minor oceanic sources. A noteworthy possibility is a significant oceanic source of highly reactive butenes ($\approx 1$ Tg/yr). Benzene and toluene are also products of fuel combustion with a major sources from biomass burning of about 2–3 Tg/yr for benzene and 1–1.5 Tg/yr for toluene. Use of toluene and xylenes as industrial solvents is also a significant source. Inventories of isoprene and terpenes have been discussed in the previous sections. It bears repeating that in many cases these estimates are educated guesses and the numbers presented have large but unquantifiable uncertainties. Although key NMHC species are included in Table 18, these do not necessarily add up to the total NMHC source, in part because many other volatile species are also emitted (Middleton and Stockwell, 1990; NAPAP, 1990).

## 5.  SUMMARY AND FUTURE NEEDS

NMHC are ubiquitous components of the atmosphere with large man-made and natural sources. They are intricately involved with many important aspects of tropospheric chemistry and are themselves key tools in validating models of the global troposphere. There are major uncertainties in our present knowledge of the distributions, sources, and sinks of these species and future research efforts are required to bridge important gaps.

One of the foremost problem stems from the lack of reliable and complete measurements. In all compartments of the atmosphere, some 15–40% of the NMHC burden is made up of unidentified species. Many of the "known" species are probably incorrectly identified since in a large number of cases identifications are based on GC retention times only. Problems are particularly challenging in remote environments where concentrations are extremely low and in studies of biogenic emission contributions to urban air where a huge range of compounds and concentrations are present. Thus, there is a clear need for basic research to identify the unknown species, confirm the identity of known species, and perform ambient intercomparison studies of complex NMHC mixtures to ensure proper separation and quantitation.

In the global atmosphere, light alkenes $(C_2-C_6)$ may play an important role but there are major uncertainties in the atmospheric and oceanic data, with the result that no firm conclusions can be drawn. Accurate oceanic measurements are hampered by the fact that the FID, the most commonly used detector, is probably pushed to the limit of its sensitivity to accomplish this task. There is a continuing need for more sensitive and specific detectors, and for fast continuous instruments that can be used to measure fluxes of specific classes of hydrocarbons. While we recognize that our knowledge of NMHC sources is quite poor, it is imperative that at least the atmospheric measurements be placed on a reliable footing before quantitatively meaningful budgets can be derived.

Small amounts of highly reactive species can have a significant impact on the chemistry of the remote troposphere. Very little effort has gone into the detection of heavier than $C_5$ NMHC species, which are also expected to be highly reactive. Enormous atmospheric inhomogeneities exist for such species (e.g., alkenes), requiring the collection of large bodies of statistically meaningful data. In many instances disagreement between models and measurements are due to the presence of such inhomogeneities and the inability of models to deal with these. While the complexity of natural emissions is evident, a great deal more research is needed to define and understand the principle mechanisms of the formation and release of these species. Although significant progress has been made during the last two decades in determining the burden and fate of NMHC in the atmosphere, the efforts to put our knowledge on a solid quantitative basis is only beginning.

### Acknowledgment

*This research is supported by the NASA Global Tropospheric Experiment and the National Center for Atmospheric Research (NCAR). NCAR is sponsored by the National Science Foundation. We are thankful to many colleagues for providing us with preprints of their unpublished work.*

## REFERENCES

Aiken, A. C., Gallagher, C., Spicer, C., and Holdren, M. (1987). "Measurement of methane and other light hydrocarbons in the troposphere and lower stratosphere." *J. Geophys. Res.* **92**, 3135–3138.

Altshuller, A. P. (1983). "Natural volatile organic substances and their effect on air quality in the United States." *Atmos. Environ.* **17**, 2131–2165.

Altshuller, A. P., and Bufalini, J. J. (1971). "Photochemical aspects of air pollution: A review." *Environ. Sci. Technol.* **5**, 36–64.

Andreae, M. O., Ferek, R. J., Bermond, F., Byrd, K. P., Engström, R. T., Hardin, S., Houmere, P. D., LeMarrec, F., and Raemdonk, H. (1985). "Dimethyl sulfide in the marine atmosphere." *J. Geophys. Res.* **90**, 12,891–12,900.

Atkinson, R. (1990). "Gas-phase tropospheric chemistry of organic compounds: A review." *Atmos. Environ.* **24A**, 1–41.

Blake, D. R., and Rowland, F. S. (1986a). "Global atmospheric concentration and source strength of ethane." *Nature (London)* **321**, 231–233.

Blake, D. R., and Rowland, F. S. (1986b). "World-wide increase in tropospheric methane." *J. Atmos. Chem.* **4**, 43–46.

Bonsang, B., and Lambert, G. (1985). "Nonmethane hydrocarbons in an oceanic atmosphere." *J. Atmos. Chem.* **2**, 257–271.

Bonsang, B., Kanakidou, M., Lambert, G., and Monfray, P. (1988). "The marine source of $C_2$–$C_5$ aliphatic hydrocarbons." *J. Atmos. Chem.* **6**, 3–20.

Bufalini, J. J., and Arnts, R. (1981). *Atmospheric Biogenic Hydrocarbons.* Ann Arbor Sci. Publ., Ann Arbor, MI.

Calvert, J. G. (1976). "Hydrocarbon involvement in photochemical smog formation in Los Angeles atmosphere." *Environ. Sci. Technol.* **10**, 256–262.

Calvert, J. G., and Madronich, S. (1987). "Theoretical study of the initial products of the atmospheric oxidation of hydrocarbons." *J. Geophys. Res.* **92**, 2211–2220.

Carsey, T. P. (1987). "Non-methane hydrocarbons in the remote marine troposphere: Results of an intercomparison exercise." *EOS, Trans. Am. Geophys. Union* **68**, 1220.

Chameides, W., Lindsay, R. W., Richardson, J., and Kiang, C. S. (1988). "The role of biogenic hydrocarbons in urban photochemical smog: Atlanta as a case study." *Science* **241**, 1473–1475.

Chatfield, R., and Crutzen, P. (1987). "Sulfur dioxide in remote oceanic air: cloud transport of reactive precursors." *J. Geophys. Res.* **89**, 7111–7132.

Ciccioli, P., Brancaleoni, E., Possanzini, M., Brachetti, A., and DiPalo, C. (1984). "Sampling, identification and quantitative determination of biogenic and anthropogenic hydrocarbons in forested areas." In *Proceedings of the Physico-Chemical Behavior of Atmospheric Pollutants, Varese, Italy.*

Colbeck, I., and Harrison, R. M. (1985). "The concentrations of specific $C_2$–$C_6$ hydrocarbons in the air of NW England." *Atmos. Environ.* **19**, 1899–1904.

Cronn, D. R., and Nutmagul, W. (1982). "Analysis of atmospheric hydrocarbons during winter MONEX." *Tellus* **34**, 159–165.

Cronn, D. R., and Robinson, E. (1979). "Tropospheric and lower stratospheric vertical profiles of ethane and acetylene." *Geophys. Res. Lett.* **6**, 641–644.

Crutzen, P. (1979). "The role of NO and $NO_2$ in the chemistry of the troposphere and stratosphere." *Annu. Rev. Earth Planet. Sci.* **7**, 443–472.

Demerjian, K. L., Kerr, J. A., and Calvert, J. G. (1974). "The mechanism of photochemical smog formation." *Adv. Environ. Sci. Technol.* **4**, 1–262.

Dimitriades, B. (1981). "The role of natural organics in photochemical air pollution: Issues and research needs." *J. Air Pollut. Control Assoc.* **31**, 229–235.

Donahue, N. M., and Prinn, R. G. (1990). "Non-methane hydrocarbon chemistry in the remote marine boundary layer." *J. Geophys. Res.* **95**, 18,387–18,411.

Duce, R. A., Mohnen, V. A., Zimmerman, P. R., Grosjean, D., Cautreels, W., Chatfield, R., Jaenicke, R., Ogren, J. A., Pellizzari, E. D., and Wallace, G. T. (1983). "Organic material in the global troposphere." *Rev. Geophys.* **21**, 921–952.

Ehhalt, D. H., Rudolph, J., Meixner, F., and Schmidt, U. (1985). "Measurements of selected $C_2$–$C_5$ hydrocarbons in the background troposphere: Vertical and latitudinal variations." *J. Atmos. Chem.* **3**, 29–52.

Ehhalt, D. H., Rudolph, J., and Schmidt, U. (1986). "On the importance of light hydrocarbons in multiphase atmospheric systems." In W. Jaeschke, ed., *Chemistry of Multiphase Atmospheric Systems*. Springer-Verlag, Berlin, pp. 321–350.

Ehhalt, D. H., Schmidt, U., Zander, R., Demoulin, P. H., and Rinsland, C. P. (1991). "Seasonal cycle and secular trend of the total and tropospheric column abundance of ethane above the Jungfraujoch." *J. Geophys. Res.* (96, 4985–4994.)

Eichmann, R., Ketseridis, G., Schebeske, G., Jaenicke, R., Hahn, J., Warneck, P., and Junge, C. (1980). "N-alkane studies in the troposphere. II. Gas and particular concentrations in Indian Ocean air." *Atmos. Environ.* **14**, 695–703.

Greenberg, J. P., and Zimmerman, P. R. (1984). "Nonmethane hydrocarbons in remote trophsphere." *J. Geophys. Res.* **89**, 4767–4778.

Greenberg, J. P., Zimmerman, P. R., Heidt, L., and Pollock, W. (1984). "Hydrocarbon and carbon monoxide emissions from biomass burning in Brazil." *J. Geophys. Res.* **89**, 1350–1354.

Greenberg, J. P., Zimmerman, P. R., and Chatfield, R. B. (1985). "Hydrocarbons and carbon monoxide in African savannah air." *Geophys. Res. Lett.* **12**, 112–116.

Greenberg, J. P., Zimmerman, P. R., and Haagenson, P. (1990). "Tropospheric hydrocarbon and CO profiles over the U.S. West Coast and Alaska." *J. Geophys. Res.* 14,015–14,026.

Greenberg, J. P., Zimmerman, P. R., Pollock, W. F., Lueb, R. A., and Heidt, L. (1991). "Diurnal variability of atmospheric methane, non-methane hydrocarbons and carbon monoxide at Mauna Loa." *J. Geophys. Res.* (submitted for publication).

Guenther, A., Monson, R., and Fall, R. (1991). "Isoprene and monoterpene emission rate variability: Observations with Eucalyptus and emission rate algorithm development." *J. Geophys. Res.* **96**, 10,799–10,808.

Haagen-Smit, A. J. (1952). "Chemistry and physiology of Los Angeles smog." *Ind. Eng. Chem.* **44**, 1342.

Hahn, J., and Matuska, P. (1990). "Nonmethane hydrocarbon measurements at two mountain stations in the Bavarian Alps." In *Physico-Chemical Behavior of Atmospheric Pollutants, CIEMAT Centre, Madrid, Spain, March, 1990.*

Hampton, C. V., Pierson, W. R., Harvey, T. M., Updegrove, W. S., and Marano, R. S. (1982). "Hydrocarbon gases emitted from vehicles on the road. I. A qualitative gas chromatography/mass spectrometry survey." *Environ. Sci. Technol.* **16**, 287–298.

Hampton, C. V., Pierson, W. R., Schuetzel, D., and Harvey, T. M. (1983). "Hydrocarbon gases emitted from vehicles on the road. 2. Determination of emission rates from diesel and spark-ignition vehicles." *Environ. Sci. Technol.* **17**, 699–708.

Harborne, J. B. (1988). *Introduction to Ecological Biochemistry*. Academic Press, San Diego, CA, 3rd ed.

Hegg, D. A., Radke, L. F., Hobbs, P. W., Rasmussen, R. A., and Riggan, P. J. (1990). "Emissions of some trace gases from biomass fires." *J. Geophys. Res.* **95**, 5669–5675.

Holdren, M. W., Westberg, H. H., and Hill, H. H. (1979). *Analytical Methodology for the Identification and Quantitation of Vapor Phase Organic Pollutants*. CRC-APRAC, Coordinating Research Council Air Pollution Research Advisory Committee, New York, Interim Report, Project CAPA-11-71.

Hov, φ., Penkett, S. A., Isaksen, I. S. A., and Semb, A. (1984). "Organic gases in the Norwegian Arctic." *Geophys. Res. Lett.* **11**, 425–428.

Hov, φ., Schmidbauer, N., and Oehme, M. (1990). "$C_2–C_5$ hydrocarbons in rural South Norway." *Atmos. Environ.* (in press).

Isidorov, V. A., Zenkench, I. G., and Iofe, B. V. (1985). "Volatile organic compounds in the atmosphere of forests." *Atmos. Environ.* **19**, 1–8.

Kanakidou, M., Bonsang, B., Le Roulley, J. C., Lambert, G., Martin, D., and Sennequier, G. (1988). "Marine source of atmospheric acetylene." *Nature (London)* **333**, 51–52.

Kanakidou, M., Bonsang, B., and Lambert, G. (1989). "Light hydrocarbons, vertical profiles and fluxes in a French rural area." *Atmos. Environ.* **23**, 921–927.

Kanakidou, M., Singh, H., Valentin, K., and Crutzen, P. (1991). "A 2-D study of ethane and propane oxidation in the troposphere." *J. Geophys. Res.* (in press).

Kasting, J. F., and Singh, H. B. (1986). "Nonmethane hydrocarbons in the troposphere: Impact on the odd hydrogen and odd nitrogen chemistry." *J. Geophys. Res.* **91**, 13,239–13,256.

Kritz, M. A., Rulley, J. L., and Danielsen, E. F. (1990). "The China Clipper—fast advective transport of radon-rich air from Asian boundary layer to the upper troposphere near California." *Tellus* **42B**, 46–61.

Lamb, B., Guenther, A., Gay, D., and Westberg, H. (1987). "A national inventory of biogenic hydrocarbon emissions." *Atmos. Environ.* **21**, 1695–1705.

Lamontagne, R. A., Swinnerton, J. W., and Linnenbom, V. J. (1974). "$C_1–C_4$ hydrocarbons in the North and South Pacific." *Tellus*, 71–77.

Liss, P. S., and Slater, P. G. (1974). "Flux of gases across the air-sea interface." *Nature (London)* **L47**, 181–184.

Lobert, J. M., Scharffe, D. H., Kuhlbusch, T. A., Seuwen, R., and Crutzen, P. J. (1991). "Experimental evaluation of biomass burning emissions: Nitrogen and carbon containing compounds." In J. S. Levine, ed., *Global Biomass Burning*. MIT Press, Cambridge, MA.

Logan, J. A. (1985). "Tropospheric ozone: Seasonal behavior trends and anthropogenic influences." *J. Geophys. Res.* **91**, 7875–7881.

Lonneman, W. A., Seila, R. L., and Meeks, S. A. (1986). "Non-methane organic composition in the Lincoln tunnel." *Environ. Sci. Technol.* **20**, 790–796.

Lovelock, J. E., Maggs, R. J., and Rasmussen, R. P. (1972). "Atmospheric dimethyl sulfide and the natural sulfur cycle." *Nature (London)* **237**, 452–453.

Matuska, P., Koval, M., and Seiler, W. (1986). "A high resolution GC-analysis method for determination of $C_2$–$C_{10}$ hydrocarbons in air samples." *J. High Resolut. Chromatogr. Chromatogr. Commun.* **9**, 577–583.

McKeen, S. A., Trainer, M., Hsie, E. Y., Tallamraju, R. K., and Liu, S. C. (1990). "On the indirect determination of atmospheric OH radical concentrations from reactive hydrocarbon measurements." *J. Geophys. Res.* **95**, 7493–7500.

Middleton, P., and Stockwell, W. (1990). "Aggregation and analysis of volatile organic compound emissions for regional modeling." *Atmos. Environ.* **24A**, 1107–1133.

Monson, R., Guenther, A., and Fall, R. (1991). "Physiological reality in relation to ecosystem level inventories of isoprene emissions." In T. Sharkey, E. Holland, and H. Mooney, eds., *Trace Gas Emissions from Plants*. Academic Press, San Diego, CA.

National Academy of Sciences (NAS) (1976). *Vapor-Phase Organic Pollutants: Volatile Hydrocarbons and Their Oxidation Products*. National Academy of Sciences Press, Washington, D.C.

National Acid Precipitation Assessment Program (NAPAP). (1990). *NAPAP State of Science and Technology*. U.S. Government Printing Office, Washington, DC, Vol. 1.

Nelson, P. F., Quigley, S. M., and Smith, M. Y. (1983). "Sources of atmospheric hydrocarbons in Sydney: a quantitative determination using a source reconciliation technique." *Atmos. Environ.* **17**, 439–449.

Niki, H., Daby, E. E., and Weinstock, B. (1972). "Mechanisms of smog reactions." *Adv. Chem. Ser.* **113**, 116–176.

Nutmagul, W., and Cronn, D. P. (1985). "Determination of Selected aromatic hydrocarbons at remote continental and oceanic locations using photoionization/flame-ionization detection." *J. Atmos. Chem.* **2**, 415–433.

Plass, C., Johner, F. J., Koppmann, R., and Rudolph, J. (1989). "The latitudinal distribution of NMHC in the Atlantic and their fluxes into the atmosphere." *Physico-Chemical Behavior of Atmospheric Pollutants, Varnese, Italy, September, 1989*.

Rasmussen, R. A. (1970). "Isoprene: Identified as a forest type emission to the atmosphere." *Environ. Sci. Technol.* **4**, 669–670.

Rasmussen, R. A. (1972). "What do the hydrocarbons from trees contribute to air pollution?" *J. Air Pollut. Control Assoc.* **22**, 537–543.

Rasmussen, R. A., and Khalil, M. A. K. (1983a). "Altitudinal and temporal variations of hydrocarbons and other gaseous tracers of arctic haze." *Geophys. Res. Lett.* **10**, 144–147.

Rasmussen, R. A., and Khalil, M. A. K. (1983b). "Atmospheric benzene and toluene." *Geophys. Res. Lett.* **10**, 1096–1099.

Rasmussen, R. A., and Khalil, M. A. K. (1988). "Isoprene over the Amazon Basin." *J. Geophys. Res.* 1,417–1,421.

Rinsland, C. P., Levine, J. S., and Miles, T. (1985). "Free tropospheric carbon monoxide concentrations in 1950 and 1951 deduced from infrared total column amount measurements." *Nature (London)* **318**, 250–254.

Rinsland, C. P., Zander, R., Farmer, C. B., Norton, R. H., and Russell, J. M., III. (1987). "Concentrations of ethane ($C_2H_6$) in the lower stratosphere and upper troposphere and acetylene ($C_2H_2$) in the upper troposphere deduced from atmospheric trace molecule spectroscopy/Spacelab 3 spectra." *J. Geophys. Res.* **92**, 11,951–11,964.

Ripperton, L. A., Jeffries, H., and Worth, J. J. B. (1971). "Natural synthesis of ozone in the troposphere." *Environ. Sci. Technol.* **5**, 246–248.

Roberts, J. M., Fehsenfeld, F. C., Albritton, D. L., and Sievers, F. C. (1983). "Measurement of monoterpene hydrocarbon at Niwot Ridge, Colorado." *J. Geophys. Res.* **88**, 10,667–10,678.

Roberts, J. M., Fehsenfeld, F. C., Liu, S. C., Bollinger, M. J., Hahn, C., Albritton, D. L., and Sievers, R. E. (1984). "Measurements of aromatic hydrocarbon ratios and $NO_x$ concentrations in the rural troposphere: Observation of air mass photochemical aging and $NO_x$ removal." *Atmos. Environ.* **181**, 2421–2432.

Robinson, E. (1978). "Hydrocarbons in the atmosphere." *Pure Appl. Geophys.* **116**, 372–384.

Robinson, E., and Robbins, R. (1969). *Source, Abundance and Fate of Gaseous Atmospheric Pollutants Supplement.* Stanford Research Institute, Menlo Park, CA.

Roemer, M. G. M., and Hout, K. D. V. D. (1988). *Nonmethane Hydrocarbons in the Troposphere and Their Impact on Global Ozone Concentrations* (in Dutch). TNO Rapport, The Netherlands, R88/315.

Rudolph, J. (1988). "Two-dimensional distribution of light hydrocarbons: results from the STRATOZ III experiment." *J. Geophys. Res.* **93**, 8367–8377.

Rudolph, J., and Ehhalt, D. H. (1981). "Measurements of $C_2$–$C_5$ hydrocarbons over the North Atlantic." *J. Geophys. Res.* **86**, 11,959–11,964.

Rudolph, J., and Johnen, F. J. (1990). "Measurements of light atmospheric hydrocarbons over the Atlantic in regions of low biological productivity." *J. Geophys. Res.* **95**, 20,583–20,592).

Rudolph, J., and Koppmann, R. (1989). "Sources and atmospheric distribution of light hydrocarbons." In *Proceedings of the 28th Liege International Astrophysical Colloquium, June, 1989.*

Rudolph, J., Ehhalt, D. H., and Tonnissen, A. (1981). "Vertical profiles of ethane and propane." *J. Geophys. Res.* **86**, 7267–7272.

Rudolph, J., Ehhalt, D. H., Schmidt, V., and Khedim, A. (1982). "Vertical distribution of some $C_2$–$C_5$ hydrocarbons in the nonurban troposphere." In *Proceedings of the Second Symposium on Composition of the Nonurban Troposphere, Williamsburg, VA, May, 1982.*

Rudolph, J., Ehhalt, D. H., and Khedim, A. (1984a). "Vertical profiles of acetylene in the troposphere and stratosphere." *J. Atmos. Chem.* **2**, 117–124.

Rudolph, J., Jobsen, C., Khedim, A., and Johnen, F. (1984b). "Measurements of the distribution of light hydrocarbons and halocarbons over the Atlantic." In *Physico-Chemical Behavior of Atmospheric Pollutants, Third European Symposium, Varese, Italy, April, 1984.*

Rudolph, J., Khedim, A., and Wagenbach, D. (1989). "The seasonal variations of light nonmethane hydrocarbons in the antarctic troposphere." *J. Geophys. Res.* **94**, 13,039–13,044.

Sanadze, G. A., and Kalandadze, A. N. (1966). "Light and temperature curves on the evolution of isoprene." *Sov. Plant. Physiol. (Engl. Transl.)* **13**, 411–413.

Sawada, S., and Tatsuka, T. (1986). "Natural and anthropogenic sources and fate of atmospheric ethylene." *Atmos. Environ.* **20**, 821–832.

Seila, R. L. (1984). "Atmospheric volatile hydrocarbon composition at five remote sites in Northwestern North Carolina." In V. Aneja, ed., *Environmental Impact of Natural Emissions*. pp. 125–140.

Seila, R. L., Lonneman, W., and Meeks, S. (1989). "Determination of $C_2$ to $C_{12}$ ambient air hydrocarbons in 39 U.S. cities, from 1984 through 1986. *U.S. Environ. Prot. Agency, Off. Res. Dev. [Rep.]* **EPA/600/S3-89/058**.

Seiler, W., and Crutzen, P. J. (1980). "Estimates of gross and net fluxes of carbon between the biosphere and the atmosphere from biomass burning." *Clim. Change* **2**, 207–247.

Sharkey, T., Loretta, F., and Delwiche, C. (1991). "The biochemistry of isoprene emissions from leaves during photosynthesis." In T. Sharkey, E. Holland, and H. Mooney, eds., *Trace Gas Emissions from Plants*. Academic Press, San Diego, CA.

Shaw, R. W., Jr., Crittenden, A. L., Stevens, R. K., Cronn, D. R., and Titov, V. S. (1983). "Ambient concentrations of hydrocarbons from conifers in atmospheric gases and aerosol particles measured in Soviet Georgia." *Environ. Sci. Technol.* **17**, 389–395.

Sigsby, J. E., Tejada, S., and Ray, W. (1987). "Volatile organic compound emissions from 46 in-use passenger cars." *Environ. Sci. Technol.* **21**, 466–475.

Singh, H. B. (1977). "Atmospheric halocarbons: Evidence in favor of reduced average hydroxyl radical concentrations." *Geophys. Res. Lett.* **4**, 101–104.

Singh, H. B. (1980). "Guidance for the collection and use of ambient hydrocarbon species data in development of ozone control strategies." *U.S. Environ. Prot. Agency, Off. Air Qual. Plann. Stand. [Tech. Rep.]* **EPA 450/14-80-008**.

Singh, H. B. (1987). "Reactive nitrogen in the troposphere." *Environ. Sci. Technol.* **21**, 320–327.

Singh, H. B., and Hanst, P. L. (1981). "Peroxyacetyl nitrate (PAN) in the unpolluted atmosphere: An important reservoir for nitrogen oxides." *Geophys. Res. Lett.* **8**, 941–944.

Singh, H. B., and Kasting, J. F. (1988). "Chlorine-hydrocarbon photochemistry in the marine troposphere and lower stratosphere." *J. Atmos. Chem.* **7**, 261–285.

Singh, H. B., and Salas, L. J. (1982). "Measurement of selected light hydrocarbons over the Pacific Ocean: Latitudinal and seasonal variations." *Geophys. Res. Lett.* **9**, 842–845.

Singh, H. B., Martinez, J. R., Hendry, D. J., Jaffe, R. J., and Johnson, W. B. (1981). "Assessment of the oxidant forming potential of light saturated hydrocarbons." *Environ. Sci. Technol.* **15**, 113–119.

Singh, H. B., Salas, L. J., Cantrell, B. K., and Redmond, R. M. (1985). "Distribution of aromatic hydrocarbons in the ambient air." *Atmos. Environ.* **19**, 1911–1919.

Singh, H. B., Viezee, W., and Salas, L. J. (1988). "Measurements of selected $C_2$–

$C_5$ hydrocarbons in the troposphere: Latitudinal, vertical and temporal variations." *J. Geophys. Res.* **93**, 15861–15878.

Swinnerton, J. W., and Lamontagne, R. A. (1974). "Oceanic distribution of low molecular weight hydrocarbons, baseline measurements." *Environ. Sci. Technol.* **8**, 657–663.

Swinnerton, J. W., and Linnenbom, V. J. (1967). "Determination of the $C_1$–$C_4$ hydrocarbons in sea water by gas chromatography." *J. Gas Chromatogr.* **5**, 570–573.

Tille, K. J. W., and Bachmann, K. (1986). "Natural and anthropogenic hydrocarbons over the Atlantic Ocean latitudinal distributions of mixing ratios and sources." In *Proceedings of the Fourth European Symposium, Stresa, Italy, September, 1986.*

Tille, K. J. W., Saelsberg, M., and Bachmann, K. (1985). "Airborne measurements of nonmethane hydrocarbons over western Europe: Vertical distributions, seasonal cycles of mixing ratios and source strengths." *Atmos. Environ.* **19**, 1751–1760.

Tingey, D. (1981). "The effect of environmental factors on the emission of biogenic hydrocarbons from live oak and slash pine." In J. Bufalini and R. Arnts, eds., *Atmospheric Biogenic Hydrocarbons.* Ann Arbor Sci. Publ., Ann Arbor, MI, pp. 53–72.

Tingey, D., Turner, D., and Weber, U. (1991). "Factors controlling the emissions of monoterpinesh) and other volatile organics." In T. Sharkey, E. Holland, and H. Mooney, eds., *Trace Gas Emissions from Plants.* Academic Press, San Diego, CA.

Trainer, M., Williams, E. J., Parrish, D. D., Buhr, M. P., Allwine, E. J., Westberg, H. H., Fehsenfeld, F. C., and Liu, S. C. (1987). "Models and observations of the impact of natural hydrocarbons on rural ozone: Modeling and observations." *Nature (London)* **329**, 705.

U.S. Department of Health, Education and Welfare (USDHEW) (1970). *Air Quality Criteria for Photochemical Oxidants.* U.S. Department of Health, Education, and Welfare, Washington, DC, NAPCA Publ. AP-63.

U.S. Environmental Protection Agency (USEPA) (1984). *Air Quality Criteria for Ozone and Other Photochemical Oxidants.* USEPA, Washington, DC, EPA-600/8-84-020A.

U.S. Environmental Protection Agency (USEPA) (1990). *National Air Pollutant Emission Estimates 1940–1988.* USEPA, Washington, DC, EPA-450/4-90-001.

Volz, A., and Kley, D. (1988). "Ozone measurements in the 19th Century: Evolution of the Mantsouris series." *Nature (London)* **332**, 240–242.

Volz, A., Ehhalt, D. H., and Derwent, R. G. (1981). "Seasonal and latitudinal variations of $^{14}CO$ and the tropospheric concentrations of OH radicals." *J. Geophys. Res.* **86**, 5163–5171.

Wagner, J. K., Walters, R. A., Maiocco, L. J., and Neal, D. R. (1986). "Development of the 1980 NAPAP emissions inventory." *U.S. Environ. Prot. Agency, Off. Res. Dev. [Rep.]* **EPA600/7-86-57a**.

Warneck, P. (1988). *Chemistry of the Natural Atmosphere.* Academic Press, San Diego, CA.

Went, F. W. (1960a). "Organic matter in the atmosphere and its possible relation to petroleum formation." *Proc. Natl. Acad. Sci. U.S.A.* **46**, 212–221.

Went, F. W. (1960b). "Blue haze in the atmosphere." *Nature (London)* **187**, 641–643.

Yokouchi, Y., Okaniwa, M., Ambe, Y., and Fuwa, K. (1983). "Seasonal variation of monoterpenes in the atmosphere of a pine forest." *Atmos. Environ.* **17**, 743–750.

Zimmerman, D., Tax, W., Smith, M., Demmy, J., and Battye, R. (1988). "Anthropogenic emissions data for the 1985 NAPAP inventory." *U.S. Environ. Prot. Agency, Off. Res. Dev. [Rep.]* **EPA-600/7-88-022**.

Zimmerman, P. R. (1979). "Testing of hydrocarbon emissions from vegetation, leaf litter and aquatic surfaces and development of a methodology for compiling biogenic emission inventories." *U.S. Environ. Prot. Agency, Off. Air Qual. Plann. Stand. [Tech. Rep.]* **EPA-450/4-79-004**.

Zimmerman, P. R., Chatfield, R. B., Fishman, J., Crutzen, P. J., and Hanst, P. L. (1978). "Estimates on the production of CO and $H_2$ from the oxidation of hydrocarbon emissions from vegetation." *Geophys. Res. Lett.* **5**, 679–682.

Zimmerman, P. R., Greenberg, J. P., and Westberg, C. (1988). "Measurement of atmospheric hydrocarbon and biogenic emission fluxes in the Amazon boundary laer." *J. Geophys. Res.* **93**, 1407–1416.

# 6

# THE CHEMISTRY OF ATMOSPHERIC HYDROGEN PEROXIDE IN SOUTHERN CALIFORNIA

*H. Sakugawa and I. R. Kaplan*

*Institute of Geophysics and Planetary Physics, University of California, Los Angeles, California*

## 1. INTRODUCTION

The earliest reported study on atmospheric hydrogen peroxide ($H_2O_2$) was conducted in the last century by Schöne (1874) in Russia, who measured the

Institute of Geophysics and Planetary Physics Contribution 3409.

*Gaseous Pollutants: Characterization and Cycling,* Edited by Jerome O. Nriagu.
ISBN 0-471-54898-7  © 1992 John Wiley & Sons, Inc.

concentration of $H_2O_2$ in rainwater in the range of 1.2–29 $\mu M$ using a colorimetric method. $H_2O_2$ concentration in rainwater (with a range of 10–25 $\mu M$) was also measured by Matsui (1949) in Japan. During the 1970s, Buffalini et al. (1972) and Kok et al. (1978b) first measured the concentration of gas phase $H_2O_2$ in Los Angeles using either colorimetry or chemiluminescence after gaseous $H_2O_2$ was stripped into aqueous solution from the air. Except for the above investigations, only limited studies of atmospheric $H_2O_2$ were conducted until the late 1970s, because the role of $H_2O_2$ in tropospheric chemistry was unknown. However, during the late 1970s and through the 1980s, a number of modeling studies demonstrated that $H_2O_2$ is a major oxidant for $SO_2$ in the formation of sulfate in the atmosphere (Penkett et al., 1979; Möller, 1980; Rodhe et al., 1981; Graedel and Goldberg, 1983; Hov, 1983; Kunen et al., 1983; Chameides, 1984; Martin, 1984; Seigneur and Saxena, 1984, 1988; Calvert et al., 1985; Jacob, 1986; Graedel et al., 1986; Kumar, 1986; Saxena et al., 1986; Nair and Peters, 1989; Pandis and Seinfeld, 1989; Tsai et al., 1990; Walcek et al., 1990), and thus may play a significant role in the formation of acidic rain, fog, and cloud waters. With worldwide concern about acid rain problem growing during the 1980s, many field studies to measure atmospheric $H_2O_2$ were performed by a number of investigators to elucidate the degree of involvement of $H_2O_2$ in acid rain formation (National Research Council (NRC), 1983; National Acid Precipitation Assessment Program (NAPAP), 1987). To date, the distribution pattern of $H_2O_2$ and its formation, decomposition, and deposition processes are still not clearly understood.

In this chapter, we review and evaluate previous studies on the chemistry of atmospheric $H_2O_2$, particularly focusing on studies conducted in the Los Angeles Air Basin, where extensive field and modeling investigations have been performed on all major primary and secondary air pollutants. The review will include the distribution pattern of $H_2O_2$ and modeling and field studies regarding the production, decomposition, and deposition processes of $H_2O_2$ in southern California. Finally, we discuss possible harmful effects of atmospheric $H_2O_2$ on forest trees in the mountains surrounding Los Angeles.

## 2.    ATMOSPHERIC LEVELS OF $H_2O_2$

### 2.1.    Gaseous $H_2O_2$

Reported concentrations of $H_2O_2$ in gas, rain, cloud, and dew phases in Los Angeles and its vicinity atmospheres are presented in Table 1. Observed concentrations of gas phase $H_2O_2$ range from 0.02 to 4.9 ppb except for a few points (>10 ppb) reported by Buffalini et al. (1972) and Kok et al. (1978b). It is likely that the previous measurements by Buffalini et al. (1972) and Kok et al. (1978b) suffered from the artifact formation of $H_2O_2$ within collectors due to the presence of $O_3$ in ambient air (Heikes et al., 1982; Zika and Saltzman, 1982; Heikes, 1984; Yoshizumi et al., 1984; Sakugawa and Kaplan,

1987). Gaseous $H_2O_2$ was analyzed by either chemiluminescent or fluorescent methods following a collection of $H_2O_2$ by aqueous stripping from the air (Kok et al., 1978a; Dasgupta et al., 1986; Lazrus et al., 1986; Tanner et al., 1986), a fluorescent method following cold trapping of $H_2O_2$ (Sakugawa and Kaplan, 1987), and tunable diode laser absorption spectroscopy (Slemr et al., 1986). Reported values of gaseous $H_2O_2$ concentrations in Los Angeles are comparable with those (0.01–7.0 ppb) measured in other regions of the United States and in other countries (Jacob et al., 1986, 1990; Slemr et al., 1986; Tanner et al., 1986; Heikes et al., 1987; Van Valin et al., 1987, 1990; Keuken et al., 1988; Olszyna et al., 1988; Possanzini et al., 1988; Barth et al., 1989; Boatman et al., 1989; Dolland et al., 1989; Lee and Busness, 1989; Mohnen and Kadlecek, 1989; Daum, 1990; Daum et al., 1990).

A clear diurnal variation of gaseous $H_2O_2$ concentrations was observed by Downs et al. (1989) in southern California by simultaneous measurements at Long Beach, downtown Los Angeles, Claremont, and Rubidoux (Fig. 1) during the Southern California Air Quality Study (SCAQS) conducted in the summer of 1987. These results indicate that, at all the sampling sites, the concentrations of gaseous $H_2O_2$ are low (0.16–0.37 ppb as the average values) in the early morning and increase in the afternoon with the maximum peak (1.49–2.09 ppb) occurring at 2–5 p.m. (local time), whereas the concentrations significantly decreased at night (0.22–0.57 ppb) (Fig. 2). Maximum $H_2O_2$ (and $O_3$) concentrations were observed in the early afternoon (at 2 p.m.) at the coast site (Long Beach), in the mid-afternoon (at 3 p.m.) in downtown Los Angeles, and in the late afternoon (at 4–5 p.m.) further inland (at Claremont and Rubidoux). Similar diurnal patterns of gaseous $H_2O_2$ and $O_3$ concentrations in southern California also observed by Lawson et al. (1988), Drummond et al. (1989), and Sakugawa and Kaplan (1989a) and elsewhere by Possanzini et al. (1988), Dolland et al. (1989), and Jacob et al. (1990).

In contrast to the studies above, a mixed diurnal pattern of gaseous $H_2O_2$ concentrations was observed at high elevation (at Lake Gregory, 1400 m above sea level) in the San Bernardino Mountains (Fig. 1) by Sakugawa et al. (1990c) during July–November 1989. The concentration of $H_2O_2$ in the region was generally highest at night and lowest during day time for the study period, whereas the $O_3$ concentration was highest in the afternoon (Fig. 3). Based on current knowledge of $H_2O_2$ chemistry in air, this diurnal pattern of $H_2O_2$ is not expected to occur from in situ chemical reaction processes because $H_2O_2$ should be photochemically produced during the daytime. The diurnal pattern above is likely to occur from the long-range transport of $H_2O_2$ by vertical and/or horizontal movement of air masses because at night the higher elevations of the San Bernardino Mountains are exposed to "free tropo-sphere" (above the inversion layer). The long-range transport of $H_2O_2$ was suggested as a major process for regulating the concentration of gaseous $H_2O_2$ in high-altitude atmospheres over the central and eastern United States (Heikes et al., 1987; Van Valin et al., 1987). A similar diurnal pattern of $H_2O_2$ concentrations showing highest concentration at night was decribed for high elevations at Mauna Loa on the island of Hawaii (Heikes, 1989).

**240**

**Table 1 Reported Concentrations of Atmospheric Hydrogen Peroxide in Gas Phase, Rain, Clouds, and Dew[a]**

| Range of Concentration | Mean | Sampling Site | Samping Period | Collection and Analytical Methods | Remarks | References |
|---|---|---|---|---|---|---|
| **Gas Phase H$_2$O$_2$ (ppb)** | | | | | | |
| 40–180 | | Riverside, L.A. Basin | Summer 1970 | Tedlar bag Colorimetry | | Bufalini et al. (1972) |
| 10–30 | | Claremont, Riverside, | Summer 1977 | Impinger or scrubbing coil Chemiluminescence | | Kok et al. (1978b) |
| 0.02–1.2 | | L.A. Basin Westwood, L.A. Basin | Summer 1985 Fall 1985 | Cryogenic collection Fluorescence | | Sakugawa and Kaplan (1987) |
| ~3.0 | | Glendora L.A. Basin | August 1986 | Various methods | High in afternoon Method comparison study | Lawson et al. (1988), Kok et al. (1990), Mackay et al. (1990), Sakugawa and Kaplan (1990), Tanner and Shen (1990), Dasgupta et al. (1990) |
| ~4.85 | | Rubidoux, Claremont, Los Angeles, Long Beach, L.A. Basin | Summer 1987 Fall 1987 | Diffusion scrubber, fluorescence, or laser absorption spectroscopy | | Downs et al. (1989) Drummond et al. (1989) |
| 0.03–2.04 | | Southern California (including L.A. Basin, Mojave Desert, San Bernardino Mountains) | 1985–1988 | Cryogenic method Fluorescence | High in afternoon High in summer | Sakugawa and Kaplan (1989a) |

| Concentration range (μM) | Mean (μM) | Location | Period | Method | Diurnal variation | Reference |
|---|---|---|---|---|---|---|
| 0.2–2.8 | | San Bernardino Mountains (1400 m above sea level) | 1989 | Cryogenic method Fluorescence | High at night | Sakugawa et al. (1990c) |
| **H$_2$O$_2$ in Rains (μM)** | | | | | | |
| 0.03–4.7 (typical range) max. 46.8 | | Claremont, L.A. Basin | 1978–1979 | Chemiluminescence | | Kok (1980) |
| 0.1–95 | 4.8 (n = 34) | Westwood, L.A. Basin | 1985–1988 | Fluorescence | High in afternoon | Sakugawa and Kaplan (1989b) |
| 0.01–145 | 4.4 (n = 71) | Westwood, L.A. Basin | 1985–1990 | Fluorescence | High in afternoon | Sakugawa et al. (1990c) |
| 0.1–6.6 | 4.1 (n = 7) | Hawthorne, L.A. Basin | Winter 1989 | | | |
| 2.2–70 | 4.5 (n = 4) | Duarte, L.A. Basin | Winter 1989 | | | |
| 9.0–90 | 33 (n = 5) | San Bernardino Mountains | May–October 1988, 1989 | | | |
| **H$_2$O$_2$ in Clouds (μM)** | | | | | | |
| 0.9–88 | 33 | L.A. Basin (<2 km altitude) | May 1982 | Fluorescence | | Richards et al. (1983) |
| 16–127 | 55 | L.A. Basin (<2 km altitude) | June 1984 | Fluorescence | | Richards (1989) |
| 12–167 | 57 | | May–June 1985 | | | |
| 9.0–62 | 25 (n = 4) | San Bernardino Mountains (1400 m above sea level) | May–October 1989 | Fluorescence | | Sakugawa et al. (1990c) |
| **H$_2$O$_2$ in Dew (μM)** | | | | | | |
| 0.01–3.0 | | Hawthorne, L.A. Basin, San Bernardino Mountains (1400 m above sea level) | 1985–1990 | Fluorescence | | Sakugawa et al. (1990c) |
| ~0.62 | | Glendora, L.A. Basin | August 1986 | Fluorescence | | Pierson and Brachaczek (1990) |

[a]Data for hydrogen peroxide concentrations were collected only in Los Angeles and its vicinity. n, number of samples; L.A. Basin, Los Angeles Basin.

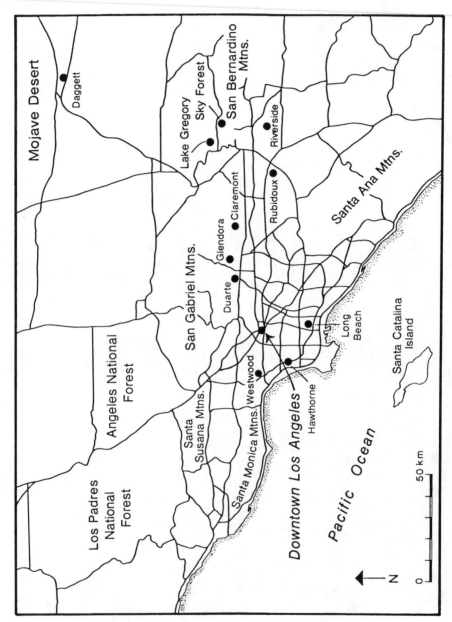

**Figure 1.** Map of Los Angeles and its adjacent areas. Solid lines indicate the major highway system in the region.

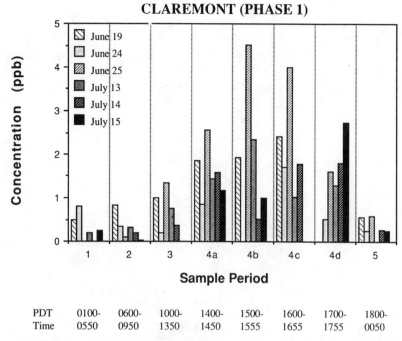

**Figure 2.** Diurnal variation of gaseous H$_2$O$_2$ concentrations at Claremont during the Southern California Air Quality Study in the summer of 1987. (Modified from Downs et al. (1989).)

Seasonal variation of gaseous H$_2$O$_2$ levels in Los Angeles was studied by Sakugawa and Kaplan (1989a). It was observed that the afternoon average 4-h concentration (12 p.m.–4 p.m.) of gaseous H$_2$O$_2$ was highest (as high as 1.35 ppb) in summer and lowest (<0.3 ppb) in the winter at Westwood during 1985–1988 (Fig. 4). During SCAQS, Drummond et al. (1989) observed that the daytime concentrations of gaseous H$_2$O$_2$ in the fall of 1987 were generally lower than those in the summer of 1987 in the Los Angeles Basin. Maximum values of gaseous H$_2$O$_2$ concentrations observed during the fall study of SCAQS (November–December) were in the 0.3- to 0.5-ppb range at Long Beach, whereas maximum values during the summer study (June–July) were in the 0.4- to 1.2-ppb range at Claremont. These results of Sakugawa and Kaplan (1989a) and Drummond et al. (1989) strongly suggest that the concentration of gaseous H$_2$O$_2$ in southern California is highest in the summer. Similar seasonal variation of gaseous H$_2$O$_2$ levels was observed at the top of Whitetop Mountain, VA (Olszyna et al., 1988) and over the central United States (Boatman et al., 1989).

The regional variation of gaseous H$_2$O$_2$ concentrations in southern California was studied by Downs et al. (1989) and Sakugawa and Kaplan (1989a, 1990). During the summer study of the SCAQS, Downs et al. (1989) compared ambient concentrations of gaseous H$_2$O$_2$ at various sites in the Los Angeles

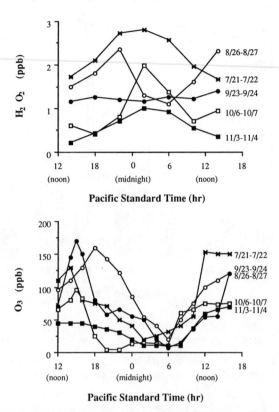

Pacific Standard Time (hr)

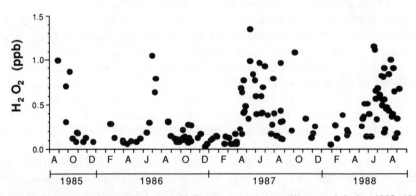

Pacific Standard Time (hr)

**Figure 3.** Diurnal variations of gaseous $H_2O_2$ and $O_3$ concentrations at Lake Gregory in the San Bernardino Mountains during July through November 1989. (From Sakugawa et al. (1990c).)

**Figure 4.** Seasonal variation of gaseous $H_2O_2$ concentrations at Westwood during 1985–1988. (Modified from Sakugawa and Kaplan (1989a).)

Air Basin. The results of their 11-day measurements during the SCAQS, indicate that the daytime maximum $H_2O_2$ concentration was highest (2.1 ppb) at Rubidoux (Fig. 1) and that the maximum $H_2O_2$ at three other sampling sites were similar (1.5 ppb at Claremont, 1.5 ppb at downtown Los Angeles, and 1.6 ppb at Long Beach). Sakugawa and Kaplan (1990) studied the regional variation of gaseous $H_2O_2$ in the Los Angeles Basin at Westwood, Duarte, and Glendora during the Carbonaceous Species Methods Comparison Study (CSMCS) in August 18–21, 1986, when a heavy photochemical smog episode occurred in the Los Angeles Basin. The results of their four-day study of gaseous $H_2O_2$ measurement showed that $H_2O_2$ concentration was highest at Glendora (the most interior site studied) and lowest at Westwood (a site near the coast).

Sakugawa and Kaplan (1989) compared the concentration levels of gaseous $H_2O_2$ in the Los Angeles Basin (at Westwood and Duarte), the San Bernardino Mountains (at Lake Gregory and Sky Forest) and the Mojave Desert (at Daggett) during the warm months (April–October) in 1985–1988. The results of simultaneous measurement of gaseous $H_2O_2$ levels at the above sites indicate that the midday (12 p.m.–4 p.m.) average concentrations of $H_2O_2$ is generally higher in the San Bernardino Mountains and the Mojave Desert than those in the Los Angeles Basin (Fig. 5), whereas the average concentrations of $H_2O_2$ were not much different within the Los Angeles Basin (at Westwood and Duarte). These results suggest that $H_2O_2$ concentrations may be higher in rural areas than those in pollutant-impacted urban areas like Los Angeles. More field studies should be conducted at various locations in Los Angeles and its vicinity before general conclusions on the regional variation of gaseous $H_2O_2$ are made.

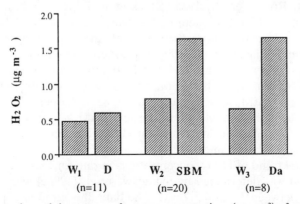

**Figure 5.** Comparison of the average afternoon concentrations ($\mu$g m$^{-3}$) of gaseous $H_2O_2$ at Westwood (W$_{1-3}$) and Duarte (D) in the Los Angeles Basin, in the San Bernardino Mountains (SBM), and Daggett (Da) in the Mojave Desert during the warm months of 1985–1988. $n$, number of observatons. (Modified from Sakugawa and Kaplan (1989a).)

## 2.2.    Aqueous $H_2O_2$

The concentration of $H_2O_2$ in rainwater was measured by Kok (1980), Sakugawa and Kaplan (1989b), and Sakugawa et al. (1990c) in Los Angeles and its vicinity (Table 1). The $H_2O_2$ concentrations in rainwater at Claremont for the majority of the samples in 1979 were within a range of 0.03–4.7 µM and the maximum measured concentration of $H_2O_2$ was 46.8 µM (Kok, 1980). The $H_2O_2$ concentrations in rainwater collected at Westwood ranged from 0.1 to 95 µM during 1985–1988 (Sakugawa and Kaplan, 1989b) and ranged from 0.01 to 145 µM during 1985–1990 (Sakugawa et al., 1990c). The concentration of $H_2O_2$ in rainwater was also measured at Hawthorne and Duarte, in the Los Angeles Basin and at Lake Gregory in the San Bernardino Mountains (Table 1). The concentrations of $H_2O_2$ in rainwater were determined by the chemiluminescent (Kok, 1980) or fluorescent methods (Sakugawa and Kaplan, 1987).

The daily variation of $H_2O_2$ concentration in rainwater was observed at Westwood by Sakugawa et al. (1990c) as shown in Fig. 6. In this study, rainwater samples were collected every 1–2 h during the daytime and the concentration of $H_2O_2$ in rainwater was measured immediately after the collection. The rainwater data at Westwood indicated that the highest concentration of $H_2O_2$ occurred in the afternoon, probably due to rapid scavenging of photochemically produced gaseous $H_2O_2$ by rain droplets. Similar patterns of daily variation of $H_2O_2$ concentrations in rain, with highest concentrations occurring in the afternoon was reported by other investigators at Long Island, NY (Lee et al., 1986) and in other countries (Yoshizumi et al., 1984; Klockow and Jacob, 1986; Jacob et al., 1990).

It was observed that the concentration of $H_2O_2$ in rain was highest in the summer and lowest in the winter in the eastern United States (Bufalini et al., 1979; Kelly et al., 1985; Lee et al., 1986) and in other countries (Yoshizumi et al., 1984; Römer et al., 1985; Klockow and Jacob, 1986). In southern California, it is difficult to examine the seasonal variation of $H_2O_2$ concentrations in rain because more than 90% of rain is precipitated during the winter season. Sakugawa et al. (1990c) collected nine summer rains (in June–September) at Westwood during 1985–1990 and measured $H_2O_2$ concentrations. $H_2O_2$ concentrations higher than 25 µM were found only in the summer rains (average 43 µM, $n = 9$) during the 1985–1990 period, whereas $H_2O_2$ concentrations in the winter rains (average 4.3 µM, $n = 53$) were mostly less than 10 µM (Fig. 7). This indicates that $H_2O_2$ concentration in Los Angeles rain is highest in the summer.

The annual trend of $H_2O_2$ concentrations in rain for the 1985–1990 period at Westwood was determined by Sakugawa et al. (1990c) with average concentrations between 3.2 and 5.2 µM. The data illustrated in Fig. 8 shows no statistically significant trend of $H_2O_2$ concentrations.

The measurement of $H_2O_2$ in clouds (nighttime stratus clouds) at various locations in the Los Angeles Basin during May 1982, June 1984, and May–

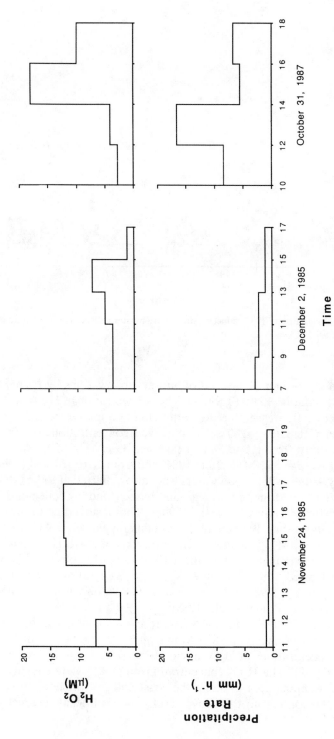

**Figure 6.** Daily variation of $H_2O_2$ concentrations in rainwater collected at Westwood. (From Sakugawa et al. (1990c).)

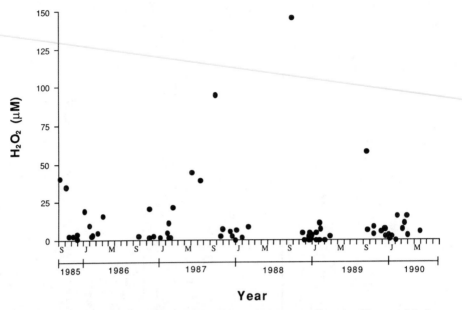

**Figure 7.** Seasonal variation of H₂O₂ concentrations in rainwater collected at Westwood during 1985–1990. (From Sakugawa et al. (1990c).)

June 1985 was performed by a series of aircraft observations by Richards et al. (1983) and Richards (1989) (Table 1). The concentration of $H_2O_2$ in clouds was measured by a fluorescent method after cloud samples were collected by a grab sampler (Richards, 1989). The observed concentrations of $H_2O_2$ in cloud waters ranged from 0.9 to 88 μM (average 33 μM) in May 1982, from 16 to 127 μM (average 55 μM) in June 1984, and from 12 to 167 μM (average 57 μM) in May–June 1985. Cloud waters were also collected at Lake Gregory in the San Bernardino Mountains by a ground-based cloud/fog collector during May–October 1989 (Sakugawa et al., 1990c). The observed concentrations of cloud $H_2O_2$ in the San Bernardino Mountains ranged from 9.0 to 62 μM ($n = 4$, average 25 μM). The above concentrations of cloud $H_2O_2$ observed in Los Angeles are comparable with those (0.01–247 μM) observed in other regions of the United States (Daum et al., 1984; Kelly et al., 1985; Lazrus et al., 1985; Kadlecek, 1986; Olszyna et al., 1988; Mohnen and Kadlecek, 1989) and in Europe (Römer et al., 1985).

Richards et al. (1987) found that offshore clouds contained the highest $H_2O_2$ concentrations (maximum 163 μM in clouds collected off Long Beach) and that the concentrations of $H_2O_2$ in clouds decreased with increasing distance inland (Fig. 9). The $H_2O_2$ concentrations in clouds were roughly 30% as great in the Fontana area as along the coast and just offshore. It should be noted that, in contrast to the above results, no significant regional vari-

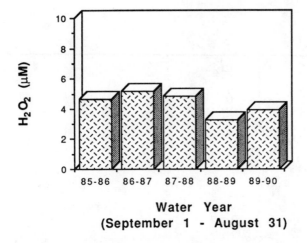

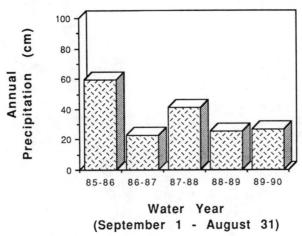

**Figure 8.** Annual volume-weighted mean concentrations of H$_2$O$_2$ in rainwater at Westwood during 1985–1990. (From Sakugawa et al. (1990c).)

ations of the concentrations of other chemical species (such as sulfate and nitrate) were found during the May–June 1985 observations.

Dew was collected at Hawthorne in the Los Angeles Basin and at Lake Gregory in the San Bernardino Mountains by Sakugawa et al. (1990c) (Table 1). In their study, dew samples were collected only at night through early morning (before sunrise) by condensing moisture from the air onto a 1.0 × 1.5-m Teflon sheet mounted on a 5-cm-thick Styrofoam pad and rested on a bench, which was set up 1.0 m above the ground surface. The result of dew water analysis indicated that, at these sampling sites, the concentrations of

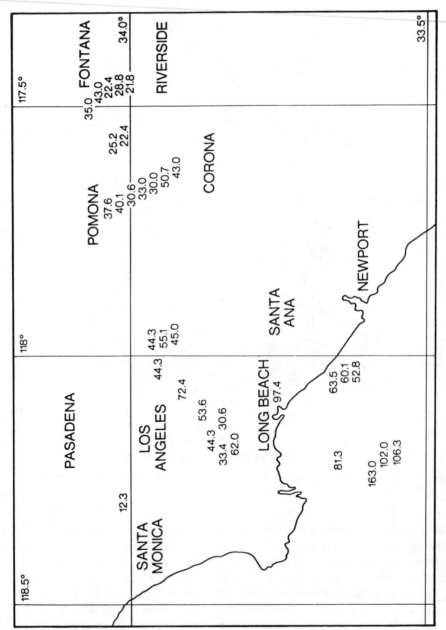

**Figure 9.** Spatial distribution of H$_2$O$_2$ concentratons ($\mu$M) in Los Angeles clouds. (Modified from Richards et al. (1987).)

H$_2$O$_2$ in dew were low (0.01–3.0 μM). Pierson and Brachaczek (1990) also measured peroxide concentrations (H$_2$O$_2$ plus organic peroxides) in dew collected at Glendora during CSMCS in August 1986. The result of their measurement indicated that peroxide concentrations in dew were less than 0.62 μM and were typically <0.13 μM (the detection limit). This low concentration of H$_2$O$_2$ in Los Angeles dew observed can be explained by the rapid reaction of H$_2$O$_2$ with dry deposited SO$_2$ (and possibly also with organic compounds).

The major source of aqueous phase (e.g., rain, cloud, fog, and dew) H$_2$O$_2$ is believed to be gaseous H$_2$O$_2$, because H$_2$O$_2$ is a highly soluble compound in water. For example, ppb levels of gaseous H$_2$O$_2$ can equilibrate with μM levels of H$_2$O$_2$ in water droplets, as calculated from the Henry's law constant for H$_2$O$_2$ (7.4 × 10$^4$ M atm$^{-1}$ at 25 °C) (Lind and Kok, 1986). A similar trend in the diurnal and seasonal variations of aqueous H$_2$O$_2$ with those of gas phase H$_2$O$_2$ (Bufalini et al., 1979; Yoshizumi et al., 1984; Kelly et al., 1985; Römer et al., 1985; Klockow and Jacob, 1986; Lee et al., 1986; Sakugawa and Kaplan, 1989b; Jacob et al., 1990) provides partial support to the claim that aqueous H$_2$O$_2$ mostly originates from gas phase H$_2$O$_2$. It has also been shown that in situ formations of H$_2$O$_2$ occur in atmospheric water droplets (e.g., HO$_2$ (aq) + O$_2^-$ $\xrightarrow{\text{H}_2\text{O}}$ H$_2$O$_2$ (aq) + O$_2$ + OH$^-$) or on the surface of aerosal particulates (Chameides and Davis, 1982; Graedel and Goldberg, 1983; Chameides, 1984; Schwartz, 1984; Seigneur and Saxena, 1984; Calvert et al., 1985; Lind and Kok, 1986; McElroy, 1986; Bahnemann et al., 1987; Sakugawa and Kaplan, 1987). The significance of these formation processes for the atmospheric production of H$_2$O$_2$ is not yet clearly understood and this area deserves further attention.

## 2.3. Air Quality Study

In Los Angeles, two major field studies of measurement of ambient H$_2$O$_2$ concentrations were conducted by the California Air Resources Board during the 1980s. CSMCS was performed in the summer of 1986 at Glendora (Fig. 1) to compare various collection and analytical methods of gaseous H$_2$O$_2$, as well as several carbonaceous species. SCAQS was performed in the summer and fall of 1987 at various locations in the Los Angeles Basin to develop a comprehensive and properly archived air quality and meteorological data base for the Basin. The determination of gaseous H$_2$O$_2$ was an important task to elucidate the oxidizing capacity of the Los Angeles Air Basin. Because the results of SCAQS are not yet fully reported, we present only the results of CSMCS in this chapter. However, some experimental results of SCAQS appear in earlier sections in this chapter.

CSMCS was conducted during August 11–21, 1986, when a photochemical smog event occurred in the Los Angeles Basin. The 1-hr maximum O$_3$ concentration observed was very high (150–255 ppb) at Glendora during CSMCS because a low and strong inversion layer had developed in the Los Angeles Basin due to the occupation of an upper-level, high-pressure ridge over the southwestern United States. The measurement of $H_2O_2$ concentrations was conducted by various groups using different collection and analytical methods of gaseous $H_2O_2$ (Dasgupta et al., 1990; Kok et al., 1990; Mackay et al., 1990; Sakugawa and Kaplan, 1990; Tanner and Shen, 1990). Details of CSMCS were reported in the special issue of *Aerosol Science & Technology* (Vol. 12, No. 1, 1990).

The results of $H_2O_2$ measurement in CSMCS can be summarized as follows:

1. There was a significant difference in the gaseous $H_2O_2$ values determined by various methodologies. The difference in $H_2O_2$ values that were determined by the various methods varied by a factor of two (Lawson et al., 1988). The above variations in the performance of different analytical methodologies suggest that further improvement in $H_2O_2$ sampling and analytical methods are required.

2. The data of $H_2O_2$ concentrations for all groups indicated that there is a clear diurnal variation of gaseous $H_2O_2$ level, with the highest and lowest concentrations occurring in the afternoon and at night, respectively.

3. The data of all the participating groups showed that concentrations of $H_2O_2$ were highest during the latter part of the study period (especially August 17–19) when a heavy photochemical smog episode occurred, suggesting that photochemical processes significantly affect the formation of $H_2O_2$.

4. Organic hydroperoxides (ROOH), which are formed by photochemical reactions including the reaction of hydroperoxyl radical with organic peroxyl radical ($HO_2 + RO_2 \rightarrow ROOH + O_2$) were measured to be up to about half of the level of gaseous $H_2O_2$ concentrations (Kok et al., 1990). Formation processes of organic hydroperoxides and their role in the oxidations of $SO_2$ and organic compounds in the air, however, still need to be explored.

## 3. PRODUCTION, DECOMPOSITION, AND DEPOSITION OF $H_2O_2$

### 3.1. Photochemical Production of $H_2O_2$

It is thought that atmospheric $H_2O_2$ is dominantly produced by the coupling reaction of two hydroperoxyl radicals ($HO_2 + HO_2 \rightarrow H_2O_2 + O_2$) (NRC, 1983; NAPAP, 1987). Since $HO_2$ radicals are formed by photochemical reactions of volatile organic compounds (VOC) (e.g., methane and other hydrocarbons, aldehydes, ketones), nitrogen oxide ($NO_x$), and carbon monoxide (CO) in the atmosphere, major air pollutants that affect the formation of gaseous $H_2O_2$ are considered to be VOC, $NO_x$, and CO in urban and rural atmospheres (NAPAP, 1987; Sakugawa et al., 1990d). A number of modeling studies demonstrated that high concentrations of VOC and CO would favor the formation of $H_2O_2$, whereas high $NO_x$ concentration would inhibit the formation of $H_2O_2$ (NRC, 1983; Calvert and Stockwell, 1983; Calvert et al., 1985; Kleinman, 1986; Stockwell, 1986; Hough, 1988; Dodge, 1989; Thompson et al., 1989a,b; Sakugawa et al., 1990a; Tsai et al., 1991). It was also demonstrated that meteorological conditions that favor the formation of gaseous $H_2O_2$ are high solar intensity, high air temperature, and high water vapor concentration (Calvert and Stockwell, 1983; Calvert et al., 1985; Kleinman, 1986; Dodge, 1989; Sakugawa et al., 1990a).

Sakugawa et al. (1990a) examined the formation mechanism of gaseous $H_2O_2$ in southern California air using a one-dimensional photochemical model. The results of their sensitivity analysis for $H_2O_2$ formation indicated that the concentration and emission rate of $NO_x$, the concentration ratio of $NO_x$/NMHC, solar intensity, temperature, relative humidity, and the height of the inversion layer significantly determine the concentration of gaseous $H_2O_2$. Among these factors, solar radiation along with temperature and relative humidity were predicted by the model to largely determine the seasonal variation of gaseous $H_2O_2$ levels in Los Angeles. Tsai et al. (1991) also studied the factors affecting the ambient levels of gaseous $H_2O_2$ in Los Angeles air. It was suggested that the concentration and emission rate of $NO_x$, the ratio of VOC/$NO_x$, the height of inversion layer, and the deposition velocity of $H_2O_2$ significantly affect ambient concentrations of gaseous $H_2O_2$. It should be noted that a series of studies conducted by Sakugawa and Kaplan (1989a) on gaseous $H_2O_2$ levels in southern California air during the 1985–1988 period showed that the concentration of gaseous $H_2O_2$ was best correlated with solar intensity, whereas $O_3$ was correlated with temperature, $NO_x$ content, and the height of inversion layer. The results of the above modeling and field studies conducted for the Los Angeles Air Basin partially support the claim that the concentrations of primary pollutants such as $NO_x$, VOC, and CO, along with meteorological parameters such as solar radiation, temperature, and water vapor concentration are the primary factors determining the atmospheric formation and content of gaseous $H_2O_2$.

## 3.2. Decomposition of $H_2O_2$

In gas phase, the reaction with OH radical ($H_2O_2$ + OH → $H_2O$ + $HO_2$) and the photolysis of $H_2O_2$ ($H_2O_2 \xrightarrow{h\nu} 2$ OH) are thought to be major destruction pathways for atmospheric $H_2O_2$ (Kleinman, 1986). In addition, atmospheric $H_2O_2$ is consumed by aqueous phase reactions (including the reactions with S(IV) and OH radical, and the photolysis of $H_2O_2$ (Chameides, 1984; Jacob, 1986; Lind et al., 1987; Pandis and Seinfeld, 1989) following the absorption of gas phase $H_2O_2$ into water droplets and subsequent dissociation. The photolysis of $H_2O_2$ ($H_2O_2$ (aq) $\xrightarrow{h\nu} 2$ OH (aq)) is an important pathway for generating aqueous OH radicals (Graedel and Goldberg, 1983; Chameides, 1984; Jacob, 1986; Lind et al., 1987; Pandis and Seinfeld, 1989), which are the initiators for aqueous phase generation of organic acids from aldehydes.

The reaction of $H_2O_2$ with S(IV) (e.g., $HSO_3^- + H_2O_2$ (aq) $+ H^+ \rightarrow SO_4^{2-} + 2 H^+ + H_2O$) in cloud and rain droplets is fast (Lind et al., 1987). The production of sulfate by this reaction is not significantly affected by the change of pH of the droplets, because the catalytic effect of $H^+$ ion for the reaction offsets the decrease in the dissociation of gaseous $SO_2$ into droplets with decreasing pH (Lind et al., 1987), although the production rates of sulfate by other oxidants (such as $O_3$, $O_2$) decrease with decreasing pH of water droplets. Thus, $H_2O_2$ is considered to be the dominant oxidant at pH values typical of cloud and rain droplets (<5) (Penkett et al., 1979; Möller, 1980; Rodhe et al., 1981; Graedel and Goldberg, 1983; Hov, 1983; Kunen et al., 1983; Chameides, 1984; Martin, 1984; Seigneur and Saxena, 1984; Calvert et al., 1985; Graedel et al., 1986; Jacob, 1986; Kumar, 1986; Saxena et al., 1986; Seigneur and Saxena, 1988; Nair and Peters, 1989; Pandis and Seinfeld, 1989; Tsai et al., 1990), whereas $O_3$ and $O_2$ catalyzed by trace metals are dominant oxidants at high pH values (>5). Evidence supporting this hypothesis was obtained by field measurements of the concentrations of gas phase or aqueous phase $H_2O_2$ and $SO_2$ in cloud level atmospheres in the United States and other countries (Daum et al., 1984; Kelly et al., 1985; Römer et al., 1985; Heikes et al., 1987; Chandler et al., 1988; Gervat et al., 1988; Barth et al., 1989; Mohnen and Kadlecek, 1989). It was found that there is an inverse relationship between the concentrations of $H_2O_2$ and $SO_2$ (Daum et al., 1984; Kelly et al., 1985; Heikes et al., 1987; Mohnen and Kadlecek, 1989) in cloud environments. Also, it was observed that $SO_2$ was rapidly oxidized by $H_2O_2$ in cloud droplets (Chandler et al., 1988; Gervat et al., 1988). Hence, $H_2O_2$ has been found to play an important role in the generation of sulfate in the atmosphere.

The significant role of $H_2O_2$ as an oxidant of $SO_2$ in the Los Angeles atmosphere was suggested by modeling and field studies for cloud and fog chemistries as described below. Jacob and Hoffmann (1983) studied the oxidation rates of $SO_2$ by various oxidants in the Los Angeles fog, assuming 1, 10, and 20 ppb of $H_2O_2$, $O_3$, and $SO_2$, respectively, as ambient gas phase concentrations and 0.5 and 0.02 $\mu$g m$^{-3}$ as the aqueous concentrations of $Fe^{3+}$ and $Mn^{2+}$, respectively, in 0.1 g m$^{-3}$ of liquid water at 10 °C. Their simulation for sulfate formation (Fig. 10) suggested that $H_2O_2$ would be the dominant oxidant for $SO_2$ during the first several minutes in fog droplets due to the rapid reaction rate of aqueous $H_2O_2$ with S(IV). However, since no photochemical generation of $H_2O_2$ occurs at night, $H_2O_2$ could be depleted rapidly. Later, $O_2$ catalyzed by trace metals would be the dominant reagent for the $SO_2$ oxidation. Other oxidants such as $O_3$, $O_2$ catalyzed soot, and N(III) could also contribute to the formation of sulfate in fog droplets, but to a lesser extent. Similar results that address the importance of $H_2O_2$ and $O_2$ (catalyzed by $Fe^{3+}$ and $Mn^{2+}$) in the $SO_2$ oxidation were reported from

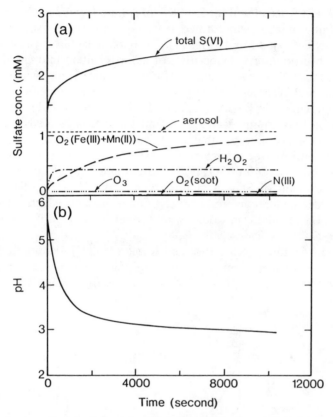

**Figure 10.** Aqueous phase formation of sulfate in fogwater by the reaction of aqueous SO₂ with various oxidants (*a*) and the resulting change of pH of fogwater (*b*). (Modified from Jacob and Hoffmann (1983).)

modeling studies by Seigneur and Saxena (1984) and Seigneur and Wegrecki (1990) for Los Angeles fog.

Seigneur and Saxena (1984) studied the formation process of sulfuric acid in Los Angeles nonprecipitating clouds using a mathematical model that includes gas and liquid phase chemistries and interfacial mass transport of chemical species. They assumed that the initial concentrations of $SO_2$, $H_2O_2$, and $O_3$ are 13–59 ppb, 0.12–1.8 ppb, and 28–68 ppb, respectively, in the cloud-level atmosphere. The aqueous concentrations of $Fe^{3+}$ and $Mn^{2+}$ in clouds were assumed to be 0.8–1.5 and 0.026–0.094 µg m⁻³, respectively, and water content in air was 0.26–0.4 g m⁻³. The concentrations of these chemical species and other air pollutant concentrations as well as meteorological conditions were based on the data collected by aircraft observations on May 22, 1982, over the Los Angeles Air Basin (Richards et al., 1983). The results of this modeling study indicated that the aqueous reaction of

S(IV) with $H_2O_2$ accounted for 75–98% of sulfate formed in clouds and that the remaining amount was formed from aqueous phase oxidation of $SO_2$ catalyzed by $Fe^{3+}$ and $Mn^{2+}$ (Fig. 11). "The initial sulfate" which is derived from the absorption and dissociation of aerosol sulfate into cloud droplets accounted for >50% of total sulfate (formed + initial) in clouds.

Seigneur and Wegrecki (1990) performed a similar simulation for sulfate formation in clouds in Los Angeles Air Basin, based on field data of cloud chemistry on May 25 and 26, 1985 collected by Richards (1989). Their simulation also clearly indicated that $H_2O_2$ would be the dominant oxidant for $SO_2$ in clouds in the Los Angeles Air Basin. In two case studies that they simulated, the oxidation of $SO_2$ by $H_2O_2$ accounted for 95 and 83% of total aqueous oxidation of $SO_2$ in cloudwater for the May 25, 1985, and the May 26, 1986, simulations, respectively. Other oxidants which contribute to the formation of sulfate in clouds would be $O_2$ (catalyzed by trace metals) and aqueous OH radicals.

Field observation by Sakugawa and Kaplan (1989a) identified an inverse correlation between ambient concentrations of gaseous $H_2O_2$ and $SO_2$ during haze events (visibility <5 km) at Westwood during 1985–1988. They also

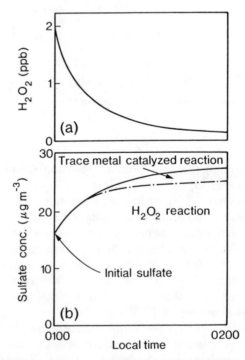

**Figure 11.** Simulation for the formation of sulfate in Los Angeles clouds. The temporal change of the concentrations of (a) $H_2O_2$ and (b) sulfate in cloud droplets. (Modified from Seigneur and Saxena (1984).)

observed that $H_2O_2$ concentrations were extremely low ($<0.05$ ppb) during nighttime fog events observed at Westwood. These results suggest that $H_2O_2$ is utilized for $SO_2$ oxidation in very humid air.

Observations of minimum concentrations of gaseous and aqueous $H_2O_2$ in winter (Figs. 4 and 7) (due to low photochemical production of $H_2O_2$) suggest that $H_2O_2$ may be a limiting reagent for the production of sulfate in the winter atmosphere of southern California where typical concentration of $SO_2$ are 3–10 ppb. This level of $SO_2$ far exceeds the concentration of gaseous $H_2O_2$. A modeling study for sulfate formation in the Los Angeles air conducted by Sakugawa and Kaplan (1989a) estimated that in Los Angeles the production rate of sulfate would be lower in the winter than in the summer because of the limited availability of $H_2O_2$ in the winter. This was also predicted for the northeastern United States (Calvert et al., 1985). It is likely that the limited production of sulfate in the winter atmosphere due to lack of sufficient $H_2O_2$ for $SO_2$ oxidation would result in a nonlinearity in the relation between $SO_2$ emission and sulfate wet deposition (NRC, 1983; Calvert et al., 1985; NAPAP, 1987). For example, a given reduction in $SO_2$ emission from industrial sources (resulting from emission control legislation) may not necessarily lead to an equivalent reduction of sulfate in wet deposition, but could result in a reduced production. The nonlinearity between $SO_2$ emission and sulfate in wet deposition complicates the development of efficient acid deposition control strategy (Schwartz, 1989) and thus has been the subject of a number of modeling and field studies (as reviewed by Hales and Renne, 1989).

When the concentration of $H_2O_2$ in the droplets is low, excess S(IV) may form hydroxymethanesulfonate (HMSA), which is formed by the reaction of S(IV) with formaldehyde in atmospheric water droplets (Richards et al., 1983; Munger et al., 1986). HMSA is stable at a normal range of pH values in cloud, fog, and rainwater (Boyce and Hoffmann, 1984; Kok et al., 1986). HMSA was found to resist rapid decomposition by $H_2O_2$ and $O_3$ (Richards et al., 1983; Hoigné et al., 1985; Kok et al., 1986) and therefore HMSA might be an important and relatively stable reservoir of $SO_2$ in the atmosphere (Munger et al., 1986). Aircraft observations by Richards et al. (1983) suggested that HMSA exists at a relatively high concentration in Los Angeles cloud water. Modeling studies for cloud and fog chemistries in Los Angeles air also predicted that HMSA is likely to exist in nonprecipitating clouds and nighttime fog, based on the data of chemical kinetics and equilibrium of formaldehyde with $SO_2$ at low pH (Jacob and Hoffmann, 1983; Seigneur and Saxena, 1984; Seigneur and Wegrecki, 1990). Field observations in California and other regions have shown that this compound exists in fog, rainwater, snow, and dew with concentrations up to 100 $\mu$M (Munger et al., 1986; Keuken et al., 1987; Ang et al., 1987). In addition, it was suggested that adducts of S(IV) with several aldehydes and organic acids such as acetaldehyde, glyoxal, glyoxylic acid, and methylglyoxal are potentially important reservoirs of S(IV) (Olson et al., 1988; Olson and Hoffmann, 1989). The fate of HMSA and other adducts of S(IV) in the atmosphere is yet unknown.

These adducts may be transported long distances and in turn they may release S(IV) by decomposing into aldehyde (or organic acid) and S(IV) (Munger et al., 1986). The released S(IV) can then be oxidized by $H_2O_2$ and $O_3$ to produce sulfate. Although HMSA is resistant to oxidation by $H_2O_2$ and $O_3$, HMSA may be oxidized by OH radical (Martin et al., 1989), which is a very powerful oxidant in the atmosphere. Currently, little information is available for the long-range transport of HMSA and other adducts of S(IV) and their possible participation in sulfate formation in water droplets, but such adduct formation and transport might be a key to the long-distance transport and formation of acid rain in high latitudes (e.g., Scandinavia, Nova Scotia) from $SO_2$ sources in the industrial south.

## 3.3.  Deposition of $H_2O_2$

$H_2O_2$ is removed from the atmosphere by wet deposition including rain scavenging and fog and mist droplet scavenging and by dry deposition to surfaces. Currently, these deposition rates of $H_2O_2$ have not yet been estimated in the United States because of lack of sufficient field data. Based on field data of $H_2O_2$ concentrations in rainwater and ambient gaseous $H_2O_2$ concentrations, and the deposition velocity of gaseous $H_2O_2$ (taken from literature), we estimated the annual deposition rates of $H_2O_2$ in the Los Angeles Basin. The rainwater data of Sakugawa et al. (1990c) indicated that annual wet deposition rate of $H_2O_2$ in Los Angeles is 0.43 kg $ha^{-1}$ $yr^{-1}$ (Table 2), assuming that the annual average $H_2O_2$ concentration in rainwater is 4.4 $\mu M$ and that the annual precipitation rate is 290 mm. The annual rate of dry deposition of $H_2O_2$ (2.68 kg $ha^{-1}$ $yr^{-1}$) was calculated by multiplying annual average concentration of gaseous $H_2O_2$ in Los Angeles (0.6 ppb, based on the data of Sakugawa and Kaplan (1989a) by the deposition velocity of gaseous $H_2O_2$ (1.0 cm $s^{-1}$, based on a theoretical estimation by Walcek (1987)). From the above results, we can determine that the annual rate of dry deposition of $H_2O_2$ is about six times larger than that of wet deposition. For comparison, reported values of rates of deposition of inorganic and organic acids such as $SO_4$, $NO_3$, formic acid, and acetic acid are also given in Table 2. These data clearly indicate that in southern California dry deposition is a much more important process for deposition of air pollutants than wet deposition, largely due to its semi arid climate.

## 4.  HARMFUL EFFECTS OF $H_2O_2$ ON VEGETATION

Harmful effects of $H_2O_2$ on plant cells are well known (Forti and Gerola, 1977; Kaiser, 1979; Robinson et al., 1980; Tanaka et al., 1982, 1985; MacRae and Ferguson, 1985). European scientists (Masuch et al., 1986; Möller, 1989) suggested that $H_2O_2$ may play a contributory role in recent decline of forests in North America and central Europe. Masuch et al. (1986) studied the effects

**Table 2 Estimated Wet and Dry Deposition Rates of Hydrogen Peroxide, Sulfate, Nitrate, Formic Acid, and Acetic Acid in the Los Angeles Air Basin**

| Chemical Species | Wet Deposition (kg ha$^{-1}$ yr$^{-1}$) | Dry Deposition (kg ha$^{-1}$ yr$^{-1}$) | Wet/(Wet + Dry) × 100 (%) |
|---|---|---|---|
| $H_2O_2$ | 0.43[a] | 2.68[b] | 14 |
| $SO_4$ | 5.11[c] | 23.0[d] | 18 |
|  | 2.01[c] | 6.84[f] | 23 |
| $NO_3$ | 2.92[e] | 70.1[f] | 4 |
| HCOOH | 1.34[g] | 28.5[g] | 5 |
| $CH_3COOH$ | 3.53[g] | 37.5[g] | 9 |

[a]Calculated in this study, based on rainwater data collected at Westwood during 1985–1990.

[b]Calculated in this study. Assumed deposition velocity of $H_2O_2$ in 1.0 cm s$^{-1}$. Annual average concentration of gaseous $H_2O_2$ in Los Angeles is assumed to be 0.6 ppb (or 0.85 μg m$^{-3}$).

[c]Calculated by Lawson and Wendt (1982), based on rainwater data collected at Orange County during 1974–1981.

[d]Calculated by Lawson and Wendt (1982), based on total deposition data collected at Orange County during 1967–1968.

[e]Calculated by Young et al. (1986) using a semi-empirical model, based on rainwater data collected at Canoga Park, Pasadena, Duarte, Riverside, downtown Los Angeles, and Corona del Mar.

[f]Calculated by Young et al. (1986) using a semi-empirical model.

[g]Calculated by Grosjean (1989).

of $H_2O_2$ containing acidic fog on young trees of Norway spruce (*Picea abies*) and red beech (*Fagus* sp.). The results of microscopic examination indicate that $H_2O_2$ causes cellular damage to young trees when the trees are exposed for 3 h per day to acidic fog with 30–300 μM of $H_2O_2$ for 6 weeks. American investigators (Hanson and McLaughlin, 1989) conducted a similar experiment for the seedlings of red spruce (*Picea rubens*) by exposing the trees to an acid mist containing 0–235 μM of $H_2O_2$ for 16 weeks. However, they failed to obtain any evidence regarding the damage of plant tissues through the experiment. Thus, at present, it is still unclear whether atmospheric $H_2O_2$ significantly damages trees.

In southern California, it is recognized that air pollution is significantly involved in the recent decline of forests in areas surrounding Los Angeles, such as the San Bernardino Mountains and San Gabriel Mountains (Miller, 1983, 1989). Since the early 1950s, the damage to pine trees (Ponderosa pine and Jeffrey pine) and other conifer trees (e.g., the occurrence of foliar injury, premature leaf fall, decreased photosynthetic capacity, and reduced radial growth) has been observed in these mountain regions (Miller, 1983, 1989). Extensive field and laboratory studies conducted during the 1960s and 1970s suggested that $O_3$ is the most likely compound that damages trees (Miller, 1983, 1989; Miller et al., 1986). However, detailed information about the concentration levels and role of other air pollutants in the mountain region is lacking.

Sakugawa and Kaplan (1989a) and Sakugawa et al. (1990c) measured ambient $H_2O_2$ concentrations in the forests of San Bernardino Mountains during the 1980s. They observed that the concentrations of $H_2O_2$ can be as high as 3 ppb in the forests and that average afternoon concentrations of $H_2O_2$ are higher than those in Los Angeles. The concentration of $H_2O_2$ in clouds was in range of 9.0–62 µM during the warm months (May–October) in the mountain region, whereas the $H_2O_2$ concentration in rain was in the 9.0–90 µM range. Sakugawa et al. (1990b), on the other hand, assessed the historic trend of ambient gaseous $H_2O_2$ concentrations in Los Angeles during the 1960s–1980s using a one-dimensional photochemical model. Their results showed that the level of gaseous $H_2O_2$ (as well as $O_3$) during the 1960s–1970s, when damage to trees was significant in the surrounding mountains of Los Angeles (Miller et al., 1989), were likely to have been higher than those during the 1980s. From these results, Sakugawa et al. (1990b) suggested that $H_2O_2$ may play a contributory role in damaging forest trees in the San Bernardino Mountains.

Although the harmful effect of $H_2O_2$ on Norway spruce, red spruce, or red beech has been examined as stated earlier, no study has been conducted on pine trees such as Ponderosa pine and Jeffrey pine, which are sensitive to ozone exposure (Williams, 1980, 1983; Miller, 1983). Because it has been demonstrated that olefines (terpenes and isoprenes) liberated from trees can rapidly react with atmospheric ozone to produce peroxides (Becker et al., 1990; Hewitt et al., 1990), the $H_2O_2$ formed by the reaction of $O_3$ with tree-emitted olefins in near surface atmospheres of tree leaves could be harmful, especially if it enters the vascular system.

## 5. SUMMARY AND CONCLUSION

Measurement of atmospheric $H_2O_2$ started in the early 1970s in southern California and a number of measurements of atmospheric $H_2O_2$ were conducted, mostly during 1980s. Observed concentrations of gaseous $H_2O_2$ are 0.02–4.9 ppb in Los Angeles and its vicinity. Observed concentrations of $H_2O_2$ in rain and clouds range from 0.01 to 145 µM and 1 to 167 µM, respectively, whereas the $H_2O_2$ concentrations in dew were found to be very low (mostly in the submicromolar range). These $H_2O_2$ concentration ranges were comparable with those measured in other regions of the United States and in other countries. Diurnal and seasonal variations of gas phase $H_2O_2$ were observed, with highest concentrations occurring in the afternoon and summer. Limited field data suggest that the concentration of $H_2O_2$ in rainwater may have similar diurnal and seasonal patterns. Regional variation of atmospheric $H_2O_2$ concentration is partially characterized for Los Angeles and its vicinity and more field studies need to be performed.

A number of modeling studies for the atmospheric formation of gaseous $H_2O_2$ have suggested that the concentrations of primary pollutants such as VOC, $NO_x$, and CO, along with solar radiation, temperature, and water vapor content in air, may be major factors controlling the formation of gaseous $H_2O_2$. It is likely that these factors are responsible for observed temporal and spatial variations of gaseous $H_2O_2$ levels in the atmosphere. The few modeling and field studies conducted for the formation of $H_2O_2$ in the Los Angeles air partially supported the above hypothesis. Additional field and modeling studies would be helpful to clarify the detailed role of these primary pollutants and meteorological parameters for the formation of gaseous $H_2O_2$ in southern California.

It is generally believed that the aqueous phase reaction of $H_2O_2$ with S(IV) is a dominant reaction for the formation of sulfate in the atmosphere. In Los Angeles, all the modeling and field studies on cloud and fog chemistries conducted during the 1980s support this conclusion. Thus, $H_2O_2$ would be a key oxidant for the production of sulfate in the southern California atmosphere.

It was estimated that the annual rate of dry deposition of atmospheric $H_2O_2$ is about six times greater than that of wet deposition. Therefore, dry deposition is a much more important process than wet deposition is removing $H_2O_2$ from the air in Los Angeles.

Field and modeling studies for the chemistry of Los Angeles clouds indicated that HMSA exists in clouds at a relatively high concentration. Due to limited reactivity of this compound with $H_2O_2$ and $O_3$ in water droplets, it is expected that HMSA is an important and relatively stable reservoir of $SO_2$ in the atmosphere.

It was suggested that atmospheric $H_2O_2$ may play a contributory role in damaging forest trees in the mountains surrounding Los Angeles. Due to large uncertainties about the harmful effect of $H_2O_2$ on trees and lack of sufficient field data, we cannot, at present, make any conclusion about the above hypothesis.

Finally, it was implied that $H_2O_2$ may be a limiting reagent for the production of sulfate in the winter atmosphere of Los Angeles. Given the fact that in Los Angeles more than 90% of annual precipitation occurs during the winter, the atmospheric level of $H_2O_2$, in addition to $SO_2$, may significantly affect the atmospheric production of sulfate. This further implies that to control the wet deposition of sulfate (and thus to control acid rain formation) not only the reduction of $SO_2$ emission but also the reduction of atmospheric $H_2O_2$ level (through the reduction of VOC and CO emissions) may need to be undertaken.

*Acknowledgments*

*We thank the U.S. Environmental Protection Agency and the National Center for Intermedia Transport Research at the University of California, Los Angeles, for sup-*

*porting this project. Although the information in this document has been funded by the Environmental Protection Agency under assistance agreement CR-812771 to the National Center for Intermedia Transport Research at UCLA, it does not necessarily reflect the views of the Agency and no official endorsement should be inferred. We also thank W. Tsai, Chemical Engineering Department, UCLA, for giving us critical comments.*

## REFERENCES

Ang, C. C., Lipari, F., and Swarin, S. J. (1987). "Determination of hydroxymethanesulfonate in wet deposition samples." *Environ. Sci. Technol.* **21**, 102–105.

Bahnemann, D. W., Hoffmann, M. R., Hong, A. P., and Kormann, C. (1987). "Photocatalytic formation of hydrogen peroxide." In R. W. Johnson and G. E. Gordon, eds., *The Chemistry of Acid Rain: Sources and Atmospheric Processes.* American Chemical Society, Washington, DC, pp. 120–132.

Barth, M. C., Hegg, D. A., Hobbs, P. V., Walega, J. G., Kok, G. L., Heikes, B. G., and Lazrus, A. L. (1989). "Measurements of atmospheric gas-phase and aqueous-phase hydrogen peroxide concentrations in winter on the east coast of the United States." *Tellus* **41B**, 61–69.

Becker, K. H., Brockmann, K. J., and Bechara, J. (1990). "Production of hydrogen peroxide in forest air by reaction of ozone with terpenes." *Nature (London)* **346**, 256–258.

Boatman, J. F., Wellman, D. L., Van Valin, C. C., Gunter, R. L., Ray, J. D., Sievering, H., Kim, Y., Wilkinson, S. W., and Luria, M. (1989). "Airborne sampling of selected trace chemicals above the central United States." *J. Geophys. Res.* **94**, 5081–5093.

Boyce, S. D., and Hoffmann, M. R. (1984). "Kinetics and mechanism of the formation of hydroxymethanesulfonic acid at low pH." *J. Phys. Chem.* **88**, 4740–4746.

Bufalini, J. J., Gay, B. W., Jr., and Brubaker, K. L. (1972). "Hydrogen peroxide formation from formaldehyde photooxidation and its presence in atmospheres." *Environ. Sci. Technol.* **6**, 816–821.

Bufalini, J. J., Lancaster, H. T., Namie, G. R., and Gay, B. W., Jr. (1979). "Hydrogen peroxide formation from the photooxidation of formaldehyde and its presence in rainwater." *J. Environ. Sci. Health* **A14**, 135–141.

Calvert, J. G., and Stockwell, W. R. (1983). "Acid generation in the troposphere by gas-phase chemistry." *Environ. Sci. Technol.*, **17**, 428A–443A.

Calvert, J. G., Lazrus, A., Kok, G. L., Heikes, B. G., Walega, J. G., Lind, J., and Cantrell, C. A. (1985). "Chemical mechanisms of acid generation in the troposphere." *Nature (London)* **317**, 27–35.

Chameides, W. L. (1984). "The photochemistry of a remote marine stratiform cloud." *J. Geophys. Res.* **89**, 4739–4755.

Chameides, W. L., and Davis, D. D. (1982). "The free radical chemistry of cloud droplets and its impact upon the composition of rain." *J. Geophys. Res.* **87**, 4863–4877.

Chandler, A. S., Choularton, T. W., Dollard, G. J., Eggleton, A. E. J., Gay, M. J., Hill, T. A., Jones, B. M. R., Tyler, B. J., Bandy, B. J., and Penkett, S. A. (1988).

"Measurements of $H_2O_2$ and $SO_2$ in clouds and estimates of their reaction rate." *Nature (London)* **336**, 562–565.

Dasgupta, P. K., McDowell, W. L., and Rhee, J. S. (1986). "Porous membrane-based diffusion scrubber for the sampling of atmosphere gases." *Analyst (London)* **111**, 87–90.

Dasgupta, P. K., Dong, S., and Hwang, H. (1990). "Diffusion scrubber-based field measurements of atmospheric formaldehyde and hydrogen peroxide." *Aerosol Sci. Technol.* **12**, 98–104.

Daum, P. H. (1990). "Observations of $H_2O_2$ and S(IV) in air, cloudwater and precipitation and their implications for the reactive scavenging of $SO_2$." *Atmos. Res.* **25**, 89–102.

Daum, P. H., Kelly, T. J., Schwartz, S. E., and Newman, L. (1984). "Measurements of the chemical composition of stratiform clouds." *Atmos. Environ.* **18**, 2671–2684.

Daum, P. H., Kleinman, L. I., Hills, A. J., Lazrus, A. L., Leslie, A. C. D., Busness, K., and Boatman, J. F. (1990). "Measurement and interpretation of concentrations of $H_2O_2$ and related species in the upper midwest during summer." *J. Geophys. Res.* **95**, 9857–9871.

Dodge, M. C. (1989). "A comparison of three photochemical oxidant mechanisms." *J. Geophys. Res.* **94**, 5121–5136.

Dolland, G. J., Sandalls, F. J., and Derwent, R. G. (1989). "Measurements of gaseous hydrogen peroxide in southern England during a photochemical episode." *Environ. Pollut.* **58**, 115–124.

Downs, J., Lin, C. C., Lev-On, M., Tanner, R., and Ferreri, E. (1989). "Atmospheric hydrogen peroxide measurements during the 1987 Southern California Air Quality Study." In *Proceedings of the 82nd Annual Meeting of the Air and Waste Management Association, Anaheim, CA*, Paper No. 89-139.5.

Drummond, J. W., Schiff, H. I., Karecki, D. R., and Mackey, G. I. (1989). "Measurements of $NO_2$, $NO_x$, $O_3$, PAN, $HNO_3$, $H_2O_2$ and $H_2CO$ during the Southern California Air Quality Study." In *Proceedings of the 82nd Annual Meeting of the Air and Waste Management Association, Anaheim, CA*, Paper No. 89-139.4.

Forti, G., and Gerola, P. (1977). "Inhibition of photosynthesis by azide and cyanide and the role of oxygen in photosynthesis." *Plant Physiol.* **59**, 859–862.

Gervat, G. P., Clark, P. A., Marsch, A. R. W., Teasdale, I., Chandler, A. S., Choularton, T. W., Gay, M. J., Hill, M. K., and Hill, T. A. (1988). "Field evidence for the oxidation of $SO_2$ by $H_2O_2$ in cap clouds." *Nature (London)* **333**, 241–243.

Graedel, T. E., and Goldberg, K. I. (1983). "Kinetic studies of raindrop chemistry. I. Inorganic and organic processes." *J. Geophys. Res.* **88**, 10865–10882.

Graedel, T. E., Mandich, M. L., and Weschler, C. J. (1986). "Kinetic model studies of atmospheric droplet chemistry. 2. Homogeneous transition metal chemistry in raindrops." *J. Phys. Chem.* **93**, 5205–5221.

Grosjean, D. (1989). "Organic acids in southern California air: Ambient concentrations, mobile source emissions, in situ formation and removal processes." *Environ. Sci. Technol.* **23**, 1506–1514.

Hales, J. M., and Renne, D. S. (1989). "Approaches to determining source-receptor relationship." In *Proceedings of the 82nd Annual Meeting of the Air and Waste Management Association, Anaheim, CA*, Paper No. 89-133.2.

Hanson, P. J., and McLaughlin, S. B. (1989). "Growth, photosynthesis, and chlorophyll concentrations of red spruce seedlings treated with mist containing hydrogen peroxide." *J. Environ. Qual.* **18**, 499–503.

Heikes, B. G. (1984). "Aqueous $H_2O_2$ production from $O_3$ in glass impinger." *Atmos. Environ.* **18**, 1433–1445.

Heikes, B. G. (1989). "Hydrogen peroxide, formaldehyde and organic peroxide signal during MLOPEX: Values, characteristic behavior, and relationship to other species." *EOS, Trans. Am. Geophys. Union* **70**, 1013.

Heikes, B. G., Lazrus, A. L., Kok, G. L., Munen, S. M., Gandrud, B. W., Gitlin, S. N., and Sperry, P. D. (1982). "Evidence for aqueous phase hydrogen peroxide synthesis in the troposphere." *J. Geophys. Res.* **87**, 3045–3051.

Heikes, B. G., Kok, B. L., Walega, J. G., and Lazrus, A. L. (1987). "$H_2O_2$, $O_3$ and $SO_2$ measurements in the lower troposphere over the eastern United States." *J. Geophys. Res., Solid Earth Planets* **92**, 915–931.

Hewitt, C. N., Kok, G. L., and Fall, R. (1990). "Hydroperoxides in plants exposed to ozone mediate air pollution damage to alkene emitters." *Nature (London)* **344**, 56–58.

Hoigné, J., Ader, H., Haag, W. R., and Staehelin, J. (1985). "Rate constant of reactions of ozone with organic and inorganic compounds in water. III." *Water Res.* **19**, 993–1004.

Hough, A. M. (1988). "An intercomparison of mechanisms for the production of photochemical oxidants." *J. Geophys. Res.* **93**, 3789–3812.

Hov, Ø. (1983). "One-dimensional vertical model for ozone and other gases in the atmospheric boundary layer." *Atmos. Environ.* **17**, 535–549.

Jacob, D. J. (1986). "Chemistry of OH in remote clouds and its role in the production of formic acid and peroxymonosulfate." *J. Geophys. Res.* **91**, 9807–9826.

Jacob, D. J., and Hoffmann, M. R. (1983). "A dynamic model for the production of $H^+$, $NO_3^-$, and $SO_4^{2-}$ in urban fog." *J. Geophys. Res.* **88**, 6611–6621.

Jacob, P., Tavares, T. M., and Klockow, D. (1986). "Methodology for the determination of gaseous hydrogen peroxide in ambient air." *Fresenius Z. Anal. Chem.* **325**, 359–364.

Jacob, P., Tavares, T. M., Rocha, V. C., and Klockow, D. (1990). "Atmospheric $H_2O_2$ field measurements in a tropical environment: Bahia, Brazil." *Atmos. Environ.* **24A**, 377–382.

Kadlecek, J. A. (1986). "Cloud water chemistry, winter-summer comparison at Whiteface Mountain." In J. Laznow and G. J. Stensland, eds., *Meteorology of Acidic Deposition.* Air Pollution Control Association, Pittsburgh, PA, pp. 333–351.

Kaiser, W. M. (1979). "Reversible inhibition of the Calvin cycle and activation of oxidative pentose phosphate cycle in isolated intact chloroplasts by hydrogen peroxide." *Planta* **145**, 377–382.

Kelly, T. J., Daum, P. H., and Schwartz, S. E. (1985). "Measurements of peroxides in cloudwater and rain." *J. Geophys. Res.* **90**, 7861–7871.

Keuken, M. P., Bakker, F. P., Lingerak, W. A., and Slanina, J. (1987). "Flow injection analysis of hydrogen peroxide, sulfite, formaldehyde and hydroxymethenesulfonic." *Int. J. Environ. Anal. Chem.* **31**, 263–279.

Keuken, M. P., Schoonebeek, C. A. M., van Wensveen-Louter, A., and Slanina, J.

(1988). "Simultaneous sampling of $NH_3$, $HNO_3$, HCl, $SO_2$ and $H_2O_2$ in ambient air by a wet annular denuder system." *J. Atmos. Environ.* **22**, 2541–2548.

Kleinman, L. I. (1986). "Photochemical formation of peroxides in the boundary layer." *J. Geophys. Res.* **91**, 10889–10904.

Klockow, D., and Jacob, P. (1986). "The peroxyoxalate chemiluminescence and its application to the determination of hydrogen peroxide in precipitation." In W. Jaeschke, ed., *Chemistry of Multiphase Atmospheric Systems*, NATO ASI Ser. G6. Springer-Verlag, Heidelberg, pp. 119–130.

Kok, G. L. (1980). "Measurements of hydrogen peroxide in rainwater." *Atmos. Environ.* **653**, 653–656.

Kok, G. L., Holler, T. P., Lopez, M. B., Nachtrieb, H. A., and Yuan, M. (1978a). "Chemiluminescent method for determination of hydrogen peroxide in the ambient atmosphere." *Environ. Sci. Technol.* **12**, 1072–1076.

Kok, G. L., Darnall, K. R., Winer, A. M., Pitts, J. N., Jr., and Gay, B. W. (1978b). "Ambient air measurements of hydrogen peroxide in the California south coast air basin." *Environ. Sci. Technol.* **12**, 1077–1080.

Kok, G. L., Gitlin, S. N., and Lazrus, A. L. (1986). "Kinetics of the formation and decomposition of hydroxymethanesulfonate." *J. Geophys. Res.* **91**, 2801–2804.

Kok, G. L., Walega, J. G., Heikes, B. G., Lind, J. A., and Lazrus, A. L. (1990). "Measurements of hydrogen peroxide and formaldehyde in Glendora, California." *Aerosol Sci. Technol.* **12**, 49–55.

Kumar, S. (1986). "Reactive scavenging of pollutants by rain: A modeling approach." *Atmos. Environ.* **20**, 1015–1024.

Kunen, S. M., Lazrus, A. L., Kok, G. L., and Heikes, B. G. (1983). "Aqueous oxidation of $SO_2$ by hydrogen peroxide." *J. Geophys. Res.* **88**, 3671–3674.

Lawson, D. R., and Wendt, J. G. (1982). "Acid deposition in California." *SAE Tech. Pap. Ser.* **821246**.

Lawson, D. R., Dasgupta, P. K., Kaplan, I. R., Sakugawa, H., Kok, G. L., Heikes, B. G., Mackay, G. I., Schiff, H. I., and Tanner, R. L. (1988). "Interlaboratory comparison of ambient $H_2O_2$ measurements in Los Angeles." In *Proceedings of the 196th American Chemical Society National Meeting, Los Angeles, CA*, Division of Environmental Chemistry, Vol. 28, p. 107.

Lazrus, A. L., Kok, G. L., Gitlin, S. N., Lind, J. A., and McLaren, S. E. (1985). "Automated fluorometric method for hydrogen peroxide in atmospheric precipitation." *Anal. Chem.* **57**, 917–922.

Lazrus, A. L., Kok, G. L., Lind, J. A., Gitlin, S. N., Heikes, B. G., and Shetter, R. E. (1986). "Automated fluorometric method for hydrogen peroxide in air." *Anal. Chem.* **58**, 594–597.

Lee, R. N., and Busness, K. M. (1989). "The concentration and vertical distribution of peroxides over the Pacific Ocean during the Pacific Stratus Investigation (PSI)." *EOS Trans. Am. Geophys. Union* **70**, 1030.

Lee, Y.-N., Shen, J., and Klotz, P. J. (1986). "Chemical composition of precipitation at Long Island, NY." *Water, Air Soil Pollut.* **30**, 143–152.

Lind, J. A., and Kok, G. L., (1986). "Henry's law determinations for aqueous solutions of hydrogen peroxide, methylhydroperoxide, and peroxyacetic acid." *J. Geophys. Res.* **91**, 7889–7895.

Lind, J. A., Lazrus, A. L., and Kok, G. L. (1987). "Aqueous phase oxidation of sulfur (IV) by hydrogen peroxide, methylhydroperoxide and peroxyacetic acid." *J. Geophys. Res.* **92**, 4171–4177.

Mackay, G. I., Mayne, L. K., and Schiff, H. I. (1990). "Measurements of $H_2O_2$ and HCHO by tunable diode laser absorption spectroscopy during the 1986 Carbonaceous Species Methods Comparison Study in Glendora, California." *Aerosol Sci. Technol.* **12**, 56–63.

MacRae, E. A., and Ferguson, I. B. (1985). "Changes in catalase activity and hydrogen peroxide concentration in plants in response to low temperature." *Physiol. Plant.* **65**, 51–56.

Martin, L. R. (1984). "Kinetic studies of sulfite oxidation in aqueous phase solutions." In J. G. Calvert, ed., *$SO_2$, NO and $NO_2$ Oxidation Mechanisms: Atmospheric Considerations*. Butterworth, Boston, MA, pp. 63–100.

Martin, L. R., Easton, M. P., Foster, J. W., and Hill, M. W. (1989). "Oxidation of hydroxymethanesulfonic acid by fenton's reagent." *Atmos. Environ.* **23**, 563–568.

Masuch, G., Kettrup, A., Mallant, R. K. A. M., and Slanina, J. (1986). "Effects of $H_2O_2$-containing acidic fog on young trees." *Int. J. Environ. Anal. Chem.* **27**, 183–213.

Matsui, H. (1949). "On the content of hydrogen peroxide of atmospheric precipitates." *J. Meteorol. Soc. Jpn.* **2**, 380–381.

McElroy, W. J. (1986). "Sources of hydrogen peroxide in cloudwater." *Atmos. Environ.* **20**, 427–438.

Miller, P. R. (1983). "Ozone effects in the San Bernardino National Forest." In D. D. David, A. A. Miller, and L. Dochinger, eds., *Air Pollution and the Productivity of the Forest*. Isaac Walton League of America, Arlington, VA, pp. 161–197.

Miller, P. R. (1989). "Concept of forest decline in relation to western U.S. forests." In J. J. MacKenzie and M. T. El-Ashry, eds., *Air Pollution's Toll on Forest and Crops*. Yale University Press, New Haven, CT, pp. 75–112.

Miller, P. R., Taylor, O. C., and Poe, M. P. (1986). "Spatial variaton of summer ozone concentrations in the San Bernardino Mountains." In *Proceedings of the 79th Annual Meeting of the Air Pollution Control Association, Minneapolis, MN*, Paper No. 86-39.2.

Miller, P. R., Schilling, S. L., Gomez, A., and Mcbride, J. R. (1989). "Trend of ozone damage to conifer forests between 1974 and 1988 in the San Bernardino Mountains of southern California." In *Proceedings of the 82nd Annual Meeting of the Air and Waste Management Association, Anaheim, CA*, Paper No. 89-129.6.

Mohnen, V. A., and Kadlecek, J. A. (1989). "Cloud chemistry research at Whiteface Mountain." *Tellus* **41B**, 79–91.

Möller, D. (1980). "Kinetic model of atmospheric $SO_2$ oxidation based in published data." *Atmos. Environ.* **14**, 1067–1076.

Möller, D. (1989). "The possible role of $H_2O_2$ in new-type forest decline." *Atmos. Environ.* **23**, 1625–1627.

Munger, J. W., Tiller, C., and Hoffmann, M. R. (1986). "Identification of hydroxymethanesulfonate in fogwater." *Science* **231**, 247–249.

Nair, S. K., and Peters, L. K. (1989). "Studies on non-precipitating cumulus cloud acidification." *Atmos. Environ.* **23**, 1399–1423.

National Acid Precipitation Assessment Program (NAPAP) (1987). *The Causes and Effects of Acidic Deposition: Interim Assessment*, Vols. I–IV. Washington, DC.

National Research Council (NRC) (1983). *Acid Deposition: Atmospheric Processes in Eastern North America*. National Academy Press, Washington, DC.

Olson, T. M., and Hoffmann, M. R. (1989). "Hydroxyalkylsulfonate formation: Its role as an S(IV) reservoir in atmospheric water droplets. *Atmos. Environ.* **23**, 985–997.

Olson, T. M., Torry, L. A., and Hoffmann, M. R. (1988). "Kinetics of the formation of hydroxyacetaldehyde-sulfur (IV) adducts at low pH." *Environ. Sci. Technol.* **22**, 1284–1289.

Olszyna, K. J., Meagher, J. F., and Bailey, E. M. (1988). "Gas-phase, cloud and rain-water measurements of hydrogen peroxide at a high-elevation site." *Atmos. Environ.* **22**, 1699–1706.

Pandis, S. N., and Seinfeld, J. H. (1989). "Sensitivity analysis of a chemical mechanism for aqueous-phase atmospheric chemistry." *J. Geophys. Res.* **94**, 1105–1126.

Penkett, S. A., Jones, B. M. R., Brice, K. A., and Eggleton, A. E. J. (1979). "The importance of atmospheric ozone and hydrogen peroxide in oxidizing sulphur dioxide in cloud and rainwater." *Atmos. Environ.* **13**, 123–137.

Pierson, W. R., and Brachaczek, W. W. (1990). "Dew chemistry and acid deposition in Glendora, California, during the 1986 Carbonaceous Species Methods Comparison Study." *Aerosol Sci. Technol.* **12**, 8–27.

Possanzini, M., Di Palo, V., and Liberti, A. (1988). "Annular denuder method for determination of $H_2O_2$ in the ambient atmosphere." *Sci. Total Environ.* **77**, 203–214.

Richards L. W. (1989). "Airborne chemical measurements in nighttime stratus clouds in the Los Angeles Basin." In R. K. Olson and A. S. Lefohn, eds., *Effects of Air Pollution on Western Forests*, Air Pollut. Control Assoc. Trans. Ser. Air and Waste Management Association, Pittsburgh, PA, pp. 51–72.

Richards L. W., Anderson, J. A., Blumenthal, D. L., McDonald, J. A., Kok, G. L., and Lazrus, A. L. (1983). "Hydrogen peroxide and sulfur (IV) in Los Angeles cloud water." *Atmos. Environ.* **17**, 911–914.

Richards, L. W., Anderson, J. A., Alexander, N. L., Blumenthal, D. L., Knuth, W. R., and McDonald, J. A. (1987). *Characterization of Reactants, Mechanisms, and Species in South Coast Air Basin Cloudwater*. Final Report to California Air Resources Board, STI 94120-606-12FR. Sonoma Technology, Inc., Santa Rosa, CA.

Robinson, J. M., Smith, M. G., and Gibbs, M. (1980). "Influence of hydrogen peroxide upon carbon dioxide photoassimulation in the spinach chloroplast." *Plant Physiol.* **65**, 755–759.

Rodhe, H. P., Crutzen, P., and Vanderpool, A. (1981). "Formation of sulfuric and nitric acids in the atmosphere during long-range transport." *Tellus* **33**, 132–141.

Römer, F. G., Viljeer, J. W., Van den Beld, L., Slangewal, H. J., Veldkamp, A. A., and Reijnders, H. F. R. (1985). "The chemical composition of cloud and rainwater. Results of preliminary measurements from an aircraft." *Atmos. Environ.* **19**, 1847–1858.

Sakugawa, H., and Kaplan, I. R. (1987). "Collection of atmospheric $H_2O_2$: Comparison of cold trap method with impinger bubbling method." *Atmos. Environ.* **21**, 1791–1798.

Sakugawa, H., and Kaplan, I. R. (1989a). "$H_2O_2$ and $O_3$ in the atmosphere of Los Angeles and its vicinity: Factors controlling their formation and their role as oxidants of $SO_2$." *J. Geophys. Res.* **94**, 12,957–12,973.

Sakugawa, H., and Kaplan, I. R. (1989b). "Distribution pattern of atmospheric $H_2O_2$ in Los Angeles and its vicinity and its controlling role as an oxidant of $SO_2$." In D. T. Allan, Y. Cohen, and I. R. Kaplan, eds., *Intermedia Pollutant Transport: Modeling and Field Measurements*. Plenum, New York, pp. 53–72.

Sakugawa, H., and Kaplan, I. R. (1990). "Observation of the diurnal variation of gaseous $H_2O_2$ in Los Angeles air using a cryogenic collection method." *Aerosol Sci. Technol.* **12**, 77–85.

Sakugawa, H., Tsai, W., Kaplan, I. R., and Cohen, Y. (1990a). "Factors controlling the photochemical generation of gaseous $H_2O_2$ in Los Angeles atmosphere." *Geophys. Res. Lett.* **17**, 93–96.

Sakugawa, H., Tsai, W., Kaplan, I. R., and Cohen, Y. (1990b). "Historic trend of the levels of atmospheric $H_2O_2$ during 1960's–1980's in Los Angeles." *Geophys. Res. Lett.* **17**, 937–940.

Sakugawa, H., Tsai, W., Kaplan, I. R., and Cohen, Y. (1990c). *Atmospheric Levels of $H_2O_2$ and $O_3$ in Los Angeles and Surrounding Mountains*. Presented at the NAPAP International Conference for Acidic Deposition: State of Science and Technology, Hilton Head Island, SC, 1990.

Sakugawa, H., Tsai, W., Kaplan, I. R., and Cohen, Y. (1990d). "Atmospheric hydrogen peroxide." *Environ. Sci. Technol.* **24**, 1452–1462.

Saxena, P., Hudischewskyj, A. B., and Seigneur, C. (1986). "The effect of meteorology on the $SO_2$/sulfate relationship." In J. Laznow and G. J. Stensland, eds., *Meteorology of Acidic Deposition*. Air Pollution Control Association, Pittsburgh, PA, pp. 211–225.

Schöne, E. (1874). "Uber das atmosphärische wasserstoffhyperoxyd." *Ber. Dtsch. Chem. Ges.* **7**, 1693–1708.

Schwartz, S. E. (1984). "Gas- and aqueous-phase chemistry of $H_2O_2$ in liquid-water clouds." *J. Geophys. Res.* **89**, 11589–11598.

Schwartz, S. E. (1989). "Acid deposition: Unraveling a regional phenomenon." *Science* **243**, 753–763.

Seigneur, C., and Saxena, P. (1984). "A study of atmospheric acid formation in different environments." *Atmos. Environ.* **18**, 2109–2124.

Seigneur, C., and Saxena, P. (1988). "A theoretical investigation of sulfate formation in clouds." *Atmos. Environ.* **22**, 101–115.

Seigneur, C., and Wegrecki, A. M. (1990). "Mathematical modeling of cloud chemistry in the Los Angeles Basin." *Atmos. Environ.* **24A**, 989–1006.

Slemr, F., Harris, G. W., Hastie, D. R., Mackay, G. I., and Schiff, H. I. (1986). "Measurement of gas phase hydrogen peroxide in air by tunable diode laser absorption spectroscopy." *J. Geophys. Res.* **91**, 5371–5378.

Stockwell, W. R. (1986). "A homogenous gas phase mechanism for use in a regional acid deposition model." *Atmos. Environ.* **20**, 1615–1632.

Tanaka, K., Otsubo, T., and Kondo, N. (1982). "Participation of hydrogen peroxide in the inactivation of Calvin-cycle SH enzymes in $SO_2$ fumigated spinach leaves." *Plant Cell Physiol.* **23**, 1009–1018.

Tanaka, K., Suda, Y., Kondo, N., and Sugihara, K. (1985). "$O_3$ tolerance and the ascorbate-dependent $H_2O_2$ decompositing system in chloroplasts." *Plant Cell Physiol.* **26**, 1425–1431.

Tanner, R. L., and Shen, J. (1990). "Measurement of hydrogen peroxide in ambient air by impinger and diffusion scrubber." *Aerosol Sci. Technol.* **12**, 86–97.

Tanner, R. L., Markovits, G. Y., Ferren, E. M., and Kelly, T. J. (1986). "Sampling and determination of gas-phase hydrogen peroxide following removal of ozone by gas-phase reaction with nitric oxide." *Anal. Chem.* **58**, 1857–1865.

Thompson, A. M., Owens, M. A., and Stewart, R. W. (1989a). "Sensitivity of tropospheric hydrogen peroxide to global chemical and climate change." *Geophys. Res. Lett.* **16**, 53–56.

Thompson, A. M., Stewart, R. W., Owens, M. A., and Herwehe, J. A. (1989b). "Sensitivity of tropospheric oxidants to global chemical and climate change." *Atmos. Environ.* **23**, 519–532.

Tsai, W., Altwicker, E. R., and Asman, W. A. H. (1990). "Numerical simulation of wet scavenging of air pollutants. Part II. Modeling of rain composition at the ground." *Atmos. Environ.* **24A**, 2485–2498.

Tsai, W., Cohen, Y., Sakugawa, H., and Kaplan, I. R. (1991). "Hydrogen peroxide levels in Los Angeles: A screening-level evaluation." *Atmos. Environ.* **25B**, 67–78.

Van Valin, C. C., Ray, J. D., Boatman, J. F., and Gunter, R. L. (1987). "Hydrogen peroxide in air during winter over the south-central United States." *Geophys. Res. Lett.* **14**, 1146–1149.

Van Valin, C. C., Luria, M., Ray, J. D., and Boatman, J. F. (1990). "Hydrogen peroxide and ozone over the northeastern United States in June 1987." *J. Geophys. Res.* **95**, 5689–5695.

Walcek, C. J. (1987). "A theoretical estimate of $O_3$ and $H_2O_2$ dry deposition over the northeast United States." *Atmos. Environ.* **21**, 2649–2659.

Walcek, C. J., Stockwell, W. R., and Chang, J. S. (1990). "Theoretical estimates of the dynamic, radiative and chemical effects of clouds in tropospheric trace gases." *Atmos. Res.* **25**, 53–69.

Williams, W. T. (1980). "Air pollution disease in the California forests. A base line fog smog disease on Ponderosa and Jeffrey pines in the Sequoia and Los Padres National Forests." *Environ. Sci. Technol.* **14**, 179–182.

Williams, W. T. (1983). "Tree growth and smog disease in the forests of California: Case study, Ponderosa pine in the southern Sierra Nevada." *Environ. Pollut. Ser. A* **30**, 59–75.

Yoshizumi, K., Aoki, K., Nouchi, I., Okita, T., Kobayashi, T., Kamakura, S., and Tajima, M. (1984). "Measurements of the concentration in rainwater and of the Henry's law constant of hydrogen peroxide." *Atmos. Environ.* **18**, 395–401.

Young, J. R., Collins, J. F., and Coyner, L. C. (1986). *Analysis of the Southern California Edison Precipitation Chemistry Data Base for Southern California*, Doc. P-D578-300. Environmental Research and Technology, Newberry Park, CA.

Zika, R. G., and Saltzman, E. S. (1982). "Interaction of ozone and hydrogen peroxide in water: Implications for analysis of $H_2O_2$ in air." *Geophys. Res. Lett.* **9**, 231–234.

# 7

# AIR POLLUTION BY GASEOUS POLLUTANTS IN ATHENS, GREECE

*Loizos G. Viras*

*Directory of Air and Noise Pollution Control, Ministry of Environment, Physical Planning and Public Works, Athens, Greece*

*And*

*Panayotis A. Siskos*

*Laboratory of Analytical Chemistry, Department of Chemistry, University of Athens, Athens, Greece*

*Gaseous Pollutants: Characterization and Cycling,* Edited by Jerome O. Nriagu.
ISBN 0-471-54898-7    © 1992 John Wiley & Sons, Inc.

## 1.   INTRODUCTION

Athens became the capital of Greece in 1834. The first city planning was scheduled in 1833 for about 40 000 inhabitants. In the last forty years, due to the rapid industrialization of the Athens area, the city population has grown enormously and now contains over 3.5 million people. This overpopulation caused a sharp increase in all types of pollution, including air pollution. Systematic measurements of air pollution levels were started in 1974 and the first measure against air pollution was the prohibition of the use of heavy oil for central heating in 1977. Since then a number of measures were taken with some effect on the air pollution levels. In the last decade, due to the increase of the number of cars as a result of the increase of the income of the population and due to the favorable meteorological conditions of the area (intense sunshine), the levels of the photochemical pollutants have increased. This has resulted, during some days of the year, in the reduction of atmospheric visibility in Athens, a phenomenon the local people call *nephos*, which means cloud. The status of air pollution in Athens by gaseous pollutants is presented in this chapter and the effect of some meteorological parameters on the intensity of *nephos* is examined.

## 2.   BASIC CONSIDERATIONS

### 2.1.   Geography

The greater Athens area (GAA) is located in the Attica peninsula, which is situated in the central part of Greece and includes the Athens basin, the

Thriassio field, and the area of Messogion (Fig. 1). The Athens basin has a NE–SW direction and is surrounded by the mountains of Hemittos (1026 m), Penteli (1107 m), Parnis (1413 m), and Egaleo (468 m) and by Saronicos Gulf in the south. Narrow geographical openings exist between the mountains. In the Athens basin there are also several hills, the most important of which are the Lycabettus hill (250 m) and the Acropolis hill (120 m).

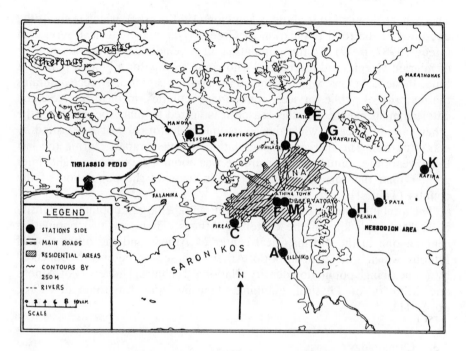

| Code letter | Name of the site |
|---|---|
| A | Elliniko |
| B | Elefsis |
| C | Pireous |
| D | Filadelfia |
| E | Tatoi |
| F | Athens town |
| G | Anavrita |
| H | Peania |
| I | Spata |
| K | Rafina |
| L | Magoula |
| M | Observatory |

**Figure 1.** The topography of the greater Athens area. The major peaks are indicated by their altitude (m) above mean sea level. Contours of 250 m are shown. The meteorological stations are shown by the letters A–M.

The Thriassio field is situated west of the Athens basin and is surrounded by the mountains Egaleo, Parnis, Kitheron (1409 m), and Patera (1911 m), while to the south is the Elefsis Gulf. The Messogion area lies east of the Athens basin and is surrounded by the mountains of Penteli and Hemittos, as well as by the gulfs of Evoikos and Saronikos.

## 2.2. Population

The Athens population is about 3.1 million (1981 census), which is about the one-third of the total population of Greece (National Statistical Service of Greece, 1987, p. 17). To this population must be added a significant number of migrant Greeks and foreigners, who come to stay for a limited period of time. Indeed, Athens, which is the administrative center of Greece, contains the majority of public services, commercial activities, medical services, and entertainment events. For this reason it is visited by many people from the other parts of Greece, while due to its unique historical past it is visited by many tourists. The increase in population was very rapid. During the last thirty years the annual rate of increase was about 4%, which is eight times greater than the corresponding rate of increase of the population of the country (Table 1).

The total inhabited area is over 350 km$^2$ and the largest population density is 32 000 people/km$^2$ in the center of Athens. There are two population centers: one in the old historical center of Athens and the other around Pireous, which is the main harbor of Athens. These two centers are connected by a zone of high population density (National Technical University of Athens and Ministry of Physical Planning, Housing and Environment, 1983, pp. 8–11).

## 2.3. Climatology

The greater Athens area has 12 meteorological stations. Seven of the stations are within the Athens basin, two are in the Thriassio field, and two are in the Messogion area (Fig. 1). The period of measurements for the various stations varies between 96 years (station M) and 18 years (station K) (National Meteorological Service, 1983).

**Table 1 Historical Changes in the Population of Athens and Greece (values in thousands of inhabitants)**

| Year | 1951 | 1961 | 1971 | 1981 | Mean annual increase (%) |
|---|---|---|---|---|---|
| Athens | 1379 | 1853 | 2540 | 3027 | 4 |
| Greece | 7633 | 8389 | 8769 | 9740 | 5 |

In Fig. 2 the mean annual frequency of wind direction at site M for the period 1949–1983 is given. It can be seen that the main wind direction is N and NE. The second most frequent wind direction is S and SW. These winds are attributed to the sea influence (land–sea breeze). Winds from the E and W directions have low frequency, which is determined by the topography of the area. The monthly variation of the wind directions is characterized by an increase in the frequency of the south winds during the months of April and May and an increase in the frequency of the north winds from June to October.

The wind speed shows remarkable differences from station to station. Thus, the maximum frequency of calms is observed inland (site D, 44% of the time) and the minimum is found near the sea (site C, only 5% of the time), (Gagaoudakis, 1979, p. 69). The monthly wind speed variation at site M, which is similar for all sites, is given in Fig. 3. From this figure it is obvious that lower wind speeds are observed during spring and autumn.

The vertical temperature gradient is measured at site A twice a day at 00:00 and 12:00 SGT. The monthly variation of the number of the temperature inversions according to their thickness, during the period 1969–1981, is given in Table 2. The highest frequency of inversions with a thickness of 0–100 m is observed during April and May (Tselepidakis et al., 1983).

In Fig. 4 the mean monthly values for precipitation are given for station M during the period of 1951–1980. The amount of precipitation in Athens is low, especially during summer months. From 1951 to 1980, the maximum mean temperature is observed during July and August and the minimum during January (Fig. 5).

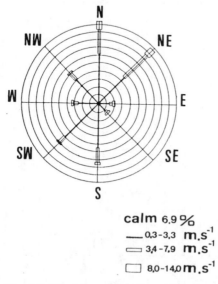

calm 6.9%

— 0,3-3,3 m.s$^{-1}$

▭ 3,4-7,9 m.s$^{-1}$

▢ 8,0-14,0 m.s$^{-1}$

**Figure 2.** Wind rose at site M for 1956–1980.

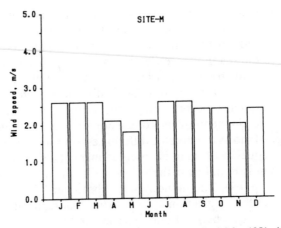

**Figure 3.** Monthly variation of wind speed at site M for 1951–1980.

**Table 2 Monthly Variation of the Number of Temperature Inversions According to Their Thickness**

| Thickness (m) | Jan | Feb | Mar | Apr | May | Jun | Jul | Aug | Sep | Oct | Nov | Dec |
|---|---|---|---|---|---|---|---|---|---|---|---|---|
| 0–100 | 24 | 29 | 25 | 48 | 48 | 34 | 44 | 42 | 8 | 9 | 14 | 21 |
| 101–200 | 87 | 75 | 96 | 96 | 88 | 76 | 72 | 96 | 120 | 109 | 93 | 96 |
| 201–300 | 24 | 10 | 4 | 4 | 1 | 2 | 0 | 2 | 0 | 19 | 44 | 33 |

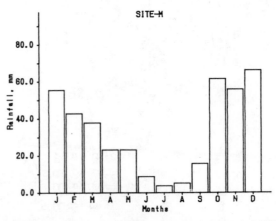

**Figure 4.** Monthly variation of rainfall at site M for the period 1951–1980.

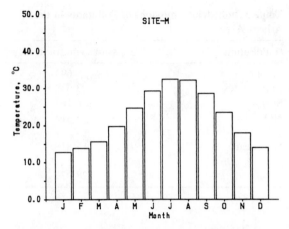

**Figure 5.** Monthly variation of temperature at site M for the period 1951–1980.

## 2.4. Pollution Sources

The main air pollution sources in Athens area are industry, traffic, and central heating. In the following sections some details of the emissions from these sources are given.

### 2.4.1. Industry

The Athens area has industries that were responsible during 1981 for about 50% of the total industrial production of the country. The main industrial activities in the area are concentrated in two zones:

1. *The Thriassio field.* In this area are installed 20 out of 80 industries, with fuel consumption for each one of over 1000 tons per year of heavy oil. The total heavy oil consumption for these industries in 1985 was about 376 000 tons, which represents about 52% of the total fuel used in Athens area by industry. The main industrial installations are oil refineries, metallurgical works, and cement plants.
2. *The area of Drapetsona near the Pireous Harbor.* There are two large industries (with annual consumption over 1000 tons of fuel) in this area, with total annual fuel consumption of about 154 000 tons (21% of the total annual consumption of heavy oil for industrial activities in GAA). The main industrial installations are fertilizer and cement plants.

The remaining industrial activities are scattered mainly in the zone between Athens and Pireous. These are mainly food processing plants, secondary metallurgical works and so on. Estimates of the air pollutants emitted by industrial activities in the year 1985 are given in Table 3.

**Table 3 Industrial Emissions of Pollutants in the Athens Area**

| Pollutant | Emissions (metric tons/yr) |
|-----------|---------------------------|
| Smoke | 693 |
| Particulates | 3 348 |
| $SO_2$ | 4 163 |
| $NO_x$ | 3 816 |
| CO | 241 |
| Hydrocarbons | 15 818 |
| Total | 28 079 |

### 2.4.2.  Traffic

The main categories of vehicles that are important for traffic air emissions in the GAA are (1) private cars, using mainly leaded gasoline (0.15 g/L) (in recent years the number of cars using unleaded gasoline has increased); (2) motorcycles, using mostly gasoline; (3) taxis, buses, and trucks, using diesel oil with low sulfur content (0.3%). The number of vehicles for each category for the year 1987 is given in Table 4. The mean age of private cars in 1986 was 11 years. The mean yearly increase of private cars in Athens is about 6% (Technical Chamber of Greece, 1990, pp. 37–51). The estimate of the air pollutants emitted by traffic in the year 1985 is given in Table 5 (Pattas et al., 1985, p. 21).

### 2.4.3.  Central Heating

Central heating systems in Athens are operated mostly by diesel oil with low sulfur content. Central heating is used from 15 Nov to 15 Apr. The amount

**Table 4 Number of Cars per Vehicle Category in the Athens Area**

| Type of Vehicle | Number of Vehicles |
|-----------------|--------------------|
| Private cars | 760 000 |
| Motorcycles | 284 000 |
| Taxis | 13 800 |
| Buses | 2 000 |
| Gasoline-powered trucks | 100 000 |
| Diesel-powered trucks | 20 000 |
| Tourist buses | 1 500 |
| Total | 497 300 |

**Table 5 Emissions of Pollutants in the Athens Area by Vehicles**

| Pollutant | Emissions (tons/yr) |
|---|---|
| Smoke | 3 370 |
| $SO_2$ | 1 410 |
| $NO_x$ | 18 405 |
| CO | 388 100 |
| Hydrocarbons | 48 790 |
| Total | 460 075 |

of pollutants emitted in 1985 from central heating using the emission factors in the National Technical University Report (1983, p. 37) is given in Table 6. The relative contributions from the various sources in the Athens area are given in Table 7.

## 2.5. Air Pollution Monitoring Network

The first measurements on air quality in Athens were done in 1965 (Alevisatos et al., 1965). A more systematic air pollution network was started in 1969 by

**Table 6 Emissions of Pollutants in the Athens Area by Central Heating**

| Pollutant | Emissions (tons/yr) |
|---|---|
| Smoke | 859 |
| $SO_2$ | 3 690 |
| $NO_x$ | 1 391 |
| CO | 380 |
| Hydrocarbons | 190 |
| Total | 6 510 |

**Table 7 Relative (%) Contribution by Emission Sources in the Athens Area**

| Pollutant | Traffic | Central Heating | Industry |
|---|---|---|---|
| Smoke | 64 | 17 | 19 |
| Particulates | 0 | 0 | 100 |
| $SO_2$ | 7 | 21 | 72 |
| $NO_x$ | 67 | 5 | 28 |
| CO | 100 | 0 | 0 |
| Hydrocarbons | 68 | 0 | 32 |

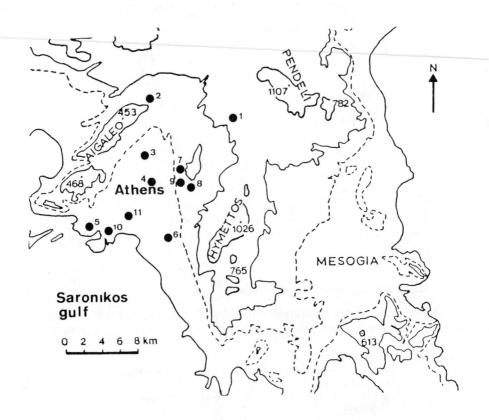

| Code number | Name of the site |
|---|---|
| 1 | Marousi |
| 2 | Liosia |
| 3 | Peristeri |
| 4 | Geoponiki |
| 5 | Drapetsona |
| 6 | Smirni |
| 7 | Patission |
| 8 | Athinas |
| 9 | Ministry |
| 10 | Pireous |
| 11 | Rentis |

**Figure 6.** The topography of Athens basin, showing the locations of air pollution monitoring sites, labeled 1 to 11. Contours of 100 m (---) and 200 m (—) are also shown.

Table 8 Monitoring Sites of the Air Pollution Network of Athens

| Sampling Site | Type of Site | Sampling Height (m) | Parameters Measured |
|---|---|---|---|
| 1 | Residential | 4 | Smoke, $SO_2$, $NO_x$, CO, $O_3$ |
| 2 | Rural | 8 | $SO_2$, $O_3$ |
| 3 | Residential–commercial | 4 | Smoke, $SO_2$, $NO_x$, CO, $O_3$ |
| 4 | Light industrial | 4 | Smoke, $SO_2$, $NO_x$, CO, $O_3$ |
| 5 | Industrial | 4 | Smoke |
| 6 | Residential | 6 | Smoke, $SO_2$, $NO_x$, CO, $O_3$ |
| 7 | Commercial | 8 | Smoke, $SO_2$, $NO_x$, CO, $O_3$ |
| 8 | Commercial | 5 | Smoke, $SO_2$, $NO_x$, CO, $O_3$ |
| 9 | Commercial | 6 | Smoke |
| 10 | Commercial | 4 | Smoke, $SO_2$, $NO_x$, CO, $O_3$ |
| 11 | Light industrial | 8 | Smoke |

the Meteorological Institute of Athens Observatory (National Observatory of Athens, 1970). A network by the Greek government in collaboration with the United Nation Development Project (UNDP) (Environmental Pollution Control Project, 1975) was started in 1973. The number of measuring sites and air pollution parameters have been expanded since 1984. The current air pollution monitoring network consists of 11 measuring sites, which are shown in Fig. 6. The pollution parameters measured at these sites as well some information on the character of the sites is given in Table 8. The methods used for measuring the various pollution parameters are given in Table 9.

## 2.6. Air Quality Limit Values and Long-Term Goals

As a member state, Greece follows the EEC directives on air quality limit values for suspended particulates, $SO_2$, and $NO_2$ (Table 10) (EEC, 1980, 1985). For the remaining pollutants no legislative limit values currently exist. As indicative values, the long-term goals recommended by the World Health Organization (WHO, 1987) are taken into account (Table 11).

Table 9 Methods Used in Measuring Air Pollution in Athens

| Pollution Parameter | Method of Measuring |
|---|---|
| Smoke | Photometric (OECD) |
| $SO_2$ | Fluorometric |
| $NO_x$ | Chemiluminometric |
| CO | IR–ND absorption |
| $O_3$ | UV absorption |

**Table 10  Air Quality Limit Values ($\mu/m^3$) for Smoke, $SO_2$, $NO_2$, and Pb According to the EEC Directives**

| | |
|---|---|
| Median of daily mean values of smoke taken in reference period 1 Apr–31 Mar (one year) | 80 |
| Median of daily mean values of smoke taken in reference period 1 Oct–31 Mar (winter) | 130 |
| 98 percentile of all daily mean values of smoke taken in reference period 1 Apr–31 Mar (one year) | 250 |
| Median of daily mean values of $SO_2$ taken in reference period 1 Apr–31 Mar (one year) | 80 |
| Median of daily mean values of $SO_2$ taken in reference period 1 Oct–31 Mar (winter) | 130 |
| 98 percentile of all daily mean values of $SO_2$ taken in reference period 1 Apr–31 Mar (one year) | 250 |
| 98 percentile of hourly mean values of $NO_2$ taken in reference period 1 Jan–31 Dec | 200 |
| Arithmetic mean of lead values taken in reference period 1 Jan–31 Dec | 2 |

**Table 11  Long-term Goals Suggested by WHO for $O_3$ and CO Concentrations in Air**

| | |
|---|---|
| Maximum 1-h mean value for $O_3$ | 150–200 $\mu g/m^3$ |
| Maximum 1-h value for CO | 30 $mg/m^3$ |
| Maximum 8-h value for CO | 10 $mg/m^3$ |

## 3.  PRIMARY POLLUTANTS

### 3.1.  Sulfur Dioxide

Figure 7 shows the yearly variation of sulfur dioxide concentrations during the period 1984–1989 for two measuring sites, one in the center and the other on the periphery of the city. There has been an increase of the yearly mean values during the last two years in the center, but not on the periphery. This increase cannot be explained by the variation of fuel consumption or by the variation of meteorological factors during the study period, and at the moment is under examination (Ministry of Environment, 1989c, pp. 178–184).

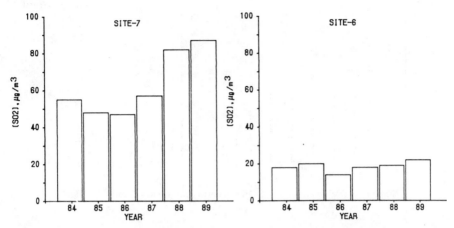

**Figure 7.** Yearly variation of SO$_2$ concentrations at site 7 (urban) and site 6 (rural) for 1984–1989.

In Fig. 8 the monthly variation of sulfur dioxide concentrations for the year 1989 is given for two monitoring sites. In both sites similar variation is observed. During winter months (November to March), when central heating is in operation, the sulfur dioxide concentrations are higher.

Figure 9 shows that the daily variation of sulfur dioxide concentrations in 1989 for two monitoring sites are similar. There is no significant difference between the sulfur dioxide concentrations measured on various days of the week, since central heating, which is the main pollution source for sulfur dioxide during winter does not show any significant daily variation.

In Fig. 10 the hourly variation of sulfur dioxide concentrations at sites 7 and 6, representative of the center and peripheral area of the city, are given. This figure shows maximum values from 9:00 to 10:00 in the morning and 21:00 to 22:00 at night. These peaks coincide with the hours when central

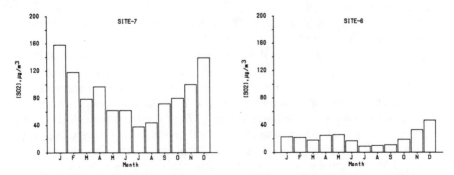

**Figure 8.** Monthly variation of SO$_2$ concentrations at sites 7 and 6 for 1989.

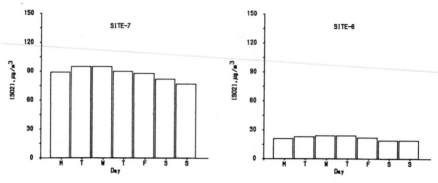

**Figure 9.** Daily variation of SO$_2$ concentrations at sites 7 and 6 for 1989.

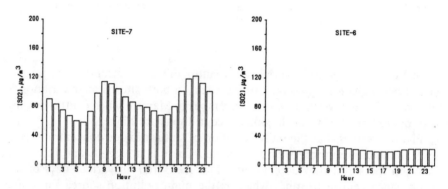

**Figure 10.** Hourly variation of SO$_2$ concentrations at sites 7 and 6 for 1989.

heating is in operation as well as with the peaks of traffic, especially at site 7 (Ministry of Environment, 1989c, pp. 205–217).

In Table 12 the yearly mean, the median, the 98 percentile, and the maximum values of 1-h sulfur dioxide measurements at all sampling sites for the year 1989 are compared. Sites operating in the central area of Athens have higher values than the other sites. This is due to the more intense operation of pollution sources (central heating, traffic) in this area of the city.

Table 13 is a comparison of sulfur dioxide concentrations measured at the site 7, which have the highest values, with the air quality limits enforced in Greece for the period 1984–1989. No violation of the limits has been observed during this 6-year period. The relatively low sulfur dioxide concentrations can be attributed to the policy followed since 1978 of reducing the sulfur content of the fuels, especially those used for central heating. This is very important for a place like Athens which has valuable historical monuments made of marble which is known to be very sensitive to sulfur dioxide (Ministry of Environment, 1989c, pp. 229–235).

**Table 12 The Yearly Mean, the Median, the 98 Percentile and the Maximum 1-h Sulfur Dioxide Values Measured during 1989 at Various Monitoring Sites in Athens (values in µg/m³)**

| Site | Mean Value | Median Value | 98 Percentile Value | Maximum Value |
|------|-----------|--------------|---------------------|---------------|
| 7 | 87 | 71 | 299 | 787 |
| 8 | 42 | 31 | 155 | 424 |
| 10 | 59 | 46 | 204 | 496 |
| 4 | 25 | 13 | 119 | 406 |
| 6 | 22 | 19 | 65 | 262 |
| 2 | 53 | 47 | 134 | 256 |

**Table 13 Comparison of Sulfur Dioxide Concentrations at Site 7 with EEC Limits for the Period 1984–1989 (values in µg/m³)**

| Year | Annual Median Value | 98 Percentile Value | Winter Median Value |
|------|---------------------|---------------------|---------------------|
| 1984 | 43 | 127 | 51 |
| 1985 | 38 | 111 | 41 |
| 1986 | 46 | 171 | 64 |
| 1987 | 53 | 185 | 73 |
| 1988 | 75 | 238 | 95 |
| 1989 | 71 | 217 | 90 |

### 3.2. Carbon Monoxide

The yearly variation of carbon monoxide concentrations during the period 1984–1989 for two measuring sites, one in the center and one on the periphery of the city, is given in Fig. 11. On the periphery of the city the mean yearly values of carbon monoxide have remained almost constant, while in the center of the city there was a decrease of carbon monoxide concentrations during 1984–1986. There has been an increase of the yearly mean values during the last 3 years but still the 1989 levels are lower than those of 1984, in spite of the fact that traffic volume has increased during those years. This is due to the enforcement of control measures on the reduction of carbon monoxide emissions from vehicles (Ministry of Environment, 1989c, pp. 115–120).

Figure 12 shows that the monthly levels of carbon monoxide concentrations for the year 1989 are similar at the two sampling stations. During summer months (July–September) when most people are on vacation there is a reduction of traffic, which leads to a decrease in carbon monoxide concentrations. Another factor affecting the carbon monoxide concentrations is the

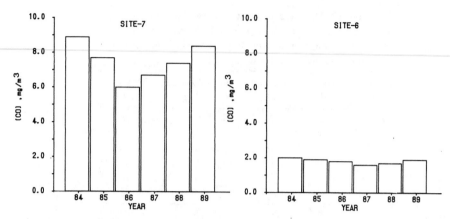

**Figure 11.** Yearly variation of CO concentrations at sites 7 and 6 for 1984–1989.

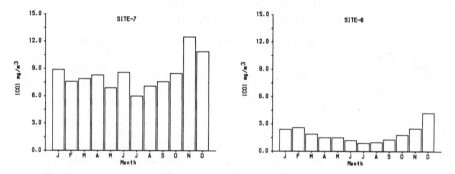

**Figure 12.** Monthly variation of CO concentrations at sites 7 and 6 for 1989.

cold start of the vehicle engines, which can explain the higher concentrations during winter months (Ministry of Environment, 1989c, pp. 121–126).

In Fig. 13, the daily variation of carbon monoxide concentrations for the year 1989 is given. For both sites similar daily variation is observed. On Sundays when traffic is reduced, especially in the center of the city, carbon monoxide concentrations are minimum. The hourly variation of carbon monoxide concentrations at two sites is given in Fig. 14. At site 7, the highest values were measured between 22:00 and 23:00 at night when the traffic was low and the meteorological conditions were unfavorable for pollution dispersion. At site 6 the highest values were observed in the morning hours between 9:00 and 10:00, when the traffic was high.

In Table 14 the yearly mean, the median, the 98 percentile, and the maximum values of 1-h carbon monoxide measurements at all sampling sites for the year 1989 are compared. Sites in the central area of Athens have higher values than the other sites. This is due to differences in the traffic density, the main source of carbon monoxide.

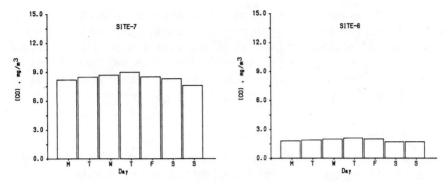

**Figure 13.** Daily variation of CO concentrations at sites 7 and 6 for 1989.

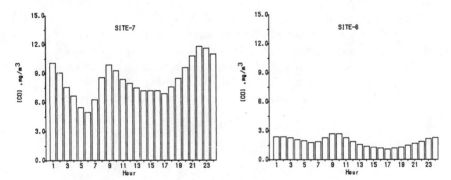

**Figure 14.** Hourly variation of CO concentrations at sites 7 and 6 for 1989.

**Table 14 The Yearly Mean, the Median, the 98 Percentile, and the Maximum 1-h Carbon Monoxide Values Measured during 1989 at Various Monitoring Sites in Athens (values in mg/m³)**

| Site | Mean Value | Median Value | 98 Percentile Value | Maximum Value |
|------|------------|--------------|---------------------|---------------|
| 7 | 8.4 | 7.3 | 21.5 | 38.6 |
| 8 | 4.9 | 4.0 | 15.9 | 32.4 |
| 10 | 5.2 | 4.4 | 12.4 | 23.1 |
| 4 | 1.8 | 1.2 | 8.4 | 21.7 |
| 6 | 1.9 | 1.3 | 7.4 | 18.4 |

Figure 15 compares the 8-h maximum values with the WHO long-term goal at the site that has the highest carbon monoxide values during the period 1984–1989. It is obvious that more intense measures should be taken to reduce carbon monoxide values below the WHO limit. This has been started by giving financial incentives to the people to buy new technology cars equipped with catalytic devices.

### 3.3.  Nitrogen Monoxide

Because nitrogen monoxide measurements were not recorded before 1987, it is impossible to make a statement about the trend of the annual mean values of nitrogen monoxide concentrations (Fig. 16). For the two monitoring sites, the monthly variation of nitrogen monoxide concentrations for the year 1989 is similar (Fig. 17). Since nitrogen monoxide and carbon monoxide have the same main source (vehicles), their concentrations show similar monthly variation. The comments made above are also valid for the monthly variation of nitrogen monoxide concentrations. Likewise, the daily variation of nitrogen monoxide concentrations for the year 1989 for two monitoring sites is similar (Fig. 18). On Sundays when traffic is reduced, especially in the center of the city, nitrogen monoxide concentrations show their minimum values.

The hourly variation of nitogen monoxide concentrations at sites 7 and 6 (center and peripheral) are highest in the morning hours 7:00–9:00 when traffic, the main source of nitrogen monoxide, is also high (Fig. 19).

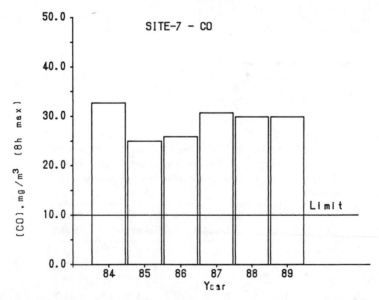

**Figure 15.** Comparison of 8-h maximum values at site 7 with the WHO guide value for 1984–1989.

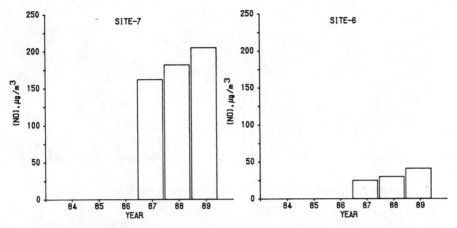

**Figure 16.** Yearly variation of NO concentrations at sites 7 and 6 for 1987–1989.

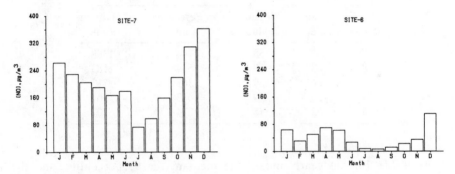

**Figure 17.** Monthly variation of NO concentrations at sites 7 and 6 for 1989.

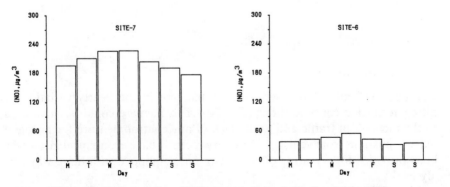

**Figure 18.** Daily variation of NO concentrations at sites 7 and 6 for 1989.

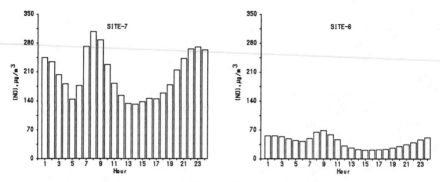

**Figure 19.** Hourly variation of NO concentrations at sites 7 and 6 for 1989.

**Table 15 The Yearly Mean, the Median, the 98 Percentile, and the Maximum 1-h Nitrogen Monoxide Values Measured during 1989 at Various Monitoring Sites in Athens (values in $\mu g/m^3$)**

| Site | Mean Value | Median Value | 98 Percentile Value | Maximum Value |
|------|-----------|--------------|---------------------|---------------|
| 7 | 205 | 158 | 744 | 1388 |
| 8 | 88 | 29 | 634 | 1183 |
| 10 | 66 | 39 | ⌐53 | 713 |
| 4 | 65 | 29 | 413 | 967 |
| 6 | 43 | 15 | 274 | 707 |

In Table 15, the yearly mean, the median, the 98 percentile, and the maximum values of 1-h measurements of nitrogen monoxide measurements at all sampling sites for 1989 are compared. Sites in the central area of Athens generally have higher values than the other sites. This is due to the heavier traffic, which is the main source for nitrogen monoxide.

## 4. SECONDARY POLLUTANTS

### 4.1. Nitrogen Dioxide

Between 1984 and 1989, there was an increase in the yearly variation of nitrogen dioxide concentrations (Fig. 20). This increase could be attributed to the increase in traffic and to the lack of any control measure for nitrogen oxides emissions from vehicles. Only very recently have economical incentives been provided for replacing old cars with ones equipped with three-way catalysts.

The monthly variation of nitrogen dioxide concentratons for the year 1989

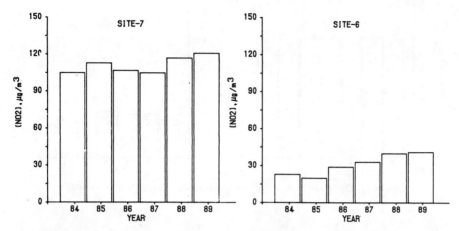

**Figure 20.** Yearly variation of NO$_2$ concentrations at sites 7 and 6 for 1984–1989.

at the two monitoring sites are similar (Fig. 21), with the spring months having the higher values. This can be explained by the fact that during the spring there is still enough traffic to emit nitrogen monoxide, which is converted to nitrogen dioxide in the intense sunshine of this period. During summer, although the sunshine is more intense, the traffic is reduced due to people going away on summer vacations.

There is no significant difference between the nitrogen dioxide concentrations measured at the two sites in 1989 for the various days of the week (Fig. 22). Nitrogen dioxide concentrations depend strongly on the intensity of sunshine, which is independent of the weekday. The hourly variation of nitrogen dioxide concentrations for two monitoring sites in 1989 show higher values occurring during morning hours between 10:00 and 12:00, that is, about 2 h after the corresponding nitrogen monoxide maximum values (Fig. 23). This time hysteresis is related to the conversion of nitrogen monoxide to nitrogen dioxide through the photochemical cycle.

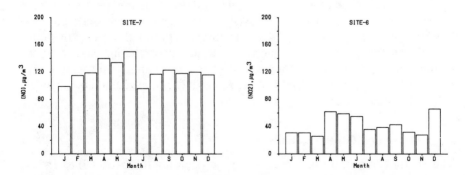

**Figure 21.** Monthly variation of NO$_2$ concentrations at sites 7 and 6 for 1989.

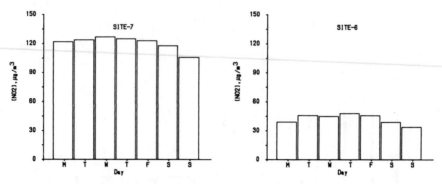

**Figure 22.** Daily variation of NO$_2$ concentrations at sites 7 and 6 for 1989.

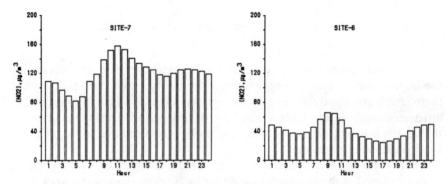

**Figure 23.** Hourly variation of NO$_2$ concentrations at sites 7 and 6 for 1989.

Table 16 compares the yearly mean, the median, the 98 percentile, and the maximum values for the 1-h nitrogen dioxide measurements for all sampling sites for the year 1989. At sites operating in the central Athens area, where traffic is intense, the nitrogen dioxide concentratons are higher than at the other sites.

In Fig. 24, a comparison is made of nitrogen dioxide concentrations measured at the site that presents the highest values, with the air quality limit enforced in Greece for the period 1984–1989. This site is also the only one in which violation of the limit value was observed during the period of study. The 98 percentile value has an increasing trend due to the increase of traffic volume. It is clear that strict measures for the reduction of nitrogen oxides emissions from vehicles should be taken to reduce the nitrogen dioxide levels below the limit.

## 4.2.  Ozone and Other Oxidants

The yearly variation of ozone concentrations during the period 1987 (when measurement of ozone was started) to 1989 for two monitoring sites are similar

**Table 16 The Yearly Mean, the Median, the 98 Percentile, and the Maximum 1-h Nitrogen Dioxide Values Measured during 1989 at Various Monitoring Sites in Athens (values in µg/m³)**

| Site | Mean Value | Median Value | 98 Percentile Value | Maximum Value |
|------|------------|--------------|---------------------|---------------|
| 7 | 121 | 112 | 270 | 634 |
| 8 | 87 | 79 | 199 | 574 |
| 10 | 75 | 70 | 159 | 317 |
| 4 | 66 | 53 | 185 | 396 |
| 6 | 41 | 29 | 157 | 338 |

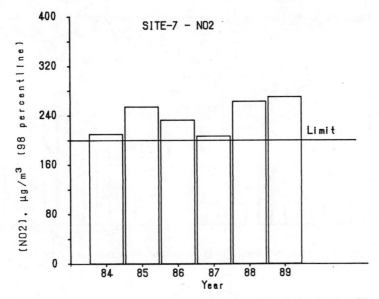

**Figure 24.** Comparison of the 98 percentile at site 7 with the EEC limit value for 1984–1989.

(Fig. 25). At both sites a rapid increase of ozone concentrations is evident. This increase should be attributed to the increase of ozone precursors, that is, hydrocarbons and nitrogen oxides (Ministry of Environment, 1989d, pp. 120–124).

The monthly variation is similar at both sites, with higher values occurring during the period from June to September when the sunshine intensity is at its highest (Fig. 26). For both sites the daily variation is negligible (Fig. 27) since, as in the case of nitrogen dioxide, ozone is strongly dependent on the sunshine intensity, which has no daily variation (Ministry of Environment, 1989d, pp. 131–136). For both sites similar hourly variation is also observed

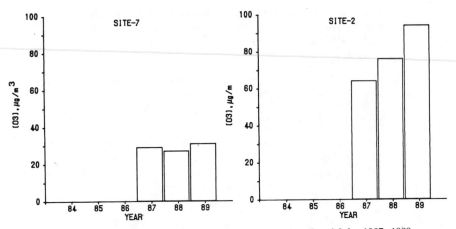

**Figure 25.** Yearly variation of $O_3$ concentrations at sites 7 and 2 for 1987–1989.

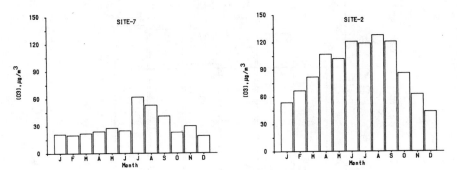

**Figure 26.** Monthly variation of $O_3$ concentrations at sites 7 and 2 for 1989.

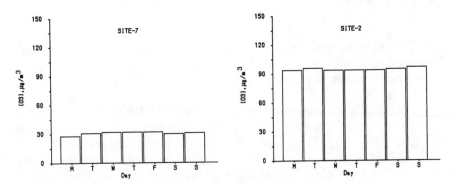

**Figure 27.** Daily variation of $O_3$ concentrations at sites 7 and 2 for 1989.

(Fig. 28), with higher values occurring during the afternoon hours (12:00–16:00) when the sunshine intensity is high.

In Table 17, the yearly mean, the median, the 98 percentile, and the maximum values for the 1-h ozone measurements for all sampling sites for 1989 are compared. Contrary to the other pollution parameters, ozone concentrations are higher on the periphery of the city. This is because ozone is destroyed by the nitrogen monoxide emitted by vehicles, so it cannot be found in the central city sites where nitrogen monoxide concentrations are high.

For the period 1987–1989 the 1-h maximum values indicate violations of the WHO long-term goal (200 $\mu g/m^3$) at almost all the sampling sites, especially at sites located at the periphery of the city (Fig. 29). The problem with ozone concentrations becomes more serious each year. For this reason special consideration should be given to reducing the precursors of ozone.

Some preliminary measurements of peroxyacetyl nitrate (PAN) were done in Athens region (Tsani-Bazaca et al., 1988) from Feb 1985 to Nov 1985, on a hill 250 m above ground level at the center of the city between 12:30 and 13:30 LST. The highest recorded PAN value was 3.7 ppb and the overall

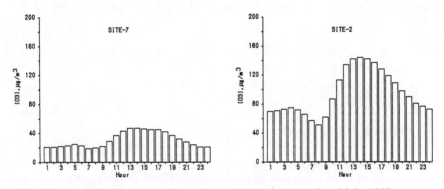

**Figure 28.** Hourly variation of $O_3$ concentrations at sites 7 and 2 for 1989.

Table 17 The Yearly Mean, the Median, the 98 Percentile, and the Maximum 1-h Ozone Values Measured during 1989 at Various Monitoring Sites in Athens (values in $\mu g/m^3$)

| Site | Mean Value | Median Value | 98 Percentile Value | Maximum Value |
|------|------------|--------------|---------------------|---------------|
| 7 | 31 | 21 | 115 | 183 |
| 10 | 55 | 43 | 160 | 265 |
| 4 | 52 | 40 | 160 | 316 |
| 6 | 46 | 32 | 135 | 247 |
| 2 | 94 | 90 | 225 | 376 |

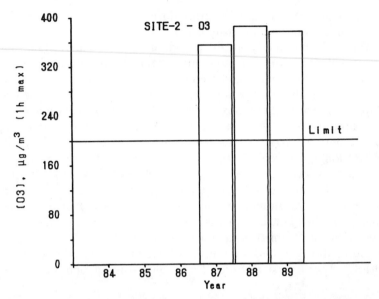

**Figure 29.** Comparison of 1-h maximum values at site 2 with the WHO guide value for 1987–1989.

mean was 0.65 ppb. Over 80% of the PAN concentrations greater than 1 ppb were associated with southwesterly winds of the sea breeze from Saronikos Gulf, which was characterized by low wind velocities.

## 5.   OTHER POLLUTANTS

In addition to the gaseous pollutants, other pollutants are also measured in the Athens area. A summary of the results of those measurements are given in the following sections.

### 5.1.   Smoke

The levels of smoke have decreased during the last few years due to the enforcement of some control measures in the emissions from diesel-powered vehicles (mainly taxis) and from the burning of fuels in the industries and central heating. In spite of this, the levels of smoke in the center of Athens are over the EEC limit values. Lower values are measured on the periphery of the city. Smoke pollution levels are higher in the winter and during working days of the week (Ministry of Environment, 1989c, pp. 21–75).

### 5.2.   Polynuclear Aromatic Hydrocarbons (PAH)

These compounds are known to have mutagenic and carcinogenic effects on human health. Measurements of PAH were done during the years 1984 and

1985 at four sites in Athens including the central and industrial parts of the city. It was found that PAH levels were higher in the center of the city during winter and during the working days of the week. The main pollution sources of PAH were estimated to be the burning of fuel for central heating (in the center of the city) or for industrial use (in the industrial area of the city) during winter. During summer the main source was traffic, especially the diesel-powered vehicles. The PAH levels in the Athens area were comparable to or lower than the levels reported in other cities (Viras et al., 1987; Viras and Siskos, 1990).

### 5.3.  Lead

The levels of atmospheric lead have decreased in the last few years due to the gradual reductions of the lead content of gasoline from 0.40 g/L in 1981 to 0.15 g/L in 1983. The annual mean levels of lead during 1989 were well below the EEC limit value (Ministry of Environment, 1989c, pp. 99–110).

### 5.4.  Acid Rain

The chemistry of wet and dry deposition have been studied in four sampling stations in Athens area since 1986. Twenty to 60% of the samples at various sites have pH values below 5.60. It seems that in Athens there is no serious acid rain problem because the primary acids in rain are neutralized by dust and sea aerosols (Dikaiakos et al., 1990).

## 6.  EFFECT OF SOME METEOROLOGICAL PARAMETERS ON AIR POLLUTION

### 6.1.  Effect of Wind Direction

The effect of wind direction on two pollution parameters, one representative of primary pollutants (CO) and one representative of secondary pollutants ($O_3$), is examined in this section. The data for pollution were taken from the sites that have the corresponding highest values (site 7 for CO and site 2 for $O_3$). The hourly wind direction observations were taken from the Meteorological Institute of Athens Observatory. For each one of the 16 wind directions the corresponding mean value of pollution was calculated for the two periods of the year, winter (January–April) and summer (May–October), and for the two years, 1987 and 1988. Since the pollution levels are affected by the wind speed, the relation between pollution and wind direction was examined for two wind speed ranges: 0.0–5.0 m/s (light and moderate winds) and >5.0 m/s (strong winds). The results in the form of circular diagrams (pollution roses) are given in Figs. 30 and 31.

In general, higher values were observed with S-SW winds (Figs. 30 and 31). Several reasons can explain this observation (Ministry of Environment, 1989c, pp. 160–165; 1989d, pp. 164–168):

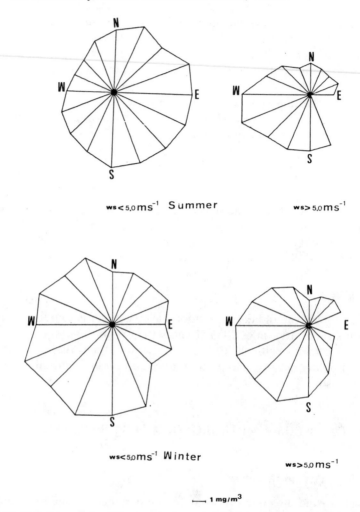

**Figure 30.** Pollution rose for CO concentrations at site 7 for summer and winter and for wind speed <5.0 and >5.0 m/s, 1987–1988.

1. The general topography of the area prevents wind ventilation in a S-SW direction, since the air masses cannot pass over the high mountain regions situated in the northern part of the city (see Fig. 1).
2. For CO, since site 7 is situated north of the center of the city, during S-SW wind flows, pollution is transported from the center of the area to the site under examination.
3. For $O_3$, the S-SW winds blow during evening hours when the ozone concentrations are high.

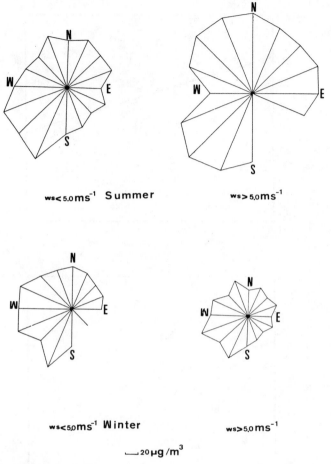

**Figure 31.** Pollution rose for $O_3$ concentrations at site 2 for summer and winter and for wind speed <5.0 and >5.0 m/s, 1987–1988.

## 6.2.  Effect of Wind Speed

The data for examining the effects of wind speed on pollution were taken from the sites that show the corresponding highest values (site 7 for CO and site 2 for $O_3$). The hourly wind speed observations were taken from the records of the Meteorological Institute of Athens Observatory. The wind speeds were divided into groups with a range of 0.5 m/s and for each group the corresponding mean concentration of the pollutant was calculated.

In Figs. 32 and 33, the mean values of CO (for summer and winter) and $O_3$ (for summer only) are plotted against the wind speed for 1987 and 1988. From these figures the following conclusions are drawn:

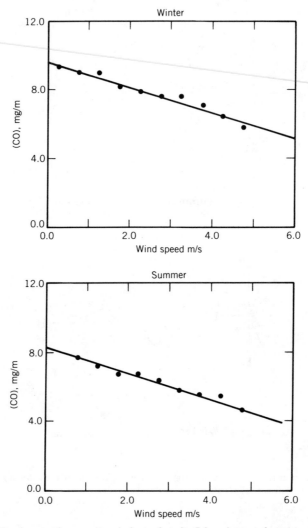

**Figure 32.** CO concentrations versus wind speed at site 7 for winter and summer of 1987–1988.

1. CO concentrations are reduced as wind speed is increased, as a result of the better pollution dispersion. From the linear relation between CO concentrations and wind speed it is estimated that the decrease of the wind speed by 0.5 m/s has as effect of increasing the mean CO concentration by 0.4 mg/m$^3$.

2. $O_3$ concentrations, by contrast, increased as wind speed increased to the value of 5 m/s, while a reduction of 25–30% was observed for wind speeds higher than 5 m/s. The reduction of $O_3$ concentrations with wind speeds higher than 5 m/s may be due to the intense dispersion of pollution by strong winds.

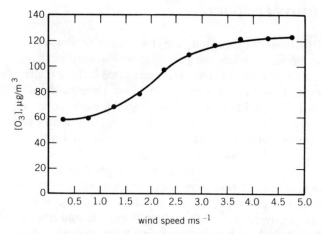

**Figure 33.** $O_3$ concentrations versus wind speed at site 2 for summer 1987–1988.

## 6.3. Effect of Temperature Inversions

The data for inversions were obtained from the National Meteorological Service and include only the night inversions (02:00 a.m. local time). These observations were correlated with the corresponding CO concentrations measured at site 7. The cases of inversions were divided into three categories according to their lapse rate. The first category includes the cases with positive lapse rate independent of its value, the second one is for the inversions with zero lapse rate, and the third one pertains to the cases with negative lapse rate (no inversion). In Table 18 the mean concentration of CO for each case of inversion is given for winter and summer of 1987. In the same table the corresponding mean wind speed value is given. From this table it is clear that CO concentrations increased as the lapse rate was changed from negative to positive, while the mean wind speed decreased at the same time. The latter indicates that during inversions the wind speed is reduced and the resulting increase in pollution is due to the prevention of horizontal and vertical dispersion (Ministry of Environment, 1989c, pp. 155–156).

**Table 18 Effect of Temperature Inversions on Carbon Monoxide Concentrations, Winter and Summer 1987**

| Period | Group I (Positive Lapse Rate) | Group II (Zero Lapse Rate) | Group III (Negative Lapse Rate) |
|---|---|---|---|
| Winter | [CO] = 8.9 ± 5.3 | [CO] = 6.4 ± 3.6 | [CO] = 4.6 ± 3.8 |
|  | WS = 1.7 ± 1.0 | WS = 2.3 ± 1.8 | WS = 4.0 ± 2.3 |
| Summer | [CO] = 5.7 ± 5.3 | [CO] = 4.0 ± 1.9 | [CO] = 2.7 ± 1.2 |
|  | WS = 1.7 ± 1.4 | WS = 2.7 ± 2.8 | WS = 3.5 ± 2.7 |

*Note.* [CO], mean concentration of CO $(mg/m^3)$; WS, mean wind speed (m/s).

## 7. POLLUTION EPISODES

Since 1981 (Ministerial Decision, 1984) a system of alarm measures was established to be taken when the pollution levels were near or exceeded the values indicated in Table 19. The alarm measures include restrictions in traffic and reductions in industrial activity, and they have been taken 19 times as described in Table 20. Most of the cases lasted for one or two days. The least number of days when measures were taken was the summer since during this period the operation of the pollution sources, mainly traffic, is reduced and the wind speed is also relatively high (Fig. 3).

Table 21 shows the number of days per each year for the period 1984–1989 when levels of carbon monoxide or nitrogen dioxide in at least one station reached or exceeded the alarm values of 20 mg/m$^3$ in 8 h or 350 μg/m$^3$ in 1 h, respectively. For both carbon monoxide and nitrogen dioxide the number of days with high pollution levels have increased during the recent years.

**Table 19 Values for Taking Alarm Measures by the Ministry of Environment**

| Pollutant | Tolerable | Alarm I | Alarm II |
|---|---|---|---|
| Smoke (24 h, μg/m$^3$) | 200 | 400 | 600 |
| SO$_2$ (24 h, μg/m$^3$) | 200 | 400 | 500 |
| NO$_2$ (1 h, μg/m$^3$) | 250 | 500 | 700 |
| O$_3$ (1 h, μg/m$^3$) | 250 | 300 | 500 |
| CO (8 h, mg/m$^3$) | 15 | 25 | 35 |

**Table 20 Cases When Alarm Measures Were Taken**

| Spring (n = 5) | Summer (n = 2) | Autumn (n = 6) | Winter (n = 6) |
|---|---|---|---|
| 9 Apr 82 | 17 Jul 84 | 25–26 Nov 81 | 12–13 Jan 82 |
| 11–13 May 82 | 31 Aug–1 Sep 88 | 9 Oct 84 | 3–4 Jan 84 |
| 24–25 May 82 | | 21 Sep 87 | 8–9 Dec 87 |
| 7 May 88 | | 10 Nov 87 | 30–31 Dec 87 |
| 31 Mar–1 Apr 89 | | 31 Oct–1 Nov 89 | 15–16 Dec 89 |
| | | 2–3 Nov 89 | 17–18 Jan 90 |

*Note. n* = number of cases.

**Table 21 Number of Violation Days per Year for the Period 1984–1989**

| Pollutant | 1984 | 1985 | 1986 | 1987 | 1988 | 1989 |
|---|---|---|---|---|---|---|
| CO | 28 | 10 | 3 | 13 | 11 | 33 |
| NO$_2$ | 5 | 12 | 8 | 5 | 20 | 20 |

A study on meteorological conditions for high sulfur dioxide levels (Lalas et al., 1982) found that in all episode cases, Greece was under a stationary high-pressure system, a situation known to produce light winds accompanied by nighttime radiation as well as subsidence cooling, which enhance the stability of the air near the ground. No episodes occurred with a daily average wind speed greater than 1.5 m/s and only two episodes occurred with daily rain duration greater than 1 h.

In another study (Lalas, 1983) it was found that days with photochemical pollution were accompanied by increased sulfur dioxide and smoke levels. In general, the same meteorological conditions described previously are valid for days with high ozone and nitrogen dioxide levels. Preliminary measurements indicated that high photochemical oxidant levels, over 100 ppb, were reached in summer months, especially during calm periods which favor the occurrence of sea breezes. Because Athens is located on the coast and because of the large number of cloudless days, sea breezes appear frequently and during the summer months occur on nearly 50% of all days (Lalas et al., 1983). During sea breeze days with basic wind direction from the north, which is usually the case, the early morning traffic emits pollutants that are advected to the sea. The same is true of all emissions that are released at night after the sea breeze circulation dies down. The pollutants over the sea produce ozone, which is then advected back over the city. The advected ozone enhances, but not by a large amount, the maximum ozone concentrations at the ground in the city and definitely assists in keeping them high in the evening hours (Lalas et al., 1985).

# 8. CONCLUSIONS

Athens, like all large cities, has its own air pollution problems. The position of the city within a basin surrounded by mountains that prevent ventilation of the air in the city, the high population density of the city, and the concentration of large industrial units in the area have led to increased air pollution. In addition, the high sunshine intensity results in the formation of photochemical pollution, especially in the summer months.

Moderate to high pollution values of carbon monoxide, sulfur dioxide, and nitrogen dioxide are found in the center of the city, while concentrations of the same pollutants outside the city are low to moderate. By contrast, ozone has its highest levels on the periphery of the city. Higher values for the primary pollutants are found in winter, while in summer the levels of photochemical pollutants are higher.

Some antipollution measures have been taken since 1977 that have resulted in the reduction of the levels of sulfur dioxide, smoke, and lead. A new plan of action is now under consideration for reducing the photochemical pollutants, mainly through the use of cars equipped with three-way catalysts.

*Acknowledgments*

*The authors express their gratitude to N. Spyropoulos, head of the section of air quality, and G. Ziros, head of the section of traffic pollution, both sections belonging in the Directorate of Air and Noise Pollution Control of the Ministry of Environment, Planning and Public Works.*

# REFERENCES

Alevisatos, P. G., Bacas, V., Alexopoulos, J., and Vericokakis, E. (1965). "The problem of atmospheric pollution in Athens and its surroundings." *Hyg. Arch.* **10-12**, 330–399.

Dikaiakos, C. G., Tsitouris, P. A., Siskos, P. A., Melissos,   , and Nastos, P. (1990). "The rainwater composition in Athens Greece." *Atmos. Environ.* **24B**, 171–176.

EEC (1980). *Directive on Air Quality Limit Values for Sulfur Dioxide and Particulates*, Directive 80/779.

EEC (1985). *Directive on Air Quality Limit Values for Nitrogen Dioxide*, Directive 85/203.

Environmental Pollution Control Project (1975). *Interim Report*, Vol. I. Ministry of Social Services and UNDP, Athens, 100 pp.

Gagaoudaki, C. (1979). "The surface wind as a climatological factor affecting the transport and dispersion of pollutants in Athens region." *National Meteorological Service* Study No. 7, p. 69.

Lalas, P. D. (1983). *Selection, Adjustment and Evaluation of a Model for Predicting Meteorological Conditions which Favor Pollution Episode*, Final Report. Athens, 45 pp.

Lalas, P. D., Veirs, R. V., Karras, G., and Kallos, G. (1982). "Analysis of the $SO_2$ concentrations levels in Athens, Greece." *Atmos. Environ.* **16**, 531–544.

Lalas, P. D., Asimakopoulos, N. D., and Deligiorgi, G. D. (1983). "Sea-breeze circulation and photochemical pollution in Athens, Greece." *Atmos. Environ.* **17**, 1621–1632.

Lalas, P. D., Tombrou-Tsela, M., Asimakopoulos, N. D., Helmis, C., and Petrakis, M. (1985). *A Study of the Horizontal and Vertical Distribution of Ozone over Athens*. Lamda Technical Ltd., Athens, pp. 19–22.

Ministerial Decision (1984). *Official Journal of Greek Democracy*, 339/49.

Ministry of Environment, Physical Planning and Public Works (1989a). *The Atmospheric Pollution in Athens Region*, Vol. I. Athens, 244 pp.

Ministry of Environment, Physical Planning and Public Works (1989b). *The Atmospheric Pollution in Athens Region*, Vol. II. Athens, 399 pp.

Ministry of Environment, Physical Planning and Public Works (1989c). *The Atmospheric Pollution in Athens Region*, Vol. III. Athens, 345 pp.

Ministry of Environment, Physical Planning and Public Works (1989d). *The Atmospheric Pollution in Athens Region*, Vol. IV. Athens, 176 pp.

National Meteorological Service (1983). *Development of a Comprehensive System for*

*Monitoring of Meteorological Parameters affecting Air Pollution in Athens*, Final Report. NMS, Athens.

National Observatory of Athens (1970). *Bulletin of Air Pollution in Athens*, Vol. 1. NOA, Athens, 10 pp.

National Observatory of Athens (1988). *Climatological Bulletin*. NOA, Athens, p. 179.

National Statistical Service of Greece (1987). *Statistical Yearbook of Greece*. NSSG, Athens.

National Technical University of Athens (1983). *Techno-economical Study of the Means for Reducing the Emissions from Central Heating*, Final Report. Natl. Tech. Univ., Athens.

National Technical University of Athens and Ministry of Physical Planning, Housing and Environment (1983). *Evaluation of Traffic and Traffic Conditions and Their Contribution to the Air Pollution*. Final Report. Natl. Tech. Univ., Athens.

Pattas, K. N., Kyriaki, N. A., and Samara, Z. X. (1985). *Exhaust Emission Study of the Current Vehicle Fleet in Athens*, Phase II, Vol. II, Final Report. Athens.

Technical Chamber of Greece (1990). Antipollution Technology—Maintenance, Control and Repair of Vehicles. Final Report. Tech. Chamber of Greece, Athens.

Tsani-Bazaca, E., Glavas, S., and Gusten, H. (1988). Peroxyacetyl Nitrate (PAN) in Athens, Greece. *Atmos. Environ.* **22**, 2283–2286.

Tselepidakis, H., Katsoulis, B., and Lalas, D. (1983). Some aspects on the temperature inversion phenomenon over Athens area. *Proceedings of the 2nd Congress on the Environmental Fluid Mechanics, Athens*, pp. 17–39.

Viras, L. G., and Siskos, P. A. (1990). "Spatial and time variation and effect of some meteorological parameters in polycyclic aromatic hydrocarbons in Athens Greece." *J. Polycyclic Aromatic Compounds* (in press).

Viras, L. G., Siskos, P. A., and Stephanou, E. (1987). "Determination of polycyclic aromatic hydrocarbons in Athens atmosphere." *J. Environ. Anal. Chem.* **28**, 71–85.

World Health Organization (WHO) (1987). *Air Quality Guidelines for Europe*, WHO Reg. Publ., Eur. Ser. No 23. WHO, Copenhagen, pp. 210–220 and 315–326.

# 8

# CHEMICAL MECHANISMS AND PROCESS PARAMETERS OF FLUE GAS CLEANING BY ELECTRON BEAM

*H. Mätzing and H.-R. Paur*

*Kernforschungszentrum Karlsruhe, Laboratorium für Aerosolphysik und Filtertechnik I, Karlsruhe, Germany*

1. **Introduction**
2. **Theory**
3. **Experimental Facilities**
4. **Dose Dependence of Removal Yields**
    4.1. Dose Dependence of $NO_x$ Removal
    4.2. Dose Dependence of $SO_2$ Removal
    4.3. Effects of Ammonia Addition
    4.4. Dose Dependence of Particle Formation
5. **Effects of Relative Humidity and Heterogeneous Reactions**
6. **Dependence of Removal Yields on Initial Concentrations**
7. **Material Balance**
8. **Improvements of Removal Efficiencies**
9. **Summary and Conclusions**
    **References**

*Gaseous Pollutants: Characterization and Cycling,* Edited by Jerome O. Nriagu.
ISBN 0-471-54898-7   © 1992 John Wiley & Sons, Inc.

## 1. INTRODUCTION

The combustion of fossil fuels is usually associated with the emission of trace gases like $NO_x$ and $SO_2$, which are known to damage buildings, crops, and human health through smog and acid formation. The reduction of pollutant emissions is being forced worldwide and some countries have already forwarded corresponding legislation. To some extent, pollutant emissions can be reduced by primary measures (optimization of the combustion process, low sulfur fuels), but additional off-gas cleaning technologies are required to achieve sufficiently low emission standards. The existent emission control technologies usually remove the $SO_2$ and $NO_x$ separately in two stages (wet scrubbers, selective catalytic reduction (SCR)). The transformation products, such as molecular nitrogen and gipsum, are of low economic value. In some cases, the handling of the waste water requires costly installations. The EBDS (electron beam dry scrubbing) process offers an economic and waste water-free alternative. In this novel technology, the flue gas is irradiated with energetic electrons (300–800 keV). This leads to the buildup of radical concentrations that are high enough to degrade the $NO_x$ and $SO_2$ traces simultaneously. The major part of the $NO_x$ and $SO_2$ is oxidized to nitric and sulfuric acids. Upon addition of ammonia, the acids form an aerosol consisting of ammonium nitrate and sulfate, which can be filtered from the main gas stream and can be used as high-quality fertilizer. Only a minor portion of the nitrogen oxides is reduced to molecular nitrogen. Obviously, the chemical transformation of $NO_x$ and $SO_2$ in the EBDS process is, in principle, comparable to the fate of these substances in the free atmosphere, but the process is run at a much higher energy level and under controlled conditions.

A simplified processing scheme is shown in Fig. 1. In the first step, the fly

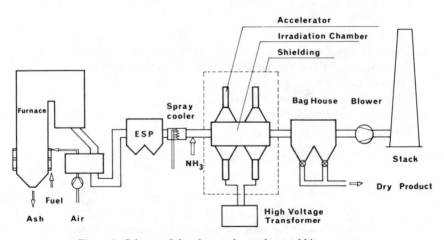

**Figure 1.** Scheme of the electron beam dry scrubbing process.

ash is removed from the flue gas to prevent its mix-up with the product fertilizer. Thereafter, water addition cools down the flue gas to the appropriate process temperature. The ammonia is added just upstream of the irradiation chamber, which is equipped with a shielding to protect the environment against x rays. The solid product is collected in a bag filter or electrostatic precipitator. There is no need for a final heat-up of the purified gas.

The EBDS process was first investigated in Japan (e.g., Suzuki and Tokunaga, 1981; Tokunaga and Suzuki, 1984; Tokunaga et al., 1978). During the 1980s, further research and development work was pursued in the United States and in Europe. Test plants with capacities of several hundred $Nm^3/h$ were installed at the nuclear research centre of Karlsruhe, KfK (Baumann et al., 1987), and at the university of Karlsruhe, ITS (Wittig et al., 1988a,b). Pilot plant investigations were performed by several institutions (KfK, ITS, Badenwerk AG) (Platzer et al., 1990) in Karlsruhe and by the Ebara corporation (Frank et al., 1988; Frank and Hirano, 1990) in Indianapolis with flow rates of 20 000 and 25 000 $Nm^3/h$, respectively. The research and development topics covered the parametric dependence of removal yields (e.g., their dependence on dose and initial concentrations), the formation, filtration, and quality of the product, the mass balance, the long-term operation of the electron accelerators, and the response to load changes. Some specific questions were addressed, in particular, the dependence of removal yields on the acceleration voltage and on the dose rate and the contribution of heterogeneous chemical reactions to the removal yields and to the product formation.

Detailed chemical kinetic models have been developed to simulate the EBDS process numerically (Busi et al., 1987; Mätzing, 1989a; Person and Ham, 1988). The model calculations have improved the understanding of the various physicochemical mechanisms involved in the EBDS process, like the absorption of the electron energy, the ionic and radical reactions in the homogeneous gas phase, the gas–solid phase transition associated with particle formation, and the role of heterogeneous gas–surface reactions. Since hitherto there are no direct measurements of intermediate species available, the importance of single reactions or reaction types must be assessed indirectly by comparison of numerical and experimental results. Consequently, some laboratories have carried out experimental and theoretical work in parallel (Baumann et al., 1987; Wittig et al., 1988a) to establish the optimum performance conditions from a sound recognition of the chemical pathways and practical experience. Only recently, Namba et al. (1988) have started laboratory studies in which selected input species are labeled with the stable $^{15}N$ isotope. Thereby, a complete nitrogen balance can be established experimentally, with particular emphasis on the formation of reduced nitrogen species such as $N_2$ and $N_2O$. Since these products are practically inert under EBDS conditions, they can be considered as favorable candidates with respect to a model verification.

This chapter reviews the most important characteristics of the electron beam dry scrubbing process. The experimental results from laboratory studies, test plants, and pilot plants are compared. The measured removal yields and product distribution depend on a variety of input conditions, such as the initial $NO_x$ and $SO_2$ concentrations, the irradiation dose, and the amounts of water vapor and ammonia additions. The experimental findings are interpreted by kinetic modeling and the mechanisms that determine the process efficiency are discussed.

## 2. THEORY

The interaction of electrons with matter depends on the electron energy and on certain target properties. In the EBDS process, combustion flue gas is irradiated with electrons of 300–800 keV incident energy. This energy is too low to permit close electron–nuclei interactions, the origin of bremsstrahlung. Rather, the electrons transfer their energy to the electron shells of the gas molecules by numerous successive inelastic collisions and become scattered throughout the gas volume that passes the irradiation chamber. The penetration depth between the electron accelerator and the opposite chamber wall is approximately 0.5 m for 300-keV electrons and 2.2 m for 800-keV electrons (Henglein et al., 1969; Lohrmann, 1983). The energy absorbed by the gas molecules leads to excitation, homolytic dissociation into neutral radicals, and heterolytic dissociation into ionic fragments. If more energy is transferred to a molecule than required for dissociative ionization, the excess energy will mainly be taken up by the secondary electron released in the ionization process. The secondary (and higher order) electrons may cause the same excitation and dissociation processes as the incident primary electrons, until finally they "cool down" to thermal energy. The thermalized electrons may directly recombine with positive ions or indirectly via the intermediate formation of negative ions through electron–molecule attachment (Armstrong, 1987; Chatterjee, 1987).

In the way described above, the incident electron energy is distributed in the flue gas components statistically. So the various flue gas components absorb parts of the energy which are proportional to their mass fraction. Therefore, more than 99% of the incident energy is transferred to the main flue gas components, that is, to $N_2$, $O_2$, $H_2O$, and $CO_2$. The primary radiolysis products formed from these gases have been measured quantitatively by Willis and Boyd (1976). The data may be represented by the following stoichiometric equations that show the primary radiolytic events per 100 eV absorbed energy:

$$4.43 \, N_2 \xrightarrow{100 \text{ eV}} 0.29 \, N_2{}^* + 0.885 \, N \, (^2D) + 0.295 \, N \, (^2P)$$
$$+ 1.87 \, N \, (^4S) + 2.27 \, N_2{}^+ + 0.69 \, N^+ + 2.96 e^-$$

$$5.377 \, O_2 \xrightarrow{100 \text{ eV}} 0.077 \, O_2{}^* + 2.25 \, O \, (^1D) + 2.8 \, O \, (^3P)$$
$$+ 0.18 \, O^* + 2.07 \, O_2{}^+ + 1.23 \, O^+ + 3.3 e^-$$

$$7.33 \text{ H}_2\text{O} \xrightarrow{100 \text{ eV}} 0.51 \text{ H}_2 + 0.46 \text{ O } (^3\text{P}) + 4.25 \text{ OH} + 4.15 \text{ H}$$
$$+ 1.99 \text{ H}_2\text{O}^+ + 0.01 \text{ H}_2^+ + 0.57 \text{ OH}^+ + 0.67 \text{ H}^+$$
$$+ 0.06 \text{ O}^+ + 3.3e^-$$

$$7.54 \text{ CO}_2 \xrightarrow{100 \text{ eV}} 4.72 \text{ CO} + 5.16 \text{ O } (^3\text{P}) + 2.24 \text{ CO}_2^+$$
$$+ 0.51 \text{ CO}^+ + 0.07 \text{ C}^+ + 0.21 \text{ O}^+ + 3.03e^-$$

For simplification, excited ionized states have been neglected and the excited states of $N_2^*$ and $O_2^*$ can be summarized by $N_2(A)$ and $O_2(^1\Delta_g)$, respectively. $O^*$ denotes a highly excited O atom above the $O(^1S)$ level. The "fingerprints" of these primary radiolytic events will get lost readily, for example, through very fast charge transfer reactions between the positive ions and surrounding molecules. So these charge transfer reactions serve to redistribute the absorbed energy and contribute to the relaxation of the system from its energetically instable state toward the final state. It is of practical interest here that many charge transfer reactions involve a partial dissociation of the reactants and produce neutral radicals thereby. For example, the major source of the OH radical is constituted by the following ion–molecule reactions:

$$O_2^+ (H_2O) + H_2O \rightarrow H_3O^+ + OH + O_2$$
$$O_2^+(H_2O) + H_2O \rightarrow H_3O^+(OH) + O_2$$
$$H_3O^+(OH) + H_2O \rightarrow H_3O^+ + OH + H_2O$$

The radical production from positive ion–molecule reactions has recently been shown to be one to two orders of magnitudes higher than the primary radical production from radiolysis (Mätzing, 1989a). The chemistry of negative ions, including the mutual ion neutralization, does not lead to substantial radical formation at dose rates below about 100 kGy/s.

While energy absorption and ion–molecule reactions preferentially involve the most abundant flue gas components, the radicals specifically react with the trace components to be removed from the flue gas. An outstanding example is the oxidation of NO by the $HO_2$ radical, $NO + HO_2 \rightarrow NO_2 + OH$, which practically constitutes the only $HO_2$ sink in the system under consideration. The OH radical reacts substantially with all trace components and plays a key role in the simultaneous oxidation of $NO_x$ and $SO_2$ to nitric and sulfuric acid. The vapor pressures of these acids differ by many orders of magnitudes at the typical process temperatures ($95 \pm 25$ °C): The sulfuric acid undergoes spontaneous nucleation and particle formation due to its low vapor pressure, while the nitric acid does not. Only in the presence of ammonia, solid ammonium nitrate particles are formed which transfer the nitric acid from the gas phase into the condensed phase. The sulfuric acid droplets, of course, react with ammonia also to form ammonium sulfate. The mixture of ammonium sulfate/nitrate particulates represents the final product output obtained from electron beam treatment of combustion flue gas. It can be filtered from the main gas stream and can be sold as high-quality fertilizer.

Many of the model results reported in this paper have been obtained with the AGATE code, which was developed at KfK for the numerical simulation of the chemical kinetics during the e-beam processing of flue gas (Mätzing, 1989a). Besides the flue gas composition, the input to the code are the primary radiolytic events (see above) and a large reaction mechanism involving 730 reactions and 95 different species. The dose distribution experienced by the flue gas can be preset to be homogeneous or inhomogeneous and measured dose distribution data can be included also. The system of 95 differential equations is solved by employing a Gear algorithm. A commercial subroutine (Harwell, 1979) is used that has been updated to reduce the time required for the matrix computations. The numerical accuracy, based on atomic mass balance, is $10^{-11}\%$ or better.

## 3.  EXPERIMENTAL FACILITIES

As outlined above, an EBDS installation includes the following three basic utilities: a stage in which the desired operating conditions are established and controlled, a reaction vessel for the electron beam irradiation, and a filtration unit. In the following, the AGATE test plant at the nuclear research center of Karlsruhe (KfK) is described.

The KfK test facility (AGATE-1 plant) is designed to investigate the e-beam treatment of combustion flue gas at variable temperature, relative humidity, input trace gas concentrations, and total mass flow. Figure 2 gives a sketch of the AGATE-1 plant, showing the preconditioning of the flue gas, the irradiation device, and the filter section. From a crude oil fired power plant, a partial volume flow between 100 and 1000 $Nm^3/h$ is fed into the test facility. The gas stream is humidified by the injection of water or steam and is cooled down to the desired process temperature by means of a water-cooled graphite heat exchanger. The prefilter is used for dedusting the gas (mainly from soot particles). In most experiments, the ammonia addition was upstream of the irradiation chamber (as shown in Fig. 2), but some experiments were run with ammonia injection directly into the irradiation chamber. The scanner of the accelerator (300 keV, 12 mA maximum, High-Voltage Eng.) is attached directly to the irradiation vessel. The accelerator window consists of two thin titanium foils (25 and 12 μm thickness) that are cooled by an intermediate air flow. The size of the radiation field and the absorbed radiation dose were measured with NBS-certified film dosimeters (Baumann et al., 1987). The irradiation chamber is designed according to the shape of the radiation field (see Fig. 3). The gas enters the irradiation vessel from two sides near the accelerator window, as shown in Fig. 3, passes the irradiation vessel along the major direction of the electron beam, and leaves it at the

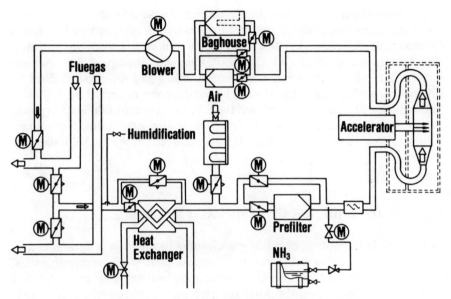

**Figure 2.** Sketch of the AGATE-1 test installation at KfK.

two far sides. Thereby, a 100% irradiation of the total gas flow is ensured. The irradiation vessel is housed in a lead chamber for protection against x rays. The aerosol particles produced in the irradiation chamber are collected in a baghouse equipped with a Dralon T felt (10 m²) or in another test filter. The scrubbed gas is re-introduced into the stack by means of a blower.

Throughout the system, probes and sensors are installed to measure the gas temperature, the operation pressure, and the pressure drops over the

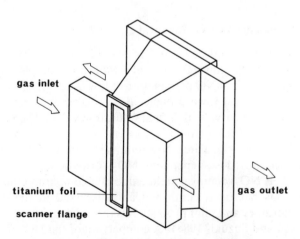

**Figure 3.** Sketch of the irradiation chamber.

filters. Temperatures and mass flow are adjusted automatically. The gas composition is determined before and after the irradiation using infrared analyzers (Spectran, Bodenseewerke) that were calibrated with known gas mixtures (for details, see Baumann et al., 1987; Paur and Jordan, 1988). Aerosol samples are taken between the irradiation vessel and the product filter by isokinetic sampling from the duct. The samples are collected on Millipore Teflon filters (0.2 µm pore size) and the chemical composition is determined by ion chromatography (Dionex). The particle size distribution is measured with a calibrated Anderson Mark III impactor according to VDI standard protocols.

## 4. DOSE DEPENDENCE OF REMOVAL YIELDS

The energy consumption by a flue gas cleaning installation is determined by the sum of energies used to operate all necessary devices, such as blowers, duct heaters, process controllers, and measuring instrumentation. In the electron beam process, a novel central component is required: the electron accelerator, which causes approximately one-third to one-half of the investment and operation costs. The effective energy consumed for the operation of the accelerator is, of course, higher than its output energy (number times voltage of the electrons), since internal losses and energy losses in the window will arise. These factors, however, are determined by the construction and design of a particular accelerator and hence vary between different machines. For a general consideration, it is more convenient to report the results as function of the energy absorbed per amount of gas, that is, the definition of dose. In this section, the basic observations about the dose dependence of removal yields and product formation are presented and discussed.

### 4.1. Dose Dependence of $NO_x$ Removal

Since the first research and development activities on the EBDS process, it has been recognized that the NO removal yield is not a simple linear function of dose. Rather, the initial NO degradation at low dose is much more efficient than the further removal toward higher dose so that the removal–dose relationship resembles a curve with a final saturation value. The early work has been summarized by Machi et al. (1977) and a representative example is given in Fig. 4, which shows calculated NO removal yields as a function of dose. Due to intermediate, dose-dependent $NO_2$ formation, the resulting $NO_x$ (= sum of NO + $NO_2$) removal yield exhibits a comparable nonlinear dose dependence. This has been confirmed and discussed in the course of all subsequent laboratory and pilot plant studies (e.g., Fuchs et al., 1988; Jordan, 1988; Tokunaga and Suzuki, 1984). A comparison of the $NO_x$ removal yields obtained with different installations is shown in Fig. 5. The shapes of the

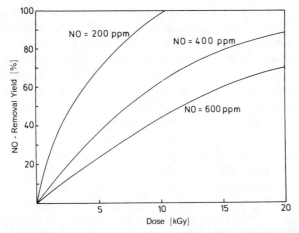

**Figure 4.** NO removal yields as a function of dose. AGATE code calculation with following input: $N_2 = 78.8\%$; $CO_2 = 10\%$; $O_2 = 6\%$; $H_2O = 5\%$; $SO_2 = 500$ ppm; NO = 200, 400, and 600 ppm; $NH_3 = 600, 700, 800$ ppm, respectively; $T = 70$ °C.

removal curves show a comparable curvature, although the input conditions are somewhat different.

The nonlinearity of the $NO_x$ versus dose relationship has been thought to arise from a particular superposition of the radiation field and the velocity distribution in the flue gas that determines the effective (three-dimensional) distribution of dose and dose rate. The homogeneity of the radiation field

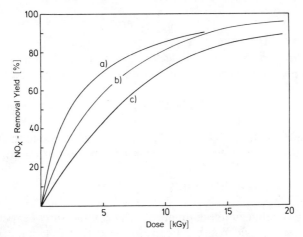

**Figure 5.** $NO_x$ removal versus dose. Experimental curves for $350 \pm 50$ ppm input $NO_x$ concentration; input $SO_2 = 500-1000$ ppm, 0.5–1 stoichiometric ammonia addition, relative humidity $\approx 60\%$. (Data from (*a*) Baumann et al. (1987), (*b*) Frank et al. (1988), and (*c*) Platzer et al. (1990).)

depends mainly on the acceleration voltage and increases with rising voltage. The velocity field of the gas flow can be controlled by the design of the irradiation chamber and ducts and also by stirring installations. Some corresponding experiments were run in which the overall dose rate was varied (Kawamura et al., 1979) and, more recently, spatially resolved measurements of the removal yields and product formation were performed under inhomogeneous irradiation conditions (Wittig et al., 1988a,b). As result, the $NO_x$ removal yield was found to be widely independent of the distribution of dose or dose rate in the flue gas, such that no corresponding recommendation could be derived for technical applications. These experimental results have also been confirmed with numerical calculations simulating one- and three-dimensional geometries (Busi et al., 1988; Mätzing et al., 1991). From these investigations, the conclusion has been derived that at given input conditions the $NO_x$ removal yield depends on the overall dose input only, but not on the spatial distribution of dose or dose rate. A dose rate effect can be expected to occur at dose rates above about 1000 kGy/s and to reduce the removal efficiency (Gentry et al., 1988); however, this regime of dose rates has not been of practical importance in previous or current investigations.

So far, there has not been any report explaining the nonlinear dose dependence of NO and $NO_x$ removal yields directly from experiments. The chemical models suggest that the oxidation of NO and $NO_2$ to nitric acid is not a simple straightforward process, but competes with "back reactions" in which partly oxidized intermediates become reduced (Person and Ham, 1988). The underlying mechanism has been investigated by sensitivity analysis with the AGATE code and the major NO oxidation reactions are found to be

$$NO + HO_2 \rightarrow NO_2 + OH \tag{1}$$

$$NO + O_3 \rightarrow NO_2 + O_2 \tag{2}$$

(ozone is formed from the termolecular addition of $O_2$ and O). The $NO_2$ is found to react extensively with O atoms and to regenerate NO:

$$NO_2 + O \rightarrow NO + O_2 \tag{3}$$

so that the intermediate $NO_x$ is trapped in the cycle of oxidation and reduction reactions until it enters the product channel to form nitric acid:

$$NO_2 + OH + M \rightarrow HNO_3 + M \tag{4}$$

and hence particulate ammonium nitrate. By comparison of experimental and calculated results, evidence has been provided that reaction 3 is the major back reaction opposing the oxidation of $NO_x$ and causing the curvature of the removal versus dose dependence (Mätzing, 1989b). An inhibition of reaction 3 is therefore expected to effect an improved linear dose dependence of the $NO_x$ removal. Since the rate constant of reaction 3 has a negligible temperature dependence (Geers-Müller and Stuhl, 1987), the only practical way to affect its importance is to control the concentrations of the reactants ($NO_2$ and/or O) appropriately. This can be achieved, for instance, by additives which inhibit the O atom source or offer alternative O atom sinks (directly or indirectly). So far, no additive has been proposed that is both economic, easy to handle, and without disadvantage to the quality of the final product. The use of high ammonia concentrations leads to increased O atom consumption via the following fast radical–radical reactions:

$$NH_2 + O \rightarrow HNO + H \tag{5a}$$

$$\rightarrow NH + OH \tag{5b}$$

($NH_2$ is formed from $NH_3$ and OH, see below). However, the ammonia concentration should not be chosen too high, to minimize the ammonia leakage. Another possibility to suppress reaction 3 by multiple irradiation is discussed in Section 8.

## 4.2. Dose Dependence of $SO_2$ Removal

There are only a few reports about the degradation of $SO_2$ in $NO_x$-free gas matrices. From these it is known that in unhydrous environments only comparatively small amounts of $SO_2$ can be decomposed by electron irradiation (e.g., Tokunaga and Suzuki, 1984). Therefore, it has been proposed to increase the $SO_2$ decomposition by the application of static electric fields to promote the electron capture by $SO_2$ and other components (Leonhardt, 1986). The focus of research, however, has been on the increase of $SO_2$ removal by water vapor and ammonia additions which markedly affect the removal efficiency of $SO_2$ (e.g., Jordan, 1988; Paur et al., 1989). At elevated irradiation temperatures (120 °C) and low relative humidity, the $SO_2$ removal is found to depend linearly on the absorbed dose (Fig. 6). Toward decreasing temperature, higher relative humidity and higher ammonia concentrations, the $SO_2$ removal increases substantially, however, $SO_2$ removal still does not exhibit a pronounced dose dependence, as shown in the comparative plot in Fig. 6. Moreover, a large part of the $SO_2$ removal appears to take place in the absence of irradiation, as is obvious from the offset of the removal curves at zero dose.

The kinetics of gas phase ion and radical reactions involving sulfur-containing species has not been investigated as extensively as the chemistry of $NO_x$. The gas phase oxidation of $SO_2$ by O atoms, $O_3$, $HO_2$, and $NO_2$ is known to be very slow around room temperature and to be unimportant in the atmospheric chemistry of sulfur compounds (Calvert et al., 1985; Filby, 1988). So the OH radical was expected to be the only important oxidizer of $SO_2$ in the gas phase, and the relevant mechanism has very recently been

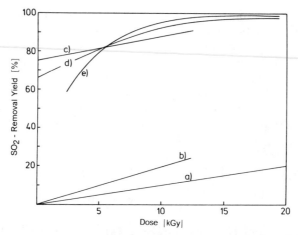

**Figure 6.** $SO_2$ removal versus dose. Experiments without $NH_3$ addition at low relative humidity (<10%) are from (a) Tokunaga and Suzuki (1984) and (b) Baumann et al. (1987). Experiments with 0.5–1 stoichiometric $NH_3$ addition at high relative humidity ($\approx 60\%$) are from (c) Baumann et al. (1987), (d) Platzer et al., (1990), and (e) Frank et al., (1988).

shown to involve the following steps:

$$SO_2 + OH + M \rightarrow HSO_3 + M \qquad (6)$$

$$HSO_3 + O_2 \rightarrow SO_3 + HO_2 \qquad (7)$$

followed by

$$SO_3 + H_2O \rightarrow H_2SO_4 \qquad (8)$$

(Gleason et al., 1987; Gleason and Howard, 1988). The three-body reaction 6 is the rate-determining step in the gas phase oxidation of $SO_2$ under both atmospheric and EBDS conditions. The pressure dependence of reaction 6 has been studied with He, Ar, $N_2$, $O_2$, $CO_2$, and $SO_2$ as the third body (Barnes et al., 1986; Martin et al., 1987, and references therein); the recommended rate expression refers to the temperature range $T = 200$–$300$ K (Atkinson et al., 1989). So the use of this rate expression in EBDS models implies some extrapolation up to temperatures around 400 K, which still needs experimental verification. With this restriction, the radiation induced oxidation of $SO_2$ by OH is calculated to be approximately 4 ppm/kGy absorbed dose independent of the initial conditions (Paur et al., 1990a). The experimental results obtained with low relative humidity and low ammonia concentrations are well represented by this modeling approach. The increase of $SO_2$ removal with rising water vapor and ammonia additions cannot yet be explained from the available kinetic data base. It appears very probable, however, that water vapor plays an important part as the third body in reaction 6. This is discussed in more detail in Section 4.4.

### 4.3. Effects of Ammonia Addition

The major part of the added ammonia is consumed for the formation of nitrate and sulfate salts, but significant amounts of the ammonia also take part in the radical reactions. Thus, ammonia provides a connecting link between the homogeneous gas phase chemistry and the heterogeneous gas–particle conversion in the EBDS process. It is convenient to relate the amount of ammonia addition to the input $NO_x$ and $SO_2$ concentrations in the following way:

$$[NH_3] = s \cdot ([NO_x] + 2 \cdot [SO_2])$$

where $s$ is the ammonia stoichiometry. For typical $s$ values between 0.5 and 1, the ammonia thus becomes the most abundant trace gas component.

Figure 7a shows a comparison of experimental and model results about the effect of increasing ammonia stoichiometry on the NO concentration. In the experiment, the NO concentration has been measured to decrease linearly with rising ammonia stoichiometry and the calculation gives almost the same result. However, the corresponding amount of the removed NO cannot be found in particulates, but rather as stable gaseous products. The reason for this is the attack of OH on ammonia:

$$NH_3 + OH \rightarrow NH_2 + H_2O \tag{9}$$

in which the oxidizing OH radical becomes substituted by the amidogen radical ($NH_2$). The amidogen radical reduces NO and $NO_2$ to molecular nitrogen and nitrous oxide:

$$NO + NH_2 \rightarrow N_2 + H_2O \tag{10}$$

$$NO_2 + NH_2 \rightarrow N_2O + H_2O \tag{11}$$

Although there are still some questions concerning the mechanism of these reactions, enough data are available to formulate the above given overall kinetics and to assess the importance of reactions 10 and 11 in the EBDS process (also see Mätzing, 1989a, 1991). According to the AGATE code calculation, reaction 11 is the only important source of nitrous oxide in the EBDS process. This becomes also evident from Fig. 7b which shows the concentration of $N_2O$ to increase linearly with rising ammonia stoichiometry. The agreement between measurement and calculation is quantitative. A small amount of nitrous oxide is calculated at zero ammonia stoichiometry. This amount is due to reaction 12:

$$NO_2 + N \rightarrow N_2O + O \tag{12}$$

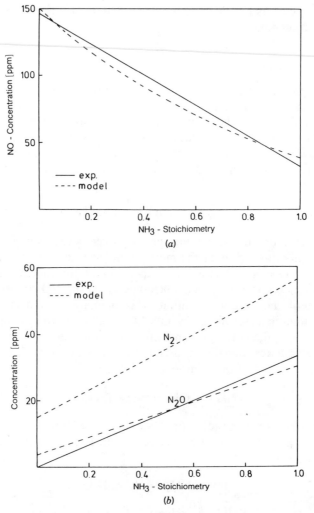

**Figure 7.** (*a*) NO removal as function of ammonia stoichiometry. Initial conditions: 360 ppm NO, 540 ppm $SO_2$, r.h. = 27%, $D$ = 12 kGy. (*b*) $N_2$ and $N_2O$ formation as function of ammonia stoichiometry. (Conditions same as in part *a*.)

which obviously is of negligible importance for $N_2O$ formation. Therefore, the nitrous oxide formation can be controlled by proper adjustment of the ammonia addition.

Figure 7*b* also shows the calculated amount of molecular nitrogen formation which is proportional to the ammonia stoichiometry and corresponds to roughly 15 ± 5% of the input NO at ammonia stoichiometries between 0.5 and 1. A substantial fraction of the molecular nitrogen is already formed at zero ammonia stoichiometry and is due to reaction 13:

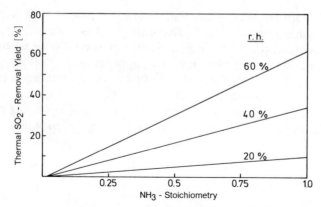

**Figure 8.** Thermal $SO_2$ removal versus ammonia stoichiometry at 20, 40, and 60% r.h. (Evaluated from experimental data by Baumann et al. (1987), Machi et al. (1987), and Wittig et al. (1988a).)

$$NO + N \rightarrow N_2 + O \qquad (13)$$

which is the major sink for N atoms under EBDS conditions. With conventional measurements, however, the additional molecular nitrogen is not easily detectable and usually shows up as a deficit in the N balance. The apparent experimental deficit corresponds well to the calculated amount of molecular nitrogen formation (Paur et al., 1988). A direct experimental measurement of the nitrogen source has been presented by Namba et al. (1988) who used isotope-labeled $^{15}NO$ as input and detected the resulting $^{29}N_2$ by mass spectroscopy. The reported result has been recalculated with the AGATE code and the agreement between experiment and model is good (Mätzing, 1991). Very recently, a complete experimental N balance has been reported by the same working group (Namba et al., 1990) for the conditions of flue gas from an iron-ore sintering furnace with a comparatively high oxygen concentration. As mentioned above, ammonia addition substantially promotes the removal of sulfur dioxide and a considerable fraction of the $SO_2$ can already be removed by the addition of ammonia and water vapor in the absence of electron irradiation, that is, by a purely thermal process. Figure 8 shows the efficiency of the thermal $SO_2$ removal as a function of the ammonia stoichiometry for three different relative humidities. The nature of this removal process is not well known, but it could be shown to take place at surfaces preferentially, mostly across the filter surface and also partly in the ducts (Paur et al., 1989, 1990b).

### 4.4.  Dose Dependence of Particle Formation

The first extensive studies on the aerosol formation were conducted at relative humidities below about 30% and the sulfate and nitrate mass concentrations have been reported to increase linearly with the absorbed dose up to about

$D$ = 12 kGy (Jordan et al., 1986; Paur et al., 1986, 1987). This is also shown by the calculated results in Fig. 9. According to these, the sulfate mass concentration continues to increase linearly with absorbed dose up to 20 kGy. The nitrate mass concentration, however, appears to level off toward higher doses, which reflects both the removal of $NO_x$ and its saturation behavior.

The average mass median aerodynamic particle diameter has been measured to be in the submicrometer range under typical operating conditions (Paur et al., 1986). This also becomes obvious from Fig. 10, which shows that

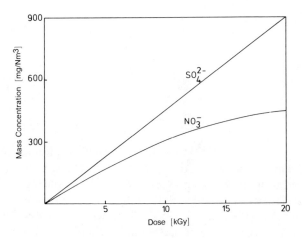

**Figure 9.** Sulfate and nitrate mass concentration as function of dose. AGATE code calculation with following input: 300 ppm NO, 1000 ppm $SO_2$, 1150 ppm $NH_3$, 6% $O_2$, 5% $H_2O$, 12% $CO_2$, rest $N_2$, $T$ = 70 °C (r.h. = 16%).

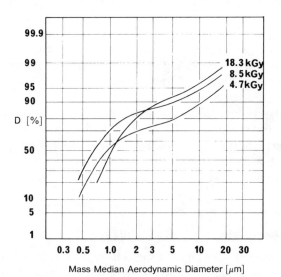

**Figure 10.** Measured particle size distribution.

50–70% of the particles are in the size range below 1 μm. Accordingly, the specific surface of the EBDS aerosol is expected to be rather high and to permit a variety of heterogeneous gas–particle reactions. Calculations about the nucleation of sulfuric acid in the system have given a specific surface around 30 $m^2/g$ at the instant of gas–particle conversion, which drops to around 3 $m^2/g$ within 6 s after irradiation (Mätzing et al., 1988). The importance of heterogeneous chemistry is widely related to the effects of relative humidity, which are discussed in the following section.

## 5. EFFECTS OF RELATIVE HUMIDITY AND HETEROGENEOUS REACTIONS

It has been pointed out above that the removal yield of $SO_2$ is markedly affected by the relative humidity of the irradiated flue gas. The $NO_x$ removal is also influenced by the relative humidity, but in a much less pronounced way. In particular, no measurable $NO_x$ removal takes place in the absence of electron irradiation. The radiation-induced effects of relative humidity on $NO_x$ and $SO_2$ removal can be compared best from the associated effects on aerosol formation.

The measured nitrate mass concentration increases linearly with rising relative humidity and is approximately doubled between 12 and 45% relative humidity (Paur and Jordan, 1988). According to the AGATE code, this fact cannot be explained in terms of radical mechanisms, but requires the consideration of ionic mechanisms, which involve the clustering of positive ions with water molecules. The key reaction to explain the relative humidity effect on nitrate formation is

$$NO_2^+(H_2O)_2 + H_2O \rightarrow HNO_3 + H_3O^+(H_2O) \tag{14}$$

This gas phase reaction proceeds in close analogy to aqueous phase chemistry (Fehsenfeld et al., 1975) and obviously requires sufficient amounts of water vapor to be effective.

The amount of sulfate contained in the aerosol has been shown to increase substantially with rising relative humidity (Baumann et al., 1987; Paur and Jordan, 1988). Together with the effect of relative humidity on the thermal $SO_2$ removal, this observation has stimulated discussion about heterogeneous sulfate formation mechanisms. Under atmospheric conditions, the oxidation of $SO_2$ by $H_2O_2$, $O_3$, $HNO_2$, $O_2$, and selected radicals in cloud droplets is of great importance (Calvert et al., 1985; Seinfeld, 1986). However, it has recently been shown that the gas to particle transport of $H_2O_2$, $O_3$, $HO_2$, and OH is too slow to permit a measurable amount of sulfate formation by heterogeneous reactions under EBDS conditions, as long as only one-step mechanisms are considered (Mätzing et al., 1988). Molecular oxygen and nitrous acid are present in sufficient amounts, but the heterogeneous oxidation of

$SO_2$ by these species usually occurs at time scales above half an hour (Chang et al., 1981; Seinfeld, 1986). Hence, some very effective chain reactions or/ and catalysis would have to be postulated, if these mechanisms are assumed effective under EBDS conditions, that is, within a few seconds of reaction time. Moreover, the heterogeneous oxidation of $SO_2$ by nitrous acid is associated with $N_2O$ formation and the model results by Busi et al. (1988) appear to predict a production of nitrous oxide on account of this mechanism that is too high (Baumann et al., 1987).

For these reasons, it presently appears much more probable that the termolecular addition of $SO_2$ and OH (reaction 6) is very sensitive to water vapor as a third body and therefore leads to increased sulfate formation at elevated relative humidity. A recent kinetic measurement supports this assumption (Nielson et al., 1987). The calculated intermediate OH concentration is high enough to permit a more extensive $SO_2$ oxidation under EBDS conditions than derived from literature data on reaction 6 and, therefore, this assumption has been investigated quantitatively with the AGATE code (Mätzing et al., 1988). If the rate constant is chosen to be $k = 4.4 \times 10^{-34} \exp(+2400K/T)$ $cm^6$/s for M = $H_2O$, close agreement between measured and calculated sulfate concentrations can be obtained. This is shown in Fig. 11, where experimental and model results are compared for the case of zero ammonia stoichiometry. The validity of the fitted rate expression needs further investigation.

The discussion above has also indicated the particular importance of nitrous acid formation and consumption in developing chemical kinetic models on the EBDS process. The major sources of nitrous acid are the termolecular addition of NO and OH and an additional ionic mechanism similar to the nitric acid formation by reaction 14:

$$NO + OH + M \rightarrow HNO_2 + M \tag{15}$$

$$NO^+(H_2O)_3 + H_2O \rightarrow HNO_2 + H_3O^+(H_2O)_2 \tag{16}$$

According to presently available literature data, there are no ionic or radical reactions that could effectively oppose these sources in homogeneous gas phase. On the other hand, nitrous acid is highly unstable against heterogeneous decomposition (Kaiser and Wu, 1977). In analogy to aqueous phase chemistry, the primary decomposition products are NO, $NO_2$, and $HNO_3$. In the AGATE code, the major heterogeneous decomposition reactions are assumed to be

$$HNO_2 \rightarrow \tfrac{1}{3} HNO_3 + \tfrac{2}{3} NO + \tfrac{1}{3} H_2O \tag{17}$$

$$HNO_2 \rightarrow 0.5\ NO_2 + 0.5\ NO + 0.5\ H_2O \tag{18}$$

The rates of these reactions are determined by the concentration of nitrous acid, by the available particle surface, and by a reaction probability between

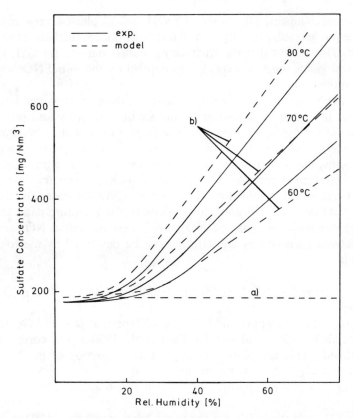

**Figure 11.** Particulate sulfate concentrations as a function of relative humidity. Initial SO₂ concentration 500 ppm, no ammonia addition, dose = 10 kGy. Experimental data from Baumann et al. (1987) and from Paur and Jordan (1988). Calculated results (*a*) based on literature data on reaction 6, and (*b*) obtained with fitted rate constant (see text).

0 and 1. The reaction probabilities were fitted by comparison with experimental results and were found to be 0.1 and 0.01 for reactions 17 and 18, respectively (Mätzing, 1989a). This modeling approach implies a complicated relation betweeen the various oxides of nitrogen and it would certainly be of interest to have direct experimental information concerning the intermediate formation and consumption of nitrous acid.

## 6.    DEPENDENCE OF REMOVAL YIELDS ON INITIAL CONCENTRATIONS

In previous sections, the NOₓ and SO₂ removal yields have been shown to depend on the irradiation dose and on the initial water and ammonia concentrations. Most of the experimental results can be interpreted quantitatively by radical mechanisms. Obviously, the OH radical plays a very important part, since it reacts substantially with all trace components including the added ammonia. This means that the trace components compete for OH. Hence, the removal yields must be expected to depend on the initial $NO_x$ and $SO_2$ concentrations also.

The dependence of $NO_x$ and $SO_2$ removal on the corresponding inlet concentrations has been measured in a number of laboratory and pilot plant studies under various operation conditions (e.g., Frank et al., 1988; Paur and Jordan, 1989; Platzer et al., 1990; Tokunaga and Suzuki, 1984). In all these investigations, the $NO_x$ removal has been shown to increase with rising initial $SO_2$ concentration at constant dose and ammonia stoichiometry. With rising $NO_x$ inlet concentration, the amount of removed $NO_x$ increases and the $NO_x$ removal yield decreases. From a comparison of the literature data, the parametric dependence of removed $NO_x$, $\Delta NO_x$, on the initial $NO_x$ and $SO_2$ concentrations (denoted by subscript 0) can be described by the following expression:

$$\Delta NO_x = 184 \log [NO_x]_0 - 266 + (0.129 \log [NO_x]_0 - 0.27)[SO_2]_0$$

for $[NO_x]_0 \approx 100-1000$ ppm and $[SO_2]_0 \leq 4000$ ppm at $D \approx 15$ kGy, relative humidity (r.h.) $\approx 60\%$, and $s \approx 0.9$ (Paur et al., 1990a). The corresponding $NO_x$ removal yields are plotted in Fig. 12 as a function of the $NO_x$ inlet concentration with $[SO_2]_0$ as parameter.

The $NO_x$ removal is supported by $SO_2$ for two reasons:

1. The $SO_2$ oxidation via reactions 6 and 7 serves to convert OH into $HO_2$, which selectively oxidizes NO with regeneration of OH (reaction 1).

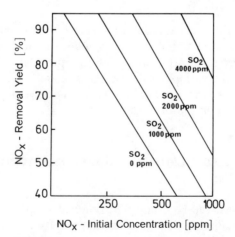

**Figure 12.** $NO_x$ removal yields in dependence of $NO_x$ and $SO_2$ inlet concentration.

This mechanism shifts the aforementioned oxidation–reduction cycle between $NO_2$ and $NO$ toward the oxidation path.

2. The total ammonia concentration is increased markedly when the $SO_2$ concentration is raised at constant ammonia stoichiometry. This favors the removal of $NO_x$ through formation of stable reduced nitrogen species ($N_2$ and $N_2O$). Indeed, Paur and Jordan (1989) have recently shown that at low relative humidity, $D = 10$ kGy, the net effect is an inhibition of particulate nitrate formation. Further work is in progress to quantify the ratio of oxidized and reduced nitrogen species as function of relative humidity and dose.

The $SO_2$ removal yield is primarily determined by the relative humidity, by the ammonia stoichiometry and by the irradiation dose. From Figs. 6 and 8, the $SO_2$ removal yield $\eta$ can be fitted by the following expression (Paur et al., 1990a):

$$\eta\,[\%] = (2.73s - 0.88) \times \text{r.h.}[\%] - 16 + \frac{4\,D[\text{kGy}]}{[SO_2]_0}$$

for $s \gtrsim 0.33$ and r.h. $\gtrsim 10\%$. A slight dependence of the $SO_2$ removal on the initial $NO_x$ concentration is known for low relative humidity conditions only (Paur and Jordan, 1989; Tokunaga and Suzuki, 1984). According to these studies, the $SO_2$ removal is slightly improved upon increasing $NO_x$ concentrations. This effect is of minor importance at high relative humidity and is therefore of low practical interest. Note, however, that this may be taken as an argument against substantial $SO_2$ oxidation by nitrogen-containing species, such as nitric or nitrous acid.

## 7. MATERIAL BALANCE

One of the first reports about the material balance in the EBDS process was given by Kawamura et al. (1980) from their pilot plant operation with flue gas from a sintering furnace. The oxygen concentration was comparatively high (15.5 vol%) under these conditions and the $NO_x$ and $SO_2$ concentrations were both around 200 ppm. The sulfur balance has been found complete within the range of experimental error, but somewhat surprisingly the $NO_x$ balance has been found to be around 125%. Since the experimental ammonia balance was nearly complete, the authors concluded that molecular nitrogen must have been oxidized into $NO_x$ and nitrate. This conclusion is confirmed by Namba et al. (1990) who indeed find some $NO_x$ and nitrate formation from molecular nitrogen (and/or ammonia). The amount, however, does not suffice to explain an $NO_x$ balance above 100% and the authors point out that the previously reported high balance value includes the nitrous oxide.

At the AGATE test plant, the balance of nitrogen compounds was investigated with flue gas of much lower oxygen concentrations between 6 and 9 vol%. The mass concentration of particulate nitrate was found to pass a maximum at 0.5 stoichiometric ammonia addition (Paur and Jordan, 1989) and the overall balance of nitrogen compounds was measured to be 80% at stoichiometric ammonia addition (Jordan, 1990). This experimental deficit in the N balance is explained quantitatively by the model calculation, which predicts the lacking amount to be contained in undetected molecular nitrogen (Mätzing, 1989a). This view is supported by the recent direct determination of molecular nitrogen formation (Namba et al., 1988; see also Section 4.3).

The material balance of sulfur species has also been determined for the conditions of high and low oxygen concentrations performed by the JAERI and AGATE groups. In the JAERI studies, 50–60% of the removed $SO_2$ was collected as sulfate by electrostatic precipitation and the remainder was found as deposits in the ducts. So the total sulfur balance was complete in these experiments (Kawamura et al., 1980; Namba et al., 1990). At the AGATE test plant, 26% of the sulfate formation was identified in the aerosol particles and 45% was shown to occur across the bag filter (Jordan, 1990). The remaining 29% of the removed $SO_2$ was oxidized to sulfate at the duct walls. These results are in good agreement with those obtained at the Indianapolis plant (Kawamura, 1988) and emphasize the necessity to keep the ducts as short as possible. In addition, it is advisable to heat the duct walls to temperatures around 80 °C to minimize deposit formation.


## 8.   IMPROVEMENTS OF REMOVAL EFFICIENCIES

Since it became clear that the dose required for EBDS processing of combustion flue gases is mainly determined by the inlet $NO_x$ concentrations, discussion has focused on the improvement of the $NO_x$ removal efficiency. The most obvious goal in this task is to achieve a linearization of the removal versus dose characteristic. The solution to this has been sought in a change of the irradiation step. Some practical experience exists in this field, but no definite agreement on recommendable means has been achieved. The present section describes items of current research and future developments.

In Section 4.1, the importance of the irradiation geometry was discussed and found negligible with respect to $NO_x$ removal as long as only one irradiation step is considered. Other situations have been considered in which the irradiation dose is not introduced totally in one step, but partly in successive steps, allowing for intermediate energy relaxation and product formation. The idea behind this is to make effective use of the low-dose regime where the slope of the $NO_x$ versus dose curve is highest and fairly linear. Practically, this can be realized by recirculating a part of the once irradiated gas stream into the irradiation zone again or by placing several accelerators along the flow line with an appropriate distance in between.

Some recirculation experiments have been performed at the KfK test plant

(Paur et al., 1987, 1989). The amount of recirculated gas was varied and also the intermediate residence time. The measured $NO_x$ removal was found to increase by a factor of approximately 1.2 in a dose range up to 14 kGy, which means that the observed dose dependence became more linear than without recirculation. Also, the $N_2O$ formation markedly decreased in the recirculation experiment. So the average $NO_2$ concentration in the irradiation chamber must have been lower due to recirculation, since practically all $N_2O$ is formed by reaction of $NO_2$ with $NH_2$ (Section 4.3). Whether the reduction of $NO_2$ is due to gas phase reactions or to heterogeneous scavenging of $NO_2$ is not yet known.

To some degree, the recirculation of flue gas simulates a genuine multiple irradiation, which, of course, requires at least two accelerators. Such an investigation has never been reported in the literature. With the AGATE code, a triple irradiation has been modeled assuming typical input concentrations and a total dose of 12.3 kGy for low relative humidity conditions. A mere triple irradiation gave the same calculated result as obtained for single irradiation. However, when all the $NO_2$ was assumed to be removed between the successive irradiation steps, the calculated dose dependencies of both the NO and $NO_x$ removals turned out perfectly linear (Paur et al., 1989). This coincides well with the observations in the recirculation experiment and indicates the need to remove the $NO_2$ efficiently between successive irradiation steps.

# 9.  SUMMARY AND CONCLUSIONS

The chemistry and the performance characteristics of the EBDS process have been reviewed. The experimental results from laboratory, test plant, and pilot plant studies agree very well and can be understood from detailed kinetic models. The parametric dependencies of the $NO_x$ and $SO_2$ removal yields on the input conditions have been discussed and formulated quantitatively. The process is best suited for flue gas with high $SO_2$ loadings. The operation conditions, such as dose, ammonia, and water additions, can be adjusted fast upon load changes. The process works waste water free and the major product is a mixture of ammonium nitrate and sulfate that can be used as fertilizer. The up-to-date results show that the EBDS technology is safe and competitive with other already well-established technologies.

Due to these interesting features, the electron beam process has gained much international recognition. Demonstration units of 100 MW have been proposed in the United States and Japan. Further pilot plants are under construction in Poland and China, countries that make abundant use of high-sulfur coal. Additional research activities are under way to further improve the energy efficiency of the process, and accelerator prices have been decreasing during the past 10 years. So the EBDS process has a good chance to start a new generation of emission-control technologies.

*Acknowledgments*

*Helpful discussions with Dr. H. Namba are gratefully acknowledged. Part of this work was funded by PEF (Projekt Europäisches Forschungszentrum für Maßnahmen zur Luftreinhaltung) under Contract 86/006/3.*

*Note added in proof*

Results from multiple irradiation experiments were reported at the Third Int. Symp. Adv. Nucl. Energy Research in Mito, Japan, March 1991 and at the IAEA Research Coordination Meeting on Radiation Processing of Flue Gases in Warsaw, Poland, May 1991. The dose requirements were reported to be up to 50% less compared to single step irradiation.

# REFERENCES

Armstrong, D. A. (1987). "The radiation chemistry of gases." In Farhataziz and M. A. J. Rogers, eds., *Radiation Chemistry*. Verlag Chemie, Weinheim, pp. 263–319.

Atkinson, R., Baulch, D. L., Cox, R. A., Hampson, R. F., Jr., Kerr, J. A., and Troe, J. (1989). "Evaluated kinetic and photochemical data for atmospheric chemistry." *J. Phys. Chem. Ref. Data* **18**, 881–1095.

Barnes, I., Bastian, V., Becker, K. H., Fink, E. H., and Nelson, W. (1986). "Oxidation of sulphur compounds in the atmosphere: I. Rate constants of OH radical reactions with sulphur dioxide, hydrogen sulphide, aliphatic thiols and thiophenol." *J. Atmos. Chem.* **4**, 445–466.

Baumann, W., Jordan, S., Mätzing, H., Paur, H.-R., Schikarski, W., and Wiens, H. (1987). "Simultane Rauchgasreinigung durch Elektronenstrahl." *Kernforschungszentr. Karlsruhe KFK-PEF* **17**, 318 pp.

Busi, F., D'Angelantonio, M., Mulazzani, Q. G., Raffaelli, V., and Tubertini, O. (1987). "A kinetic model for radiation treatment of combustion gases." *Sci. Total Environ.* **64**, 231–238.

Busi, F., D'Angelantonio, M., Mulazzani, Q. G., and Tubertini, O. (1988). "Radiation induced $NO_x/SO_2$ emission control for industrial and power plants flue gas." *Radiat. Phys. Chem.* **31**, 101–108.

Calvert, J. G., Lazrus, A., Kok, G. L., Heikes, B. G., Walega, J. G., Lind, J., and Cantrell, C. A. (1985). "Chemical mechanisms of acid generation in the troposphere." *Nature (London)* **317**, 27–35.

Chang, S. G., Toossi, R., and Novakov, T. (1981). "The importance of soot particles and nitrous acid in oxidizing $SO_2$ in atmospheric aqueous droplets." *Atmos. Environ.* **15**, 1287–1292.

Chatterjee, A. (1987). "Interaction of ionizing radiation with matter." In Farhataziz and M. A. J. Rogers, eds., *Radiation Chemistry*. Verlag Chemie, Weinheim, pp. 1–28.

Fehsenfeld, F. C., Howard, C. J., and Schmeltekopf, A. L. (1975). "Gas phase ion chemistry of $HNO_3$." *J. Chem. Phys.* **63**, 2835–2841.

Filby, W. G. (1988). "Participation of free radicals in atmospheric chemistry." In Z. B. Alfassi, ed., *Chemical Kinetics of Small Organic Radicals*, Vol. IV. CRC Press, Boca Raton, FL, pp. 31–57.

Frank, N., and Hirano, Sh. (1990). "The electron-beam FGT process." *Radiat. Phys. Chem.* **35**, 416–421.

Frank, N., Hirano, Sh., and Kawamura, K. (1988). "EBARA electron beam process for flue gas clean-up." *Radiat. Phys. Chem.* **31**, 57–82.

Fuchs, P., Roth, B., Schwing, U., Angele, H., and Gottstein, J. (1988). "Removal of $NO_x$ and $SO_2$ by the electron beam process." *Radiat. Phys. Chem.* **31**, 45–56.

Geers-Müller, R., and Stuhl, F. (1987). "On the kinetics of the reactions of oxygen atoms with $NO_2$, $N_2O_4$ and $N_2O_3$ at low temperatures." *Chem. Phys. Lett.* **135**, 263–268.

Gentry, J. W., Paur, H.-R., Mätzing, H., and Baumann, W. (1988). "A modelling study on the dose rate effect on the efficiency of the EBDS-process." *Radiat. Phys. Chem.* **31**, 95–100.

Gleason, J. F., and Howard, C. J. (1988). "Temperature dependence of the gas-phase reaction $HOSO_2 + O_2 \rightarrow HO_2 + SO_3$." *J. Phys. Chem.* **92**, 3414–3417.

Gleason, J. F., Sinha, A., and Howard, C. J. (1987). "Kinetics of the gas-phase reaction $HOSO_2 + O_2 \rightarrow HO_2 + SO_3$." *J. Phys. Chem.* **91**, 719–724.

Henglein, A., Schnabel, W., and Wendenburg, J. (1969). *Einführung in die Strahlenchemie*. Verlag Chemie, Weinheim, 400 pp.

Jordan, S. (1988). "Progress in the electron beam treatment of stack gases." *Radiat. Phys. Chem.* **31**, 21–28.

Jordan, S., (1990). "On the state of the art of flue gas cleaning by irradiation with fast electrons." *Radiat. Phys. Chem.* **35**, 409–415.

Jordan, S., Paur, H.-R., Cherdron, W., and Lindner, W. (1986). "Physical and chemical properties of the aerosol produced by the 'ES-Verfahren'." *J. Aerosol Sci.* **17**, 669–675.

Kaiser, E. W., and Wu, C. H. (1977). "A kinetic study of the gas phase formation and decomposition reactions of nitrous acid." *J. Phys. Chem.* **81**, 1701–1706.

Kawamura, Ke. (1988). *Testing Conducted on the Ebara E-Beam Flue Gas Treatment System Process Demonstration Unit at Indianapolis, Indiana: Final Report*, DOE/PC/60259-T74, Vol. 1, 253 pp.

Kawamura, Ke., Hirasawa, A., Aoki, S., Kimura, H., Fujii, T., Mizutani, S., Higo, T., Ishikawa, R., Adachi, K., and Hosoki, S. (1979). "Pilot plant experiment of $NO_x$ and $SO_2$ removal from exhaust gases by electron beam irradiation." *Radiat. Phys. Chem.* **13**, 5–12.

Kawamura, Ke., Aoki, S., Kimura, H., Adachi, K., Kawamura, Ka., Katayama, T., Kengaku, K., and Sawada, Y. (1980). "Pilot plant experiment on the treatment of exhaust gas from a sintering machine by electron beam irradiation." *Environ. Sci. Technol.* **14**, 288–293.

Leonhardt, J. W. (1986). "Radiation induced cleaning of exhaust gases—current status, prospects and tasks for basic research." *Radiat. Phys. Chem.* **28**, 559–568.

Lohrmann, E. (1983). *Einführung in die Elementarteilchenphysik*. Teubner Studienbücher, Stuttgart, 148 pp.

Machi, S., Tokunaga, O., Nishimura, K., Hashimoto, S., Kawakami, W., Washino,

M., Kawamura, K., Aoki, S., and Adachi, K. (1977). "Radiation treatment of combustion gases." *Radiat. Phys. Chem.* **9**, 371–388.

Machi, S., Namba, H., and Suzuki, N. (1987). "Research and development of electron beam treatment of combustion flue gases." In *Electron Beam Processing of Combustion Flue Gases*, IAEA-TECDOC-428. IAEA, Vienna, pp. 13–20.

Martin, D., Jourdain, J. L., Laverdet, G., and Le Bras, G. (1987). "Kinetic studies of oxidation reactions of sulfur compounds." *Comm. Eur. Communities* [*Rep.*] *EUR* **EUR 10832**, 212–217.

Mätzing, H. (1989a). "Chemical kinetics of flue gas cleaning by electron beam." *Kernforschungszentr. Karlsruhe KFK* **4494**, 110 pp.

Mätzing, H. (1989b). "The kinetics of O atoms in the radiation treatment of waste gases." *Radiat. Phys. Chem.* **33**, 81–84.

Mätzing, H. (1991). "Molecular nitrogen formation by e-beam processing of flue gas." *Radiat. Phys. Chem.* **37**, 245–248.

Mätzing, H., Paur, H.-R., and Bunz, H. (1988). "Dynamics of particulate formation in the electron beam dry scrubbing process." *J. Aerosol Sci.* **19**, 883–885.

Mätzing, H., Bauman, W., and Paur, H.-R. (1991). "E-beam treatment of waste gas: Dose distribution measurements used in model calculations." *Radiat. Phys. Chem.* **37**, 249–255.

Namba, H., Aoki, Y., Tokunaga, O., Suzuki, R., and Aoki, S. (1988). "Experimental evidence of $N_2$ formation from NO in simulated coal-fired flue gas by electron beam irradiation." *Chem. Lett.*, 1465–1468.

Namba, H., Tokunaga, O., Suzuki, R., and Aoki, S. (1990). "Material balance of nitrogen and sulfur components in simulated flue gas treated by an electron beam." *Appl. Radiat. Isotopes* **41**, 569–573.

Nielson, O. J., Pagsberg, P., and Sillesen, A. (1987). "Absolute rate constant for the reaction of OH with $SO_2$ in the presence of water at atmospheric pressure." *Comm. Eur. Communities* [*Rep.*] *EUR* **EUR 10832**, 338–344.

Paur, H.-R., and Jordan, S. (1988). "Aerosol formation in the electron beam dry scrubbing process (ES-Verfahren)." *Radiat. Phys. Chem.* **31**, 9–13.

Paur, H.-R., and Jordan, S. (1989). "The influence of $SO_2$ and $NH_3$ concentrations on the aerosol formation in the electron beam dry scrubbing process." *J. Aerosol Sci.* **20**, 7–11.

Paur, H.-R., Jordan, S., Baumann, W., Cherdron, W., Lindner, W., and Wiens, H. (1986). "The influence of flue gas humidity and reaction time on the aerosol formation process in the electron beam dry scrubbing process (ES-Verfahren)." In *AEROSOLS: Formation and Reactivity*, Proc. 2nd Int. Aerosol Conf., Berlin, 1986. Pergamon Press, Oxford, pp. 1024–1027.

Paur, H.-R., Jordan, S., and Schikarski, W. (1987). "Simultane Rauchgasreinigung durch Elektronenstrahl." *Kernforschungszentr. Karlsruhe KFK-PEF* **12**, 561–605.

Paur, H.-R., Baumann, W., Mätzing, H., and Jordan, S. (1988). "Aerosolbildung und heterogene Reaktionen bei der simultanen Rauchgasreinigung durch Elektronenstrahl." *Kernforschungszentr. Karlsruhe KFK-PEF* **35**, 739–754.

Paur, H.-R., Jordan, S., Baumann, W., and Mätzing, H. (1989). *Flue Gas Treatment by Electron Beam Irradiation.* Paper presented at RadTech Europe 89, Florence, Italy, 1989.

Paur, H.-R., Baumann, W., Mätzing, H., and Jordan, S. (1990a). "Aerosolbildung und heterogene Reaktionen bei der simultanen Rauchgasreinigung durch Elektronenstrahl." *Kernforschungszentr. Karlsruhe, KFK-PEF* **61**, 665–681.

Paur, H.-R., Namba, H., Tokunaga, O., and Mätzing, H. (1990b). "Aerosol formation and material balance in the electron beam dry scrubbing process." In *AEROSOLS: Science, Industry, Health and Environment*, Vol. 2, Proc. 3rd Int. Aerosol Conf., Kyoto, 1990. Pergamon Press, Oxford, p. 745–748.

Person, J. C., and Ham, D. O. (1988). "Removal of $SO_2$ and $NO_x$ from stack gases by electron beam irradiation." *Radiat. Phys. Chem.* **31**, 1–8.

Platzer, K.-H., Willibald, U., Gottstein, J., Tremmel, A., Angele, H.-J., and Zellner, K. (1990). "Flue gas cleaning by the electron beam process (II): Recent activities at the RDK-7 pilot plant, Karlsruhe." *Radiat. Phys. Chem.* **35**, 427–431.

Seinfeld, J. H. (1986). *Atmospheric Chemistry and Physics of Air Pollution.* Wiley, New York, 738 pp.

Suzuki, N., and Tokunaga, O. (1981). "Radiation chemical studies on electron beam treatment of exhaust gases." In S. Masuda, S. Koda, S. Tsuchiya, M. Taniguchi, O. Tokunaga, and N. Suzuki, eds., *Basic Research on Electron Beam Desulfurization and Denitrification Process*, Final Report of Research Project. Tokyo, pp. 30–47.

Tokunaga, O., and Suzuki, N. (1984). "Radiation chemical reaction in $NO_x$ and $SO_2$ removals from flue gas." *Radiat. Phys. Chem.* **24**, 145–165.

Tokunaga, O., Nishimura, N., Suzuki, N., and Washino, M. (1978). "Radiation treatment of exhaust gases. IV. Oxidation of NO in the moist mixture of $O_2$ and $N_2$." *Radiat. Phys. Chem.* **11**, 117–122.

Willis, C., and Boyd, A. W. (1976). "Excitation in the radiation chemistry of inorganic gases." *Radiat. Phys. Chem.* **8**, 71–111.

Wittig, S., Spiegel, G., Platzer, K.-H., and Willibald, U. (1988a). "Simultane Rauchgasreinigung (Entschwefelung, Denitrierung) durch Elektronenstrahl." *Kernforschungszentr. Karlsruhe KFK-PEF* **45**, 111 pp.

Wittig, S., Spiegel, G., Platzer, K.-H., and Willibald, U. (1988b). "The performance characteristics of the electron-beam-technique: Detailed studies at the (ITS) flue gas facility." *Radiat. Phys. Chem.* **31**, 83–94.

# 9

# VOLATILE ORGANICS IN THE INDOOR ENVIRONMENT: SOURCES AND OCCURRENCE

*Rein Otson*

Bureau of Chemical Hazards, Environmenal Health Directorate, Health Protection Branch, Health & Welfare Canada, Ottawa, Ontario, Canada

*Philip Fellin*

Concord Environmental Corporation, Toronto, Ontario, Canada

*Gaseous Pollutants: Characterization and Cycling,* Edited by Jerome O. Nriagu.
ISBN 0-471-54898-7  © 1992 John Wiley & Sons, Inc.

# 1. INTRODUCTION

## 1.1. Background

The significance of air hygiene in buildings was recognized by Max von Pettenkofer and others more than 100 years ago (Meyer, 1983). However, over the intervening years, air quality health concerns and research have been directed mainly at outdoor air pollution resulting from transportation and industrial sources. Since the 1960s, a relatively weak association has been found between outdoor and indoor pollutants. An increased awareness of the importance of indoor sources and their potential impact on human health (Molhave, 1982; Molhave et al., 1986) has stimulated an interest in indoor air quality (IAQ). There is now evidence that exposure to airborne volatile organic compounds (VOCs) may cause discomfort or more serious health effects (Molhave et al., 1986; Bureau of National Affairs (BNA), 1988; U.S. Environmental Protection Agency (USEPA), 1989; Norback et al., 1990). A New York architect, Richard Stern, has concluded (BNA, 1988) that the end of World War II was the beginning of indoor air deterioration in America and noted that the post war era ushered in the mechanically ventilated building. A further incentive to construct tight buildings was provided by the energy crisis in the 1970s. It is ironic that the "sick building" problems (Molhave et al., 1986; Wallingford, 1988; BNA, 1988) reported in the last few decades may largely be due to the lack of adequate building ventilation by what was previously considered as relatively "polluted" outdoor air.

The scientific evidence now indicates that air within homes and other structures occupied by humans can be more seriously polluted than the outdoor air. Indeed, levels of some volatile organic compounds may be several times higher indoors than outdoors (Wallace et al., 1987a; Yocom, 1982). Also, time–activity studies have demonstrated that most people spend the greater part of their time indoors. Studies in the United States (USEPA, 1989; Meyer, 1983) indicate that approximately 93% of time is spent indoors, 5% of time is spent in transit, and only 2% of time outdoors. Others have reported similar values (World Health Organization (WHO), 1986; Moschandreas, 1981). The combination of potentially greater concentrations and proportion of time spent indoors than outdoors suggests that exposure to airborne organics, and hence health risks, can be greater indoors than outdoors.

Air contaminant levels in indoor environments are the result of complex interactions of interrelated variables, which include contaminant source characteristics, building characteristics, contaminant removal mechanisms, outdoor air quality, and meteorological conditions. Increases in indoor airborne pollutant levels in recent years have, in part, been attributed to efforts to reduce heating and cooling costs for buildings (Wallingford, 1988; BNA, 1988). Although building energy conservation measures, such as reduction of outdoor air infiltration, are important, it has been suggested (Wallace, 1987) that the presence of a large number of emission sources with significant emission strengths is even more important in the development of high indoor air pollutant concentrations. In the last 50 years, the development and use of a large number of synthetic products and materials that represent potential sources of indoor air pollutants has grown. It is now common practice to use unvented combustion appliances and synthetic materials for building furnishings and structural components. All of these factors—increased number of sources, building energy conservation measures, and the shift to indoor activities up to 80 or 90% of the time—have led to a potential increase in human exposure to indoor pollutants.

In 1984, further research on indoor VOCs, their sources, and decay rates was identified as a high priority by the World Health Organization (WHO, 1986). Recent advances in analytical techniques have allowed selective and accurate determination of VOCs and have contributed to provide a rapidly increasing amount of data on the occurrence of both indoor and outdoor VOCs. There has also been an increasing awareness of the need to identify the origin of airborne VOCs. As a result, the potential sources and occurrence of airborne VOCs have been examined in numerous studies, the most comprehensive being the Total Exposure Assessment Methodology (TEAM) study (Wallace, 1987) in the United States. In other developed countries, notably some in Europe, considerable efforts have also been expended on monitoring and research of indoor VOCs.

There are several excellent text books and articles that review and discuss various aspects of IAQ. These include Beat Meyer's exhaustive review of the state of knowledge in 1983, a text on "Indoor Air Pollution: Characterization, Prediction and Control" (Wadden and Scheff, 1983), and a more recent book on indoor air pollution control and related topics (Godish, 1989c). A 1981 publication by the Committee on Indoor Pollutants (National Research Council (NRC), 1981) and a 1984 WHO meeting report on indoor air quality research (WHO, 1986) appraised the available knowledge and provided recommendations for further research. Information on selected IAQ related topics can also be found in texts edited by Walsh et al. (1984) and Perry and Kirk (1988). IAQ control and microenvironments, respectively, are discussed in books by Fisk et al. (1987) and by Jokl (1989). There are also several books and articles that provide excellent reviews of tobacco (Hoffman and Wynder, 1986; International Agency for Research on Cancer (IARC), 1987) and formaldehyde (Meyer, 1979; Turoski, 1985) related issues and IAQ research and do this in greater detail than is possible here.

## 1.2.  Scope of Review

This review intends to inform the reader about recent advances in research and understanding of factors that affect IAQ. Current knowledge of sources and their contribution to airborne indoor VOC levels is described first. Then, much of the available information on the occurrence of indoor airborne VOCs in a variety of nonindustrial environments is reviewed and also summarized in tabular form. A separate section on characterization methods evolved because many of the procedures for characterization of source emissions and for VOCs that occur indoors are the same or similar. Then, processes that may affect the fate of indoor VOCs are described and mathematical representation of the factors affecting IAQ is briefly discussed. Finally, a summary of findings and a brief discussion of research needs are provided. Throughout, the focus is mainly on advances during the past decade.

Differences in physical, chemical, and biological behavior have led to classification of airborne organic contaminants into nonvolatile, semivolatile, and volatile categories. Several definitions have been proposed for volatile organic compounds. Lewis (1986), for example, has classified semivolatiles (SVOCs) as compounds with vapor pressures (Vp) between $10^{-1}$ and $10^{-7}$ Torr (mm Hg) at 25 °C. By inference, volatiles are compounds with Vp exceeding $10^{-1}$ Torr. More recently, Lewis (1989) classified compounds with Vp above $10^{-2}$ kPa (0.08 Torr) and between $10^{-2}$ and $10^{-8}$ kPa, respectively, as volatile and semivolatile. A practical reason for defining VOCs as compounds with Vp > $10^{-1}$ Torr is related to the fact that these compounds are normally present and measured in air at levels above 0.1 µg/m³. Such concentrations allow measurement by simpler techniques than those employed for the semivolatiles since the SVOCs are normally present in the nanogram to picogram/m³ range. This practical distinction is likely the rationale for the otherwise arbitrary definition and has been adopted for the purposes of this review.

Since the potential for human exposure provides the rationale for assessment of airborne organic compounds, physical properties other than volatility have been recognized to be important in the classification of organics. Andelman (1985a), for example, suggested that compounds with Henry's constant $k_H \geq 2.5 \times 10^{-6}$ atm m³/mol should be considered together with volatiles. This property allows human exposure to such compounds through the air route to be at least as great as through the water route, based on daily intakes of 20 m³ air and 2 L water. Applying adsorption theory to typical atmospheres, Junge (1977) estimated that for an aerosol surface area to air volume ratio of about $4 \times 10^{-5}$, compounds with Vp of $10^{-4}$ Torr or larger exist primarily (95%) in vapor form. These considerations would lead to inclusion of compounds such as polychlorinated biphenyls, for example, Arochlor 1248, Vp $4.9 \times 10^{-4}$ Torr (Andelman, 1985a), as well as the more traditional VOCs with Vp exceeding $10^{-1}$ Torr in the volatile class. Foreman and Bidleman (1987) examined the relationship between Vp and vapor/particulate phase distribution during collection of SVOCs. They considered pesticides, polychlorinated biphenyls, and polycyclic aromatic hydrocarbons as SVOCs, although a significant portion of these compounds is present in the atmosphere in the vapor form. Although these considerations have not been employed, generally, in the classification of compounds for this review, they are important in the assessment of human exposure to environmental pollutants. A few organics, such as nicotine and diethylphthalate, which have a Vp below, but near, $10^{-1}$ Torr, will be discussed due to their ubiquity and the potential for human exposure.

## 2. SOURCES AND EMISSIONS

### 2.1. Overview

Only in recent years has there been a significant effort to identify and characterize the wide range of potential sources of indoor airborne VOCs. Mostly due to an increasing health concern and advances in analytical techniques, the focus of IAQ research has shifted from characterization of particulate matter and inorganic gases toward the measurement of VOCs. There are now several review articles (Wadden and Scheff, 1983; NRC, 1981; USEPA, 1989; Levin, 1989) that categorize and identify many of the sources of indoor pollutants, including VOCs. A further indicator of progress is the development of a computerized source emissions data base (Bare, 1988). Nevertheless, there is limited published information on quantitative characterization of

VOC sources and less on emission characteristics under various conditions. Most investigations of emission characteristics have been limited to laboratory tests using controlled atmosphere chambers and other devices.

Sources of indoor airborne VOCs can be categorized into those that are located outdoors and those within the building or structure "envelope". For the purpose of this review, they are arbitrarily further categorized into tobacco smoke, other combustion sources, building materials and furnishings, consumer products and other indoor sources, and outdoor sources and other considerations. Discussion is usually limited to provide examples and an indication of the state of the technology and knowledge. For a convenient overview, tables (Tables 1–4) of some sources, emission components, and rates (where available) and related references are provided.

### 2.2.  Tobacco Smoke

The impact of environmental tobacco smoke (ETS) on indoor air quality is of current public interest. Tobacco smoke is a complex mixture of gaseous and particulate pollutants that has been extensively studied compared to other indoor pollutant sources. The smoke emitted by a burning cigarette (or cigar or pipe) between puffs (inhalation) is commonly defined as sidestream (SS) smoke and is the primary tobacco source contribution to indoor air pollution and "passive smoking" exposure (IARC, 1987; Lofroth, 1989a; Guérin et al., 1987; Reasor, 1987). Mainstream (MS) smoke, which is generated at higher temperatures than SS smoke and travels through the tobacco column, is of somewhat different composition and is inhaled directly by the smoker. A portion of the MS smoke is subsequently exhaled. The chemical characteristics of tobacco smoke are largely determined by the nature of the tobacco source and the combustion conditions. Factors that affect emission characteristics and fate have been examined and discussed elsewhere (IARC, 1987; Lofroth, 1989a; Guérin et al., 1987; Reasor, 1987).

Much of the research done until the mid 1980s on tobacco smoking is reviewed and discussed in two WHO publications on tobacco and passive smoking (Hoffmann and Wynder, 1986; IARC, 1987). The chapter by Hoffmann and Wynder (1986) summarizes the literature on MS smoke with emphasis on biologically active constituents. Hoffmann and Wynder report that about 3900 compounds have been identified in MS smoke. The book on passive smoking (IARC, 1987) provides an overview on the formation and physicochemical nature of SS smoke and the available analytical methods. The occurrence of about 400 constituents has been quantitatively compared in MS and SS smoke (IARC, 1987). The commonly used procedure of calculating SS smoke constituent concentration from SS/MS ratios and MS deliveries is questioned by Guérin et al. (1987), since MS deliveries are influenced by factors such as puffing conditions and filter type, which have relatively little effect on SS emissions. In their critical review of the cigarette smoke literature Guérin et al. conclude that SS emissions are relatively constant, that direct measurement is likely to be more meaningful than the calculation method, and that further research is required to develop validated SS smoke generation and sampling devices.

A brief review (Proctor, 1988) of the contribution of ETS to indoor air concludes that it is likely that the ETS contribution of volatile compounds is much less than the contribution from other sources. In a study (Proctor, 1989) of VOC occurrence in three different indoor environments (offices, betting shops, train compartments), the author also concludes that ETS is not a major source of VOCs in indoor air and suggests that the outdoors may be a major source. These findings, based on correlations between ETS indicators (nicotine, RSP, UV-RSP) and aromatic VOC levels, are different from the TEAM study (Wallace et al., 1987a) conclusion that "tobacco smoke is a main source of exposure to aromatics." It has been suggested that use of VOCs for determining the contribution of ETS to environmental pollution is not reliable (Bayer and Black, 1987).

Volatile organic constituents of SS smoke include aromatic amines, N-nitrosamines, aldehydes, simple aromatics, carboxylic acids, and phenols. The estimated emission rates for many compounds have been summarized elsewhere (IARC, 1987) and some samples from other studies are provided in Table 1. Recently, Lofroth (1989a) has summarized what is known of the chemical composition of side (SS) and main stream (MS) ETS with an emphasis on genotoxic and carcinogenic components. Volatile hydrocarbon emissions include alkenes such as ethene, propene, 1,3-butadiene, and isoprene. In addition to nicotine, 16 gas phase organic bases were detected in a study that focused on the chemical characterization of ETS (Eatough et al., 1989a). In the examination (Lofroth, 1989b) of a potential indicator for ETS, isoprene was detected in human breath (average 216 ppb), herbal cigarettes (2 mg/g tobacco), incense (ca. 1 mg/g), and insecticidal coils (1.4 mg/g). Sidestream emissions from a commercial cigarette were 6.8 mg isoprene/g tobacco, which is somewhat higher than reported in other studies, and rates for exhaled isoprene ranged from 30 to 530 µg/h (Lofroth, 1989b). It is suggested that the discrepancies in values from different studies may be due to differences in analytical procedures.

The occurrence of formaldehyde in ETS has been reported in several studies (Godish, 1989b; IARC, 1987). In an examination of the ability of air cleaners to remove ETS components, emission rates were 0.4 mg/cigarette for HCN and 2.4 mg/cigarette for formaldehyde (Olander et al., 1988). A quantitative study (Schlitt and Knoppel, 1989) of smoke from many international and domestic cigarette brands in Italy showed acetaldehyde, acetone, butanone, propanal, and acrolein as the most abundant of 10 aldehydes and four ketones determined. Formaldehyde was a minor constituent and emission rates for selected carbonyl compounds were considerably greater in SS than in MS smoke.

Table 1 VOC Emissions in SS Smoke from Burning
Cigarettes (illustrative)

| Compound | Emission rate (μg/cigarette) | References |
|---|---|---|
| Nicotine[a] | 2700–6900 | 1 |
| Phenol[a] | 0.069–0.418 | 1 |
| Catechol[a] | 138–292 | 1 |
| Hydrogen cyanide | 53–111 | 1 |
| N-Nitrosodimethylamine | 0.140–1.900 | 1 |
| Ethene | 1200 | 2 |
| Propene | 1300 | 2 |
| 1,3-Butadiene | 400 | 2 |
| Isoprene | 3100 | 2 |
| Methyl chloride | 1300 | 3 |
| Pyridine | 90–930 | 3 |
| Formaldehyde | 2360 | 4 |
| Acetaldehyde | 4830 | 4 |
| Acrolein + furfural | 1090 | 4 |
| Acetone | 1620 | 4 |
| Propanal | 390 | 4 |
| Crotonaldehyde | 280 | 4 |
| Methyl vinyl ketone | 260 | 4 |
| Butanone | 660 | 4 |
| Methyl glyoxal | 160 | 4 |
| Formic acid | 435 | 5 |
| Acetic acid | 1241 | 5 |
| Limonene | 63 | 5 |
| Toluene[a] | 605 | 5 |

[a]Including particulate phase.
*References*: 1, Guérin et al. (1987); 2, Lofroth (1989a); 3, Wadden
(1985); 4, Schlitt and Knoppel (1989); 5, IARC (1987).

## 2.3.  Other Combustion Sources

The abundance of indoor sources of combustion pollutants other than ETS
is evident. It has been estimated (Tucker, 1987) that there were 20 million
gas stoves, 10 million kerosene- or gas-fired unvented space heaters, 50 million
gas- and oil-fired appliances and furnaces, and 20 million wood stoves and
fireplaces in the United States in 1987. Also, more than 5% of the petroleum
burned in the United States is consumed in residential oil-fired furnaces
(Braun et al., 1987). The ubiquity of indoor combustion sources very likely
holds for many other countries where heating and/or cooking occur indoors.
For example, nearly 33% of the urban households in India depend on LPG
for their cooking needs (Haraprasad and Dave, 1989).

Due to the size and nature of combustion appliances and equipment and
the variability of operational requirements, it is difficult to obtain a simple

and accurate measure of their emissions to the indoor environment. Both controlled tests in the laboratory and field tests in buildings have been conducted. Investigations have focused more often on the measurement of inorganic combustion gases and particulate emissions than on the measurement of individual VOCs emitted due to incomplete combustion or from stored or transported fuel. Often, one (typically formaldehyde) or a few VOCs have been measured in conjunction with, and sometimes secondarily to, the determination of inorganic and particulate emissions. Some examples of VOCs and their emission rates for various combustion processes are shown in Table 2. The impact of the use of vented combustion appliances on IAQ is difficult to assess since only emission rates for exhaust gases can be readily measured.

It has been demonstrated (Girman et al., 1982) in tests conducted in a 27-m$^3$ chamber that formaldehyde emissions from gas ranges (oven, 2.7 μg/kJ; top burners, 1.7 μg/kJ) are typically higher than from gas space heaters (0.8 μg/kJ). Also, it was found that formaldehyde emissions from gas appli-

**Table 2  VOC Emissions from Combustion Devices and Fuels (illustrative)**

| Source | Compound | Emission Rate | References |
|---|---|---|---|
| Gas range | Formaldehyde | 28–140  μg/min | 1 |
|  | Other VOCS | ND | 1 |
| LPG stove | Formaldehyde | 14–154  μg/min | 2 |
| Kerosene stove | Formaldehyde | 222–807  μg/min | 3 |
| Kerosene heater | Alkyl benzenes | 91–61000  μg/h | 4 |
|  | Aliphatic hydrocarbons | 6–4300  μg/h | 4 |
|  | Aliphatic alcohols | 4900–10000  μg/h | 4 |
|  | Aliphatic ketones | 1–4500  μg/h | 4 |
|  | Cyclohexane | 530  μg/h | 4 |
|  | Methoxybenzene | 3000  μg/h | 4 |
| Wood fireplace | Formaldehyde | 1–40  mg/min | 5 |
|  | Acetaldehyde | 1–26  mg/min | 5 |
|  | Furfural | ND–16  mg/min | 5 |
|  | Acrolein | ND–16  mg/min | 5 |
|  | Acetone | ND–23  mg/min | 5 |
|  | Propionaldehyde | ND–50  mg/min | 5 |
|  | Crotonaldehyde | ND–23  mg/min | 5 |
|  | p-Tolualdehyde | ND–51  mg/min | 5 |
|  | Total carbonyls | 0.61–2.26  g/kg fuel | 5 |
| Wood stove | Total hydrocarbons | 1.5  g/kg fuel | 6 |
| Wood | Total hydrocarbons | 74  ng/J | 6 |
| Coal | Total hydrocarbons | 50  ng/J | 6 |
| Oil | Total hydrocarbons | 3  ng/J | 6 |
| Gas | Total hydrocarbons | 5  ng/J | 6 |

*Note.* ND, not detectable.
*References*: 1, Moschandreas et al. (1983); 2, Haraprasad and Dave (1989); 3, Dave and Haraprasad (1989); 4, Traynor et al. (1986); 5, Lipari et al. (1984); 6, Muhlbaier (1982).

ances are significantly dependent on operating conditions. Moschandreas et al. (1983) have reported on gas range emission experiments that were done in a 512-ft³ box with emission venting into an exhaust funnel and sampling near the flame at two settings. Formaldehyde emission rates were 1654–8401 μg/h and were calculated to correspond to indoor ambient concentrations of 47–259 ppb, but with qualifying comments. They indicate that a variety of VOCs were identified, list concentrations for 20 compounds, and conclude that emission factors were relatively low. Emission rates of VOCs and other pollutants from gas-fired ranges and space heaters contained in a controlled environment chamber (31 m³) were determined by Moschandreas and collaborators (1986b). It was concluded that "the emission rates of VOCs and PAH are nearly zero for the gas used in this study," although no VOC values were provided. Also, the emission rate for formaldehyde (0.48 μg/kJ) was lower than the corresponding rate (1.7 μg/kJ) found by Girman et al. (1982).

Mean emission rates ranging from 840 to 9240 μg formaldehyde/h were found by a direct flue gas sampling method for an LPG stove (Haraprasad and Dave, 1989). Based on certain assumptions, including zero air exchange, negligible decay rates, and a calculated activity-related emission inventory, it was calculated that the resulting formaldehyde concentration in the 24-m³ kitchen would be 250 μg/m³. Also, the effect of falling reservoir levels on emissions was examined for the kerosene wick stove, which is popular among low-income families in India (Dave and Haraprasad, 1989). Formaldehyde emissions ranged from 2.02 to 7.34 μg/kJ and, in general, were inversely related to burn rate. Under assumptions similar to those in their work with the LPG stove, the investigators projected that formaldehyde concentrations in the 24-m³ kitchen would be 990 μg/m³. In another study (Leaderer et al., 1986), formaldehyde levels were found to be low (27.5 ± 17.5 μg/m³) and not related to the use of unvented combustion sources in homes with and without kerosene heaters. Tests with four portable kerosene-fired heaters (radiant and convective) showed that formaldehyde emissions to the indoor environment ranged up to about 1 μg/kJ (Traynor et al., 1982). It has been suggested (Tucker, 1987) that unvented gas-fired appliances are not important sources of organic emissions, whereas there is evidence that kerosene heaters may contribute to indoor organics levels. Traynor et al. (1986) determined emission rates for several classes of VOCs, for example, alkylbenzenes, other hydrocarbons, alcohols, and ketones, from kerosene heaters and found rates as high as 61 mg/h for alkylbenzenes. It was also concluded (Tucker, 1987) that emission rates for organic compounds from indoor combustion sources are virtually unknown and the extent of leakage from vented appliances due to spillage or back-drafting is not known.

Various investigators (Muhlbaier, 1982; Cooke et al., 1981; Lipari et al., 1984) have reported on VOC emissions in the exhaust of woodburning appliances, but few have provided data on emissions directly to the indoor environment. Direct indoor pollution can occur from woodstoves during starting, stoking, and reloading operations or pollutants can be emitted continuously from leaks in the stove or vent system. Indirect pollution may occur by infiltration of organic vapors that have been exhausted outdoors. Although Cooke et al. (1981) found a decrease of total hydrocarbon (THC) emissions with an increase in burn rate or air supply, it is evident from their and other studies that exhaust emissions are governed by many factors, such as wood-burning appliance type, design features, and operating conditions. The relative THC emission factors for wood (74 µg/kJ) were estimated (Muhlbaier, 1982) to be higher than for three other commonly used residential fuels: coal (50 µg/kJ), oil (2.9 µg/kJ), and natural gas (5 µg/kJ).

Formaldehyde was the major component among 11 aldehydes identified in fireplace exhaust gas (Lipari et al., 1984). A comparison of estimated emission rates showed that woodburning in the fireplace (0.6–2.3 g/kg fuel) yields higher aldehyde values than burning of coal (0.0002 g/kg), oil (0.1 g/kg), or natural gas (0.2 g/kg) in power plants. The contribution to indoor formaldehyde levels was negligible for three airtight stoves, but raised concentrations by almost 50 ppb in two of the tests with a nonairtight stove (Traynor et al., 1984). Matthews et al. (1985) have examined and reviewed the formaldehyde emission rates of various combustion sources and products containing formaldehyde resin and have estimated the potential contributions of these sources to the indoor formaldehyde levels. They concluded that pressed-wood products and foam insulation containing urea–formaldehyde resins were the strongest contributors and combustion sources and phenol–formaldehyde resin bonded products are generally weaker emitters.

Combustion of wood, charcoal, animal dung, and crop residues for heating and cooking in Nepal generally elevated the indoor levels of methane, other hydrocarbons, and methyl chloride (Davidson et al., 1986). Chamber tests with four combustion sources showed no alcohols, aldehydes, ketones, amines, or alkyl halides in organic vapor emissions from a gas stove and kerosene heater, whereas isopropanol, acetone, and dichloromethane were detected in cigarette smoke, and frying hamburger emitted various aldehydes, ketones, and amines (Sexton et al., 1986b). Methane, paraffins, olefins, and aromatics were among the pollutants measured in flue gas from a woodburning stove and a fireplace and a coal burning stove (Zeedijk, 1986). Although the results varied with combustion conditions, VOC emissions were similar for the wood appliances and an anthracite-fired stove, but higher with bitumen.

As part of the USEPA's Integrated Air Cancer Project (IACP), nonmethane organic carbon (NMOC) and carbonyl compounds were monitored at ten pairs of homes (Zweidinger et al., 1988). Examination of averaged data showed that total NMOC and total carbonyl levels were higher inside than outside all homes and formaldehyde levels were slightly higher inside homes with woodstoves. VOCs were determined in indoor and outdoor air and in flue gas at three homes with operating wood stoves in a pilot study to identify ranges and trends in emissions that can be attributed to indoor and outdoor sources (Highsmith et al., 1988). The VOC distribution was uniform from indoors to outdoors and their occurrence could not be attributed to wood combustion. Indoor aldehyde levels were somewhat higher than outdoor levels and the investigators suggest that this may have been due to building materials, furnishings and household activities. In summary, there seems to be no evidence that properly operated and vented combustion appliances will contribute significantly to indoor VOC levels, whereas unvented appliances may measurably contribute to VOC levels.

### 2.4. Building Materials and Furnishings

Building occupants, owners, designers, and product manufacturers are increasingly aware of the effect of emissions from construction materials and furnishings on IAQ, largely due to occupant complaints and reports by the news media. Numerous studies have demonstrated that these sources emit a wide variety of VOCs. The data are probably the most abundant among VOC emissions data from all indoor sources and it is not possible here to report results of all such published studies. Table 3 depicts the scope of research results and the nature of the investigations is briefly illustrated below by some examples. Reports on these studies and other publications (Meyer, 1979, 1983; Levin, 1989; Wadden and Scheff, 1983; NRC, 1981; Rousseau et al., 1988; Ozkaynak et al., 1987) should be consulted for additional details.

Emissions of formaldehyde and other VOCs from urea formaldehyde foam insulation, pressed wood products, and other products containing formaldehyde-based resins and bonding material have been thoroughly chronicled elsewhere (Meyer, 1979; Matthews et al., 1985). Godish (1988) has recently published a brief and excellent review of studies designed to determine concentrations of formaldehyde in different housing types and those factors that significantly affect residential airborne formaldehyde levels. He includes a list of emission rates for various products.

In a 1979 report, Molhave concluded that some of the VOCs detected at 14 locations, where occupants complained about IAQ problems, probably originated from construction materials. A total of 29 organics, including $C_6$–$C_{10}$ aromatics, $C_6$–$C_{12}$ alkanes, and terpenes, were found in air of the 14 rooms, and head space analysis of 32 materials showed emissions of 38 different compounds. Later, Molhave (1982) reported that VOC emissions from 42 (including the initial 32) commonly used building materials contained a total of 52 different compounds. Emission rates ranged from $10^{-3}$ to 270 mg $m^{-2} h^{-1}$. Miksch et al. (1982) have also listed predominant VOCs associated with various building materials and offices. Some VOCs identified in an energy-efficient building, where there had been complaints of odor and poor IAQ, were subjectively related to emissions from building materials and furniture (Hijazi et al., 1983). Monteith et al. (1984) measured the levels of 18 VOCs at more than 150 mobile homes and in head space samples of four building materials used in construction of such homes. Thirteen of 18 target compounds were detected in emissions from two plywood products, particle board, and carpeting. Dynamic head space sampling was conducted with 22

**Table 3 VOC Emissions from Building Materials (illustrative)**

| Source | Compound | Emisson Rate | References |
|---|---|---|---|
| Chipboard | TOV | 0.1 mg/m² h⁻¹ | 1 |
| Gypsum board | TOV | <0.1 mg/m² h⁻¹ | 1 |
| Wall paper | TOV | 0.1 mg/m² h⁻¹ | 1 |
| Floor wax | TOV | ND–80 mg/m² h⁻¹ | 1 |
| Wood stain | TOV | ND–10 mg/m² h⁻¹ | 1 |
| Wood finish | TOV | ND–9 mg/m² h⁻¹ | 1 |
| Floor varnish | TOV | 1 mg/m² h⁻¹ | 1 |
| Latex back carpet | 4-Phenylcyclohexene | 0.1 mg/m² h⁻¹ | 1 |
| Particle board | Formaldehyde | 0.2–2 mg/m² h⁻¹ | 1 |
| Plywood panel | Formaldehyde | 1 mg/m² h⁻¹ | 1 |
| Floor adhesive | TOV | 100–1700 mg/m² h⁻¹ | 2 |
| Silicone caulk | TOV | 2–50 mg/m² h⁻¹ | 2 |
| Spray foam (PUF) | Acetone | ND–0.02 mg/m² h⁻¹ | 3 |
| | Chlorobenzene | 0.1–0.2 mg/m² h⁻¹ | 3 |
| | Xylenes | 0.03 mg/m² h⁻¹ | 3 |
| | Cyclohexane | ND–0.03 mg/m² h⁻¹ | 3 |
| | Dichlorobenzenes | ND–0.03 mg/m² h⁻¹ | 3 |
| Foam board (PUF) | Dichloromethane | 0.01–0.08 mg/m² h⁻¹ | 3 |
| | Chloroform | ND–0.002 mg/m² h⁻¹ | 3 |
| | Acetone | ND–0.03 mg/m² h⁻¹ | 3 |
| | Chlorobenzene | 0.2–0.3 mg/m² h⁻¹ | 3 |
| | Dichlorobenzenes | ND–0.04 mg/m² h⁻¹ | 3 |
| Particle board | Formaldehyde | 95–230 μg/m² h⁻¹ | 2 |
| | Acetone | 37–41 μg/m² h⁻¹ | 2 |
| | Hexanal | 15–26 μg/m² h⁻¹ | 2 |
| Painted sheetrock | 1,1,1-Trichloroethane | <0.1 μg/m² h⁻¹ | 4 |
| | Benzene | 0.1 μg/m² h⁻¹ | 4 |
| | n-Decane | 0.2 μg/m² h⁻¹ | 4 |
| | n-Undecane | 1.5 μg/m² h⁻¹ | 4 |
| Glued wallpaper | 1,2-Dichloroethane | 0.3 μg/m² h⁻¹ | 4 |
| | 1,1,1-Trichloroethane | <0.1 μg/m² h⁻¹ | 4 |
| | n-Decane | 0.2 μg/m² h⁻¹ | 4 |
| | n-Undecane | 0.3 μg/m² h⁻¹ | 4 |
| Glued carpet | 1,2-Dichloroethane | 0.2 μg/m² h⁻¹ | 4 |
| | 1,1,1-Trichloroethane | 0.3 μg/m² h⁻¹ | 4 |
| | p-Dichlorobenzene | <0.1 μg/m² h⁻¹ | 4 |
| | n-Decane | 0.5 μg/m² h⁻¹ | 4 |
| | n-Undecane | 0.5 μg/m² h⁻¹ | 4 |
| Linoleum tile | Trichloroethylene | 3.6 μg/m² h⁻¹ | 5 |
| Telephone cable | Tetrachloroethylene | 3.8 μg/m² h⁻¹ | 5 |
| | p-Dichlorobenzene | 0.2 μg/m² h⁻¹ | 5 |
| Polystyrene foam | Chlorobenzene | 0.5 μg/m² h⁻¹ | 5 |
| | p-Dichlorobenzene | 0.7 μg/m² h⁻¹ | 5 |

*Note.* TOV, total organic vapors; ND, not detectable; PUF, polyurethane.
*References*: 1, Levin (1989); 2, Tichenor and Mason (1988); 3, Hartwell (1986); 4, Wallace et al. (1987b); 5, Sheldon et al. (1986).

solid and 10 solvent-based materials used in the construction of a new office building (Sheldon et al., 1986). Emission rates were reported for 24 compounds, including 7 halogenated compounds, 4 aliphatics, and 12 aromatics. Many other volatile organics in the emissions were also identified.

Jungers and Sheldon (1987) report on the monitoring of aromatic hydrocarbons, aliphatic hydrocarbons, chlorinated hydrocarbons, and oxygenated hydrocarbons in new and old public access buildings. New buildings were examined soon after construction and again after occupation. Emission rates of chemicals from building materials were also determined in chamber studies. The new buildings showed very high total VOC levels immediately after construction and the levels decreased substantially over 5–8 months. Aliphatic and aromatic hydrocarbon levels were highest immediately after construction and were probably due to building material emissions. Mean chlorinated hydrocarbon levels were highest for buildings after occupation and presumably due to solvent based office materials. Oxygenated compounds showed no trends. Two closely related studies (USEPA, 1988) on IAQ in 10 public buildings included examination of VOC sources and emission rates and found indoor/outdoor concentration ratios well above unity, particularly for new buildings. By comparing data from room air samples and chamber tests with building materials from floor, walls, and ceiling, Berglund et al. (1989) identified 17 compounds as representative of the materials and air in a room of a 7-year-old building. The effect of temperature was evident in a bake-out (32–39 °C) to reduce VOC emissions from building materials and furnishings in a newly renovated office building (Girman et al., 1989b). The airborne VOC concentrations increased by more than four times during bake-out and were reduced to 71% of the original concentration after bake-out.

Wallace et al. (1987b) employed results from both TEAM studies and laboratory tests to identify potential VOC sources. They identified numerous organics in glues and floor and wall covering emissions and reported on measured emission rates. Other investigations (Sanchez et al., 1987; Tichenor and Mason, 1988) for the USEPA have also identified VOCs in the emissions from a variety of building materials and examined emission rates under various conditions. Toluene, styrene, and a variety of cyclic, branched, and normal alkanes were major components in emissions from 15 solvent- and water-based adhesives used to construct and finish interiors of buildings (Girman et al., 1986). VOC emission rates ranged from below the detection limit (0.1 $\mu g \, g^{-1} \, h^{-1}$) to a total alkane emission rate of over 700 $\mu g \, g^{-1} \, h^{-1}$. Mineral wool (van der Wal et al., 1989) and polyurethane foam, PUF (Krzymien, 1989; Hartwell, 1986) insulating materials have been identified as potential sources of indoor VOCs. Moistened mineral wool emitted about 100-fold more carbonyl compounds than dry wool, but the presence of airborne VOCs in buildings could not be directly related to use of this insulating material. Over 80 compounds, most of them VOCs, were identified by dynamic head space analysis of PUF samples (Krzymien, 1989; Hartwell, 1986). There is evidence of the presence of significant levels of airborne diisocynates during and after spray application of PUF (Bilan et al., 1989). However, for the majority of the wide range of compounds, emissions from PUF would be a minor contributor to indoor VOC levels.

There are additional examples of the variety of sources. Major components in emissions from a floor adhesive and particle board were identified by Merrill et al. (1987). Various amines and short-chain carboxylic acids were found in damaged, malodorous self-leveling floor topping compounds and were associated with *Clostridial* putrefaction (Karlsson et al., 1984). Formaldehyde and other VOCs were emitted for several years after application of a Swedish floor finish (Van Netten et al., 1988). Saarela et al. (1989) investigated some of the common additives used in plasticized PVC materials that are frequently used in wall and floor coverings and found varying amounts of VOCs. Baechler et al. (1989) examined paint emissions data to predict VOC levels in indoor air and noted discrepancies between data from various field and chamber studies. Fluid in an elevator hydraulic system was the dominant source of VOCs in a poorly ventilated facility where cleaning products, floor wax, latex paint, and reentrained vehicle exhaust were also identified as sources of airborne organics (Weschler et al., 1990). Weschler and Shields (1986) measured the occurrence of vapor phase additives in telephone company offices and identified plasticizers and other commercial products as potential sources. In another study of a telephone office, these investigators determined levels of VOCs, found that the data supported the validity of a mass balance model and used the model to calculate indoor generation rates (3–25 μg/min) for VOCs associated with dominant indoor sources (Weschler and Shields, 1989).

The variety of materials that can contribute to indoor pollution is illustrated in Table 3. There is little doubt that such materials are an important consideration for the evaluation of potential sources of human exposure, particularly when recently manufactured or installed products are used in buildings. Investigations concerning health and IAQ related complaints by building occupants have often implicated materials and furnishings as causative factors (Levin, 1989; USEPA, 1989). It is also evident that building temperature, humidity, and ventilation have considerable influence on VOC emissions and their fates.

## 2.5 Consumer Products and Other Indoor Sources

The variety and number of consumer and household products change constantly and their VOC emissions and hazards are largely dependent on use patterns. Girman et al. (1987) have discussed the considerations in evaluating emissions from such products and refer to similarities in requirements for investigations of other emission sources. They proposed an eight-step investigative approach that entails defining the potential health hazards, acquiring application and usage information, establishing goals, assessing capabilities, writing protocol, pretesting, conducting the study, and post-study considerations. Consumer products that may contribute to indoor VOC levels include aerosol products, cleaning and polishing agents, hobby materials, personal hygiene products, solvents, adhesives, paints, and a number of other products (Ter Konda, 1987). A major investigation of over 5000 materials (e.g., paints, adhesives, cosmetics) was conducted for the National Aeronautics and Space Administration (NASA) to determine potential VOC emissions in spacecraft and provided an extensive data base (Ozkaynak et al., 1987). Less comprehensive lists of products and related VOCs are available from several other sources (Ter Konda, 1987; Tichenor and Mason, 1988; Knoppel and Schauenburg, 1989; NRC, 1981; Meyer, 1983). The range of products and emissions is illustrated in Table 4 and some examples are cited below.

Ozkaynak et al. (1987) made use of the large computerized NASA data base when they calculated emission rates for 10 of the more than 5000 consumer products and other materials listed. Eight groups of aerosol products (insect sprays, paints and finishes, household, personal, animal care, automotive and industrial, food, and miscellaneous products) and three groups of nonaerosol products (personal, household, garage) were examined (Rogozen et al., 1988) on behalf of the USEPA. A list of VOCs identified in many products and emission pattern estimates based on per capita emission rates was provided. Various alcohols, acetaldehyde, acetone, and halides were identified in organic vapor emissions from two aerosol spray products examined in chamber tests (Sexton et al., 1986b). Tichenor and Mason (1988) list the major organic ingredients for several household products, such as room deodorants and moth crystals. Time profiles and time-weighted average concentration values were determined by Otson et al. (1981, 1984) after application of eight household products in a test room. The major VOC components were dichloromethane and methanol in brush-on paint removers; dichloromethane, toluene, and propane in an aerosol spray remover; and propane, 1,1,1-trichloroethane, and petroleum distillates in aerosol-type fabric protector products.

Cleanser, polish, and pesticide formulations were found to emit a variety of VOCs and emission rates were reported for 17 target compounds in 10 products that had been identified as potential sources in field and laboratory investigations (Wallace et al., 1987b). Formaldehyde may be found in leather products, rubber, printing products, plastics, photographic supplies, paper, latex paint, germicides, fumigants, and dyes (Ter Konda, 1987). In an examination (Knoppel and Schauenburg, 1989) of eight waxes and polishes and two detergents, it was found that the water-containing products emitted hydrocarbons (alkanes, alkenes, terpenes) at an initial rate of 0.2–2 $\mu g$ $cm^{-2}$ $min^{-1}$, while products without water emitted oxygenated compounds (including terpene alcohols and their acetates, aliphatic alcohols and esters, alkoxy-alcohols) and at higher rates (up to 430 $\mu g$ $cm^{-2}$ $min^{-1}$). The wide variety of detected VOCs was listed. Bruun Hansen (1987) determined VOC emissions for 13 household products (paint removers, paints, lacquer, glues) applied to surfaces under various conditions.

**Table 4 VOC Components and Emissions for Consumer Products and Other Sources**

| Source | Compound | Emission Rate | References |
|---|---|---|---|
| Cleaning agents | Chloroform | 15 μg/m² h$^{-1}$ | 1 |
| and pesticides | 1,2-Dichloroethane | 1.2 μg/m² h$^{-1}$ | 1 |
| | 1,1,1-Trichloroethane | 37 μg/m² h$^{-1}$ | 1 |
| | Carbon tetrachloride | 71 μg/m² h$^{-1}$ | 1 |
| | m-Dichlorobenzene | 0.6 μg/m² h$^{-1}$ | 1 |
| | p-Dichlorobenzene | 0.4 μg/m² h$^{-1}$ | 1 |
| | n-Decane | 0.2 μg/m² h$^{-1}$ | 1 |
| | n-Undecane | 1.1 μg/m² h$^{-1}$ | 1 |
| Moth cake | p-Dichlorobenzene | 14000 mg/m² h$^{-1}$ | 2 |
| Dry-cleaned clothes | Tetrachloroethylene | 0.5–1 mg/m² h$^{-1}$ | 2 |
| Liquid floor wax | TOV | 96 g/m² h$^{-1}$ | 3 |
| | Trimethylpentene | | 3 |
| | Dodecane isomers | | 3 |
| Paste leather wax | TOV | 3.3 g/m² h$^{-1}$ | 3 |
| | Pinene | | 3 |
| | 2-Methyl-1-propanol | | 3 |
| Detergent | TOV | 240 mg/m² h$^{-1}$ | 3 |
| | Limonene, pinene | | 3 |
| | Myrcene | | 3 |
| Human emissions | Acetone | 50.7 mg/day[a] | 4 |
| | Acetaldehyde | 6.2 mg/day[a] | 4 |
| | Acetic acid | 19.9 mg/day[a] | 4 |
| | Methyl alcohol | 74.4 mg/day[a] | 4 |
| Copy paper | Formaldehyde | 0.4 μg/form | 5 |
| Steam humidifier | Diethylaminoethanol | | 6 |
| | Cyclohexylamine | | 6 |
| Wet copy machine | 2,2,4-Trimethylheptane | | 7 |
| Household solvents | Toluene, ethyl benzene | | 8 |
| Paint removers[b] | Dichloromethane, Methanol | | 9 |
| Paint removers[c] | Dichloromethane | | 9 |
| | Toluene, propane | | 9 |
| Fabric protector[c] | 1,1,1-Trichloroethane | | 10 |
| | Propane, petroleum distillates | | 10 |
| Latex paint | 2-Propanol, butanone | | 11 |
| | Ethylbenzene, toluene | | 11 |
| Room freshener | Nonane, decane | | 11 |
| | Ethylheptane, limonene | | 11 |
| Shower water | Chloroform | | 12 |
| | Trichloroethylene | | 13 |

[a]Per person.   [c]Spray.
[b]Brush on.

*Note.* TOV, total organic vapors.

*References*: 1, Wallace et al. (1987b); 2, Levin (1989); 3, Knoppel and Schauenburg (1989); 4, Wadden and Scheff (1983); 5, Wadden (1985); 6, Edgerton et al. (1989); 7, Tsuchiya et al. (1988); 8, Ter Konda (1987); 9, Otson et al. (1981); 10, Otson et al. (1984); 11, Tichenor and Mason (1988); 12, Stern and Andrews (1989); 13, Andelman (1985a).

There are a variety of sources and activities in and outside the home that can result in human exposure to VOCs. The following are examples of sources that may not be readily recognized. A considerable amount of tetrachloroethylene was detected in dry-cleaned clothes (13.6 mg/g) and emissions were detectable for several months upon storage of clothes in the home (Kawauchi and Nishiyama, 1989). Hijazi et al. (1983) detected lactic and pyruvic acids in an energy-efficient building and ascribed the presence of these compounds to human emissions. Bruun Hansen and Andersen (1986) have suggested that vapors emitted from photocopy machines may react with the ozone emissions to form various oxygenated products. Tsuchiya et al. (1988) examined wet and dry photocopying processes and found that exhaust from "wet" machines was a major source of VOCs. They reported TVOC results for the exhaust (maximum 4150 mg/m$^3$) and room air (maximum 64 mg/m$^3$) and observed that 2,2,4-trimethylpentane was a major component among $C_8$–$C_{12}$ branched alkanes in the air. Low concentrations of selected amines (cyclohexylamine, dimethylaminoethanol) found in indoor air were attributed to their use in steam humidification of buildings (Edgerton et al., 1989). Brimblecombe (1990) in a review of museum atmospheres composition indicates that formaldehyde and organic acids (e.g., formic and acetic) from glues, varnishes, wood products, and so on, and diethylaminoethanol from steam line corrosion inhibitors can be damaging to objects of art. Also, cleaning materials used in museums emit various solvent components. A large number of volatile organics emitted by waste water has been detected in the air of municipal treatment plants (Dunovant et al., 1986).

Several studies (Shehata, 1985; Andelman, 1985b; McKone, 1987; McKone and Knezovich, 1989) have indicated that inhalation exposure to VOCs in tap water may be of the same order of magnitude as exposure by ingestion. Showers, baths, running tap water, toilets, cooking, humidifiers, dishwashers, washing machines, and other activities and equipment that use tap water indoors may contribute to indoor levels of airborne VOCs. The release of VOCs from water is largely dependent on the solubility and vapor pressure of individual compounds. Mathematical models (Andelman, 1985b; Shehata, 1985; Andelman et al., 1986; McKone, 1987; McKone and Knezovich, 1989) using these and other parameters have been applied, with some success, to predict indoor airborne VOCs (e.g., chloroform, trichloroethylene) levels, but usually at relatively high waterborne VOC concentrations (e.g., 1 mg/L water). Although individual VOC levels may approach 1 mg/m$^3$ in shower stalls, levels in other rooms of buildings with normal water use patterns are usually at least an order of magnitude lower (Andelman et al., 1986; Shehata, 1985; Stern and Andrews, 1989). Clearly, the contribution of many factors, such as water temperature, waterborne VOC concentration, and method of water usage, makes prediction of emission rates difficult.

It is evident that consumer products and other potential sources of indoor VOCs are numbered in the thousands. Although the composition of many such sources has been published, or may be available from manufacturers, there are relatively few data on emission rates. Consequently, their impact on IAQ is difficult to predict, even when use and activity pattern information is available. As will be discussed in the following section (2.6), indirect, for example, statistical, methods are sometimes used to identify dominant VOC sources among the multitude of possibilities.

## 2.6. Outdoor Sources and Other Considerations

More than 300 volatile organics that have been detected in outdoor and indoor air are now listed in a computerized data base compiled for the USEPA (Shah and Heyerdahl, 1988; Shah and Singh, 1988). The concentration file for ambient measurements contains more than 122 000 records on 261 chemicals. Also, for the outdoor environment, source profiles represented by 23 VOCs have been prepared for 10 common sources (Scheff et al., 1989). These and other data in studies that provide characteristics of VOC emissions are useful in assessing the impact of outdoor air on IAQ. Several investigators (Yocom, 1982; Moschandreas et al., 1982; Cohen et al., 1989) have examined and discussed the relationship between indoor and outdoor air. The composition of outdoor air and external source emissions is extensively discussed in other publications (IARC, 1977; Altshuller, 1983; Shah and Singh, 1988; Scheff et al., 1989) and in other chapters of this book. Therefore, only the results of some recent studies will be reported to exemplify the state of knowledge of the influence of outdoor VOCs on IAQ and on personal exposure.

In 52% of 446 IAQ investigations conducted by the National Institute for Occupational Safety and Health (NIOSH) until 1987, the building ventilation was determined to be inadequate in some way (Wallingford, 1988). In this category, insufficient outdoor air supply to the office space was the largest cause of problems. An additional 11% of problems were due to exterior sources, such as motor vehicle exhaust, reentrainment of previously exhausted building contaminants, boiler gases, and outdoor construction. Contamination from building materials was the major problem in only 3% of investigations, whereas sources inside the office space (e.g., copy machines, cleaning agents, tobacco smoke) were identified in 17% of all cases. About one-half of these IAQ problems appeared to be caused or exacerbated by energy conservation attempts.

The importance of good ventilation and an adequate supply of outdoor air for buildings is now evident (Wallingford, 1988; NRC, 1981; Meyer, 1983; USEPA, 1989). Under most circumstances VOC levels are considerably lower in the outdoor source than indoors (Wallace et al., 1987a; Yocom, 1982). Nevertheless, nearby exterior sources of VOCs and reentrainment of building exhaust may have a deleterious effect on IAQ in public, commercial, residential, and other structures. In either case, that is, concern with the clean air supply or with external VOC sources, the dilution effect or emission rate of external air is governed by infiltration and exfiltration characteristics of a building. Even when accurate air exchange rates (Nagda et al., 1986) are available, it is difficult to identify external point sources or to determine the contribution (emission rate) of outdoor air since many VOCs are common to both indoor and outdoor environments.

Although there have been a relatively large number of studies on the indoor occurrence and indoor–outdoor relationships of particle phase organics, until recently, there have been only a few of such studies with VOCs. Overall, a difference in occurrence and behavior is evident in that the indoor/outdoor occurrence ratio is generally <1 for particles and >1 for VOCs (Yocom, 1982). Also, as an example of behavior differences, an infiltration factor of about 0.85 for selected VOCs and 0.5 for fine particles was found upon examination of indoor and outdoor data from a wood smoke impacted community (Lewis et al., 1989). As part of a larger study designed to investigate the impact of the chemical industry on human exposure to VOCs, an examination of the relationship between indoor and outdoor VOC concentrations was conducted in West Virginia (Cohen et al., 1989). Since the outdoor measurements for many of the 18 target VOCs were below the detection limit (1.8 to 11.1 $\mu g/m^3$), data evaluation focused on the indoor results. Statistical analyses revealed that house age, presence of an attached garage or forced air heating, the use of central air conditioning, and the frequency of open windows were good predictors for overall VOC levels in a house. Clearly, these factors are related to ventilation and air exchange.

Berglund et al. (1982) examined the influence of building ventilation on IAQ and the relationship between outdoor and indoor VOCs for an energy-efficient office building. The results indicated that some aliphatics ($C_9$–$C_{11}$ alkanes) were found almost exclusively indoors. Outdoor VOCs were also detected indoors and changes in ventilation had varying effects on the occurrence of individual compounds. It was also observed that the number and concentrations of air pollutants increased between the building fresh air intake and the air exhaust, that is, within the building. Continuous monitoring with an atmospheric pressure chemical ionization mass spectrometer (APCI-MS) identified numerous VOCs in the air of complaint-related offices and confirmed a considerable improvement in IAQ in one office after reentrainment of building exhaust had been reduced (Hijazi et al., 1983; Yocom et al., 1984). The effect of reduced ventilation on IAQ was also examined for a newly constructed office building in San Francisco (Turiel et al., 1983). Evaluation of the data showed that aldehyde levels were higher indoors than outdoors and more gas phase organics and higher concentrations were found indoors. Indoor median and maximum levels were generally higher than corresponding outdoor values for the more than 20 VOCs measured in overnight samples from several communities (Pellizzari et al., 1986). A weak correlation between concurrent indoor and outdoor overnight measurements was found and concentrations were generally higher indoors than outdoors for 12 halocarbons in a study by Hartwell et al. (1984).

Results of several additional studies support the thesis that the outdoor air contribution is unimportant compared to the effect of indoor sources on human exposure to VOCs. Several activities and sources associated with exposure to specific VOCs (20–26 target VOCs) were identified in the TEAM study (Wallace et al., 1987a,c). Some of these were tobacco smoke (benzene, styrene, ethylbenzene, *m*- and *p*-xylenes), hot tap water (chloroform), dry cleaning and dry-cleaned clothes (tetrachloroethylene), filling gas tank (benzene), moth crystals and room deodorizers (*p*-dichlorobenzene), and the work place (various VOCs). The California TEAM study (Wallace et al., 1988a) examined the occurrence of 26 VOCs and identified smoking, employment, and automobile related activities as important sources of personal exposure. Again, indoor VOC levels exceeded those outdoors except at night during February in Los Angeles. The occurrence of airborne VOCs at 14 homes and one small office building in Northern Italy was examined (De Bortoli et al., 1986) and the concentration of total volatile organic compounds (TVOC) was found to be higher indoors (mean TVOC 3 mg/m$^3$) than outdoors (mean TVOC, 0.4 mg/m$^3$). Many of the identified VOCs were believed to originate from consumer products and the investigators suggest that this source category is as important as building materials and combustion processes. Levels of 45 VOCs were determined outdoors and in living rooms of more than 300 Dutch homes (Lebret et al., 1986). Virtually all VOCs showed higher indoor than outdoor concentrations. Straight-chain and aromatic hydrocarbon levels were higher in newly occupied than in older homes and it was also concluded that most of the VOCs had both consumer products and building materials as indoor sources. In a study that compared activity records and personal exposure measurements for 17 VOCs, the results also indicated that outdoor exposure levels were considerably lower than indoor levels (Wallace et al., 1989). Major exposures were associated with use of deodorizers (*p*-dichlorobenzene), washing clothes and dishes (chloroform), visiting a dry cleaners (1,1,1-trichloroethane, tetrachloroethylene), smoking (benzene, styrene), cleaning a car engine (xylenes, ethylbenzene, tetrachloroethylene), painting and using paint remover (*n*-decane, *n*-undecane), and working in a scientific laboratory (many VOCs).

There is evidence that the influence of nearby external sources on indoor air quality cannot be disregarded. The sources of persistent VOCs in the air of selected East Tennessee residences were identified as attached garages, automobiles, and stored gasoline and motor oil (Gammage et al., 1984). In another study (Verhoeff et al., 1987), the occurrence of organic solvents was determined in a car-body repair shop, an offset printing office, and the surrounding houses and compared with personal exposure data for employees and residents. Both the repair shop and printing office proved to be sources of organics found in the surrounding houses. In an investigation of 10 small screen printing plants and surrounding houses, organic solvents from the plants were detected in some poorly sealed adjacent dwellings (Verhoeff et al., 1988).

The migration of VOCs in soil gas and groundwater from hazardous waste landfill has been demonstrated and evaluated (Stephens et al., 1986). Infiltration of VOCs from the ground into buildings has been the subject of several recent investigations. The impact of groundwater contaminants on IAQ was examined by Highland et al. (1985) using toluene (shower water) and tetrachloroethylene (groundwater) as indicator VOCs. It was concluded that both sources contributed to indoor levels of VOCs. An investigation of air quality in the basement of a large building on a former swamp area, identified potential sources of VOCs and suggested that decomposition of organic material in the soil was the major source of detected methane (Barboza et al., 1987). In a survey of IAQ in homes near a major hazardous waste disposal site, levels of target VOCs were only slightly higher in nearby residences than in control homes (Miller and Beizer, 1985). However, the measurements were obtained after the start of apparently successful remedial measures which were initiated after detection of methane, vinyl chloride, and other gases in some homes. The entry and dissipation rates of VOCs from a simulated sump well for residences with basements has been examined by means of experimental and computer simulation tests (Eichler et al., 1986). Simulation tests with $SF_6$ demonstrated that basement depressurization by 25–50 Pa (relative to ambient) resulted in effective soil air entry velocities exceeding 1 m h$^{-1}$ and considerably greater than entry by diffusion (Nazaroff et al., 1987). The relationship between IAQ and soil contaminated with VOCs (>40 VOCs) was evaluated from measurements at 77 houses in the Netherlands (Kliest et al., 1989). Models for transport of pollutants in soil gave poor prediction of indoor concentrations. The influence of soil contamination was detected in seven houses, but the contribution was variable with time. Experiments (Garbesi and Sextro, 1989; Hodgson et al., 1988) by scientists from the Lawrence Berkeley Laboratory indicated that the rate of entry of soil gas into basements can approach 1 m$^3$ h$^{-1}$ as a result of typical wind conditions and indoor–outdoor temperature differentials. The VOCs identified at the study house were predominantly halogenated and oxygenated compounds. Based on data from two shower experiments, it was estimated that soil gas was a relatively important source of exposure to VOCs compared to the household use of water.

From the foregoing results, it must be concluded that the potential impact of outdoor sources on indoor or overall exposure is not always evident and depends on many factors, some of them difficult to anticipate. Studies such as the USEPA total exposure assessment methodology (TEAM) project are valuable in providing clues to subjects for further investigation. In an evaluation of the TEAM study results, Wallace (1989) concluded that the major sources of exposure were small, close to the person, and usually inside the home. Statistical analysis of the data suggested that residence near major outdoor point sources of VOCs did not affect exposure and personal exposures were consistently higher than outdoor concentrations (Wallace et al., 1987a). However, as reported above (Miller and Beizer, 1985; Verhoeff et al., 1987; Garbesi and Sextro, 1989), there are situations where outdoor sources have a measureable impact on IAQ and, hence, indoor exposures. In addition, there is always some, albeit frequently difficult to discern, contribution to indoor pollution by outdoor air. Infiltered outdoor air contains VOCs, which add to the indoor burden, as well as some reactive chemical species (Becker and Salter, 1982; Yocom, 1982; Brimblecombe, 1990) that may affect the nature of indoor VOCs. Cleaning and recirculation of indoor air can be a desirable, but costly, option in circumstances with high outdoor pollution levels. Nevertheless, if indoor VOC levels are the most significant contributors to human exposure (Wallace, 1989), reduction and removal of indoor source emissions is most important in reducing exposure.

## 2.7. Summary

There has been a considerable increase in information on sources of indoor VOCs during the last decade. Various investigators have conducted extensive tests, reviewed the literature, and compiled emission inventories for tobacco smoke, building materials, and consumer products in their efforts to expand the understanding of the influence of these sources on IAQ. However, the development of new products and processes, as well as improvements and changes in their nature, will continue to present a never ending challenge to the analyst and the IAQ investigator.

Tables 1–4 provide a rough indication of the potential relative contribution of selected sources to indoor pollution. There are few data on VOC emission rates for some categories of potential sources. Emissions from indoor combustion of fuels for heating and cooking, various human activities, tap water use, and soil gas have proved difficult to quantitate or have not yet received much attention. There is little or no information on the contribution of pets, houseplants, hobbies, and transport vehicles to indoor VOC levels, although some of these have been implicated (USEPA, 1989; Tada and Seki, 1989; Wadden and Scheff, 1983). For example, houseplants may be a source of VOCs since it is known that a variety of terpenes and other hydrocarbons are emitted by outdoor vegetation (Altshuller, 1983). Nevertheless, it is evident from recent conferences and symposia on IAQ related topics that the number and range of VOC source investigations are increasing, and this will aid in filling some information gaps.

Absolute characterization of emissions is not feasible in many instances. The pattern and duration of activities and source use must be considered in the evaluation of their impact on IAQ. Many materials, products, and appliances are used only intermittently or their VOC emissions may vary or decay with time and conditions. Aerosol spray products, such as paint removers (Otson et al., 1981) and fabric protectors (Otson et al., 1984), are used briefly and volatile components in residues on application surfaces have half-lives of less than one day. VOCs in some products, such as fresh building materials and floor wax (Berglund et al., 1989; Jungers and Sheldon, 1987; USEPA, 1988; Levin, 1989), may undergo preferential evaporation and their emissions decay significantly over a period of days or months. The effect of building temperature and air humidity on formaldehyde emissions from particle board and foam insulation has been demonstrated in a number of studies (Meyer, 1979, 1986; Matthews et al., 1986). Thus, it is not sufficient to determine the character and rate of emissions under a limited set of laboratory conditions. To adequately predict the impact of sources on the indoor occurrence of VOCs, the many factors that influence emissions must be understood and quantified.

## 3.    OCCURRENCE AND EXPOSURE

### 3.1.    Introduction

The occurrence of indoor VOCs has been examined in numerous studies. The objectives have been to evaluate the potential for exposure to VOCs and to examine the potential sources of these compounds. Studies have employed several strategies, including statistical sampling designs, to evaluate the occurrence of VOCs for representative populations (Wallace, 1987). The sources of the VOCs found indoors have been tentatively identified by the combined use of sophisticated questionnaires and statistical techniques. An alternative strategy has been used to investigate VOC emissions directly and to correlate indoor occurrence to the emissions. A study by Berglund et al. (1989) provides an example, but few other studies have attempted this approach.

In Table 5, a summary of important studies on VOC occurrence also indicates the objectives and methods employed. In the following subsection (3.2), a few of the studies are selected to indicate the results for VOCs measured indoors and an attempt is made to relate the VOC concentrations to the major sources identified in Section 2. In most instances, the study designs did not permit clear identification of indoor sources by direct measurement; rather, statistical techniques were employed to draw inferences regarding the probable sources of VOCs. The TEAM study (Wallace, 1987) represents a good example of this approach.

It is evident from a review of Table 5 that measurements have been conducted in a number of indoor environments, including public buildings, private buildings, and personal residences, and under a large variety of conditions and situations. Study objectives have in large part defined the strategies employed for surveys of occurrence. These objectives are many and varied and somewhat dependent on logistical and resource considerations. Some literature references with suggestions regarding study strategy formulation are provided in the section on methodology (4.2).

In some instances the identification of VOCs has been performed without detailed quantitative measurement of concentrations. Jarke et al. (1981), for example, identified over 250 different organic compounds and Berglund et al. (1989) reported that more than 300 organic compounds have been detected in indoor air. Over 90 organic compounds were identified by Monteith et al. (1984) in field samples from 150 mobile homes. The VOCs included aldehydes $(C_7-C_{10})$, terpenes, alkanes $(C_6-C_{16})$, alkenes $(C_6)$, alcohols $(C_6-C_8)$, aro-

**Table 5 Summary of Important IAQ Studies**

| Study Title | Description | Sampling and Analysis Methods |
|---|---|---|
| **VOC Studies** | | |
| The Atmospheric Environment in Modern Danish Dwellings— Measurements in 39 Flats (Molhave and Moller, 1979) | Organic gases; main compounds alkanes, alkylbenzenes and terpenes. | Active GC-FID GC-MS |
| Indoor Air Pollution Due to Building Materials (Molhave, 1979) | Organic vapors, dust in indoor air of new offices where occupants had IAQ complaints. | Active GC-FID GC-MS |
| Organic Contaminants in Indoor Air and Their Relation to Outdoor Contaminants (Jarke et al., 1981) | 36 homes in Washington DC surveyed—118 compounds quantified indoors and 29 outdoors. | Active TD-GC-MS |
| Monitoring Individual Exposure. Measurements of Volatile Organic Compunds in Breathing-Zone Air, Drinking Water and Exhaled Breath (Wallace et al., 1982) | 15 VOCs To field test the candidate methods to be used in the TEAM study for estimating personal exposures to VOCs. | Active/personal Active/breath GC-MS |
| Survey of Selected Organics in Office Air (Otson et al., 1983) | Thirty offices surveyed for VOC. Maximum of 5.9 ppm DCM, 1.1 ppm TTCE, 0.3 ppm toluene found. | Passive GC-FID |
| Indoor Air Quality in a Large Hospital Building (Vanderstraeten et al., 1984) | Monitored inorganic compounds mainly but grab samples for TVOC on two occasions. | Bag samples GC-MS |

**Table 5** (*Continued*)

| Study Title | Description | Sampling and Analysis Methods |
|---|---|---|
| Personal Exposure to Volatile Organic Compounds, 1. Direct Measurements in Breathing-Zone Air, Drinking Water, Food and Exhaled Breath (Wallace et al., 1984) | To conduct a pilot study of methods for estimating personal exposure to VOCs. | Passive/area Active/area Active/breath Active/personal GC-MS |
| Integrating "Real Life" Measurements of Organic Pollution in Indoor and Outdoor air of Homes in Northern Italy (De Bortoli et al., 1984b) | 14 dwellings monitored for 33 VOCs and TVOC. | Passive Active DNPH impingers for aldehydes Extraction or TD GC-FID GC-ECD HPLC |
| Sources and Characterization of Organic Air Contaminants Inside Manufactured Housing (Monteith et al., 1984) | To identify VOCs present in manufactured housing and to compare indoor/outdoor levels. | Active GC-MS |
| Results of a Forty-Home Indoor Air Pollutant Monitoring Study (Hawthorne et al., 1983) | HCHO, selected VOCs and other criteria pollutants. | Active Passive Pararosaniline GC-FID |
| Residential Measurements of High Volatility Organics and Their Sources (Gammage et al., 1984) | To measure the levels of VOCs inside homes in east Tennessee and to identify their sources. | Portable GC Active GC-FID Active/PID |
| Air Quality in Residences Adjacent to an Active Hazardous Waste Disposal Site (Miller and Beizer, 1985) | 72 homes monitored in W. Covina, California within 250 ft of waste site. Found many VOCs at low levels. | Tedlar bags Active GC-MS |

**Table 5** (*Continued*)

| Study Title | Description | Sampling and Analysis Methods |
|---|---|---|
| Inhalation Exposure in the Home to Volatile Organic Contaminants of Drinking Water (Andelman, 1985b) | To demonstrate the volatilization of TCE from domestic water and inhalation routes of exposure from chemicals in potable water. | Continuous real-time monitor |
| Surveys of Selected Organics in Residential Air (Otson and Benoit, 1986) | To study the effects of energy conservation measures on total organic matter and respirable organics in homes. | Active Passive GC-MS GC-FID |
| Sample Integrity of Trace Level Volatile Organic Compounds in Ambient Air Stored in Summa Polished Canisters (Oliver et al., 1986) | To test new and used Summa polished stainless steel canisters for storage stability of 15 VOCs. | Canister sampling GC-FID GC-ECD |
| Indoor and Outdoor Air Pollution in the Himalayas (Davidson et al., 1986) | To measure typical light hydrocarbon concentrations in four Nepalese residences. Mainly $C_1$–$C_5$ found, also benzene and toluene. | Canister sampling GC-FID |
| Volatile Organic Compounds in Dutch Homes (Lebret et al., 1986) | To determine the distribution and variability levels of 45 VOCs in typical homes. | Active GC-FID GC-MS |
| Indoor Pollutants in 70 Houses in The Tennessee Valley Area: Study Design and Measurement | Survey of 70 homes in Tennessee Valley area for HCHO, PAH | Active/ multisorbent sampling tube Passive |

**Table 5** (*Continued*)

| Study Title | Description | Sampling and Analysis Methods |
|---|---|---|
| Methods (Dudney et al., 1986) | and other pollutants. | Pararosaniline |
| Methodologies for Evaluating Sources of Volatile Organic Chemicals (VOC) in Homes (Seifert and Ullrich, 1987) | To develop a classification of indoor sources of VOC and testing of VOC sampling procedures. | Active Passive GC-FID GC-ECD |
| The Comprehensive TEAM Study: Volume 1 (Wallace, 1987) | To measure 26 VOCs in the air, drinking water, and exhaled breath of 650 persons in four states. | Passive/area Active/area Active/breath Active/personal GC-MS |
| Indoor Air Quality in Health Care Facilities (Morey and Woods, 1987) | VOCs in Hospital Data Analysis Centre had values of 0.5–10 mg/m³. Ventilation modification resolved problems. | Active Passive GC-MS |
| Intercomparison of Sampling Techniques for Toxic Organic Compounds in Indoor Air (Spicer et al., 1987a) | To compare several VOC sampling techniques in an indoor environment. | Passive (high and low flow rates) Canister Distributive air sampling GC-MS |
| Participant Evaluation Results for Two Indoor Air Quality Studies (Hawthorne et al., 1987) | VOCs, HCHO To demonstrate methods for maximization of homeowner response. | Passive Active GC-FID |
| Volatile Organic Compounds in Indoor Air: Types, Sources and Characteristics (Gammage and Matthews, 1988) | Evaluation of indoor air contamination emitted by transient and persistent sources of VOC | Hand-held PID Active GC-FID Formaldehyde by SEM |

**Table 5** (*Continued*)

| Study Title | Description | Sampling and Analysis Methods |
|---|---|---|
| | and to examine HCHO emission characteristics of pressed-wood products. | |
| Comparison of Volatile Organic Levels Between Sites and Seasons for the Total Exposure Assessment Methodology (TEAM) Study (Hartwell et al., 1987) | To statistically compare VOC levels between different areas and different seasons in the same area. | Active/area Active/breath Active/personal GC-MS |
| The California TEAM Study: Breath Concentrations and Personal Exposures to 26 Volatile Compounds in Air and Drinking Water of 188 Residents of Los Angeles, Antioch and Pittsburgh, CA (Wallace et al., 1988a) | 26 VOCs. To carry out the TEAM study in California, using the same sampling methods used in Bayonne and Elizabeth, NJ, phases of the study. | Active/area Active/personal Active/breath GC-MS |
| Preliminary Results from the Baltimore TEAM Study (Wallace et al., 1988b) | To extend the findings of the TEAM study on VOCs to cities representative of major urban areas. | Active/personal Active/breath Canister/area GC-MS |
| Volatile Organic Compounds in Indoor Air (Tsuchiya, 1988) | To study TVOC and fingerprint of the compounds in indoor air. | Active GC-MS GC-FID |
| Residential Indoor Air Contamination by Screen Printing Plants (Verhoeff et al., 1988) | 17 VOCs found in air of screen printing plants in Amsterdam. Lesser number found in adjacent homes. | Active GC-FID |

**Table 5** (*Continued*)

| Study Title | Description | Sampling and Analysis Methods |
|---|---|---|
| Joint Representation of Physical Locations and Volatile Organic Compounds in Indoor Air From a Healthy and Sick Building (Noma et al., 1988) | Identification of major indoor VOCs, 158 compounds identified. | Active TD-GC-FID TD-GC-MS |
| Volatile Organic Contaminants in Indoor Air (Daisey and Hodgson, 1987) | VOCs, extractable organic matter, THC, and inorganic gases | Active GC-MS |
| Measurement of Volatile Organic Compounds (VOC) in Eight East Tennessee Homes (Gammage et al., 1988) | VOCs measured in 8 east Tennessee homes | Passive |
| Sampling and Analysis Techniques for Trace Volatile Organic Emissions from Consumer Products (Bayer et al., 1988) | To compare two techniques for the trace analysis of VOC emissions from consumer products | Direct on-line sampling and analysis GC-FID Active GC-MS |
| Bakeout of a New Office Building to Reduce Volatile Organic Concentrations (Girman et al., 1989a) | To examine effect of bakeout of an office building. During bakeout, VOC levels were four times pre-bake-out levels. | Active Active/ impingers TD-GC-MS CA |
| The Use of a Passive Sampler for the Simultaneous Determination of Long-Term Ventilation Rates and VOC Concentrations (Mailahn et al., 1989) | VOCs measured in 12 homes in Berlin over 26 two-week periods Air exchange rates measured via a perfluorocarbon tracer. (same as Seifert et al., 1989) | Passive GC-FID GC-ECD |

**Table 5** (*Continued*)

| Study Title | Description | Sampling and Analysis Methods |
|---|---|---|
| The Effects of Ventilation, Filtration and Outdoor Air on the Composition of Indoor Air at a Telephone Office Building (Weschler and Shields, 1989) | To examine selected VOCs, in an office building. | Passive GC-MS |
| Indoor/Outdoor Air Quality. Reference Pollutant Concentrations' in Complaint-free Residences. (Montgomery and Kalman, 1989) | 17 residences in Washington State monitored for typical VOCs. Winter values were higher than summer values. | Active GC-MS |
| Temporal Variability Measurement of Specific Volatile Organic Compounds (Pleil et al., 1989) | VOCs monitored in bathroom during showering, automobile while driving, residential and office air. | Syringe sequential sampler Cryotrapping GC-MSD |
| Determination of Organic Contaminants in Residential Indoor Air Using an Adsorption/Thermal Desorption Technique (Chan et al., 1990) | 12 homes tested in fall and winter for aromatics, 10–50 $\mu g/m^3$; other compounds, 1–10 $\mu g/m^3$. | Active/ multisorbent sampling tube TD-GC-MS |

**Formaldehyde Studies**

| | | |
|---|---|---|
| A Comparison of IAQ in Pacific Northwest Existing and New Energy-Efficient Homes (Grimsrud et al., 1986) | 161 houses tested for HCHO and infiltration. | Passive |
| Study Design to Relate Residential Energy | HCHO and criteria pollutants. | Mobile lab Continuous |

**Table 5** (*Continued*)

| Study Title | Description | Sampling and Analysis Methods |
|---|---|---|
| Use, Air Infiltration and Indoor Air Quality (Nagda et al., 1983) | To present the design of a study to investigate infiltration, IAQ, and energy use. | analytical devices, CA |
| A Survey of Typical Exposures to Formaldehyde in Houston Area Residences (Stock and Mendez, 1985) | 78 homes in Houston monitored during summer of 1980; 0.04 ppm HCHO found in conventional homes; 0.18 ppm in energy efficient residences. | Active/ impinger CA |
| Formaldehyde Concentrations in Wisconsin Mobile Homes (Hanrahan et al., 1985) | 137 homes monitored over 4 months. Values of HCHO ranged from 0.1 to 2.8 ppm (median of 0.39). | Active/ impinger CA |
| Formaldehyde Concentrations Inside Private Residences: A Mail-Out Approach to Indoor Air Monitoring (Sexton et al., 1986a) | 766 mobile homes surveyed; mean value of HCHO was 0.072 ppm; maximum was 0.46 ppm in summer, 0.31 ppm in winter. | Passive monitor CA |
| A Comparison of Formaldehyde Monitoring Methods for Residential Environments (Monsen and Stock, 1986) | 164 mobile homes in Texas; average concentration of HCHO was 0.14 ppm in 205 measurements. | Detector tubes, Active Passive CA TGM555 continuous analyzer |
| Formaldehyde Concentrations inside Conventional | 43 homes, sampled over 1-week periods, two | Passive CA |

**Table 5** (*Continued*)

| Study Title | Description | Sampling and Analysis Methods |
|---|---|---|
| Housing (Stock, 1987) | locations in each home; found mean HCHO values of 0.08 ppm ± 0.03 ppm. | |
| Indoor Air Quality in 300 Homes in Kingston/Harriman Tennessee (Hawthorne et al., 1988) | 300 homes survey to determine HCHO levels. | Passive |
| Formaldehyde Exposures Inside Mobile Homes (Sexton et al., 1989) | 470 mobile homes monitored for 1-week periods in summer and winter; 319 exceeded 0.10 ppm. | Passive CA |
| Monitoring of Parts-per-billion Levels of Formaldehyde Using a Diffusive Sampler (Levin et al., 1989) | Monitored office, residence, factory, and textile plant; good correlation of active and passive methods found; range of values 6–45 ppb HCHO. | Active DNPH coated filters HPLC |

**Environmental Tobacco Smoke (ETS)**

| Study Title | Description | Sampling and Analysis Methods |
|---|---|---|
| Chemical Studies on Tobacco Smoke, LIX: Analysis of Volatile Nitrosamines in Tobacco Smoke and Polluted Indoor Environments (Brunnemann and Hoffmann, 1978) | Various areas monitored for NDMA levels: trains, 0.11–0.13 $\mu g/m^3$; bar, 0.24 $\mu g/m^3$; suburban residence, <0.005 $\mu g/m^3$. | Collected in a nitrate buffer impinger GC-TED |
| Concentrations of Dimethylnitrosamine in the Air of Smoke Filled Rooms | Investigated NDMA in smoking areas. Found < 0.01 | Impinger/ nitrate buffer GC-TED |

**Table 5** (*Continued*)

| Study Title | Description | Sampling and Analysis Methods |
|---|---|---|
| (Stehlik et al., 1982) | $\mu g/m^3$ in a restaurant and a maximum of 0.07 $\mu g/m^3$ in an Austrian bar. | |
| Measurement of Nicotine in Building Air as Indicators of Tobacco Smoke Levels (Williams et al., 1985) | Investigated gradients of nicotine in a building with restricted smoking areas. | Cold plate collection method with methanol extraction GC-NPD |
| Thermal Desorption/ Gas Chromatographic/ Mass Spectrometric Analysis of Volatile Organic Compounds in the Offices of Smokers and Nonsmokers (Bayer and Black, 1987) | Used VOC and nicotine measurement to differentiate smokers' from nonsmokers' offices. | Active GC-FID for VOC GC-NPD for nicotine |
| Collection and Determination of Solanesol as a Tracer of Environmental Tobacco Smoke in Indoor Air (Ogden and Maiolo, 1989) | To evaluate solanesol as a tracer for ETS compared to nicotine. | Teflon filter extracted and derivatized with BSTFA GC-FID GC-MS Active GC-NPD for nicotine |
| A Thermal Desorption Method for the Determination of Nicotine in Indoor Environments. (Thompson et al., 1989) | Used test sampler to measure nicotine in labs (10.5 $\mu g/m^3$), restaurant (0.5– 37 $\mu g/m^3$). | Active GC-FID TD-GC-NPD Triethylamine assisted |
| Portable Air Sampler for Measurements in Aircraft and Public Buildings (van der Wal, 1989) | Portable sampler described for monitoring of nicotine; 0.4– 104 $\mu g/m^3$ were measured in | Active GC-NPD |

Table 5 (*Continued*)

| Study Title | Description | Sampling and Analysis Methods |
|---|---|---|
| | cabin air during flights. | |
| Continuous Monitoring of Air Impurities in Dwellings (Keskinen and Graeffe, 1989) | To examine the effect of smoking in 19 homes. | Impinger CA |

*Abbreviations*:

| | |
|---|---|
| Active | Active sampling methods employed |
| BSTFA | *N,O*-Bis(trimethylsilyl)trifluoroacetamide |
| CA | Chromotropic acid |
| DNPH | Dinitrophenylhydrazine |
| DCM | Dichloromethane |
| ECD | Electron capture detector |
| ETS | Environmental tobacco smoke |
| FID | Flame ionization detector |
| GC | Gas chromatograph |
| HCHO | Formaldehyde |
| HPLC | High-pressure liquid chromatography |
| IAQ | Indoor air quality |
| MS | Mass spectrograph |
| MSD | Mass selective detector |
| NDMA | Nitrosodimethylamine |
| NPD | Nitrogen phosphorous detector |
| Passive | Passive sampling methods employed |
| PAH | Polyaromatic hydrocarbons |
| PID | Photoionization detector |
| SEM | Surface emission monitor |
| TD | Thermal desorption |
| TCE | Trichloroethylene |
| TED | Thermal energy detector |
| THC | Total hydrocarbons |
| TTCE | Tetrachloroethylene |
| TVOC | Total volatile organic compounds |

matics, and numerous branched alkanes. Hydrocarbons and aldehydes were measured inside and outside paired homes with and without woodstoves, in Boise, Idaho (Zweidinger et al., 1988). Average concentrations for 14 aldehydes, ethylene, isoprene, benzene, and toluene were reported and the investigators concluded that levels of carbonyl compounds were higher indoors than outdoors. In other studies, the focus has been on specific target compounds and measurement of air concentrations indoors under many conditions. Wallace (1987) reports that 26 VOCs were identified in the latter phases of the TEAM study. Lebret et al. (1986) examined the occurrence of 45 compounds in Dutch homes, while Seifert et al. (1989) determined concentrations of 60 compounds in Berlin residences. Numerous other examples of studies on occurrence will be discussed in this part of the review.

A report to the U.S. Congress (USEPA, 1989) indicated that more than 900 compounds have been identified indoors, although not all could be classified as VOCs. In addition to studies mentioned previously, 79 compounds were identified in a study of two schoolrooms by Johansson (1978); 100 compounds were identified by De Bortoli et al. (1985) in a study of 15 homes in northern Italy; Noma et al. (1988) found 158 peaks (compounds) using mass spectrometry in some Stockholm preschools; Molhave (1982) identified at least 52 compounds that originated from use of common building materials while performing chamber studies on the materials. Weschler and Shields (1986) detected significant concentrations of additives in the indoor air of a telephone exchange building and found six compounds, including compounds such as dimethyl- and diethylphthalate. Table 6 provides a summary of com-

**Table 6 Volatile Organic Compounds Found Indoors**

**Alcohols**

Methanol[1]
Ethanol[2]
Propanol[3]
Butanol[4]
Pentanol[4]
1-Hexanol[5]

**Aldehydes and Ketones**

Formaldehyde[6]
Acetaldehyde[3]
Hexenal[4]
Acetone[2]
Butanone[4]
Diethylketone[3]

**Aliphatics**

Ethane[7]
Propane[7]
Butane[8]
Pentane[5]
Hexane[5]
Heptane[5]

Octane[5]
Nonane[5]
Decane[5]
2-Methylpentane[5]
3-Methylpentane[5]
2-Methylhexane[5]
3-Methylhexane[5]
2,4-Dimethylhexane[5]
2-Methylheptane[5]
3-Methylnonane[5]
4-Methyldecane[5]
1,3-Dimethylcyclopentane[5]
Cyclohexane[5]
Methylcyclohexane[5]

**Unsaturated Aliphatics**

Ethylene[7]
Acetylene[7]
α-Pinene[4]
Carene[4]
Limonene[4]

**Aromatic Hydrocarbons**

Benzene[5]
Styrene[5]

**Table 6** (*Continued*)

| | |
|---|---|
| Toluene[5] | **Miscellaneous** |
| Xylenes[5] | |
| Ethylbenzene[5] | |
| Diethylbenzenes[5] | Ketene[10] |
| Dimethylethylbenzenes[5] | Ethyl acetate[5] |
| Ethylmethylbenzenes[5] | Acetonitrile[10] |
| 1-Propylbenzene[5] | Acrylonitrile[10] |
| Propylmethylbenzenes[5] | Carbon disulphide[10] |
| Trimethylbenzene[4] | Formic acid[1] |
| | Acetic Acid[5] |
| | Butyric acid[3] |
| **Halogenated** | Pyruvic acid[11] |
| **Hydrocarbons** | Lactic acid[11] |
| | Ethylene oxide[5] |
| Chloroform[5] | Ethylamine[11] |
| Dichloromethane[5] | Diethylamine[11] |
| Tetrachloroethylene[5] | Nicotine[12] |
| 1,1,1-Trichloroethane[5] | N-nitrosodimethylamine[13] |
| Trichloroethylene[5] | N-nitrosodiethylamine[13] |
| Trichlorofluoromethane[5] | Cyclohexylamine[14] |
| Vinylidene chloride[9] | Diethylaminoethanol[14] |
| Ethylene dibromide[9] | Phenol[3] |
| Dichlorobenzenes[5] | Diethylphthalate[15] |

*References*: 1, Pitts et al. (1989); 2, Keskinen and Graeffe (1989); 3, Brimblecombe (1990); 4, Noma et al. (1988); 5, Ember (1988); 6, Girman et al. (1989a); 7, Davidson et al. (1986); 8, Vanderstraeten et al. (1984); 9, Wallace et al. (1982); 10, Jarke et al. (1981); 11, Yocom et al. (1984); 12, Proctor 1989); 13, Stehlik et al. (1982); 14, Edgerton et al. (1989); 15, Weschler and Shields (1986).

pounds and compound classes reported in many literature sources, including some that are noted above. This brief synopsis indicates that the indoor environment is very complex and, as described in Section 2, subject to the influence of many sources.

A valuable resource for information on IAQ in residences and public buildings is available (Shah and Singh, 1988; USEPA, 1988). Detailed information on the occurrence of VOCs arising from potential sources and on their determination in a hospital room, primary school, university building, two nursing homes, two homes for the elderly, and three offices is summarized in these references. It is neither feasible nor desirable to provide comprehensive listings of compounds found in indoor air in this review since these data are more amenable to summary in a database format such as that developed for the USEPA and described by Shah and Singh (1988). In this database, the concentration file for ambient measurements contains more than 122 000 records and there are more than 52 000 records on 71 chemicals in the indoor concentration file available from studies conducted in the United States.

Table 7 has been extracted from the database and provides information on the typical VOCs identified in indoor air, the average concentrations observed in studies conducted in the United States, and the frequency of

Table 7  Daily Indoor Concentrations for Selected Compounds for Indoor Sites

| Compound | Number of Measurements | Average Concentration (ppb) |
|---|---|---|
| Acetone | 4 | 7.955 |
| Benzaldehyde | 106 | 1.578 |
| Benzene | 2128 | 5.162 |
| 2-Butoxyethanol | 14 | 0.214 |
| Carbon tetrachloride | 2120 | 0.400 |
| Chloroform | 2120 | 0.832 |
| Cumene | 103 | 0.177 |
| Cyclohexane | 4 | 1.380 |
| Decamethylcyclopentasiloxane | 25 | 0.206 |
| Decane | 710 | 0.775 |
| Dichlorobenzene | 3 | 0.900 |
| 1,3-Dichlorobenzene | 2121 | 3.988 |
| 1,4-Dichlorobenzene | 2121 | 3.988 |
| Dimethylbenzene | 2216 | 2.844 |
| Dimethylbenzene | 2305 | 8.666 |
| 1,4-Dioxane | 585 | 1.029 |
| Ethenylbenzene | 2125 | 1.413 |
| Ethylbenzene | 2278 | 2.887 |
| Formaldehyde | 315 | 49.400 |
| Methyl ethyl ketone | 4 | 9.238 |
| Nonane | 134 | 1.164 |
| Octane | 605 | 0.882 |
| Pentadecane | 112 | 0.227 |
| α-Pinene | 623 | 0.547 |
| Tetrachloroethane | 2195 | 3.056 |
| 1,1,2,2-Tetrachloroethane | 585 | 0.014 |
| Tetradecane | 119 | 0.803 |
| Toluene | 220 | 7.388 |
| Trichlorobenzene | 2 | 0.065 |
| Trichloroethane | 2120 | 48.900 |
| Trichloroethene | 2132 | 1.347 |
| Tridecane | 122 | 1.484 |
| Trimethylbenzene | 96 | 0.574 |
| Trimethylbenzene | 178 | 0.914 |
| Undecane | 706 | 0.746 |

Source. Shah and Singh (1988).

detection of each compound. Some of the studies that have provided data for this database will be used as examples to illustrate the relationship between sources and occurrence of VOCs in the subsequent section.

## 3.2.  Sources and Occurrence Relationships

### 3.2.1.  Environmental Tobacco Smoke (ETS)

Major reviews regarding ETS have been cited in the section on sources. The contribution of ETS to indoor nicotine levels has been clearly established (Proctor, 1988). However, the specific contribution of ETS, whether MS or SS smoke, to indoor VOCs is less clear and is a subject of considerable controversy. Bayer and Black (1987), for example, used a Tenax sorbent coupled with thermal desorption and GC-MS analysis to establish the relationship between ETS and VOCs. Nicotine was used as the primary indicator of ETS occurrence in this study. The findings suggested that the contribution of ETS to VOC levels (e.g., alkanes, benzene, and ethylbenzene) was confounded by the contribution of VOCs from unspecified other sources. No statistical correlation was established between nicotine (ETS) and VOCs and the authors concluded that VOC levels could not be unambiguously used as indicators of ETS.

Repace and Lowrey (1987) in a criticism of Sterling et al. (1987) suggested an important relationship between ETS and indoor air pollution, but failed to specify VOCs as indicators of ETS. In contrast, many other researchers have conducted extensive indoor VOC surveys that suggest a significant contribution of ETS to the indoor air levels of VOCs. Seifert et al. (1989), Wallace et al. (1987a), and Krause et al. (1987) all suggest enhancement of indoor levels of aromatics, specifically in smokers' homes. Krause et al. (1987), for example, in a study of 500 West German homes, divided the homes into roughly equal groups (smoking and nonsmoking homes) and found elevated levels of benzene in smokers' homes (9.3 $\mu g/m^3$ compared to 6.9 $\mu g/m^3$). Wallace et al. (1987a,c) reported on enhanced levels of benzene, ethylbenzene, and styrene in the breath of smokers and pointed to smoking as one of the major sources of light aromatic compounds indoors. Proctor (1989) in southern England examined the occurrence of 18 VOCs in offices, betting shops, and train compartments with and without tobacco smoking. The arithmetic mean, standard deviation, median, minimum, and maximum values for the VOCs were reported and the author concluded that VOC concentrations were similar in the three environments and similar to data reported in the USA TEAM study (Wallace et al., 1987a). Based on their findings, the authors concluded that no ETS contribution to VOCs was evident in these indoor environments.

A review (Holcomb, 1988) of some studies on the impact of ETS on airline cabin air quality suggests that only low levels of ETS-related substances, including nicotine, are found in smoking sections and even lower levels are found in nonsmoking sections. Effective ventilation of cabins in most airliners is suggested as a possible reason for the low ETS levels. Stehlik et al. (1982) also examined concentrations of dimethylnitrosamine and found this compound at levels of $0.02-0.05$ $\mu g/m^3$ in smoke filled rooms, at $<0.01$ $\mu g/m^3$ in a restaurant and at $0.07$ $\mu g/m^3$ in a dancing bar. Similar results were found by Brunnemann and Hoffmann (1978) for dimethylnitrosamine.

Fishbein (1985) also suggested that cigarette smoking would contribute to indoor levels of simple aliphatics, aromatics, and other VOCs. Measurements in an apartment after cigarette smoking indicated concentrations of 20 ppb of methyl chloride, while in a restaurant the levels were between 0.65 and 8 ppb. Williams et al. (1985) tested nicotine as an indicator of tobacco smoke and noted a good correlation between smoking and the presence of nicotine in selected areas of buildings.

The evidence cited in this section suggests that VOCs (specifically the aromatics) may be poor indicators of ETS because of the multiplicity of indoor sources of these compounds. However, the studies also suggest that ETS contributes to the indoor VOC burden.

### 3.2.2.  Formaldehyde

Formaldehyde has been one of the most frequently measured compounds in IAQ studies. Indoor formaldehyde arises from combustion sources, ETS, consumer products, and building materials (Miksch et al., 1982). It is apparent from the few studies cited here that there is still concern for indoor formaldehyde levels and that many sources exist that can lead to significant concentration levels.

Concern with building materials emissions has provided the rationale for many of the IAQ studies on this compound. De Bortoli et al. (1984b) found formaldehyde values between 8 and 87 $\mu g/m^3$ in a survey of 15 homes in Northern Italy; and Keskinen and Graeffe (1989) reported values of 75 and 85 $\mu g/m^3$ in Tampere, Finland. Stock (1987) surveyed conventional U.S. housing by means of a variety of passive and active measurement techniques and found that concentration levels were lognormally distributed in the homes and mean values were $0.08 \pm 0.03$ ppm. Day versus night variation in concentrations was evident and attributed to the heating of the wall cavity. Smoking was found to have little influence on the weekly average values for formaldehyde.

Konopinski (1985) also measured seasonal formaldehyde levels in an office building and indicated that warm weather results were twice as high as those attained under winter conditions. Meyer and Hermanns (1985) examined the diurnal variation of formaldehyde exposure in mobile homes and found that early morning levels were $0.25-1.25$ ppm, while values increased to 5 ppm when the walls were warmer. The test homes in this study exhibited an air exchange rate of 0.5 air changes per hour (ACH). Stock and Mendez (1985) surveyed Houston area homes during the summer of 1980. In this study, an attempt was made to correlate the formaldehyde levels with the type of housing. The maximum value observed was 0.3 ppm; different house types had different mean concentration levels, with conventional homes at 0.04 ppm, energy-efficient homes at 0.07 ppm, apartments at 0.08 ppm, condominiums at 0.09 ppm, and energy-efficient condominiums at 0.18 ppm. The small sample size made it difficult to determine if statistically significant differences exist in the data.

Mobile homes have received considerable attention in view of the use of urea formaldehyde resins and particle board in their construction. Hanrahan et al. (1985) and Sexton et al. (1986a) have conducted major surveys of mobile homes in Wisconsin and California, respectively. In the Wisconsin survey of 137 homes over 9 months, median concentration values of formaldehyde were 0.39 ppm with a range of 0.1–2.8 ppm. A more detailed analysis was conducted of the California mobile home data by segregating the data according to age of the home and time of year. The mean concentration value of formaldehyde for all (766) homes was found to be 0.072 ppm in summer and 0.078 ppm in winter. Mean concentrations found in different seasonal periods were statistically not significantly different. However, specific information on environmental variables such as insolation and temperature was not used to categorize the measurement results. When results were correlated according to the age of the home, it was found that homes built before 1981 had lower mean concentrations than homes built after 1981; 0.06 versus 0.08 ppm were comparative mean concentrations found for summer data, and 0.06 versus 0.09 ppm were mean concentrations found for winter data.

### 3.2.3.  Other Building Materials, Furnishings, and Human Activities

Building materials make a significant contribution to indoor concentrations of VOCs. Molhave (1982) has conducted some pioneering work in identification of sources and compounds and in determining the contributions by these sources to indoor levels of VOCs. His results suggested that the ten most frequently identified compounds indoors were toluene, o- and m-xylene, 1,2,4,- and 1,2,5-trimethylbenzene, ethylbenzene, n-propylbenzene, n-decane, n-undecane, and n-nonane, and that the occurrence of these VOCs was attributable to building materials, furnishings, and consumer products. Lebret et al. (1986) arrived at a similar conclusion after surveying levels of 45 VOCs in 300 Dutch homes. In this study, statistical interdependencies found among compounds suggested common sources for the compounds. In addition, a greater number of compounds with concentrations exceeding 100 $\mu g/m^3$ was found in new homes than older homes. Lebret et al. suggested that building materials and new furnishings and carpets contributed to the higher VOC concentrations found in the newer homes. In a study that focused on the effects of woodstove usage on VOC levels indoors, Highsmith et al. (1988) found that the distribution of aldehydes measured indoors differed from that found outdoors and suggested that materials, furnishings and occupant activities had more impact on indoor aldehyde levels than the wood stoves. De Bortoli et al. (1985) identified up to 100 compounds in a survey of 15 homes in northern Italy. The observation that compound identities were consistent with emissions arising from building materials and household products suggested that these were the major sources. De Bortoli found formaldehyde, acetaldehyde, and acetone in all homes at mean values of 29, 16, and 40 $\mu g/m^3$, respectively; hexanal, nonanal, and other ketones were frequently but not always found; $C_6$–$C_{13}$ paraffinic compounds were detected at levels between 10 and 100 $\mu g/m^3$; 1,1,1-trichloroethylene, carbon tetrachloride, trichloro- and tetrachloroethane, and p-dichlorobenzene were found at levels between 10 and 100 $\mu g/m^3$; also found were benzene, alkylbenzenes, toluene, xylenes, limonene, and $\alpha$-pinene.

Hawthorne et al. (1985) compared a variety of indoor structural materials and furnishings, such as carpets, and evaluated their qualitative contribution to indoor VOC levels by comparison of chromatograms obtained simultaneously for indoor and outdoor air. Edgerton et al. (1989) evaluated levels of cyclohexylamine and diethylaminoethanol, resulting from steam humidification and found concentrations of 2.4 and 0.8 ppb for these two compounds indoors. Pleil et al. (1989) employed a sequential syringe sampler, cryogenic focusing, and GC-MSD analysis to monitor the air in various indoor environments. For example, they found levels of 1–20 ppbv of benzene, toluene, xylenes, styrene, and trimethylbenzene in an automobile while driving. During use of household spray products in a home, measurable levels of xylenes, 1,1,1-trichloroethane, dichloromethane, and carbon tetrachloride were found. Measurements in a kitchen indicated slight increments in hydrocarbons, benzene (1.5 ppbv), and toluene (3.5 ppbv) during cooking. Fishbein (1985) reported on the influence of consumer products on indoor air using information obtained mainly from a Maryland study. Indoor levels of carbon tetrachloride in residences near a solvent recovery plant were found to be three to four times as high as (or higher than) outdoor levels. Freon 12 was also reported at levels of 500 ppb in homes compared to outdoor values of approximately 1 ppb. Indoor use of spray cans and refrigerants was implicated by these measurements.

Seifert et al. (1989) studied 12 homes in Berlin during 26 two-week periods and pointed to renovations and painting as factors that contribute to significant elevation of VOC levels. Concentrations were 3–10 times higher during these activities, compared with periods when renovations or painting were not occurring in the home. Concentrations of ethyl acetate, 1,1,1-trichloroethane, and limonene exceeding 20 $\mu g/m^3$ were found during specific activities. During residents' absence, VOC levels were found to decrease by approximately 50%.

Weschler and Shields (1989) investigated the air quality in a telephone office building. VOCs, specifically, toluene (0.65–0.93 $\mu g/m^3$), xylene (0.80–0.86 $\mu g/m^3$), trimethylbenzene (0.44–0.48 $\mu g/m^3$), alkylbenzene (0.19–0.31 $\mu g/m^3$), $C_9$–$C_{13}$ (0.10–0.26 $\mu g/m^3$), trichloroethane (0.11–0.13 $\mu g/m^3$), and chlorobenzene (0.02–0.13 $\mu g/m^3$), were measured in this study. Organic compound concentrations were reduced only by elimination of specific sources such as cleaning solvents, waxes, and restriction of smoking. It can be inferred that these activities and products contributed to the enhanced levels of VOCs in the office building.

Interesting viewpoints regarding the impact of sources on indoor VOC occurrence are provided by Tsuchiya (1988) and De Bortoli et al. (1985). Tsuchiya examined the differences in VOC levels in homes with and without urea formaldehyde foam insulation (UFFI). Differences between these two types of residences in terms of VOC profiles or concentrations were not detectable. It was noted that any differences were overshadowed by the variations caused by differences in materials, furnishings, and activities occurring in the different residences. De Bortoli et al. (1985), however, have claimed that great similarities exist in the profiles of volatile organic compounds in indoor air in many surveys conducted in Europe and North America, and have suggested that the use of common building materials, solvents, and specific human activities are the primary sources of indoor organics.

The studies discussed in this section, in particular the latter two cited, point to the important influence of building materials, furnishings, consumer products, and human activities on indoor VOC concentrations. These findings are consistent with one of the major conclusions of the TEAM study (Wallace, 1987) regarding the overall importance of indoor sources compared with physical factors, such as air exchange rates, on indoor air quality.

### 3.2.4. Domestic Water

Indoor use of water for many human activities has been postulated to lead to significant increases of VOC concentrations, particularly of the light chlorinated organics. While data from studies using fortified water (Andelman, 1985b) suggest that this source may have a significant impact indoors, few studies have been reported in which source contribution and direct air quality measurements have been performed. Stern and Andrews (1989), performed an investigation on indoor air concentrations of chloroform in occupied and unoccupied apartments in New York City. Outdoor air was also monitored during the study using Du Pont passive organic vapor monitors and detailed water usage logs were kept. The results of this study suggest that no significant difference was observed between the mean concentration levels during periods when residences were occupied and unoccupied. Mean values were 501 $ng/m^3$ for occupied and 569 $ng/m^3$ for unoccupied residences. This result suggests that domestic water use made no detectable contribution to indoor air levels of chloroform. The authors, however, noted that, since the study population was quite small and highly stratified, it was difficult to extrapolate the study conclusions to the general population.

Andelman (1985b) has also investigated the potential for contamination of indoor air by use of water. He noted that contaminated groundwater could lead to significant human exposure of trichloroethylene (TCE) via the air route as a result of domestic water use. This study employed a different approach compared with that of Stern and Andrews (1989) and relied on measurement of air concentrations of TCE in a model shower enclosure in which TCE spiked water was employed. The transport process in a typical residence was then modeled mathematically. Based on the results, he concluded that potential for exposure was significant and he predicted concentrations for other compounds that are likely to volatilize during domestic water use, such as dichloromethane (170–14 000 ppt v/v), chloroform (14–730 ppt v/v), and carbon tetrachloride (140–460 ppt v/v). As part of a testing program of a novel syringe-based passive sampler, Pleil et al. (1989) measured levels of trihalomethanes in a residential bathroom during a morning shower and found levels of 3 ppbv initially and a maximum of 97 ppbv by the end of the shower.

### 3.2.5.   Outdoor Air and Other Sources

A number of studies have investigated the importance of various other factors contributing to VOC contamination indoors, including the importance of specific outdoor sources. While outdoor air influence was not the only focus of such studies, the study design allowed the investigation and assessment of the importance of the outdoor sources. A number of the studies have been selected to illustrate the findings.

The TEAM study represents one of the most comprehensive of those that have reported on VOC occurrence indoors. Wallace et al. (1987a) and Wallace (1987, 1989) have described the study, its design and findings in great detail. The TEAM study investigated the potential for human exposure of 26 VOCs by using personal 12-h measurement periods in several nonindustrial (Greensboro, NC, and Grand Forks, ND) and heavily industrialized (Elizabeth, NJ, and Los Angeles, CA) areas of the United States. By adopting the personal monitoring approach, exposures during excursions into outdoor environments were also monitored. However, in this study, nighttime samples were considered representative of indoor concentrations in the selected residences. Outdoor air and breath samples were also obtained in the study. Air concentrations were determined by use of Tenax GC sorbent tubes, and thermal desorption followed by analysis with GC-MSD.

Typical concentrations and compounds measured in the various areas were reported by Hartwell et al. (1987) and are provided in Table 8. The main conclusions of the study, based on statistical analysis of the data, were that indoor sources were the most important factor influencing indoor concentrations of VOCs and that indoor air generally contained higher VOC concentrations than outdoor air by a considerable margin. The major sources of exposure were identified as occupational environments, smoking (aromatics), dry cleaning, and filling of gas tanks. The air exchange rate was suggested to be secondary in importance in influencing indoor VOC occurrence compared to the presence of sources. These findings were generally applicable even for

**Table 8 Population Weighted Median Concentrations (µg/m³)**

| Compound | Elizabeth | | Elizabeth (summer) | | Los Angeles (winter) | | Los Angeles (spring) | | Pittsburgh/ Antioch (spring) | |
|---|---|---|---|---|---|---|---|---|---|---|
| | Night | Day | Night | Day | Night | Day | Night | Day | Night | Day |
| Ethylbenzene | 5.3 | 8.2 | 4.9 | 4.2 | 7.9 | 8.0 | 2.5 | 6.0 | 1.9 | 2.0 |
| o-Xylene | 6.0 | 9.9 | 5.4 | 5.1 | 9.7 | 10 | 2.6 | 4.0 | 2.2 | 3.6 |
| m,p-Xylene | 19 | 25 | 13 | 13 | 22 | 23 | 8.7 | 18 | 6.1 | 10 |
| Benzene | 14 | 14 | 8.5 | 3.9 | 15 | 15 | 4.4 | 7.2 | 4.4 | 6.3 |
| Tetrachloroethylene | 6.6 | 9.7 | 5.5 | 5.9 | 8.3 | 8.2 | 1.9 | 3.4 | 1.8 | 2.2 |
| 1,1,1-Trichloroethylene | 17.5 | 26 | 12 | 6.6 | 26 | 29 | 7.2 | 11 | 4.3 | 5.5 |
| m,p-DCB | 4.2 | 5.8 | 3.2 | 2.3 | 2.6 | 2.2 | 0.8 | 0.9 | 0.5 | 0.9 |
| Styrene | 1.3 | 1.7 | 1.4 | 1.2 | 2.8 | 2.3 | 0.8 | 1.5 | 0.7 | 0.8 |
| Chloroform | 2.2 | 2.2 | 0.9 | 0.8 | 1.5 | 1.0 | 0.7 | 0.3 | 0.03 | 0.03 |
| Sample Size | 49 | 47 | 150–156 | 142–146 | 113–119 | 110–114 | 51 | 51 | 70 | 66–68 |

*Source.* Hartwell et al. (1987)

study areas that were heavily industrialized and where many outdoor point sources of the target VOCs existed. The lack of correlation of the indoor VOC levels with geographical areas supports the conclusion stated earlier that the major influence on indoor VOCs is the presence of indoor sources.

A corollary finding was the applicability of breath sample analysis in detecting previous exposure. Lebret et al. (1986) conducted a major survey of 300 Dutch homes in two cities between 1981 and 1983 and measured weekly average concentrations of 45 compounds. Four of the homes were subjected to repeat sampling every second week for a year. Charcoal tube active sampling at 0.1 L/min, $CS_2$ extraction and capillary GC-FID, and some GC-MS was used as the measurement method. Table 9 summarizes the results from this study. The primary conclusions were that indoor concentrations were generally higher than outdoor levels, and that smokers' homes (total volatile organic hydrocarbons, TVOC, 353 $\mu g/m^3$) exhibited both higher concentrations and a greater number of compounds than nonsmokers' homes (TVOC, 213 $\mu g/m^3$). New homes had the highest number of compounds with concentrations over 100 $\mu g/m^3$. This result suggests the importance of building materials and possibly new furnishings as sources of indoor VOCs.

Verhoeff et al. (1988), in another Dutch study, examined the influence of small screen printing shops on the air quality within local residences. Monitoring of the screen printing facilities led to identification of 14–17 major volatile organic solvents, including benzene, heptane, and ethoxyethylacetate. In households situated above the shops, fewer compounds (0–15) were found and concentrations were generally much lower. Few of the residential data showed a direct significant correlation with analytical results from screen printing plants in the neighborhood. In a few of the residences some evidence was found for infiltration of screen printing emissions. However, the significantly lower concentrations and the incidence of fewer compounds overall in the residences than in the plants suggested that the external source did not lead to significant contamination of the indoor air.

Krause et al. (1987) studied the occurrence of VOCs in the air of 500 homes in West Germany from 1985 to 1986. Fifty-seven compounds were measured by means of Gasbadge or 3M OVM passive devices, deployed over 2-week periods. Carbon disulfide extraction was coupled with GC-FID and GC-ECD analyses and peak identification was achieved using retention indices and retention index differences. Table 10 summarizes the range of VOC concentrations measured in 230 of the homes in this study. The authors suggest that aromatic VOCs are emitted from continuous sources, while most halocarbons seem to be released from intermittent sources. The basis for this conclusion seems tenuous at best since long-term integrative sampling was the only sampling strategy reported in the study. In this study, smokers' homes were found to exhibit higher concentrations of benzene compared with homes where smoking did not normally occur: smokers' homes exhibited 9.3 $\mu g/m^3$ compared with 6.9 $\mu g/m^3$ of benzene in nonsmokers' homes. Once data from this study are more fully analyzed and reported, they should provide valuable insights on indoor VOC levels.

**Table 9  Maximum Indoor and Outdoor Air Concentrations of 45 VOCs Measured in 300 Dutch Homes**

| Compound | Maximum Concentration ($\mu g/m^3$) | |
| --- | --- | --- |
| | Indoor | Outdoor |
| n-Hexane | 338 | 4 |
| n-Heptane | 556 | 3 |
| n-Octane | 553 | 1 |
| n-Nonane | 407 | 8 |
| n-Decane | 205 | 5 |
| n-Undecane | 445 | 3 |
| n-Dodecane | 118 | 1 |
| n-Tridecane | 38 | <0.3 |
| n-Tetradecane | 46 | <0.3 |
| n-Pentadecane | 5 | 0.5 |
| n-Hexadecane | 4 | <0.3 |
| 3-Methylpentane | 101 | 3 |
| 2-Methylhexane | 278 | 4 |
| 3-Methylhexane | 233 | 3 |
| Cyclohexane | 355 | 2 |
| Methylcyclohexane | 504 | 2 |
| Dimethylcyclopentane isomer | 60 | 0.4 |
| Dimethylcyclopentane isomer | 29 | 0.4 |
| Dimethylcyclopentane isomer | 59 | 0.9 |
| Limonene | 693 | 10 |
| Benzene | 148 | 7 |
| Toluene | 2252 | 17 |
| Xylenes | 757 | 30 |
| Ethylbenzene | 138 | 14 |
| n-Propylbenzene | 27 | 0.7 |
| Isopropylbenzene | 11 | 0.3 |
| o-Methylethylbenzene | 156 | 2 |
| m-Methylethylbenzene | 227 | 2 |
| p-Methylethylbenzene | 94 | 1 |
| 1,2,3-Trimethylbenzene | 40 | 0.9 |
| 1,2,4-Trimethylbenzene | 400 | 4 |
| 1,3,5-Trimethylbenzene | 99 | 1 |
| n-Butylbenzene | 40 | 0.6 |
| p-Methylisopropylbenzene | 32 | <0.3 |
| Naphthalene | 14 | <0.3 |
| 1-Methylnaphthalene | 2 | <0.3 |
| Tetrachloromethane | 25 | 20 |
| Trichloroethene | 106 | <2 |
| Tetrachloroethane | 205 | <2 |
| Chlorobenzene | 27 | <0.4 |
| m-Dichlorobenzene | 9 | <0.6 |
| p-Dichlorobenzene | 299 | <0.6 |
| 1,2,3-Trichlorobenzene | 28 | <0.8 |
| 1,2,4-Trichlorobenzene | 33 | <0.8 |
| 1,3,5-Trichlorobenzene | 5 | <0.8 |

*Source.* Lebret et al. (1986).

**Table 10 Occurrence of VOCs in West German Homes**

| Compound | Mean Concentration ($\mu g/m^3$) | Range ($\mu g/m^3$) |
|---|---|---|
| n-Hexane | 9 | 2.0–46 |
| n-Heptane | 7 | 1.5–82 |
| n-Octane | 4.9 | <1.0–64 |
| n-Nonane | 9.7 | <1.0–101 |
| n-Decane | 14 | <1.0–136 |
| n-Undecane | 10 | <1.0–88 |
| n-Dodecane | 5.6 | <1.0–35 |
| n-Tridecane | 6.3 | <1.0–79 |
| Methylcyclopentane | 2.8 | <1.0–15 |
| Cyclohexane | 7.4 | <1.0–29 |
| Methylcyclohexane | 5.5 | 1.1–51 |
| Benzene | 9.3 | <1.0–39 |
| Toluene | 76 | 11.0–578 |
| Ethylbenzene | 10 | 1.5–161 |
| m,p-Xylene | 23 | 3.3–304 |
| o-Xylene | 7.0 | 1.2–45 |
| n-Propylbenzene and isopropylbenzene | 4.5 | <1.0–89 |
| Styrene | 2.5 | <1.0–41 |
| 1-Ethyl-2-methylbenzene | 4.4 | <1.0–103 |
| 1-Ethyl-(3- and 4-)methylbenzene | 8.9 | <1.0–229 |
| 1,2,3-Trimethylbenzene | 3.5 | <1.0–83 |
| 1,2,4-Trimethylbenzene | 11 | <1.0–312 |
| 1,3,5-Trimethylbenzene | 3.9 | <1.0–111 |
| Naphthalene | 2.3 | <1.0–14 |
| 1,1,1-Trichloroethane | 7.9 | <1.0–254 |
| Trichloroethylene | 13 | <1.0–1200 |
| Tetrachloroethylene | 12 | <1.0–617 |
| 1,4-Dichlorobenzene | 22 | <1.0–1270 |
| α-Pinene | 11 | <1.0–97 |
| β-Pinene | 1.3 | <1.0–11 |
| α-terpinene | 4.1 | <1.0–37 |
| Limonene | 25 | <1.0–315 |
| Ethylacetate | 12 | 1.0–204 |
| n-Butylacetate | 6.1 | <1.0–125 |
| Isobutylacetate | 2 | <1.0–33 |
| Methylethylketone | 6.2 | <1.0–25 |
| 4-Methyl-2-pentanone | 1 | <1.0–12 |
| Hexanal | 2 | <1.0–11 |
| n-Butanol | 1 | <1.0–11 |
| Isobutanol | 3 | <1.0–20 |
| Isoamylalcohol | 1 | <1.0–10 |
| 2-Ethylhexanol | 2 | <1.0–10 |

*Source.* Krause et al. (1987).

### 3.2.6.  Summary

The results of studies reported in this section show that there is substantial indirect evidence of relationships between VOC occurrence indoors and previously identified sources. Most of the indirect evidence for relationships has been derived from analysis of survey data by statistical means rather than by direct measurement of source emission factors coupled with dilution/transport calculations and air concentration predictions. A study by Berglund et al. (1989) represents an attempt at direct measurements of source emissions and indoor air occurrence of VOCs. Few other studies linked source emissions by direct measurement to indoor air occurrence of VOCs.

Environmental tobacco smoke and certain building materials, furnishings, and human activities have been established as significant contributors to indoor VOCs. Outdoor sources and use of domestic water are of less importance and there is little information on the contribution of combustion appliances. Many of the VOCs associated with specific sources have been identified in indoor air.

## 4.  CHARACTERIZATION METHODS

### 4.1.  Introduction

Many of the traditional procedures used for measurement of VOCs indoors had their genesis as occupational hygiene monitoring methods. Although the objective to estimate human exposure is common to all types of air quality investigations, occupational hygiene investigations usually have limited objectives and requirements. Normally, the objective is the determination of one or a few pollutants with acceptable exposure limits that are usually several orders of magnitude greater than the concentrations commonly found in nonindustrial environments. IAQ investigations, on the other hand, have a greater need for sensitivity and selectivity in detection of VOCs and a variety of objectives ranging from source characterization to examination of temporal trends. This has required modification of the existing and development of new methods.

Air quality monitoring procedures may be categorized under grab (instantaneous), continuous (frequent intervals), and integrative (single sample over period of time) sampling techniques. Further classifications are area (stationary) and personal (on person) sampling strategies and direct (immediate) and indirect reading methods. In addition, techniques may be classified as active (air sampled by mechanical means) or passive (air sampled by diffusion). Examples of these classes of techniques can be found in Table 5 and in the following reports on specific methods.

There are a variety of chemical analysis techniques in use. Many of these are listed in Table 5, which shows that most studies on VOCs have applied the traditional gas chromatography techniques. Occasionally, and for special applications, other methods, such as those based on detector tubes (Draeger), infrared spectroscopy (MIRAN 1B2), and APCI-MS, have been used. For certain investigations, such as those on aldehydes occurrence, HPLC often has been favored over GC. Recent advances in capillary column technology allow improved gas chromatography of complex mixtures, polar compounds, and some relatively reactive or unstable species (e.g., amines, phenols). However, traditional wet chemistry techniques are still used for some applications, such as cigarette smoke characterization.

Specialized equipment is required for the determination of source emission characteristics in the field and the laboratory. The apparatus used varies in size and complexity, depending on the application. By and large, the chemical analysis techniques are usually the same as those used in IAQ monitoring. Some of the procedures are listed below.

The intent of this section is not to provide a comprehensive critical review, but to summarize methods commonly employed for indoor VOC investigations. Excellent reviews on measurement techniques for tobacco smoke are found in a book on passive smoking (IARC, 1987). The determination of formaldehyde has been reviewed in a book edited by Turoski (1985) and a book by Meyer (1979). Godish (1985) and Otson and Fellin (1988) reviewed methods for airborne aldehydes, Seifert (1988) and Lewis (1989) discuss VOC measurement methods, and Fishbein (1985) discusses analytical techniques for indoor halocarbons. A book edited by Keith (1984) provides valuable information on source emissions and the sampling and analysis of organic pollutants in air. Books on air sampling procedures (American Conference of Governmental Industrial Hygienists (ACGIH), 1988), instrumentation (Hering, 1989) and passive sampling (Berlin et al., 1987) are additional sources of information. These resources should be consulted by those who require details on procedures or additional information on certain topics.

## 4.2.  Study Design

In addition to other practical considerations, monitoring objectives dictate the equipment and procedures used in IAQ investigations. Important considerations that affect the choice of strategy and procedures are sampling locations, time, duration, and frequency. The need for representative sampling, supplementary and complementary measurements, the availability, suitability, and convenience of sampling methods, sample stability or storability, and quality assurance and control procedures must also be considered. Selection of procedures depends on a number of factors. For example, indoor VOC sources include building materials and furnishings that emit continuously, as well as intermittent occupant-related activities, such as the use of consumer products, fireplaces, and tobacco. Short and frequent sampling periods are more suitable for detailed documentation of temporal trends (e.g., for intermittent emissions and acute exposures), while long-term integrative sampling is applicable to relatively constant persistent emissions and assessment of average exposure (e.g., for chronic health effects). Small, unobtrusive (e.g., low noise and airflow) equipment is usually required for IAQ or personal monitoring.

For indoor environments, selection of study strategy is particularly difficult due to the abundance and variety of both known and unidentified sources and the potential for rapid temporal and spatial variations in VOC levels. Because of the complexity of the indoor environment, it is important to have clearly defined objectives and strategies before embarking on any major IAQ investigation. Although there is a lack of comprehensive, published reports on protocols and procedures to aid in planning IAQ surveys and monitoring studies for what are often quite unique microenvironments, some helpful information is available. Issues related to the design and execution of indoor studies have been discussed by Levin (1987), Seifert (1988), and Fortmann et al. (1987) and in a book edited by Nagda and Harper (1989). Also, the TEAM study (Wallace, 1987) serves as a useful template for considerations that need to be addressed in study design.

## 4.3.  Source Characterization Methods

A variety of equipment for characterization of sources and their emissions have been designed to suit individual study requirements. Apparatus for containment of sources range from sophisticated, room-sized, controlled environment chambers to sealed vials for static head space analysis. Emission rates for sources in dynamic flow-through chambers of various sizes generally have been calculated by use of mass balance models (e.g., Sanchez et al., 1987) based on the formula described in Section 5.5. Steady-state conditions, a well-mixed chamber atmosphere, and no VOC losses within the chamber have usually been assumed. A simple version that has frequently been used is

$$S = Vq(C - c_0)$$

where $S$ = source emission rate ($\mu$g h$^{-1}$), $V$ = chamber volume (m$^3$), $q$ = air change rate (h$^{-1}$), $c_0$ = background VOC concentration ($\mu$g/m$^3$) in the chamber, and $C$ = equilibrium VOC concentration ($\mu$g/m$^3$).

Emission analysis techniques have included use of direct reading instruments, active sampling with various sorbents, and grab sampling. Collection of VOCs onto Tenax followed by thermal desorption into a GC-FID or GC-MS has been the most popular method. Other sorbents, canisters, and passive samplers commonly used in IAQ investigations have also been applied. In fact, sampling and analysis procedures used in source characterization generally have been the same or similar to those described later in this section.

It is difficult to accurately measure the impact of combustion emissions on IAQ. Traditional methods for generation and collection of SS and MS smoke components from tobacco or similar products have relied on small smoking chambers of various designs (IARC, 1987; Guérin et al., 1987). These are generally made of glass and are intended for use with one or a few cigarettes. The use of large controlled environment chambers (e.g., Eatough et al., 1989b) has allowed examination of ETS, but not SS or MS smoke, under relatively realistic conditions. Only an estimation of the character of SS and MS emissions is possible in studies with the small chambers since the smoke is usually generated by mechanical means and the nature of the complex, partly unstable, smoke components may be affected by the chamber and the collection devices. It has been suggested that further research is required to develop validated SS smoke generation and sampling devices (Guérin et al., 1987). The range of sampling methods is described and emission factors and smoke formation mechanisms are discussed in several recent reviews (IARC, 1987; Guérin et al., 1987; Reasor, 1987). Chemical analysis methods for MS and SS smoke are generally the same as those for ETS and are discussed below.

It is also difficult to accurately determine the nature and amounts of emissions that enter the indoor environment during operation of combustion appliances for heating and cooking. Sampling and analysis of exhaust or flue gas has been conducted in several investigations (Cooke et al., 1981; Muhlbaier, 1982; Lipari et al., 1984; Zeedijk, 1986). Such procedures permit only estimation of the impact of emissions on IAQ for unvented appliances and there is no assurance that representative samples are obtained, even with direct probe sampling (Moschandreas et al., 1986b; Haraprasad and Dave, 1989). Available combustion testing equipment only allows simulation of actual use conditions. Mass balance models and analytical data obtained for large (ca. 15–30 m$^3$) controlled environment chambers containing combustion appliances have been used to estimate the effect of VOC emissions on IAQ (Moschandreas et al., 1986a,b; Sexton et al., 1986b). The direct sampling and mass balance methods were compared for unvented gas appliances in a 14-m$^3$ chamber and in houses (Moschandreas et al., 1986a, 1987). Based on NO and NO$_2$ data, it was concluded that the two methods provide comparable results.

Emissions from building materials, furnishings, and consumer products have been extensively examined by use of a wide variety of chambers. Dynamic systems have included a 43-mm-wide, 30-cm-long, glass tube (Krzymien, 1989), a small, custom-designed glass jar (Sheldon et al., 1986), a 1-L Teflon-lined chamber (Tichenor and Mason, 1988), a 3.8-L steel paint can (Girman et al., 1986), a 9-L desiccator (Monteith et al., 1984), 30-L (van Netten et al., 1988) and 1-m$^3$ (Molhave, 1982) stainless steel chambers, and 1.9-m$^3$ (Berglund et al., 1989) and 34-m$^3$ (Wallace et al., 1987b) controlled environment chambers. The USEPA has also tested and applied a well-equipped, 166-L, stainless steel, controlled environment chamber (Sanchez et al., 1987; Tichenor and Mason, 1988). Static head space determinations have been conducted by use of 100-mL (Saarela et al., 1989) and 250-mL (Wallace et al., 1987b) glass vials. Matthews et al. (1986) used a monitor based on passive sampling for determination of formaldehyde emissions from surfaces.

There is also a different and possibly more practical approach to examination of the effect of various sources on IAQ. For example, Weschler and Shields (1989) used a mass balance model and data on indoor and outdoor VOC concentrations and building operation conditions to calculate indoor generation rates for VOCs associated with dominant indoor sources. A useful extension to such whole-building studies is implementation of complementary chamber studies to identify and characterize likely sources of the identified VOCs. Based on these findings, mitigation measures can be implemented and their success can be monitored.

### 4.4.  Tobacco Smoke Characterization

Smoke components are representative of a wide range of chemicals found in indoor air. Thus, many of the methods for characterizing tobacco smoke are common to those used for investigations involving other VOC sources. ETS and SS and MS smoke constituents have been collected by means of Cambridge filters, cold traps, gas bubblers, solid sorbents, and other methods. Many of these and corresponding chemical analysis methods are reviewed in detail in a book on passive smoking (IARC, 1987). Also, a review (Proctor, 1988) of the contribution of ETS to indoor air briefly discusses some of the analytical methods for volatile ETS components such as nicotine, hydrogen cyanide, nitrosamines, and simple aromatics.

Nicotine has been collected (for personal exposure assessment) on Millipore filters and Cambridge filters treated with sulfuric acid, in syringes (grab sample), on glass fiber filters treated with potassium bisulfate, and by pumping contaminated air through a tube containing Uniport-S coated with OV-17 (IARC, 1987). Chemical analyses were done by means of GC methods. Tobacco alkaloids in SS smoke collected on potassium bisulfate-impregnated Cambridge filters can be desorbed with methanol containing a small amount of NaOH and determined by means of a GC equipped with an FID or an NPD (IARC, 1987). A sampling method for collection of both particulate and vapor phase nicotine onto sodium bisulfate-treated, Teflon-coated, glass fiber filters has been reported (Hammond et al., 1987). The extracted nicotine was determined on a GC equipped with a nitrogen selective detector. The method was judged efficient and sensitive (0.2 $\mu g/m^3$) by the investigators, but did not permit differentiation between vapor and particulate phase nicotine.

Volatile airborne alkaloids, including nicotine, have also been collected by means of sorbents such as XAD, Tenax, and ORBO-42, and by condensation on a cold surface. Chuang et al. (1987), for example, used a two-stage Soxhlet extraction step to recover nicotine from XAD-2 and XAD-4 and followed this with a concentration step and GC-NPD analysis. Volatile alkaloids in air contaminated by ETS or from other sources can be determined by collection on XAD-2 resin, extraction with ethyl acetate, and analysis by GC-FID or GC-NPD (IARC, 1987). In an investigation (Sterling and Mueller, 1988) of the relationship between ETS, ventilation, and smoking regulation, nicotine was collected at 1 L/min on 80 mg/40 mg of XAD-4 in tubes. The collected nicotine was extracted with ethyl acetate containing 0.01% triethylamine and determined on a GC equipped with an NPD. Thompson et al. (1989), in contrast, collected nicotine on Tenax-GC and used triethylamine-assisted thermal desorption followed by direct injection onto a packed GC column (10% wax/2% KOH on 80/100 chromosorb W-AW) with detection by NPD to overcome some of the poor detection limit difficulties encountered with the NIOSH (1977) protocol. The latter protocol employs solvent extraction of XAD-2 with GC-NPD and has detection limits of 300 $\mu$g/m$^3$. Bayer and Black (1987) employed ethyl acetate extraction of nicotine from ORBO-43 followed by GC-NPD analysis with a DB-5 column. In a study (Williams et al., 1985) aimed at estimation of the distribution of ETS in a large office complex, airborne nicotine was collected on the surface of a cooled ($-10$ °C) petri dish and subsequently recovered in methanol. The methanol solution was analyzed on a GC equipped with a DB-1 capillary column and an NPD.

N-Nitrosamines from tobacco smoke are typically collected in aqueous buffer solutions containing ascorbic acid, extracted with dichloromethane, and determined by GC-thermal energy analysis (TEA) procedures. Stehlik et al. (1982) used a low-volume impinger charged with 75 mL of 1 $m$ citrate/phosphate buffer (pH 4.5) and ascorbic acid. Brunnemann and Hoffmann (1978) and Hoffmann et al. (1983) collected volatile nitrosamines in traps charged with ascorbic acid solutions. The solutions were then extracted with dichloromethane and the extract was analyzed by GC-TEA. A procedure (IARC, 1987) for determination of aromatic amines in SS smoke entails trapping of the compounds in aqueous HCl, neutralization with NaOH, extraction with ether, cleanup by basic alumina column chromatography, derivatization with pentafluoropropionic anhydride, and analysis by GC-ECD.

Volatile aldehydes in SS smoke from a smoking apparatus are often collected in traps containing ice-cooled 2,4-dinitrophenylhydrazine (DNPH) in hydrochloric acid and chloroform (IARC, 1987). The aldehyde derivatives (hydrazones) are extracted with chloroform and determined by HPLC with UV detection. A similar procedure (IARC, 1987) has been prescribed for aldehydes in ETS and both procedures are suitable for determination of aldehydes from other sources. In an investigation (Olander et al., 1988) of the ability of various air cleaners to remove ETS components, formaldehyde was measured with a modified chromotropic acid method, HCN with a CEA TGM 555 continuous colorimetric analyzer, and hydrocarbons with a Miran 1A spectrophotometer. Carbonyl compounds in ETS were determined (Schlitt and Knoppel, 1989), essentially by reaction with DNPH solutions in impingers and determination by HPLC with a scanning UV detector.

Acids and phenols have been collected on Cambridge filters followed by 5% NaOH in gas wash bottles (IARC, 1987). Cambridge filters followed by acidic solutions in bottles were used for collection of alkaline constituents. Other constituents, including many VOCs, have been collected by means of Cambridge filters and cold traps. Volatile phenols, such as the cresols and methylated catechols, from tobacco combustion have been determined by trapping in aqueous NaOH, acidification of the solution, extraction with dichloromethane, fractionation on Sephadex LH-20, derivatization with $N,O$-bis(trimethylsilyl)trifluoroacetamide (BSTFA), and analysis by GC-FID (IARC, 1987). In their examination of potential ETS tracer compounds, Eatough et al. (1989b) collected gas phase organic bases with benzene–sulfonic acid (BSA)-coated diffusion denuders and with an XAD-2 or Tenax sorbent bed following a quartz filter. Analyses of the extracted bases were done by GC or GC-MS. In a more detailed chemical characterization of ETS generated in a chamber (Eatough et al., 1989a), gas phase acids and bases were determined by use of several different diffusion denuder samplers and ion chromatography, GC, or GC-MS. As compared to most other available methods, diffusion denuders allow segregation and differentiation of vapor and particle phase organics.

In a study (Proctor, 1989) of smoking and nonsmoking environments in England, VOCs were collected at 50 cm$^3$/min on 0.2 g Tenax TA, the sampling tube was thermally desorbed, and the VOCs were cryogenically ($-30$ °C) trapped on Tenax and then separated on a DB-5 capillary column before identification and quantitation with an ion trap detector. Detection limits were about 0.1 μg/m$^3$ and a briefcase was used to conceal the sampling equipment. A Photovac 10S50 gas chromatograph has been used for direct measurement of airborne isoprene in ETS (Lofroth, 1989b).

### 4.5. Determination of Aldehydes

The variety of methods available for airborne aldehyde determination has recently been reviewed by Otson and Fellin (1988) and Godish (1985). It is evident from these reviews and the published literature that most IAQ investigations have focused on formaldehyde determination. Examples of some commonly used methods for aldehyde determination are provided below.

Methods used for source characterization have generally been the same as those used for other IAQ investigations. Girman et al. (1989a) used midget impingers and a modified chromotropic acid method to determine formaldehyde in a building during bake-out tests. Van Netten et al. (1988) determined formaldehyde levels by means of the chromotropic acid method and DuPont dosimeters. The PF-1 diffusion tube monitor (Air Quality Research, Inc.) was used for formaldehyde determination in a survey of air in homes with and without unvented combustion sources (Leaderer et al., 1986). Matthews et al. (1986) used a novel surface monitor for passive sampling of emissions onto a molecular sieve bed. Formaldehyde was eluted with water and determined colorimetrically. Indoor and outdoor formaldehyde (para-rosaniline method) and total aliphatic aldehyde (3-methyl-2-benzothiazolane hydrazone (MBTH) method) levels were determined by use of the LBL sampler (Turiel et al., 1983). Emissions from gas appliances were analyzed for aldehydes by a modified pararosaniline colorimetric method (Moschandreas et al., 1983). Carbonyls were determined by means of DNPH (2,4-dinitrophenyl hydrazine) coated silica gel followed by HPLC analysis in an examination of air quality at wood-burning residences (Highsmith et al., 1988). Aldehyde emissions in the chimneys of wood-burning fireplaces were measured by means of a modified DNPH/acetonitrile/impingers/HPLC procedure (Lipari et al., 1984). Aldehydes in exhaust gas were determined by collection in DNPH solutions followed by HPLC analysis (Muhlbaier, 1982).

Among the active sampling methods for formaldehyde reviewed by Godish (1985), detector tubes were not recommended due to their poor detection limit (0.5 ppm) and precision ($\pm 25\%$) and it was reported that the continuous TGM555 monitor (pararosaniline method) was capable of monitoring over a concentration range of 0.05–10 ppm. Operation of the TGM was found to be very time-consuming, noisy, and messy, and the disposal of spent reagents was a problem (Godish, 1985; Monsen and Stock, 1986). The impinger-based chromotropic acid colorimetric method was found to be adequate. However, the pararosaniline method, based on different colorimetric principles, was reported to offer better sensitivity and specificity. Monsen and Stock (1986) employed Draeger detector tubes based on colorimetric principles and found them to be of little value at the concentrations monitored (average 0.14 ppm). Collection of formaldehyde in water contained in midget impingers and subsequent chromotropic acid analysis offered consistent and reliable performance, whereas diffusion badges exhibited poor detection limits and variable blank values.

Among the passive devices evaluated, Godish (1985) indicated that the DuPont Pro-Tek formaldehyde badge offered the most consistent performance, with a measurement range of 0.12–6.8 ppm for an 8-h exposure time. The 3M formaldehyde monitor was judged to be relatively poor from the perspective of blank control and reproducibility ($\pm 25\%$), while the AQRI passive device (PF-1) offered reasonable performance with detection limits of 0.01 ppm after 7 days of exposure and reproducibility in the $\pm 25\%$ range. Envirotech's dosimeter was found to exhibit serious problems due to lack of specificity for formaldehyde and was recommended only for use as a range-finding device. Lack of linear response also proved to be a drawback.

### 4.6. Determination of Other VOCs

Methods representing the range of available sampling and chemical analysis techniques used in IAQ studies are presented in Table 5. Reviews by Lewis (1989) and Seifert (1988) have also briefly described the methods typically employed in determination of airborne VOCs. Grab sampling of VOCs can be done by use of rigid containers, usually made of glass and stainless steel, or by use of bags, usually made of plastics. Chromatographic methods have usually been applied for analysis of the collected samples. Electropolished, stainless steel canisters (Davidson et al., 1986) or sequentially operated syringe samplers (Krasnec and Demaray, 1984; Pleil et al., 1989) have been used together with cryogenic preconcentration methods and GC-FID and GC-MSD instrumentation. Detection limits for such combinations with GC-MSD instrumentation are generally below 1 μg/m³ (Lewis, 1989). Analysis of whole air samples or samples from bags and other containers has also been conducted with similar instrumentation. Seifert (1988) reported that the use of bags constructed from polytetrafluoroethylene (PTFE) gives better results than the use of other materials.

Electropolished canisters have also been employed to obtain integrative samples. In this case, a mass flow control or limiting orifice device is interjected in the sampling train to permit a slower (e.g., 24 h), controlled filling cycle. Lewis (1989) has noted that electropolished canisters are generally suitable for compounds with vapor pressures greater than $10^{-2}$ kPa, that is, chiefly VOCs, including some polar compounds. Special analytical protocols are required for analysis of polar compounds since, normally, the Nafion driers utilized for water removal in capillary column analyses also tend to remove these compounds.

The canister technique has been applied in studies with quite different objectives. Stainless steel canisters and GC instrumentation were used to monitor the occurrence of VOCs in indoor and outdoor air in the Himalayas (Davidson et al., 1986). VOCs were collected in pacified stainless steel canisters and analyzed by GC-FID in a receptor modeling study (Lewis et al., 1989). Canisters were also used to determine airborne hydrocarbons (GC-MS analysis) in a wood-burning community (Zweidinger et al., 1988) and in an examination of air quality at wood-burning residences (Highsmith et al., 1988). Analyses of integrative samples were done by GC-FID and GC-MS. Volatile hydrocarbons in emissions from catalytic and noncatalytic wood heaters were analyzed in a GC-FID by injection of an aliquot of exhaust gas collected in a stainless steel cylinder (Turiel et al., 1983).

Direct reading indicator tubes may be used for grab sampling or short-term integrative sampling. An example of such devices is the Draeger tube for determination of benzene, toluene, and xylene (Seifert, 1988). Detection is based on the formation and appearance of a brownish quinoid compound and a detection limit of 0.5 ppm is reported. Indicator tubes have limited applicability since they are not specific and, generally, not sufficiently sensitive for IAQ studies.

Integrative, active sampling methods using a variety of sorbents have been employed most frequently for the determination of indoor VOCs. Among the sorbents used are charcoal coupled with carbon disulfide (Lebret et al., 1984, 1986; Verhoeff et al., 1988; Girman et al., 1986), dichloromethane (Andersson et al., 1984), or dimethylformamide (Molhave, 1979) desorption; silica gel with solvent extraction (Seifert, 1988); Porapak with thermal desorption (Berglund et al., 1982, 1989; Seifert, 1988); XAD with solvent extraction (Chuang et al., 1987); and Carbotrap and Carbosieve with thermal desorption (Seifert, 1988). Thermosorb has been used for nitrosamine determination (Dropkin, 1985). Collection on Tenax followed by thermal desorption of VOCs into a GC-FID or GC-MS has been a popular method used for source emission tests (van der Wal et al., 1989; Krzymien, 1989; Hartwell, 1986; van Netten et al., 1988) and IAQ studies (Hawthorne et al., 1986; Bayer and Black, 1987; Wallace, 1987; Miksch et al., 1982; Girman et al., 1989a). In some instances, VOCs desorbed from Tenax were reconcentrated in a solvent (Turiel et al., 1983) before analysis or collected in a cryogenic (Sheldon et al., 1986) or Tenax/charcoal trap (Sanchez et al., 1987) before redesorption and analysis. Collection of VOCs in a Tenax/Chromosorb 106 trap and thermal desorption followed by GC-MS analysis has also been applied (Monteith et al., 1984). The advantages of thermal desorption include good sensitivity, and chromatography is improved if cryofocusing is applied. Among the disadvantages is the limited thermal stability of many sorbents. Good reproducibility has been achieved by thermal desorption techniques; for example, precision values of ±3 to 9% were reported by Seifert (1988) for a range of compounds.

Traps containing several sorbents have been employed in an attempt to increase the range of compounds that can be monitored. Tsuchiya (1988) employed a combination of glass beads, Tenax GC, and Ambersorb XE-340 with thermal desorption. Chan et al. (1990) used glass beads, Tenax-GC, Ambersorb-XE-340, and charcoal with thermal desorption. A multisorbent sampler containing Tenax-TA, Ambersorb XE-340, and activated charcoal was evaluated (Hodgson and Girman, 1987) and found to effectively collect VOCs covering a wide range of boiling points (≥36 °C). Tenax alone was not effective for collecting some of the very volatile compounds.

Passive sampling based on diffusive principles has begun to achieve wide acceptance due to the convenience of the available devices. Various passive devices and associated methods have been employed in major IAQ studies. OVM 3500 badges from the 3M Company and the Gasbadge sampler from Abcor have been employed by Seifert et al. (1989), Seifert and Abraham (1983), and De Bortoli et al. (1984a). In the latter study, methods were compared over a limited range of compounds and the results obtained with the passive devices (3M OVM 3500) were found to compare favorably with those obtained for apolar compounds with the active samplers (Tenax or Porapak; thermal desorption). Weschler and Shields (1986, 1989) used the 3M OVM 3500 badges to collect long-term (2–4 weeks) samples of VOCs in indoor and outdoor air.

As a result of a literature review of personal air monitors (Lutz, 1982) conducted for the USEPA, four passive and two active monitors intended for industrial hygiene monitoring were recommended for further evaluation. Lewis (1989) described a thermally desorbable passive sampler developed by the EPA. Different sorbents may be used with this device, sampling rates are typically 75–100 mL/min, and the method sensitivity is 1 ppbv with 1- to 3-h exposure periods for typical VOCs. Other passive samplers have also been used and Seifert (1988) claims that VOCs can be accurately sampled using passive devices for detection of VOCs between 30 and 300 μg/m$^3$ with reproducibility of ±20%. Analyses are usually done by GC-MS. Spicer et al. (1987a) compared distributive air volume sampling (Tenax), high and low rate passive sampling (Tenax), and whole air collection in canisters and used cryofocusing for GC analyses for the determination of 10 target VOCs. They found generally good correlation between results obtained by the different methods, although poor results were obtained for benzene and the canister method results were relatively high in value.

A variety of other methods have proved useful in IAQ studies of VOCs. These include the use of continuous monitors. Instruments range in complexity from the photoionization detector (Hawthorne et al., 1986) to the more complex atmospheric pressure chemical ionization (APCI) methods for detection of specific compounds (Yocom et al., 1984). An APCI-MS (TAGA-3000) was used to determine VOCs in various locations within two buildings in an investigation of occupant complaints and potential sources of indoor pollutants (Hijazi et al., 1983). Other methods employed include a variety of automated GCs that utilize whole air sample injection techniques. The Scentograph PC and Photovac 10S70 are examples of such instruments (Lewis, 1989). Continuous monitoring of total hydrocarbons in wood-burning appliance exhaust gas was conducted by means of a Beckman 108A hydrocarbon analyzer (Muhlbaier, 1982) and a FID (Cooke et al., 1981). In some chamber tests (Miksch et al., 1982), VOCs were collected cryogenically and the condensed liquid was injected directly into a GC-MS.

## 4.7.  Summary

Improved sorbents and relatively new techniques, such as those based on diffusion denuders (integrative), canisters (integrative or grab), and passive (integrative) sampling devices, have provided more versatility in sampling approaches and a better understanding of the nature of airborne VOCs. The development of new techniques and devices for isolation and concentration of analytes and the introduction of high-resolution capillary columns have resulted in improved specificity and detection limits for chromatographic methods. Transportable mass spectrometers and portable gas chromatographs now permit high-quality, on-site analyses. The improved design of emission measurement methods and equipment has contributed to a better understanding of sources contributing to indoor pollution. Notwithstanding all these improvements and innovations, there are few generally accepted, standard procedures for conducting IAQ investigations or for VOC determination.

## 5.  FATE OF VOCs

### 5.1.  Introduction

The fate of indoor VOCs has received little attention compared with characterization of sources and occurrence of VOCs. Indoor processes that control the fates are analogous to those that occur outdoors. However, due to the complexity of the indoor environment, understanding of the indoor processes has been relatively slow. Brimblecombe (1990), for example, suggests that the kinds of chemical transformation "that can occur on indoor surfaces are not at all well investigated." Other processes, such as deposition, have proven to be difficult to study indoors. This is because methods generally applicable in the outdoor environment for deposition measurement (i.e, eddy correlation, eddy accumulation, gradient and surface absorption/adsorption methods) are, as a rule, not suitable for application indoors. A lack of sensitive detectors or of suitable physical characteristics of the indoor environment prevent the application of such methods. Nonetheless, the importance of establishing the fate of indoor VOCs has been recognized and is receiving increasing attention.

Indoor processes are described in this section under headings of dilution and dispersion, deposition and sorption, and transformations. The description of the behavior of VOCs in terms of these processes has been formulated by mathematical models of varying complexity and types. A short description of the types of models most frequently used is provided to illustrate the power of such a tool. Ultimately, the efforts in characterization of processes and the use of models are directed at the mitigation of indoor VOCs and other pollutants.

## 5.2.  Dilution and Dispersion

In Section 4.3 on methods, a general equation was presented that described the source emission characteristics and the relationship between source emission strengths and concentrations. In the very simplest situation, the steady-state concentration achieved in a confined space is directly related to the emission factor, the volume of the space, the rate of air exchange, and the time after introduction of any given source. Seifert (1984) has provided an example of the concentration time profile in a 23-m$^3$ room with an artificial, constant formaldehyde source. He demonstrated mathematically that concentrations rose exponentially at all dilution or air exchange rates and that the highest steady-state concentrations were achieved at the lowest air change rates. In this idealistic example, the potential influence of deposition, sorption, and transformations on indoor VOC levels has been ignored. It is apparent that the two factors that determine the indoor concentrations, particularly of VOCs, are primarily the source emission strength and the dilution or ventilation rate. Wallace (1987) has suggested that the source strength is the more important of the two factors. He based his conclusions on the observation that concentrations and, by implication, source strengths vary over several orders of magnitude, but that air exchange rates are generally maintained within a factor of 10. The latter observation is consistent with measurements of air exchange in residences. Sulatisky (1984), for example, studied 200 residences across Canada and demonstrated that most exhibited rates between 1.5 and 10.4 ACH.

De Bortoli et al. (1985), Lebret et al. (1986), Wallace et al. (1987a), and others already cited in this review have all reported that outside air generally exhibits lower concentrations of VOCs than indoor air. Infiltration of outside air causes indoor concentrations to be diluted. The degree of dilution of indoor VOCs is related to the background of VOCs present in the outside air and the infiltration rate (see Section 5.5). Rates are normally determined by a pressure equilibrium approach (Sulatisky, 1984) or a chemical tracer approach. Carbon dioxide is generated in substantial amounts in occupied dwellings and has been employed as a natural tracer of outdoor air infiltration. Because of the somewhat variable background of outdoor $CO_2$ levels, $SF_6$ or perfluorocarbon (Mailahn et al., 1989) tracers have frequently been employed to assess air exchange rates indoors. Tichenor et al. (1990) employed $SF_6$ in their test home study. Many of the surveys cited in this review have not measured the air exchange with outside air; however, Seifert and Ullrich (1987) have demonstrated the feasibility of using passive monitoring of air exchange rates by using the same collection device for determination of VOC concentrations and a perfluorocarbon injected continuously in residences by a permeation device. This allowed the determination of indoor air dilution by outside air over an extended measurement period.

Tracers have also been used to evaluate pathways and effectiveness of room to room dispersion within indoor spaces. In small residences with operating central ventilation systems, room to room variability is generally low, as is evidenced in test house experiments with p-dichlorobenzene (p-DCB) conducted by Tichenor et al. (1990). In larger buildings with multicomponent ventilation systems and isolated areas, this is not likely to be the case and considerable variability in indoor VOC concentrations would occur due to barriers to transport from area to area. Maki and Woods (1984) made a similar observation and suggested that, under nonsteady conditions, it is invalid to assume that pollutants from a local source disperse uniformly throughout a house. They suggested that further research should be directed to dynamic behavioral characteristics. Inhomogeneities that can occur indoors are illustrated by the results of a study on nicotine levels. Concentrations of nicotine and other ETS-related substances were measured in offices under different conditions of smoking regulation (Sterling and Mueller, 1988). The effect of dilution of ETS was evident from the observation that nicotine levels averaged 14.0 $\mu g/m^3$ in designated smoking areas and 6.2 $\mu g/m^3$ in nonsmoking areas in cafeterias and from the comparison of levels in other areas of buildings under different ventilation and smoking conditions.

It is evident that both dilution and dispersion are important in determining the concentration levels, as well as the homogeneity of concentrations of VOCs in indoor compartments. Techniques employing tracers combined with multicompartment models for description of dilution, dispersion, and source strengths are important factors in understanding VOC behavior indoors.

### 5.3. Sorption and Deposition

Gebefugi (1989) has suggested that only chemicals with Vp $< 10^{-2}$ millibar (ca. $10^{-2}$ mm Hg) will accumulate on indoor surfaces. However, there is evidence that many VOCs will also collect on surfaces and materials. Among the VOCs, formaldehyde is probably the most extensively examined in terms of sorption and reemission processes. Matthews et al. (1987) have briefly reviewed investigations of such processes and, based on their own experiments, described a three-parameter model for the interactions of formaldehyde with gypsum wall board. Others (van Netten et al., 1989; Godish and Guindon, 1989; Godish, 1989a) have reported on the temperature and humidity dependence of formaldehyde release from various materials including those that do not initially contain formaldehyde. It has also been demonstrated that indoor dust may collect and then slowly release airborne VOCs such as formaldehyde (Rothenberg et al., 1989).

There are other examples of removal and reemission of indoor airborne VOCs, particularly those present in ETS. After an investigation of a 7-year-old building and chamber emission tests, Berglund et al. (1989) concluded that some of the VOCs emitted from wall, floor, and ceiling materials had been adsorbed from room air. In another study (Williams et al., 1985), a gradual reduction in airborne nicotine levels in a large office complex was observed after restriction of tobacco smoking. Based on simple dilution characteristics of such a building, a rapid reduction in levels would be accounted for by air exchange processes. The slow reduction implies reemission from contaminated indoor surfaces. Also, Ogden and Maiolo (1989) suggested that although more than 90% of the nicotine in ETS aerosol exists in the vapor phase, it is removed from the environment at a faster rate than the particulate fraction of ETS due to adsorptive and chemical interactions. Recent chamber studies (Ingebrethsen and Sears, 1989) of tobacco smoke particle removal processes showed that long-time (>100 min) particle count decay rates were first order and explainable by air exchange and size-dependent surface removal. Short-term decay rates were not first order and were explained by an additional process, evaporation of volatile compounds from particles. Eatough et al. (1989b) examined the effect of residence time in a chamber on the nature of ETS with the objective to identify candidate ETS tracer components. Several of the candidate organic bases occurred predominantly in the gas phase and the results indicated that gas-phase nicotine and myosmine are removed from indoor environments at a much faster rate than 2- or 3-ethenylpyridine or particle phase nicotine or cotinine.

Some recent studies have led to detailed mathematical representations and a better understanding of the importance of sorption and desorption processes. Nazaroff and Cass (1989) have conducted a detailed examination of mass-transport aspects of pollutant removal at indoor surfaces. Dunn and Tichenor (1988) have used $p$-DCB to examine sink effects and have developed a mathematical model that accounts for such effects and the fate of $p$-DCB in a chamber. In addition, Tichenor et al. (1990) quantitated $p$-DCB deposited on chamber walls during emission tests and detected continuing emission of $p$-DCB after removal of the primary source (moth crystals) from a test house. These few examples illustrate the importance of sorptive processes indoors.

## 5.4. Transformations

There is very little information on the chemical transformations of airborne VOCs in indoor environments. Even assuming commonality in chemical composition, it is not necessarily appropriate to predict the extent of indoor reactions based on a knowledge of outdoor atmospheric chemistry since indoor conditions, such as light intensity and reactant concentrations, are quite different from those outdoors. Also, surface reactions are likely to be more important indoors due to the significantly larger surface to air volume ratio found in buildings than outdoors.

Much of the work on indoor chemical transformations has been done because of concern with museum air quality and potential damage to works of art. Brimblecombe (1990) has listed many of the chemical species, including VOCs, that are found indoors and has reviewed the types of transformations that may occur in buildings. Reactive species that occur and are of concern in museum (and other) environments are inorganic species, such as nitrogen and sulfur oxides ($NO_x$ and $SO_x$) and ozone ($O_3$). Organic species include formaldehyde, volatile organic acids (e.g., $HCOOH$, $CH_3COOH$), alkenes, and organic sulfides. Brimblecombe reported that homogeneous chemistry within buildings may be important where there are glass walls allowing for high photolysis rates and enhanced oxidant concentrations. Oxidation of hydrocarbons to aldehydes and other species is then possible. There is little known about heterogeneous reactions at surfaces. Brimblecombe (1990) also suggested that conversion of HCHO to formic acid may occur on surfaces. Significant amounts of reactive species have also been detected in other indoor environments. Pitts et al. (1989) have reported that simultaneous identification and measurement of indoor gaseous $HCOOH$ and $CH_3OH$ in a study that also included quantitation of HONO, $NO_2$, and HCHO in structures containing combustion sources.

Eatough et al. (1989a) described the chemical characterization of gas phase components in ETS and the effect of UV light on some organic bases. Gas phase $HNO_2$ was the predominant acidic inorganic species found and there was evidence of photochemical reactions of nicotine. Although gas and particle phase ETS constituents are subject to oxidation and photochemical transformation, detailed studies of the phase distribution and chemical transformations have not been reported (IARC, 1987). Bruun Hansen and Andersen (1986) suggest that vapors from photocopy machines may react with ozone, which is also produced, and that the heated zones (e.g., 150–200 °C) may aid reactions. They further suggest that decomposition products of the monomer from toner resin are unstable ozonides, diperoxides, and epoxides and more stable oxygenated compounds such as aldehydes, ketones, and acids, but provide no experimental evidence.

Indoor chemical reactions may, in some instances, proceed at rates comparable to, or even much faster than, the ventilation rate (Nazaroff and Cass, 1986). In addition to photolytic and thermal reactions occurring in the gas phase, transformations may occur on fixed surfaces such as the floor, walls, and ceiling and on or within airborne particles. Nazaroff and Cass (1986) also indicate that such processes have substantial impact on indoor pollutant concentrations. They have formulated a general mathematical model that describes the time dependence of indoor air pollutants in a chemically reactive system and use estimated, as well as measured, values for various inorganic and organic species in evaluation of the model. Accurate values for deposition velocities of reactive species are generally unavailable, although measured values for NO, $NO_2$, $O_3$, and HCHO were used by Nazaroff and Cass (1986).

Harrison et al. (1988) have done a comparative evaluation of indoor and outdoor air quality and have reviewed major environmental pathways of important primary and secondary air pollutants. They briefly discuss reactions and deposition of some reactive inorganic species ($SO_2$, NO, $NO_2$, $O_3$ and peroxyacetylnitrate (PAN)), and conclude that levels of these pollutants indoors are normally lower than those outdoors due to enhanced decomposition processes in the indoor environment. The half-life of $O_3$ may be as short as 10 min (Harrison et al., 1988) and the levels decrease rapidly in the absence of a fresh supply from the outdoors or from indoor sources, such as photocopying machines (Bruun Hansen and Andersen, 1986). Maki and Woods (1984) have discussed photolytic and catalytic reactions which may remove reactive inorganic species from indoor air. Spicer et al. (1987b) have identified the processes and materials responsible for $NO_2$ removal in indoor air and have characterized the rates, mechanisms, and factors that affect these processes. The interaction of airborne organics and inorganics with metallic surfaces has been discussed by Sinclair (1989) in the context of indoor corrosion of metals.

Although a number of reactive inorganic species have been identified and measured, there is little direct evidence of their reactions with VOCs in the indoor environment. Information on the reactions of VOCs in the outdoor atmosphere (Altshuller, 1983) may be used to speculate on indoor transformations. To quote Altshuller (1983), "Almost all volatile organic compounds will react in the presence of solar radiation at varying rates with $NO_x$ to convert NO to $NO_2$ which subsequently reacts to form nitric acid and peroxyacyl nitrates. $O_3$ is produced as a product. The volatile organic compounds react with hydroxyl radicals and $O_3$ to form other radicals and stable products including a variety of volatile oxygenated organic compounds." Rates of reaction can be expected to be different indoors than outdoors due to different conditions, including light intensity and wavelength. Nazaroff and Cass (1986) have described some of the photochemical and other reactions that may occur indoors but do not indicate the relative importance of transformations compared to deposition, dilution, and sorptive processes.

## 5.5. Mathematical Models

The relationship between indoor processes can be formulated in mathematical form by use of models. A number of such models have been developed, which vary in complexity and basic physical principles. Models based on mass balance principles have been most commonly used in studies of indoor VOCs and other pollutants. In many instances, the models have been simplified by assuming that indoor transport and mixing are instantaneous processes and by eliminating terms in the model.

More sophisticated versions of such models describe pollutant transport between various indoor compartments and incorporate loss terms for deposition and chemical transformation. Still other models are based on computational fluid dynamics, statistics, and other principles. Nonetheless, the primary objective of all models is to provide a mathematical representation of the physical and chemical processes and to provide predictive capability. Numerous examples of the use of such models exist in the literature. However, detailed specific examples, except for those based on mass balance principles, will not be provided here.

A typical mass balance model is described by Wadden (1985). In this model, the conceptual framework is described in equation 1:

Indoor pollutant accumulation

$$
= \text{Pollutant flow inbound} - \text{Pollutant flow outbound} \qquad (1)
$$
$$
+ \text{ Source emissions} - \text{Sinks Removals}
$$

The mathematical representation of equation 1 is

$$
C_1 = \frac{k[q_0\,(1 - F_0) + q_2]\,C_0 + S - R}{k(q_0 + q_1 F_1 + q_2)}\,(1 - e^{-(k/V)(q_0 + q_1 F_1 + q_2)t})
$$
$$
+ C_s e^{-(k/V)(q_0 + q_1 F_1 + q_2)t} \qquad (2)
$$

where  $t$ = time
  $C_I$ = indoor concentration; and $C_I = C_S$ at $t = 0$
  $C_0$ = outdoor concentration
  $q_0$ = volumetric flow rate for makeup air
  $q_1$ = volumetric flow rate for recirculation air
  $q_2$ = volumetric flow rate for infiltration air
  $F_0$ = filter efficiency for makeup air
  $F_1$ = filter efficiency for recirculation air
  $V$ = room volume
  $S$ = indoor source emission rate
  $R$ = indoor sink removal rate
  $k$ = a factor that accounts for inefficiency of mixing (the fraction of incoming air that completely mixes within the room volume)

A more complete treatment of the indoor environment would also incorporate a separate term for chemical transformations. But, since little is known about indoor transformation of VOCs (Brimblecombe, 1990), either on surfaces or in homogeneous processes, this added complexity seems unwarranted and has, in the main, not been incorporated in IAQ models.

An examination of the various terms in equation 1 with reference to a few literature reports on modeling demonstrates the utility of the mass balance approach. The terms in equation 1 are established to allow prediction of indoor pollutant accumulation. Equally, rearrangement of the equation would permit the evaluation of other variables.

The source terms (source emissions and pollutant flow inbound) can be determined in several ways. In a text on passive smoking (IARC, 1987), for example, a simple phenomenological model was presented relating equilibrium ETS concentrations to source (smoker) density. Other less complex source strengths have been evaluated by employing chambers or field measurements as described by Tichenor et al. (1990) in tests on $p$-DCB from moth crystals. In the tests of Tichenor et al., weight difference over specified periods gave accurate assessment of source strength. Equation 1 also indicates the necessity for determining contaminants in dilution air. In only a very few instances can source strengths be determined with accuracy since source emissions can be highly variable. Section 2 of this review has alluded to the variability and the difficulty in establishing source strength data.

Dilution and dispersion of pollutants can usually be determined by establishing air exchange rates and the background concentrations of VOCs in the dilution air. As mentioned earlier, air exchange rates can be evaluated accurately by an artificial tracer, such as a perfluorocarbon tracer, or natural tracers, such as $CO_2$. By evaluating the exponential decay in concentration of any tracer after release into a confined space, it is possible to determine the ventilation or exchange rate between the space and outdoor air. This, in turn, allows calculation of dilution rates and pollutant outbound flow.

As was described in Sections 5.3 and 5.4, the sinks/removals term in equation 1 is not well understood for VOCs. Several avenues are available for estimating this term. Lofroth (1989b), for example, suggests the use of isoprene as a potential indicator for ETS behavior indoors. However, use of the surrogate compound is problematic since it can be emitted by several other sources and since it is not known whether isoprene is absorbed on indoor surfaces or if it reacts with other indoor pollutants. Direct measurement of factors that lead to indoor losses has also proven difficult. Aging and fate of particulate and inorganic components of tobacco smoke have until recently been examined in laboratory studies more thoroughly than the fate of VOCs (IRAC, 1987). Tichenor et al. (1990) used a calculated sink term based on solution of other terms in equation 1 for $p$-DCB. Losses due to direct surface deposition are expected to be relatively small since VOCs have vapor pressures exceeding $10^{-1}$ Torr (Gebefugi, 1989). Other sorptive processes have not been evaluated in detail for VOCs and have largely been ignored in modeling studies.

Losses can also occur by chemical transformation, but for indoor VOCs, this loss mechanism has not been evaluated fully. In a theoretical study, Nazaroff and Cass (1986) have evaluated photolytic, thermal, and heterogenous reactions for inorganic compounds and formaldehyde. Although not directly applicable to all VOCs, their results suggest that photolysis is unimportant indoors as a result of the low photon flux. Depletion of $O_3$ occurs rapidly indoors and Harrison et al. (1988) also suggest that other reactive species are lower in concentrations indoors than outdoors. Thus, homogeneous reactions are expected to be slow. Thermal and heterogeneous reactions could be important indoors, but direct evidence for reactions of VOCs was not presented by Nazaroff and Cass. Information to allow accurate inputs to the loss/sink term in equation 1 is therefore incomplete.

In spite of the incomplete information base for use of equation 1, Tichenor et al. (1990) described the use of a mass balance model similar to Wadden's. They employed a controlled test house and p-DCB as a test compound. Results calculated by the model showed good agreement with measurements obtained in the house when appropriate values were used for source emissions (p-DCB from moth crystals), outside air exchange, in-house air movement, and deposition of p-DCB on surfaces. Other loss terms were ignored.

Even simple formulations can quickly lead to complex models with many variables or factors that need to be solved or measured. At this time, indoor models have developed to a reasonable degree of sophistication. The major elements cannot always be solved because of the lack of specific information regarding some of the terms in the model. Indoor sinks (or fates), for example, are not usually well characterized. When accurate data on the major factors influencing IAQ become available, such IAQ models, if further refined, will be very useful. The potential environmental impact of sources can be evaluated, emission rate limits for indoor sources based on "acceptable" indoor pollutant concentrations can be determined, and building ventilation requirements for control of pollutant concentrations can be better estimated.

## 5.6. Summary

This brief description of fates of VOCs has pointed out some important deficiencies in our understanding the behavior of VOCs indoors. Conceptual models have evolved to a sophisticated level. Currently, the implementation of the models in a comprehensive fashion is hampered by a lack of detailed information on sorptive and reemission processes and by a poor understanding of transformations. Progress has been made, particularly in the areas of model development, evaluation methods for air exchange between indoor and outdoor environments, and determining source strengths.

For a more detailed treatment, including discussion of multicompartment models, the reader should consult more specialized literature (Jokl, 1989). A brief overview of available IAQ transport models is provided in the Report to Congress (USEPA, 1989).

## 6.  SUMMARY

### 6.1.  Overview and Recommendations

In the past 10 years, there has been an increasing awareness of the need to identify and measure airborne VOCs in indoor environments due to concerns with the potential deleterious effect on human health. Health- and comfort-related complaints by building occupants are often blamed on poor IAQ (NRC, 1981; Wallingford, 1988), particularly now that there is a greater public awareness of the possible connection between poor IAQ and health effects.

Reduction of infiltration by outdoor air as a result of energy conservation measures and changes in building construction has alleviated the effect of outdoor pollution on the indoor environment. However, tightening of the building envelope has at the same time exacerbated the effect of indoor pollutant sources. It has been demonstrated by numerous research efforts that a variety of indoor sources can emit a wide range of VOCs, some of which are potential irritants, and may cause discomfort or more serious toxic effects (Molhave et al., 1986; BNA, 1988; USEPA, 1989). For those who are affected and those who are responsible for solving IAQ related complaints and problems, it is important to be informed about the likely pollutant sources and strengths of emissions. For the scientific investigator, understanding the character of source emissions and their fate can expedite research and solutions to IAQ problems.

Many research efforts have been directed toward elucidation of these problems. Advances in analytical techniques have facilitated the investigation of IAQ and the origin and behavior of pollutants. The tools available to the analyst now include sophisticated continuous monitoring instruments, such as the APCI-MS and MIRAN 1B2, as well as portable gas chromatographs and automated canister sampling instruments. In addition, there have been improvements in air sampling materials and development of procedures, such as diffusion denuders and passive sampling devices, and advances in laboratory analytical instrumentation. Developments in apparatus and techniques for source emission characterization have resulted in more accurate determination of emission rates. These advances have contributed to a rapidly increasing amount of information on the sources and the occurrence of VOCs.

Recent air quality studies have ranged from surveys involving a large number of homes to detailed examination of large public buildings and single family dwellings. To date, the number of VOCs identified in indoor air has been estimated at about 900 (USEPA, 1989) and covers many chemical classes. A selected portion of the published data has been consolidated and made available in a computerized data base format (Shah and Singh, 1988); however, most of the data is available only in print. The lack of use of standard or, at least, common methods discourages comparison of data from different studies and creates a concern for the reliability of VOC data. Validated methods that are accepted by professional analysts and organizations are required for the characterization of both indoor air and of source emissions. Similarly, protocols for conducting surveys and other studies need to be established.

Some IAQ monitoring studies have been conducted to identify potential sources of VOCs or to determine the influence of known sources on IAQ. To be successful, such studies require a strong statistical basis or unique and dominant sources, and target compounds must be present. Identification of sources of particular airborne VOCs can be difficult since the organics can be generated from a variety of indoor sources, as well as outdoor sources, such as ambient air, soil, and groundwater. Indoor sources include building materials, combustion appliances, furnishings, consumer and household products, and various occupant-related activities. There have been significant efforts in recent years to improve understanding of the contributions of such sources to indoor pollutant levels. This review suggests that the greatest efforts have been directed at measurement of VOC emissions from building materials and to a lesser extent from consumer products. Considerable effort has been expended in the examination of tobacco smoke, combustion appliances, and pressed wood products. For these sources, extensive information has been developed for formaldehyde and nicotine emissions. However, there is a need for emissions data for the wide range of conditions that may apply indoors. There has been relatively little work done on quantitation of volatile organic emissions from combustion sources and even less on a variety of occupant-related activities. Concern with the contributions of domestic water, groundwater, and soil gas is very recent and few quantitative data are available. Infiltration of VOCs from outdoor sources, including ambient air, is a complex issue and depends on factors that are difficult to control and measure.

There is little information on the fate of indoor VOCs. Dilution and dispersion processes that affect VOC levels are the best understood, largely due to the interest in specific contaminants, such as formaldehyde and ETS. Considerable work has been done on the ventilation of buildings, and some investigations have focused on the effects on airborne VOCs. The nature and effect of sorptive and transformation processes have only recently received some attention. The lack of information on certain indoor processes has hampered application of detailed modeling techniques to predict indoor VOC levels or to allow widespread use of models for determining measures to control IAQ. For most applications, only values for air exchange rates and source emission rates are available. The presence of a large number and variety of both transient and persistent VOC emitters in most buildings makes the task of source identification and, hence, implementation of mitigation measures difficult. In general, source apportionment and predictive modeling for indoor VOCs is in relative infancy compared to outdoor air research. Comprehensive, simultaneous examination of sources, occurrence, and processes that control the fate of indoor VOCs is required to allow accurate prediction of IAQ and human exposure to VOCs. In many cases, indoor VOC levels can be reduced without significant energy expenditure by increasing ventilation or by cleaning recirculated air. Nevertheless, a better understanding of sources, the fate of VOC emissions, and factors that control airborne VOC levels is necessary.

## 6.2. Concluding Remarks

A major contribution to the understanding of the sources of direct human exposure to VOCs has been the results and conclusions drawn from the USEPA TEAM studies (Wallace, 1987). Wallace (1989) has concluded that, "Probably the central finding of all of these studies has been that the major sources of exposure to all chemical groups studied have been small and close to the person, usually inside his or her home. This finding is at odds with the conventional wisdom (that the major sources are industry, autos, urban areas, incinerators, landfills, and hazardous waste sites)." He has further suggested that monitoring and research efforts should be focused more on personal exposure than on outdoor air. It should be noted that exposure assessments have generally focused on the most abundant organics, which are not necessarily the most harmful. The TEAM and similar studies have examined only a small segment of indoor and outdoor pollutants and sources and have been successful in quantitating exposure to only a few VOC sources. Considerable research remains to be done in the identification and characterization of VOC sources, determination of the fate of emissions, and the real, as well as the potential, impact of emissions and their transformation products on humans and their environment.

*Acknowledgments*

*The authors thank L. Gamble, S. Barnett, and P.D. Bothwell for their assistance with literature searches; D. T. Williams, C. S. Davis, and L. Elliott for proofreading and comments; and M. C. Spano and N. Douglas for assistance with typing.*

## REFERENCES

Altshuller, A. P. (1983). "Review: Natural volatile organic substances and their effect on air quality in the United States." *Atmos. Environ.* **17**, 2131–2165.

American Conference of Governmental Industrial Hygienists (ACGIH) (1988). *Advances in Air Sampling.* Lewis Publishers, Chelsea, MI, 409 pp.

Andelman, J. B. (1985a). "Human exposures to volatile halogenated organic chemicals in indoor and outdoor air." *Environ. Health Perspect.* **62**, 313–318.

Andelman, J. B. (1985b). "Inhalation exposure in the home to volatile organic contaminants of drinking water." *Sci. Total Environ.* **47**, 443–460.

Andelman, J. B., Meyers, S. M., and Wilder, L. C. (1986). "Volatilization of organic chemicals from indoor uses of water." In J. N. Lester, R. Perry, and R. M. Sterritt, eds., *Proceedings of the International Conference on Chemicals in the Environment.* Selper, London, pp. 323–330.

Andersson, B., Andersson, K., and Nilsson, C.-A. (1984). "Mass spectrometric identification of 2-ethylhexanol in indoor air: Recovery studies by charcoal sampling and gas chromatographic analysis at the micrograms per cubic metre level." *J. Chromatogr.* **291**, 257–263.

Baechler, M., Ozkaynak, H., Spengler, J., Wallace, L., and Nelson, W. (1989). "Assessing indoor exposure to volatile organic compounds released from paint USI—the NASA data base." In *Proceedings of the 82nd Annual Meeting of the Air and Waste Management Association, Anaheim, CA, Paper No. 89–86.7.*

Barboza, M. J., Militana, L. M., and Leifer, S. A. (1987). "Indoor air toxic due to underground sources. Case studies." In *Proceedings of the 80th Annual Meeting of the Air Pollution Control Association*, New York, NY, Vol. 5, Paper No. 87/84.4.

Bare, J. C. (1988). "Indoor air pollution source database." *J. Air Pollut. Control Assoc.* **38**, 670–671.

Bayer, C. W., and Black, M. S. (1987). "Thermal desorption/gas chromatographic/ mass spectrometric analysis of volatile organic compounds in the offices of smokers and nonsmokers." *Biomed. Environ. Mass Spectrom.* **14**, 363–367.

Bayer, C. W., Black, M. S., and Galloway, L. M. (1988). "Sampling and analysis techniques for trace volatile organic emissions from consumer products." *J. Chromatogr. Sci.* **26**, 168–173.

Becker, D. S., and Salter, R. G. (1982). "Organic contaminants in indoor air and their relation to outdoor contaminants: Phase 2—statistical analysis." *ASHRAE Trans.* **88**, 491–502.

Berglund, B., Johansson, I., and Lindvall, T. (1982). "The influence of ventilation on indoor/outdoor air contaminants in an office building." *Environ. Int.* **8**, 395–399.

Berglund, B., Johansson, I., and Lindvall, T. (1989). "Volatile organic compounds from used building materials in a simulated chamber study." *Environ. Int.* **15**, 383–387.

Berlin, A., Brown, R. H., and Saunders, K. J., eds. (1987). *Diffusive Sampling: An Alternative Approach to Workplace Air Monitoring.* Royal Society of Chemistry, London, 484 pp.

Bilan, R. A., Haflidson, W. O., and McVittie, D. J. (1989). "Assessment of isocyanate exposure during the spray application of polyurethane foam." *Am. Ind. Hyg. Assoc. J.* **50**, 303–306.

Braun, A. G., Busby, W. F., Jr., Liber, H. L., and Thilly, W. G. (1987). "Chemical and toxicological characteristics of residential oil burner emissions: II. Mutagenic, tumorogenic and potential teratogenic activity." *Environ. Health Perspect* **73**, 235–246.

Brimblecombe, P. (1990). "The composition of museum atmospheres." *Atmos. Environ.* **24B**, 1–8.

Brunnemann, K. D., and Hoffmann, D. (1978). "Chemical studies on tobacco smoke. LIX. Analysis of volatile nitrosamines in tobacco smoke and polluted indoor environments." In E. A. Walker, L. Griciute, M. Castegnaro, and R. E. Lyle, eds., *Environmental Aspects of N-Nitroso Compounds*, IARC Sci. Publ. No. 19. International Agency for Research on Cancer, Lyon, pp. 343–356.

Bruun Hansen, T. (1987). "Prediction of the air concentration caused by the use of household products containing organic solvents." In B. Seifert, H. Esdorn, M. Fischer, H. Ruden, and J. Wegner, eds., *Indoor Air '87. Vol. 1. Volatile Organic Compounds, Combustion Gases, Particles and Fibres, Microbiological Agents*, Proc. 4th Int. Conf. Indoor Air Qual. Clim. Institute for Water, Soil and Air Hygiene, Berlin, pp. 32–36.

Bruun Hansen, T., and Andersen, B. (1986). "Ozone and other air pollutants from photocopying machines." *Am. Ind. Hyg. Assoc. J.* **47**, 659–665.

Bureau of National Affairs (BNA) (1988). *Indoor Air Pollution: The Complete Resource Guide*, Vol. 1, BNA, Washington, DC, 81 pp.

Chan, C. C., Vainer, L., Martin, J. W., and Williams, D. T. (1990). "Determination of organic contaminants in residential indoor air using an adsorption-thermal desorption technique. *J. Air Waste Manage. Assoc.* **40**, 62–67.

Chuang, J. C., Kuhlman, M. R., Hannan, S. W., and Bridges, C. (1987). *Evaluation of Sampling and Analytical Methods for Nicotine and Polynuclear Aromatic Hydrocarbon in Indoor Air*, Report No. EPA/600/4-87/031. U.S. Environmental Protection Agency, Research Triangle Park, NC.

Cohen, M. A., Ryan, P. B., Yanagisawa, Y., Spengler, J. D., Ozkaynak, H., and Epstein, P. S. (1989). "Indoor/outdoor measurements of volatile organic compounds in the Kanawha Valley of West Virginia." *J. Air Waste Manage. Assoc.* **39**, 1086–1093.

Cooke, W. M., Allen, J. M., and Hall, R. E. (1981). "Characterization of emissions from residential wood combustion sources." In J. A. Cooper and D. Malek, eds., *Proceedings of the 1981 International Conference On Residential Solid Fuels Environmental Impacts and Solutions*. Oregon Graduate Center, Beaverton, pp. 139–163.

Daisey, J. M., and Hodgson, A. T. (1987). "Volatile organic contaminants in indoor air." In E. J. Cairns and D. T. Grimsrud, eds., *Lawrence Berkeley Laboratory, Applied Science Division Annual Report. Indoor Environment Program FY 1987*. Report prepared for the U.S. Department of Energy under Contract No. DE-AC03-76 S F00098, 37 pp.

Dave, J. M., and Haraprasad, V. (1989). "Effect of reservoir fuel levels on emissions from a kerosene cook stove." In C. J. Bieva, Y. Courtois, and M. Govaerts, eds., *Present and Future of Indoor Air Quality*, Proc. Brussels Conf. Elsevier, Amsterdam, pp. 257–259.

Davidson, C. I., Lin, S.-F., Osborn, J. F., Pandey, M. R., Rasmussen, R. A., and Khalil, M. A. K. (1986). "Indoor and outdoor air pollution in the Himalayas." *Environ. Sci. Technol.* **20**, 561–567.

De Bortoli, M., Knoeppel, H., Molhave, L., Seifert, B., and Ullrich, D. (1984a). *Interlaboratory Comparison of Passive Samplers for Organic Vapours with Respect to Their Applicability to Indoor Air Pollution Monitoring: A Pilot Study*, Report No. EUR-9450-EN. Office for Official Publications of the European Communities, Luxembourg.

De Bortoli, M., Artoli, M., Knoppel, H., Rogora, L., Pecchio, E., Schauenburg, H., Peil, A., and Schlitt, H. (1984b). "Integrating 'real life' measurements of organic pollution in indoor and outdoor air of homes in northern Italy." In B. Berglund, T. Lindvall, and J. Sundell, eds., *Indoor Air. Chemical Characterization and Personal Exposure*, Proc. 3rd Int. Conf. Indoor Air Qual. Clim., Vol. 4. Swedish Council of Building Research, Stockholm, pp. 21–26.

De Bortoli, M., Knoppel, H., Pecchio, E., Peil, A., Rogora, L., Schauenburg, H., Schlitt, H., and Vissers, H. (1985). *Measurements of Indoor Air Quality and Comparison with Ambient Air. A Study on 15 Homes in Northern Italy*, Report No. EUR 9656 EN. Office for Official Publications of the European Communities, Luxembourg.

De Bortoli, M., Knoppel, H., Pecchio, E., Peil, A., Rogora, L., Schauenburg, H., Schlitt, H., Vissers, H., and Schauenburg, H. (1986). "Concentrations of selected organic pollutants in indoor and outdoor air in northern Italy." *Environ. Int.* **12**, 343–350.

Dropkin, D. (1985). *Sampling of Automobile Interiors for Organic Emissions*, Report No. EPA/600/3-85/008. U.S. Environmental Protection Agency, Research Triangle Park, NC.

Dudney, C. S., Matthews, T. G., Dreibelbis, W. G., Hawthorne, A. R., Thompson, C. V., Monar, K. P., Quillen, J. L., and Hjelmfelt, A. (1986). "Indoor pollutants in 70 houses in the Tennessee Valley area: Study design and measurement methods." In *Proceedings of the 1986 EPA/APCA Symposium on Measurement of Toxic Air Pollutants*, APCA Publ. VIP-7. Air Pollution Control Association, Pittsburgh, PA, pp. 116–127.

Dunn, J. E., and Tichenor, B. A. (1988). "Compensating for sink effects in emissions test chambers by mathematical modelling." *Atmos. Environ.* **22**, 885–894.

Dunovant, V. S., Clark, C. S., Que Hee, S. S., Hertzberg, V. S., and Trapp, J. H. (1986). "Volatile organics in the wastewater and airspaces of three wastewater treatment plants." *J. Water Pollut. Control Fed.* **58**, 886–895.

Eatough, D. J., Benner, C. L., Bayona, J. M., Richards, G., Lamb, J. D., Lee, M. L., Lewis, E. A., and Hansen, L. D. (1989a). "Chemical composition of environmental tobacco smoke. 1. Gas-phase acids and bases." *Environ. Sci. Technol.* **23**, 679–687.

Eatough, D. J., Benner, C. L., Tang, H., Landon, V, Richards, G., Caka, F. M., Crawford, J., Lewis, E. A., Hansen, L. D., and Eatough, N. L. (1989b). "The chemical composition of environmental tobacco smoke. III. Identification of conservative tracers of environmental tobacco smoke." *Environ. Int.* **15**, 19–28.

Edgerton, S. A., Kenny, D. V., and Joseph, D. W. (1989). "Determination of amines in indoor air from steam humidification." *Environ. Sci. Technol.* **23**, 484–488.

Eichler, D. L., Mackey, J. H., Niblock, S. F., and Weston, A. F. (1986). "Entry and dissipation rates of VOCs for residences with basements." In *Proceedings of the 79th Annual Meeting of the Air Pollution Control Association, Minneapolis, MN*, Paper No. 86–7.3.

Ember, L. R. (1988). "Survey finds high indoor levels of volatile organic chemicals." *Chem. Eng. News* **66**(49), 23–25.

Fishbein, L. (1985). "Halocarbons in indoor environments." In L. Fishbein and I. K. O'Neill, eds., *Environmental Carcinogens Selected Methods of Analysis. Vol. 7. Some Volatile Halogenated Hydrocarbons*, IARC Sci. Publ. No. 68. International Agency for Research on Cancer, Lyon, pp. 91–106.

Fisk, W. J., Spencer, R. K., Grimsrud, D. T., Offermann, F. J., Pedersen, B., and Sextro, R. (1987). *Indoor Air Quality Control Techniques: Radon, Formaldehyde, Combustion Products*. Noyes Data Corporation, Park Ridge, NJ, 245 pp.

Foreman, W. T., and Bidleman, T. F. (1987). "An experimental system for investigating vapor-particle partitioning of trace organic pollutants." *Environ. Sci. Technol.* **21**, 869–875.

Fortmann, R. C., Nagda, N. L., and Koontz, M. D. (1987). "Indoor air quality measurements." *ASTM Spec. Tech. Publ.* **957**, 35–45.

Gammage, R. B., and Matthews, T. G. (1988). "Volatile organic compounds in indoor air: Types, sources and characteristics." *Environ. Prog.* **7**, 279–283.

Gammage, R. B., White, D. A., and Gupta, K. C. (1984). "Residential measurements of high volatility organics and their sources." In B. Berglund, T. Lindvall, and J. Sundell, eds., *Indoor Air. Chemical Characterization and Personal Exposure*, Proc. 3rd Int. Conf. Indoor Air Qual. Clim., Vol. 4. Swedish Council of Building Research, Stockholm, pp. 157–162.

Gammage, R. B., Higgins, C. E., Dreibelbis, W. G., Guérin, M. R., Buchanan, M. V., White, D. A., Olerich, G., and Hawthorne, A. R. (1988). *Measurement of Volatile Organic Compounds (VOC) in Eight East Tennessee Homes*, Report No. ORNL-6286. U.S. Department of Energy, Washington, DC.

Garbesi, K., and Sextro, R. G. (1989). "Modeling and field evidence of pressure-driven entry of soil gas into a house through permeable below-grade walls." *Environ. Sci. Technol.* **23**, 1481–1487.

Gebefugi, I. (1989). "Chemical exposure in enclosed environments." *Toxicol. Environ. Chem.* **20–21**, 121–127.

Girman, J. R., Apte, M. G., Traynor, G. W., Allen, J. R., and Hollowell, C. D. (1982). "Pollutant emission rates from indoor combustion appliances and sidestream cigarette smoke." *Environ. Int.* **8**, 213–221.

Girman, J. R., Hodgson, A. T., Newton, A. S., and Winkes, A. W. (1986). "Emissions of volatile organic compounds from adhesives with indoor applications." *Environ. Int.* **12**, 317–321.

Girman, J. R., Hodgson, A. T., and Wind, M. L. (1987). "Considerations in evaluating emissions from consumer products." *Atmos. Environ.* **21**, 315–320.

Girman, J. R., Alevantis, L. E., Petreas, M. X., and Webber, L. M. (1989a). "Bake-out of a new office building to reduce volatile organic concentrations." In *Proceedings of the 82nd Annual Meeting of the Air and Waste Management Association, Anaheim CA*, Paper No. 89–80.8.

Girman, J. R., Alevantis, L. E., Kulasingam, G. C., Petreas, M. X., and Webber, L. M. (1989b). "The bake-out of an office building: A case study." *Environ. Int.* **15**, 449–453.

Godish, T. (1985). "Residential formaldehyde sampling—current and recommended practices." *Am. Ind. Hyg. Assoc. J.* **46**, 105–110.

Godish, T. (1988). "Residential formaldehyde contamination: Sources and levels." *Comments Toxicol.* **2**, 115–134.

Godish, T. (1989a). "Effect of ambient environmental factors on indoor formaldehyde levels." *Atmos. Environ.* **23**, 1695–1698.

Godish, T. (1989b). "Formaldehyde exposures from tobacco smoke: A review." *Am. J. Public Health* **79**, 1044–1045.

Godish, T. (1989c). *Indoor Air Pollution Control*. Lewis Publishers, Chelsea, 401 pp.

Godish, T., and Guindon, C. (1989). "An assessment of botanical air purification as a formaldehyde mitigation measure under dynamic laboratory chamber conditions." *Environ. Pollut.* **62**, 13–20.

Grimsrud, D. T., Turk, B. H., Harrison, J., and Prill, R. J. (1986). "A comparison of indoor air quality in Pacific Northwest existing and new energy-efficient homes." In *Proceedings of the 79th Annual Meeting of the Air Pollution Control Association, Minneapolis, MN,* Paper No. 86–16.3.

Guérin, M. R., Higgins, C. E., and Jenkins, R. A. (1987). "Measuring environmental emissions from tobacco combustions: Sidestream cigarette smoke literature review." *Atmos. Environ.* **21**, 291–297.

Hammond, S. K., Leaderer, B. P., Roche, A. C., and Schenker, M. (1987). "Collection and analysis of nicotine as a marker for environmental tobacco smoke." *Atmos. Environ.* **21**, 457–462.

Hanrahan, L. P., Anderson, H. A., Dally, K. A., Eckmann, A. D., and Kanarek, M. S. (1985). "Formaldehyde concentrations in Wisconsin mobile homes." *J. Air Pollut. Control Assoc.* **35**, 1164–1167.

Haraprasad, V., and Dave, J. M. (1989). "Assessment of exposure to emissions from a liquefied petroleum gas stove used in urban India." In C. J. Bieva, Y. Courtois, and M. Govaerts, eds., *Present and Future of Indoor Air Quality*, Proc. Brussels Conf. Elsevier, Amsterdam, pp. 239–242.

Harrison, R. M., Colbeck, T., and Simmons, A. (1988). "Comparative evaluation of indoor and outdoor air quality—chemical considerations." *Environ. Technol. Lett.* **9**, 521–530.

Hartwell, J. M. (1986). "Off-gassing of polyurethane foam." *Proc. 30th Annu. Polyurethane Tech. Mark. Conf.* pp. 184–186.

Hartwell, T. D., Zelon, H. S., Crowder, J. H., Leininger, C. C., Pellizzari, E. D., and Clayton, C. A. (1984). "Comparative statistical analysis for volatile halocarbons in indoor and outdoor air." In B. Berglund, T. Lindvall, and J. Sundell, eds., *Indoor Air. Chemical Characterization and Personal Exposure*, Proc. 3rd Int. Conf. Indoor Air Qual. Clim., Vol. 4, Swedish Council of Building Research, Stockholm, pp. 57–61.

Hartwell, T. D., Pellizzari, E. D., Perritt, R. L., Whitmore, R. W., Zelon, H. S., and Wallace, L. (1987). "Comparison of volatile organic levels between sites and seasons for the Total Exposure Assessment Methodology (TEAM) study." *Atmos. Environ.* **21**, 2413–2424.

Hawthorne, A. R., Gammage, R. B., Dudney, C. S., Womack, D. R., Morris, S. A., Westley, R. R., White, D. A., and Gupta, K. C. (1983). "Results of a forty-home indoor air pollutant monitoring study." In *Proceedings of the 76th Annual Meeting of the Air Pollution Control Association,* Paper No. 83–9.10.

Hawthorne, A. R., Matthews, T. G., and Gammage, R. B. (1985). "Characterization and mitigation of organic vapors." In *Proceedings of the 78th Annual Meeting of the Air Pollution Control Association,* Detroit, MI, Vol. 3, Paper No. 85–30B.1.

Hawthorne, A. R., Gammage, R. B., and Dudney, C. S. (1986). "An indoor air quality study of 40 east Tennessee homes." *Environ. Int.* **12**, 221–239.

Hawthorne, A. R., Dudney, C. S., Cohen, M. A., and Spengler, J. D. (1987). *Participant Evaluation Results for Two Indoor Air Quality Studies.* Report prepared for the U.S. Department of Energy under Contract Number DE-AC05-840R21400, ORNL Report, CONF-87-0695—3, 9 pp.

Hawthorne, A. R., Aldrich, T. E., Vo-Dinh, T., Uziel, M., and Cohen, M. A. (1988). *Indoor Air Quality in 300 Homes in Kingston/Harriman, Tennessee,* Report ORNL—6401. Oak Ridge National Laboratory, Oak Ridge, TN.

Hering, S. V., ed. (1989). *Air Sampling Instruments for Evaluation of Atmospheric Contaminants.* American Conference of Governmental Industrial Hygienists, Cincinnati, OH, 612 pp.

Highland, J. H., Harris, R. H., Harrje, D. T., English, C. W., and Weiss, C. F. (1985). "The impact of groundwater contaminants on indoor air quality." *Energy Technol.* **12**, 728–740.

Highsmith, V. R., Zweidinger, R. B., and Merrill, R. G. (1988). "Characterization of indoor and outdoor air associated with residences using woodstoves: A pilot study." *Environ. Int.* **14**, 213–219.

Hijazi, N., Chai, R., Amster, M., and Duffee, R. (1983). "Indoor organic contaminants in energy efficient buildings." In E. R. Frederick, ed., *Proceedings of a Specialty Conference on Measurement and Monitoring of Non-Criteria (Toxic) Contaminants in Air.* Air Pollution Control Association, Pittsburgh, PA, pp. 471–477.

Hodgson, A. T., and Girman, J. R. (1987). "Application of a multisorbent sampling technique for investigations of volatile organic compounds in buildings." In N. L. Nagda and J. P. Harper, eds., *Design and Protocol for Monitoring Indoor Air Quality.* American Society for Testing and Materials, Philadelphia, PA, pp. 224–256.

Hodgson, A. T., Garbesi, K., Sextro, R. G., and Daisey, J. M. (1988). *Evaluation of Soil-Gas Transport of Organic Chemicals into Residential Buildings: Final Report*, Report No. LBL-25465. Lawrence Berkeley Laboratory, Berkeley, CA.

Hoffmann, D., and Wynder, E. L. (1986). "Chemical constituents and bioactivity of tobacco smoke." In D. G. Zaridze and R. Peto, eds., *Tobacco: A Major International Health Hazard*, IARC Sci. Publ. No. 74. International Agency for Research on Cancer, Lyon, pp. 145–165.

Hoffmann, D., Brunnemann, K. D., and Webb, K. S. (1983). "Volatile nitrosamines in tobacco and mainstream and sidestream smoke and indoor environments." In H. Egan, R. Preussmann, I. K. O'Neill, G. Eisenbrand, B. Spiegelhalder, and H. Bartsch, eds., *Environmental Carcinogens Selected Methods of Analysis. Vol. 6. N-Nitroso Compounds*, IARC Sci. Publ. No. 45. International Agency for Research on Cancer, Lyon, pp. 69–83.

Holcomb, L. C. (1988). "Impact of environmental tobacco smoke on airline cabin air quality." *Environ. Technol. Lett.* **9**, 509–514.

Ingebrethsen, B. J., and Sears, S. B. (1989). "Particle evaporation of sidestream tobacco smoke in a stirred tank." *J. Colloid Interface Sci.* **131**, 526–536.

International Agency for Research on Cancer (IARC) (1977). *Air Pollution and Cancer in Man*, IARC Sci. Publ. No. 16. IARC, Lyon, 331 pp.

International Agency for Research on Cancer (IARC) (1987). *Environmental Carcinogens Methods of Analysis and Exposure Measurement. Vol. 9. Passive Smoking.* IARC Sci. Publ. No. 81. IARC, Lyon, 372 pp.

Jarke, F. H., Dravnieks, A., and Gordon, S. M. (1981). "Organic contaminants in indoor air and their relation to outdoor contaminants." *ASHRAE Trans.* **87**, 153–166.

Johansson, I. (1978). "Determination of organic compounds in indoor air with potential reference to air quality." *Atmos. Environ.* **12**, 1371–1377.

Jokl, M. V. (1989). *Microenvironment: The Theory and Practice of Indoor Climate.* Thomas, Springfield, IL, 416 pp.

Junge, C. E. (1977). "Basic considerations about trace constituents in the atmosphere as related to the fate of global pollutants." In I. H. Suffet, ed., *Fate of Pollutants in the Air and Water Environments*, Part 1. Wiley, New York, pp. 7–25.

Jungers, R. H., and Sheldon, L. S. (1987). "Characterization of volatile organic chemicals in public access buildings." In B. Seifert, H. Esdorn, M. Fischer, H. Ruden, and J. Wegner, eds., *Indoor Air '87. Vol. 1. Volatile Organic Compounds, Combustion Gases, Particles and Fibres, Microbiological Agents*, Proc. 4th Int. Conf. Indoor Air Qual. Clim. Institute for Water, Soil and Air Hygiene, Berlin, pp. 144–148.

Karlsson, S., Banhidi, E., Albertsson, A.-C., and Banhidi, Z. (1984). "Accumulation of malodorous amines and polyamines due to clostridial putrefication indoors." In B. Berglund, T. Lindvall, and J. Sundell, eds., *Indoor Air. Sensory and Hyperreactivity Reactions to Sick Buildings*, Proc. 3rd Int. Conf. Indoor Air Qual. Clim., Vol. 3, Swedish Council of Building Research, Stockholm, pp. 287–293.

Kawauchi, T., and Nishiyama, K. (1989). "Residual tetrachloroethylene in dry-cleaned clothes." *Environ. Res.* **48**, 296–301.

Keith, L. H., ed. (1984). *Identification and Analysis of Organic Pollutants in Air*. Butterworth, Woburn, MA, 486 pp.

Keskinen, J., and Graeffe, G. (1989). "Continuous monitoring of air impurities in dwellings." *Environ. Int.* **15**, 557–562.

Kliest, J., Fast, T., Boley, J. S. M., van de Wiel, H., and Bloemen, H. (1989). "The relationship between soil contaminated with volatile organic compounds and indoor air pollution." *Environ. Int.* **15**, 1–6.

Knoppel, H., and Schauenburg, H. (1989). "Screening of household products for the emission of volatile organic compounds." *Environ. Int.* **15**, 413–418.

Konopinski, V. J. (1985). "Seasonal formaldehyde concentrations in an office building." *Am. Ind. Hyg. Assoc. J.* **46**, 65–68.

Krasnec, J. P., and Demaray, D. E. (1984). "Sampling and monitoring toxic gases in indoor environments." In B. Berglund, T. Lindvall, and J. Sundell, eds., *Indoor Air. Chemical Characterization and Personal Exposure*, Proc. 3rd Int. Conf. Indoor Air Qual. Clim., Vol. 4, Swedish Council of Building Research, Stockholm, pp. 191–196.

Krause, C., Mailahn, W., Nagel, R., Schulz, C., Seifert, B., and Ullrich, D. (1987). "Occurrence of volatile organic compounds in the air of 500 homes in the Federal Republic of Germany." In B. Seifert, H. Esdorn, M. Fischer, H. Ruden, and J. Wegner, eds., *Indoor Air '87. Vol. 1. Volatile Organic Compounds, Combustion Gases, Particles and Fibres, Microbiological Agents*, Proc. 4th Int. Conf. Indoor Air Qual. Clim. Institute for Water, Soil and Air Hygiene, Berlin, pp. 102–106.

Krzymien, M. E. (1989). "GC-MS analysis of organic vapors emitted from polyurethane foam insulation." *Int. J. Environ. Anal. Chem.* **36**, 193–207.

Leaderer, B. P., Zagraniski, R. T., Berwick, M., and Stolwijk, J. A. J. (1986). "Assessment of exposure to indoor air contaminants from combustion sources: Methodology and application." *Am. J. Epidemiol.* **124**, 275–289.

Lebret, E., van de Wiel, H. J., Noij, D., Bos, H. P., and Boleij, J. S. (1984). "Volatile hydrocarbons in Dutch homes." In B. Berglund, T. Lindvall, and J. Sundell, eds., *Indoor Air. Chemical Characterization and Personal Exposure*, Proc. 3rd Int. Conf. Indoor Air Qual. Clim., Vol. 4, Swedish Council of Building Research, Stockholm, pp. 169–174.

Lebret, E., van de Weil, H. J., Bos, H., Noij, D., and Boleij, J. S. M. (1986). "Volatile organic compounds in Dutch homes." *Environ. Int.* **12**, 323–332.

Levin, H. (1987). "Overview of indoor air quality sampling and analysis." *ASTM Spec. Tech. Publ.* **957**, 21–34.

Levin, H. (1989). "Building materials and indoor air quality." *Occup. Med. State Art Rev.* **4**, 667–693.

Levin, J. O., Lindahl, R., and Andersson, K. (1989). "Monitoring of parts-per-billion levels of formaldehyde using a diffusive sampler." *J. Air Pollut. Control Assoc.* **39**, 44–47.

Lewis, C. W., Dzubay, T. G., Highsmith, V. R., Stevens, R. K., and Zweidinger, R. B. (1989). "Indoor-outdoor comparisons of aerosol and VOC source tracer species in a residential woodsmoke-impacted community." In *Proceedings of the 82nd Annual Meeting of the Air and Waste Management Association, Anaheim, CA, Paper No. 89–104.6.*

Lewis, R. G. (1986). "Problems associated with sampling for semivolatile organic chemicals in air." In *Proceedings of the 1986 EPA/APCA Symposium on Measurement of Toxic Air Pollutants*, APCA Publ. VIP-7. Air Pollution Control Association, Pittsburgh, PA, pp. 134–145.

Lewis, R. G. (1989). "Advanced methodologies for sampling and analysis of toxic organic chemicals in ambient outdoor, indoor, and personal respiratory air." *J. Chin. Chem. Soc.* **36**, 261–277.

Lipari, F., Dasch, J. M., and Scruggs, W. F. (1984). "Aldehyde emissions from wood-burning fireplaces." *Environ. Sci. Technol.* **18**, 326–329.

Lofroth, G. (1989a). "Environmental tobacco smoke: Overview of chemical composition and genotoxic components." *Mutat. Res.* **222**, 73–80.

Lofroth, G. (1989b). "Isoprene—a potential indoor indicator for environmental tobacco smoke." In C. J. Bieva, Y. Courtois, and M. Govaerts, eds., *Present and Future of Indoor Air Quality*. Proc. Brussels Conf. Elsevier, Amsterdam, pp. 147–153.

Lutz, G. A. (1982). *Literature Review of Personal Air Monitors for Potential Use in Ambient Air Monitoring of Organic Compounds*, Report No. EPA-600/4-82-048, U.S. Environmental Protection Agency, Research Triangle Park, NC, 88 pp.

Mailahn, W., Seifert, B., Ullrich, D., and Moriske, H.-J. (1989). "The use of a passive sampler for the simultaneous determination of long-term ventilation rates and VOC concentrations." *Environ. Int.* **15**, 537–544.

Maki, H. T., and Woods, J. E., Jr. (1984). "Dynamic behaviour of pollutants generated by indoor combustion." In B. Berglund, T. Lindvall, and J. Sundell, eds., *Indoor Air. Buildings, Ventilation and Thermal, Climate*, Proc. 3rd Int. Conf. Indoor Air Qual. Clim., Vol. 5, Swedish Council of Building Research, Stockholm, pp. 73–78.

Matthews, T. G., Reed, T. J., Tromberg, B. J., Daffron, C. R., and Hawthorne, A. R. (1985). "Formaldehyde emission from combustion sources and solid formaldehyde-resin-containing products. Potential impact on indoor formaldehyde concentrations." In V. Turoski, ed., *Formaldehyde: Analytical Chemistry and Toxicology*, Adv. Chem. Ser. No. 210. American Chemical Society, Washington, DC, pp. 131–150.

Matthews, T. G., Fung, K. W., Tromberg, B. J., and Hawthorne, A. R. (1986). "Surface emission monitoring of pressed-wood products containing urea-formaldehyde resins." *Environ. Int.* **12**, 301–309.

Matthews, T. G., Hawthorne, A. R., and Thompson, C. V. (1987). "Formaldehyde sorption and desorption characteristics of gypsum wallboard." *Environ. Sci. Technol.* **21**, 629–634.

McKone, T. E. (1987). "Human exposure to volatile organic compounds in household tap water: The indoor inhalation pathway. *Environ. Sci. Technol.* **21**, 1194–1201.

McKone, T. E., and Knezovich, J. P. (1989). "Assessing tap-water contributions to inhalation exposure for volatile organic chemicals." In *Proceedings of the 82nd Annual Meeting of the Air and Waste Management Association, Anaheim, CA,* Paper No. 89-80.6.

Merrill, R. G., Steiber, R. S., Martz, R. F., and Nelms, L. H. (1987). "Screening methods for the identification of organic emissions from indoor air pollution sources." *Atmos. Environ.* **21**, 331–336.

Meyer, B. (1979). *Urea-Formaldehyde Resins.* Addison-Wesley, Reading, MA 423 pp.

Meyer, B. (1983). *Indoor Air Quality.* Addison-Wesley, Reading, MA, 434 pp.

Meyer, B. (1986). "Formaldehyde exposure from building products." *Environ. Int.* **12**, 283–288.

Meyer, B., and Hermanns, K. (1985). "Diurnal variations of formaldehyde exposure in mobile homes." *J. Environ. Health* **48**, 57–61.

Miksch, R. R., Hollowell, C. D., and Schmidt, H. E. (1982). "Trace organic chemical contaminants in office spaces." *Environ. Int.* **8**, 129–137.

Miller, J. J., and Beizer, M. B. (1985). "Air quality in residences adjacent to an active hazardous waste disposal site." In *Proceedings of the 78th Annual Meeting of the Air Pollution Control Association,* Vol. 6, Paper No. 85–72.7.

Molhave, L. (1979). "Indoor air pollution due to building materials." In P. O. Fanger and O. Valbjorn, eds., *Indoor Climate, Effects of Human Comfort, Performance and Health in Residential, Commercial and Light-Industry Buildings,* Proc. 1st Int. Indoor Clim. Symp. 1978. Danish Building Research Institute, Copenhagen, Denmark, pp. 89–110.

Molhave, L. (1982). "Indoor air pollution due to organic gases and vapours of solvents in building materials." *Environ. Int.* **8**, 117–127.

Molhave, L., and Moller, J. (1979). "The atmospheric environment in modern Danish dwellings—measurements in 39 flats." In P. O. Fanger and O. Valbjorn, eds., *Indoor Climate, Effects of Human Comfort, Performance and Health in Residential, Commercial and Light-Industry Buildings,* Proc. 1st Int. Indoor Clim. Symp., 1978. Danish Building Research Institute, Copenhagen, Denmark, pp. 171–186.

Molhave, L., Bach, B., and Pedersen, O. F. (1986). "Human reactions to low concentrations of volatile organic compounds." *Environ. Int.* **12**, 167–175.

Monsen, R. M., and Stock, T. H. (1986). "A comparison of formaldehyde monitoring methods for residential environments." *J. Environ. Health* **49**, 72–75.

Monteith, D. K., Stock, T. H., and Seifert, W. E. (1984). "Sources and characterization of organic air contaminants inside manufactured housing." In B. Berglund, T. Lindvall, and J. Sundell, eds., *Indoor Air. Chemical Characterization and Per-*

*sonal Exposure*, Proc. 3rd Int. Conf. Indoor Air Qual. Clim., Vol. 4, Swedish Council of Building Research, Stockholm, pp. 285–290.

Montgomery, D. D., and Kalman, D. A. (1989). "Indoor/outdoor air quality: Reference pollutant concentrations in complaint-free residences." *Appl. Ind. Hyg.* **4**, 17–20.

Morey, P. R., and Woods, J. E. (1987). "Indoor air quality in health care facilities." *Occup. Med. State Art Rev.* **2**, 547–563.

Moschandreas, D. J. (1981). "Exposure to pollutants and daily time budgets of people." *Bull. N.Y. Acad. Med.* [2] **57**, 845–859.

Moschandreas, D. J., Spengler, J. D., Biersteker, K., and Howlett, C. T. (1982). "Indoor-outdoor air quality relationships. A critical review. Discussion papers." *J. Air Pollut. Control Assoc.* **32**, 904–915.

Moschandreas, D. J., Gordon, S. M., Eisenberg, W. C., and Relwani, S. (1983). "Emission factors of volatile organic compounds and other air constituents from unvented gas appliances." In E. R. Frederick, ed., *Proceedings of a Specialty Conference on Measurement and Monitoring of Non-Criteria (Toxic) Contaminants in Air.* Air Pollution Control Association, Pittsburgh, PA, pp. 425–434.

Moschandreas, D. J., Relwani, S. M., Macriss, R. A., and Cole, J. T. (1986a). "A comparison of emission rates of unvented gas appliances measured by two different methods." *Environ. Int.* **12**, 241–246.

Moschandreas, D., Relwani, S., Johnson, D., and Billick, I. (1986b). "Emission rates from unvented gas appliances." *Environ. Int.* **12**, 247–253.

Moschandreas, D. J., Relwani, S. M., Billick, I. H., and Macriss, R. A. (1987). "Emission rates from range-top burners—assessment of measurement methods." *Atmos. Environ.* **21**, 285–289.

Muhlbaier, J. L. (1982). "A characterization of emissions from wood-burning fireplaces." In J. Cooper and D. Malik, eds., *Proceedings of the 1981 International Conference on Residential Solid Fuels.* Oregon Graduate Center, Portland, Oregon, pp. 164–187.

Nagda, N. L., and Harper, J. P., eds. (1989). *Design and Protocol for Monitoring Indoor Air Quality.* American Society for Testing and Materials, Philadelphia, PA, 309 pp.

Nagda, N. L., Koontz, M. D., Rector, H. E., Harrje, D., Lannus, A., Patterson, R., and Purcell, G. (1983). "Study design to relate residential energy use, air infiltration, and indoor air quality." In *Proceedings of the 76th Annual Meeting of the Air Pollution Control Association*, Atlanta, GA, Vol. 2, Paper No. 83–29.

Nagda, N. L., Fortmann, R. C., Koontz, M. D., and Rector, H. E. (1986). "Air infiltration and building factors: Comparison of measurement methods." In *Proceedings of the 79th Annual Meeting of the Air Pollution Control Association*, Minneapolis, MN, Vol. 1, 86-16.1.

National Institute for Occupational Safety and Health (NIOSH) (1977). *Manual of Analytical Methods*, Part II, Vol. 3. Cincinnati, OH, Method No. S293.

National Research Council, Committee on Indoor Pollutants (NRC) (1981). *Indoor Pollutants.* National Academy Press, Washington, D.C., 537 pp.

Nazaroff, W. W., and Cass, G. R. (1986). "Mathematical modeling of chemically reactive pollutants in indoor air." *Environ. Sci. Technol.* **20**, 924–934.

Nazaroff, W. W., and Cass, G. R. (1989). "Mass-transport aspects of pollutant removal at indoor surfaces." *Environ. Int.* **15**, 567–583.

Nazaroff, W. W., Lewis, S. R., Doyle, S. M., Moed, B. A., and Nero, A. V. (1987). "Experiments on pollutant transport from soil into residential basements by pressure-driven airflow." *Environ. Sci. Technol.* **21**, 459–466.

Noma, E., Berglund, B., Berglund, U., Johansson, I., and Baird, J. C. (1988). "Joint representation of physical locations and volatile organic compounds in indoor air from a healthy and a sick building." *Atmos. Environ.* **22**, 451–460.

Norback, D., Michel, I., and Widstrom, J. (1990). "Indoor air quality and personal factors related to the sick building syndrome." *Scand. J. Work Environ. Health* **16**, 121–128.

Ogden, M. W., and Maiolo, K. C. (1989). "Collection and determination of solanesol as a tracer of environmental tobacco smoke in indoor air." *Environ. Sci. Technol.* **23**, 1148–1154.

Olander, L., Johansson, J., and Johansson, R. (1988). "Tobacco smoke removal with room air cleaners." *Scand. J. Work Environ. Health* **14**, 390–397.

Oliver, K. D., Pleil, J. D., and McClenny, W. A. (1986). "Sample integrity of trace level volatile organic compounds in ambient air stored in SUMMA® polished canisters." *Atmos. Environ.* **20**, 1403–1411.

Otson, R., and Benoit, F.M. (1986). "Surveys of selected organics in residential air." In D. S. Walkinshaw, ed., *Indoor Air Quality in Cold Climates. Hazards and Abatement Measures.* Air Pollution Control Association, Pittsburgh, PA, pp. 224–233.

Otson, R., and Fellin, P. (1988). "A review of techniques for measurement of airborne aldehydes." *Sci. Total Environ.* **77**, 95–131.

Otson, R., Williams, D. T., and Bothwell, P. (1981). "Dichloromethane levels in air after application of paint removers." *Am. Ind. Hyg. Assoc. J.* **42**, 56–60.

Otson, R., Williams, D. T., Bothwell, P. D., and Doyle, E. E. (1983). "Survey of selected organics in office air." *Bull. Environ. Contam. Toxicol.* **31**, 222–229.

Otson, R., Williams, D. T., and Bothwell, P. (1984). "Fabric protectors. Part II. Propane, 1,1,1-trichloroethane, and petroleum distillates levels in air after application of fabric protectors." *Am. Ind. Hyg. Assoc. J.* **45**, 28–33.

Ozkaynak, H., Ryan, P. B., Wallace, L. A., Nelson, W. C., and Behar, J. V. (1987). "Sources and emission rates of organic chemical vapors in homes and buildings." In B. Seifert, H. Esdorn, M. Fischer, H. Ruden, and J. Wegner, eds., *Indoor Air '87. Vol. 1. Volatile Organic Compounds, Combustion Gases, Particles and Fibres, Microbiological Agents*, Proc. 4th Int. Conf. Indoor Air Qual. Clim. Institute for Water, Soil and Air Hygiene, Berlin, pp. 3–7.

Pellizzari, E. D., Hartwell, T. D., Perritt, R. L., Sparacino, C. M., Sheldon, L. S., Zelon, H. S., Whitmore, R. W., Breen, J. J., and Wallace, L. (1986). "Comparison of indoor and outdoor residential levels of volatile organic chemicals in five U.S. geographical areas." *Environ. Int.* **12**, 619–623.

Perry, R., and Kirk, P. W., eds. (1988). *Indoor Ambient Air Quality.* Selper Ltd., London, 604 pp.

Pitts, J. N., Jr., Biermann, H. W., Tuazon, E. C., Green, M., Long, W. D., and Winer, A. M. (1989). "Time-resolved identification and measurement of indoor air pollutants by spectroscopic techniques: Gaseous nitrous acid, methanol, formaldehyde and formic acid." *J. Air Pollut. Control Assoc.* **39**, 1344–1347.

Pleil, J. D., McClenny, W. A., and Oliver, K. D. (1989). "Temporal variability measurement of specific volatile organic compounds." *Int. J. Environ. Anal. Chem.* **37**, 263–276.

Proctor, C. J. (1988). "The analysis of the contribution of ETS to indoor air." *Environ. Technol. Lett.* **9**, 553–562.

Proctor, C. J. (1989). "A comparison of the volatile organic compounds present in the air of real-world environments with and without environmental tobacco smoke." In *Proceedings of the 82nd Annual Meeting of the Air and Waste Management Association, Anaheim, CA*, Paper No. 89–80.4.

Reasor, M. J. (1987). "The composition and dynamics of environmental tobacco smoke." *J. Environ. Health* **50**, 20–24.

Repace, J. L., and Lowrey, A. H. (1987). "Environmental tobacco smoke and indoor air quality in modern office work environments." *J. Occup. Med.* **29**, 628–629.

Rogozen, M. B., Baca, R. J., and Lamason, W. H., II (1988). "Development of species profiles and emission factors for 20 consumer and commercial products." In *Proceedings of the 81st Annual Meeting of the Air Pollution Control Association*, Dallas, TX, Paper No. 88–66.7.

Rothenberg, S. J., Nagy, P. A., Pickrell, J. A., and Hobbs, C. H. (1989). "Surface area, adsorption and desorption studies on indoor dust samples." *Am. Ind. Hyg. Assoc. J.* **50**, 15–23.

Rousseau, D., Rea, W. J., and Enwright, J. (1988). *Your Home, Your Health and Well-Being.* Ten Speed Press, Berkeley, CA, 300 pp.

Saarela, K., Kaustia, K., and Kiviranta, A. (1989). "Emissions from materials; the role of additives in PVC." In C. J. Bieva, Y. Courtois, and M. Govaerts, eds., *Present and Future of Indoor Air Quality*, Proc. Brussels Conf. Elsevier, Amsterdam, pp. 329–336.

Sanchez, D. C., Mason, M., and Norris, C. (1987). "Methods and results of characterization of organic emissions from an indoor material." *Atmos. Environ.* **21**, 337–345.

Scheff, P. A., Wadden, R. A., Bates, B. A., and Aronian, P. F. (1989). "Source fingerprints for receptor modeling of volatile organics." *J. Air Pollut. Control Assoc.* **39**, 469–478.

Schlitt, H., and Knoppel, H. (1989). "Carbonyl compounds in mainstream and sidestream cigarette smoke." In C. J. Bieva, Y. Courtois, and M. Govaerts, eds., *Present and Future of Indoor Air Quality*, Proc. Brussels Conf. Elsevier, Amsterdam, pp. 197–206.

Seifert, B. (1984). "Planung und durchfuhrung von luftmessungen innenraumen." *GI, Haustech. Bauphys.-Umwelttech.* **105**, 15–18.

Seifert, B. (1988). "Sampling and analysis of ambient and indoor air." In L. Fishbein and I. K. O'Neill, eds., *Environmental Carcinogens. Methods of Analysis and Exposure Measurement. Vol. 10. Benzene and Alkylated Benzenes*, IARC Sci. Publ. No. 85. International Agency for Research on Cancer, Lyon, pp. 165–178.

Seifert, B., and Abraham, H. J. (1983). "Use of passive samplers for the determination of gaseous organic substances in indoor air at low concentration levels." *Int. J. Environ. Anal. Chem.* **13**, 237–254.

Seifert, B., and Ullrich, D. (1987). "Methodologies for evaluating sources of volatile organic chemicals (VOC) in homes." *Atmos. Environ.* **21**, 395–404.

Seifert, B., Mailahn, W., Schulz, C., and Ullrich, D. (1989). "Seasonal variation of concentrations of volatile organic compounds in selected German homes." *Environ. Int.* **15**, 397–408.

Sexton, K., Liu, K.-S., and Petreas, M. X. (1986a). "Formaldehyde concentrations inside private residences: A mail-out approach to indoor air monitoring." *J. Air Pollut. Control Assoc.* **36**, 698–704.

Sexton, K. Webber, L. M., Hayward, S. B., and Sextro, R. G. (1986b). "Characterization of particle composition, organic vapor constituents, and mutagenicity of indoor air pollutant emissions." *Environ. Int.* **12**, 351–362.

Sexton, K., Petreas, M. X., and Liu, K.-S. (1989). "Formaldehyde exposures inside mobile homes." *Environ. Sci. Technol.* **23**, 985–988.

Shah, J. J., and Heyerdahl, E. K. (1988). "National ambient volatile organic compounds (VOCs) data base update." In *Proceedings of the 81st Annual Meeting of the Air Pollution Control Association*, Dallas, TX, Paper No. 88–95A.1.

Shah, J. J., and Singh, H. B. (1988). "Distribution of volatile organic chemicals in outdoor and indoor air: A national VOCs data base." *Environ. Sci. Technol.* **22**, 1381–1388.

Shehata, A. T. (1985). "A multi-route exposure assessment of chemically contaminated drinking water." *Toxicol. Ind. Health* **1**, 277–298.

Sheldon, L. S., Thomas, K. W., and Jungers, R. H. (1986). "Volatile organic emissions from building materials." In *Proceedings of the 79th Annual Meeting of the Air Pollution Control Association, Minneapolis, MN*, Paper No. 86–52.3.

Sinclair, J. D. (1989). "Interaction of the environment with metallic surfaces in indoor environments." In F. M. Doyle, S. Raghavan, P. Somasundaran, and G. W. Warren, eds., *Innovations Materials Processing Using Aqueous, Colloid and Surface Chemistry*. The Minerals, Metals & Materials Society, Warrendale, PA, pp. 55–70.

Spicer, C. W., Holdren, M. W., Slivon, L. E., Coutant, R. W., and Shadwick, D. S. (1987a). *Intercomparison of Sampling Techniques for Toxic Organic Compounds in Indoor Air*, Report No. EPA/600/4-87/008. U.S. Environmental Protection Agency, Research Triangle Park, NC.

Spicer, C. W., Coutant, R. W., Ward, G. F., Joseph, D. W., Gaynor, A. J., and Billick, I. H. (1987b). "Rates and mechanisms of $NO_2$ removal from indoor air by residential materials." In B. Seifert, H. Esdorn, M. Fischer, H. Ruden, and J. Wegner, eds., *Indoor Air '87. Vol. 1. Volatile Organic Compounds, Combustion Gases, Particles and Fibres, Microbiological Agents*, Proc. 4th Int. Conf. Indoor Air. Qual. Clim. Institute for Water, Soil and Air Hygiene, Berlin, pp. 371–375.

Stehlik, G., Richter, O., and Altmann, H. (1982). "Concentration of dimethylnitrosamine in the air of smoke-filled rooms." *Ecotoxicol. Environ. Saf.* **6**, 495–500.

Stephens, R. D., Ball, N. B., and Mar, D. M. (1986). "A multimedia study of hazardous waste landfill gas migration." *Pollut. Multimedia Environ.* **1**, 265–287.

Sterling, T. D., and Mueller, B. (1988). "Concentrations of nicotine, RSP, CO and $CO_2$ in nonsmoking areas of offices ventilated by air recirculated from smoking designated areas." *Am. Ind. Hyg. Assoc. J.* **49**, 423–426.

Sterling, T. D., Collett, C. W., and Sterling, E. M. (1987). "Environmental tobacco smoke and indoor air quality in modern office work environments." *J. Occup. Med.* **29**, 57–62.

Stern, A. H., and Andrews, L. R. (1989). "The contribution of domestic water use to indoor air concentrations of chloroform in New York City apartments: A pilot study." *Toxicol. Environ. Chem.* **24**, 71–81.

Stock, T. H. (1987). "Formaldehyde concentrations inside conventional housing." *J. Air Pollut. Control Assoc.* **37**, 913–918.

Stock, T. H., and Mendez, S. R. (1985) "A survey of typical exposures to formaldehyde in Houston area residences." *Am. Ind. Hyg. Assoc. J.* **46**, 313–317.

Sulatisky, M. T. (1984). *Airtightness Tests on 200 New Houses Across Canada: Summary of Results*, BETT Publ. No. 84.01. Prepared for Energy, Mines and Resources Canada under the Buildings Energy Technology Transfer Program (BETT).

Tada, M., and Seki, H. (1989). "Toxic diterpenes from Euphorbia trigona (saiunkaku: An indoor foliage plant in Japan)." *Agric. Biol. Chem.* **53**, 425–430.

Ter Konda, P. K. (1987). "Sources and characteristics of indoor air toxic organics from consumer products." In *Proceedings of the 80th Annual Meeting of the Air Pollution Control Association*, New York, NY, Paper No. 87–81.4.

Thompson, C. V., Jenkins, R. A., and Higgins, C. E. (1989). "A thermal desorption method for the determination of nicotine in indoor environments." *Environ. Sci. Technol.* **23**, 429–435.

Tichenor, B. A., and Mason, M. A. (1988). "Organic emissions from consumer products and building materials to the indoor environment." *J. Air Pollut. Control Assoc.* **38**, 264–268.

Tichenor, B. A., Sparks, L. A., White, J. B., and Jackson, M. D. (1990). "Evaluating sources of indoor air pollution." *J. Air Waste Manage. Assoc.* **40**, 487–492.

Traynor, G. W., Allen, J. R., Apte, M. G., Dillworth, J. F., Girman, J. R., Hollowell, C. D., and Koonce, J. F. (1982). "Indoor air pollution from portable kerosene-fired space heaters, wood-burning stoves, and wood-burning furnaces." In E. R. Frederick, ed., *Proceedings of the Residential Wood and Coal Combustion Specialty Conference*. Air Pollution Control Association, Pittsburgh, PA, pp. 253–263.

Traynor, G. W., Apte, M. G., Carruthers, A. R., Dillworth, J. F., Grimsrud, D. T., and Gundel, L. A. (1984). "Indoor air pollution to emissions from wood burning stoves." In *Proceedings of the 77th Annual Meeting of the Air Pollution Control Association*, San Francisco, CA, Vol. 2, Paper No. 84–33.4.

Traynor, G. W., Apte, M. G., Sokol, H. A., Chuang, J. C., and Mumford, J. L. (1986). "Selected organic pollutant emissions from unvented kerosene heaters." In *Proceedings of the 79th Annual Meeting of the Air Pollution Control Association, Minneapolis, MN*, Vol. 4, Paper No. 86–52.5.

Tsuchiya, Y. (1988). "Volatile organic compounds in indoor air." *Chemosphere* **17**, 79–82.

Tsuchiya, Y., Clermont, M. J., and Walkinshaw, D. S. (1988). "Wet process copying machines: A source of volatile organic compound emissions in buildings." *Environ. Toxicol. Chem.* **7**, 15–18.

Tucker, W. G. (1987). "Chairman's Summary, Session I. Characterization of emissions from combustion sources: Controlled studies." *Atmos. Environ.* **21**, 281–284.

Turiel, I., Hollowell, C. D., Miksch, R. R., Rudy, J. V., Young, R. A., and Coye,

M. J. (1983). "Effects of reduced ventilation on indoor air quality in an office building." *Atmos. Environ.* **17**, 51–64.

Turoski, V., ed. (1985). *Formaldehyde: Analytical Chemistry and Toxicology*, Adv. Chem. Ser. No. 210. American Chemical Society, Washington, DC, 303 pp.

U.S. Environmental Protection Agency (USEPA) (1988). *Indoor Air Quality in Public Buildings*, Report Nos. EPA/600/6-88/009a and b, USEPA, Washington, DC, Vol. I, 535 pp. Vol. II, 733 pp.

U.S. Environmental Protection Agency (USEPA) (1989). *Report to Congress on Indoor Air Quality. Vol. II: Assessment and Control of Indoor Air Pollution*, Report No. EPA/400/1-89/001C. USEPA, Washington, DC, 243 pp.

Vanderstraeten, P., Muylle, E., and Verduyn, G. (1984). "Indoor air quality in a large hospital building." In B. Berglund, T. Lindvall, and J. Sundell, eds., *Indoor Air. Chemical Characterization and Personal Exposure*, Proc. 3rd Int. Conf. Indoor Air Qual. Clim., Vol. 4, Swedish Council of Building Research, Stockholm, pp. 335–341.

van der Wal, J. F. (1989). "Portable air sampler for measurements in aircraft and public buildings." In C. J. Bieva, Y. Courtois, and M. Govaerts, eds., *Present and Future of Indoor Air Quality*, Proc. Brussels Conf. Elsevier, Amsterdam, pp. 371–378.

van der Wal, J. F., Moons, A. M. M., and Steenlage, R. (1989). "Thermal insulation as a source of air pollution. *Environ. Int.* **15**, 409–412.

Van Netten, C., Shirtliffe, C., and Svec, J. (1988). "Formaldehyde release characteristics from a Swedish floor finish." *Bull. Environ. Contam. Toxicol.* **40**, 672–677.

Van Netten, C., Shirtliffe, C., and Svec, J. (1989). "Temperature and humidity dependence of formaldehyde release from selected building materials." *Bull. Environ. Contam. Toxicol.* **42**, 558–565.

Verhoeff, A. P., Wilders, M. M. W., Monster, A. C., and Van Wijnen, J. H. (1987). "Organic solvents in the indoor air of two small factories and surrounding houses." *Int. Arch. Occup. Environ. Health* **59**, 153–163.

Verhoeff, A. P., Suk, J., and van Wijnen, J. H. (1988). "Residential indoor air contamination by screen printing plants." *Int. Arch. Occup. Environ. Health* **60**, 201–209.

Wadden, R. A. (1985). "Indoor air quality and emissions." *Carcinog. Mutagens Environ.* **4**, 167–175.

Wadden, R. A., and Scheff, P. A. (1983). *Indoor Air Pollution*. Wiley, New York, 213 pp.

Wallace, L. A. (1987). *The Total Exposure Assessment Methodology (TEAM) Study: Summary and Analysis*, Vol. 1, Report No. EPA/600/6-87/002a. U.S. Environmental Protection Agency, Washington, DC, 209 pp.

Wallace, L. A. (1989). "The total exposure assessment methodology (TEAM) study: An analysis of exposures, sources, and risks associated with four volatile organic chemicals." *J. Am. Coll. Toxicol.* **8**, 883–895.

Wallace, L. A., Zweidinger, R., Erickson, M., Cooper, S., Whitaker, D., and Pellizzari, E. (1982). "Monitoring individual exposure. Measurements of volatile organic compounds in breathing-zone air, drinking water, and exhaled breath." *Environ. Int.* **8**, 269–282.

Wallace, L. A., Pellizzari, E., Hartwell, T., Rosenzweig, M., Erickson, M., Sparacino, C., and Zelon, H. (1984). "Personal exposure to volatile organic compounds. 1. Direct measurements in breathing-zone air, drinking water, food and exhaled breath." *Environ. Res.* **35**, 293–319.

Wallace, L. A., Pellizzari, E. D., Hartwell, T. D., Sparacino, C., Whitmore, R., Sheldon, L., Zelon, H., and Perritt, R. (1987a). "The TEAM study: Personal exposures to toxic substances in air, drinking water, and breath of 400 residents of New Jersey, North Carolina, and North Dakota." *Environ. Res.* **43**, 290–307.

Wallace, L. A., Pellizzari, E., Leaderer, B., Zelon, H., and Sheldon, L. (1987b). "Emissions of volatile organic compounds from building materials and consumer products." *Atmos. Environ.* **21**, 385–393.

Wallace, L. A., Pellizzari, E., Hartwell, T. D., Perritt, K., and Ziegenfus, R. (1987c). "Exposures to benzene and other volatile compounds from active and passive smoking." *Arch. Environ. Health* **42**, 272–279.

Wallace, L. A., Pellizzari, E. D., Hartwell, T. D., Whitmore, R., Zelon, H., Perritt, R., and Sheldon, L. (1988a). "The California Team Study: Breath concentrations and personal exposures to 26 volatile compounds in air and drinking water of 188 residents of Los Angeles, Antioch, and Pittsburgh, CA." *Atmos. Environ.* **22**, 2141–2163.

Wallace, L. A., Nelson, W. C., Hartwell, T. D., Perritt, R., Smith, D., Sebestik, J., Keever, J., Michael, L., and Pellizzari, E. (1988b). "Preliminary results from the Baltimore TEAM study." In *Proceedings of the 81st Annual Meeting of the Air Pollution Control Association*, Dallas, TX, Paper No. 88–115.6.

Wallace, L. A., Pellizzari, E. D., Hartwell, T. D., Davis, V., and Michael, L. C. (1989). "The influence of personal activities on exposure to volatile organic compounds." *Environ. Res.* **50**, 37–55.

Wallingford, K. M. (1988). "Indoor air quality investigations in office buildings." *Energy Technol.* **15**, 436–444.

Walsh, P. J., Dudney, C. S., and Copenhaver, E. D., eds. (1984). *Indoor Air Quality.* CRC Press, Boca Raton, FL, 207 pp.

Weschler, C. J., and Shields, H. C. (1986). "The accumulation of "additives" in office air." In *Proceedings of the 79th Annual Meeting of the Air Pollution Control Association, Minneapolis, MN*, Paper No. 86–52.2.

Weschler, C. J., and Shields, H. C. (1989). "The effects of ventilation, filtration, and outdoor air on the composition of indoor air at a telephone office building. *Environ. Int.* **15**, 593–604.

Weschler, C. J., Shields, H. C., and Rainer, D. (1990). "Concentrations of volatile organic compounds at a building with health and comfort complaints." *Am. Ind. Hyg. Assoc. J.* **51**, 261–268.

Williams, D. C., Whitaker, J. R., and Jennings, W. G. (1985). "Measurement of nicotine in building air as indicators of tobacco smoke levels." *Environ. Health Perspect.* **60**, 405–410.

World Health Organization (WHO) (1986). *Indoor Air Quality Research*, EURO Reports and Studies 103. WHO Regional Office for Europe, Copenhagen, 64 pp.

Yocom, J. E. (1982). "Indoor-outdoor air quality relationships. A critical review." *J. Air Pollut. Control Assoc.* **32**, 500–520.

Yocom, J. E., Hijazi, N. H., and Zoldak, J. J. (1984). "Use of direct analysis mass spectrometry to solve indoor air quality problems." In B. Berglund, T. Lindvall, and J. Sundell, eds., *Indoor Air. Chemical Characterization and Personal Exposure*, Proc. 3rd Int. Conf. Indoor Air Qual. Clim., Vol. 4, Swedish Council of Building Research, Stockholm, pp. 245–250.

Zeedijk, H. (1986). "Emissions by combustion of solid fuels in domestic stoves." In H. F. Hartmann, ed., *Clean Air Congress '86*, Proc. 7th World Clean Air Congr., Vol. IV. Clean Air Society of Australia and New Zealand, Eastwood, pp. 78–85.

Zweidinger, R., Tejada, S., Highsmith, R., Westburg, H., and Gage, L. (1988). "Distribution of volatile organic hydrocarbons and aldehydes during the IACP Boise, Idaho residential study." In *Proceedings of the 1988 EPA/APCA International Symposium on Measurement of Toxic and Related Air Pollutants*. Air Pollution Control Association, Pittsburgh, PA, pp. 814–820.

# 10

# THORON AND ITS PROGENY IN THE ATMOSPHERIC ENVIRONMENT

*S. D. Schery and D. M. Grumm*

*Department of Physics, New Mexico Institute of Mining and Technology, Socorro, New Mexico*

*Gaseous Pollutants: Characterization and Cycling,* Edited by Jerome O. Nriagu.
ISBN 0-471-54898-7   © 1992 John Wiley & Sons, Inc.

## 1. INTRODUCTION

Air quality issues involving the radon isotope $^{222}$Rn (radon) are well known. Less has been heard about the isotope $^{220}$Rn (thoron), even though it is a major radioactive gas in the earth's crust produced at rates comparable to that of radon and its chemical properties are the same as radon. Thoron comes from the decay series of $^{232}$Th, pervasive in trace amounts throughout most soils, rocks, and inorganic building materials.

Early health investigations into the radon isotopes centered on the mining industry where the effects of thoron were estimated to be small compared with radon. Since then, radon has received more attention as a pollutant, although on an absolute basis, or in comparison with other pollutants, thoron can be important. Our understanding of thoron as a pollutant is in an early stage, and probably more than one surprise awaits us. Thoron has long been studied in the earth and atmospheric sciences, and we can draw on that experience to provide the background required to understand thoron's dissemination in the environment and potential as a pollutant.

This chapter discusses thoron in the indoor and outdoor atmospheres and the physical processes by which it is brought there from the earth's surface. After a brief historical perspective and review of physical properties, we discuss the distribution of the parent isotope, thorium, in the earth's surface and the escape of thoron from soil to the air. The mixing of thoron in the outdoor atmosphere is then explored, followed by a discussion of thoron in the indoor environment where current estimates suggest it has the potential to be a significant pollutant. Concluding sections deal with health effects from thoron, and experimental procedures for measuring airborne thoron and its progeny and reducing their concentration in indoor air. The airborne progeny from decay of thoron are responsible for most of the health effects associated with thoron, so they will be an important part of our discussion.

## 2. HISTORICAL BACKGROUND

Thoron was discovered in 1899 by R. B. Owens at McGill University in collaboration with Ernest Rutherford. Thoron was initially called *thorium emanation* and it took extensive investigation by Rutherford, Madame Curie, and others to fully enumerate its physical properties. Most of this early work focused on the fundamental physical properties of natural radioactivity, but some of it is still relevant to modern environmental considerations. For example, one early paper, "Einfluss der Temperatur auf die Emanationen radioactiver Substanzen" by Rutherford (1901), dealt with the release of radon isotope gases from solids, a subject still important today in an environmental context.

Another important step in thoron research occurred in the atmospheric sciences when it was realized that thoron and its progeny are a major source of atmospheric ions near the earth's surface. These ions are important to a wide range of atmospheric processes, including nucleation of water drops necessary for rain and formation of thunderstorms. Thoron and its progeny have been found useful as tracers in studies of atmospheric transport processes, such as eddy diffusion. Much of this early atmospheric research was by Hans Israel and his colleagues (Israel, 1971, 1972; Israel et al., 1968) and the field has continued to be active (Burchfield et al., 1983).

An early paper dealing with health hazards associated with airborne thoron was that of Evans and Goodman (1940). Based on an understanding of the underlying dosimetry for airborne alpha-particle emitters developed in their work, an early recommendation for limits on radon (but, surprisingly, not thoron) appeared (National Bureau of Standards (NBS), 1941). Occupational health concerns about radon isotopes and their progeny gained momentum during the 1950s in connection with the mining industry, particularly uranium mining. Most of the attention was focused on radon progeny where it was becoming clear that, at higher doses, the progeny were causing deaths due to lung cancer. The potential for a health problem due to thoron was recognized, but almost always thoron in mine air was accompanied by radon, which then dominated as a health concern.

There are a few industrial situations where thoron can be found in isolation from radon. Most of these are connected with industrial applications of thorium. Thorium has been used for light bulb filaments and is still used for gas mantles. Thorium is a component in certain alloys, such as magnesium, and plays a small role in the nuclear fuel industry. However, health problems associated with these applications have not been striking (although due to the small number of people involved it would be easy to overlook low-level or latent effects). The mining of thorium ores is usually done in well-ventilated open pits. The wastes and tailings from thorium-bearing ores (such as monazite) processed for metals other than thorium can potentially release significant thoron.

The most recent concern about health effects associated with airborne radon isotopes and their progeny centers on the indoor air of ordinary housing and buildings. Work in this area intensified in the late 1980s. Current analysis indicates that inhalation of the progeny of radon isotopes typically provides the single largest effective dose to the body of all forms of natural ionizing radioactivity (Committee on Biological Effects of Ionizing Radiation, 1990; United Nations Scientific Committee on the Effects of Atomic Radiation (UNSCEAR), 1988). Efforts have been under way, particularly in the United States and Europe, to set guidelines for indoor exposure and to identify and remediate houses with high indoor levels. As with mining, the isotope of priority so far has been radon, since it clearly is the major source of dose.

## 3.  REVIEW OF PHYSICAL PROPERTIES

Table 1 shows the nuclear decay scheme for the $^{232}$Th series. Thoron has a half-life of 55.6 s and upon decay emits an alpha particle with energy of 6.29

**Table 1  Principal Members of the $^{232}_{90}$Th Series**

| Nuclide | Half-life | Alpha | Beta | Gamma and X Rays |
|---|---|---|---|---|
| $^{232}_{90}$Th | $1.4 \times 10^{10}$ y | 3.95 MeV (23%)<br>4.01 (77) | | L |
| $^{228}_{88}$Ra | 5.8 y | | 39 KeV (100%) | |
| $^{228}_{89}$Ac | 6.13 h | | 1170 (32)<br>1740 (12) | L<br>338.3 KeV (11%)<br>911.1 (28)<br>969.0 (17) |
| $^{228}_{90}$Th | 1.91 y | 5.34 (26.7)<br>5.42 (72.7) | | L<br>84.4 (1.2) |
| $^{224}_{88}$Ra | 3.62 d | 5.45 (5)<br>5.69 (95) | | L<br>241.0 (3.9) |
| $^{220}_{86}$Rn | 55.6 s | 6.29 (100) | | |
| $^{216}_{84}$Po | 0.15 s | 6.78 (100) | | |
| $^{212}_{82}$Pb | 10.6 h | | 334 (85)<br>573 (10) | 238.6 (45)<br>300.1 (3.4) |
| $^{212}_{83}$Bi | 60.6 m | 6.05 (25.2)<br>6.09 (9.6) | 1520 (8.0)<br>2250 (48.4) | L<br>727.2 (11.8)<br>785.5 (2.0)<br>1620.6 (2.7) |
| $^{212}_{84}$Po | $2.98 \times 10^{-7}$ s | 8.78 (100) | | |
| $^{208}_{81}$Tl | 3.05 m | | 1280 (23)<br>1520 (23)<br>1790 (49) | L<br>510.8 (21.6)<br>583.1 (84.2)<br>860.4 (12.5)<br>2614.7 (99.8) |
| $^{208}_{82}$Pb | Stable | | | |

64.1%   35.9%

[a]Parentheses enclose intensities (percentage emissions accompanying decay). Maximum beta energy listed. L denotes L x rays.

MeV. Table 2 lists other properties based mostly on measurements of radon (UNSCEAR, 1982; National Council on Radiation Protection and Measurements (NCRP), 1988). (Except in very special circumstances, differences in chemical properties between radon and thoron should be negligible for the purposes of environmental science.) Radon isotopes are inert noble gases which can form chemical compounds only with difficulty (Stein, 1987). The thoron progeny in the decay sequence up to $^{208}$Tl are the most important for airborne dosimetry, particularly the alpha-particle emitters. All progeny are chemically reactive heavy metals that readily oxidize and attach to surfaces

**Table 2 Properties of Thoron**

| | |
|---|---|
| Boiling point | −61.8 °C |
| Melting point | −71 °C |
| Solubility[a] in | |
|    Water | 0.51 at 0 °C |
| | 0.25 at 20 °C |
| | 0.14 at 50 °C |
|    Acetone | 8.0 at 0 °C |
| Diffusion coefficient in | |
|    Air | 0.1 cm$^2$ s$^{-1}$ at STP |
|    Water | 1.1 × 10$^{-5}$ cm$^2$ s$^{-1}$ at 18 °C |

[a]Volumetric concentration in solution relative to volumetric concentration in overlying air.
*Sources.* UNSCEAR (1982), Tanner (1964).

(such as walls or the surface of aerosols). Immediately after decay the recoiling nucleus of these progeny is most frequently in a positive charge state. If unattached to aerosols, these progeny (usually existing as molecular clusters) have a diffusion coefficient in air on the order of 0.05 cm$^2$ s$^{-1}$, with the exact value depending on properties of the air such as moisture content and the presence of trace gases.

The international unit for activity is the becquerel (Bq), with one becquerel equal to one decay per second. The concentration of airborne radioisotopes is customarily given in units of becquerel per cubic meter (Bq m$^{-3}$). However, the health effects of thoron progeny are more closely related to the energy released by alpha-particle decays. This is quantified by the concept of PAEC (potential alpha-particle energy concentration), which is the amount of alpha-particle energy ultimately to be released through decay of the thoron progeny present in the air per unit volume of air. The international units are joules per cubic meter. If PAEC is in nJ m$^{-3}$ and progeny concentrations $C$ are in Bq m$^{-3}$, then PAEC for thoron, PAEC (Tn), is given by (International Commission on Radiological Protection (ICRP), 1981)

$$\text{PAEC (Tn)} = 5.32 \times 10^{-4} C(^{216}\text{Po}) + 69.1 C(^{212}\text{Pb}) + 6.56 C(^{212}\text{Bi}) \quad (1)$$

neglecting the much smaller contribution of $^{212}$Po.

## 4. THORON IN THE EARTH'S SURFACE

### 4.1. The Distribution of Thorium and Thoron

Since thorium ($^{232}$Th) is the ultimate progenitor of thoron, its distribution in the earth's crust is important for controlling the production of thoron. Trace amounts of thorium permeate almost all soils and rocks, in part because of the influence of groundwater from which thorium can precipitate over geological time scales. Thorium usually exists in the plus four valence state. Thorium is not highly soluble itself, but forms complex ions that are more soluble (Langmuir and Herman, 1980). Thus, thorium can be leached from primary source rock under proper conditions of acidity (pH) and oxidation potential (Eh). It then can be carried by water to other locations where it is precipitated when pH and Eh change to conditions unfavorable for remaining in solution. Although thorium is not as soluble as uranium, there is some similarity in their geochemistry, and soils enhanced in uranium are often enhanced in thorium. In magmas and hard rocks there is likely to be an even stronger correlation between thorium and uranium deposits since here the respective geochemical processes have a greater similarity yet.

The average trace concentration of thorium in soils is about 25 Bq kg$^{-1}$, with the majority of samples having a concentration within a factor of two of this value. Due to association with organic matter, there is some tendency for thorium to be enhanced in the top layer (A horizon) of soils. Minerals with unusually high thorium content include monazite, thorite, zircon, sphene, and allanite. Rocks composed of granite or black shale are likely to have a high content of thorium. Monazite and zircon sands have an especially high concentration of thorium. In contrast, basalts, limestone, and sandstone typically have a below average concentration. A few representative values are shown in Table 3.

A number of locations with higher surface content of thorium have been identified (Fig. 1). Best known, perhaps, are the monazite sands along the southeast coast of Brazil, in Sri Lanka (Ceylon), and on the south tip of India. In the United States, the Triassic Conway granite of New Hampshire and

**Table 3  Common Values for Thorium Content and Thoron Flux Density**

| Material | $^{232}$Th (Bq kg$^{-1}$) | Flux Density (Bq m$^{-2}$ s$^{-1}$) |
| --- | --- | --- |
| Soil | 10–70 | 0.5–5 |
| Limestone | 5 | 0.04 |
| Pumice stone (thick) | 100 | 0.5 |
| Black shales | Up to 400 | |
| Granites | 100–200 | |
| Sandstone | 5 | 0.05 |
| Basalts | 2–15 | |
| Concrete | 25 | 0.04 |
| Gypsum | 10 | 0.1 |
| Monazite sands | $4 \times 10^4$ to $3 \times 10^5$ | |

*Sources.* Schery et al. (1989), Keller et al. (1982), Overstreet (1967), Wedepohl (1978), Durrance (1986), Folkerts et al. (1984).

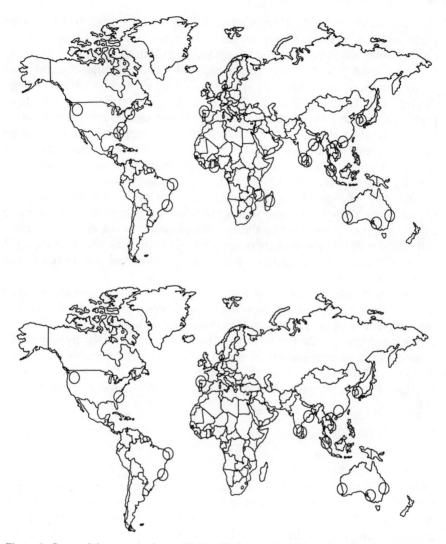

**Figure 1.** Some of the areas in the world identified as having high surface deposits of thorium. Although areas such as Sri Lanka and southern India are well known, many parts of the world have been poorly surveyed and available information is incomplete. (*Principle sources*: Overstreet (1967), Organization for Economic Co-operation and Development (OECD) (1965), Luxin et al. (1988), Staatz et al. (1979), Wang (1990).)

coastal areas of the southeast have large surface deposits of thorium. In contrast, the thorium content of the oceans far from freshwater discharge is typically quite low, less than $1 \times 10^{-4}$ Bq kg$^{-1}$ (Wedepohl, 1978).

The immediate parent of thoron is $^{224}$Ra. Although this isotope is not always in equilibrium with thorium, particularly in groundwater, in broad terms its concentration in soils and rocks will correlate well with thorium.

Upon decay of $^{224}$Ra, the thoron atom will experience recoil. If decay takes place within a mineral, the recoil range is of the order of 30 nm. Hence, thoron atoms might be expected to remain trapped in the grains for the short time they exist before decay. Surprisingly, in soils, a significant number of thoron atoms escape to the pore space where they are free to diffuse through the pore air. The ratio of atoms that escape into the pore space to the total produced is called the emanation coefficient (or escape coefficient) and is typically about a third (Schery et al., 1989). The explanation for this large coefficient rests in part with a preferential deposit of thorium minerals on the surface of grains when precipitated from groundwater. Another factor believed to be important is pervasive moisture in soils, which can provide a diffuse layer of material on the surface of grains preventing penetration of the recoiling nucleus into the dense mineral body where trapping is more likely. A typical value for the thoron concentration in the pore air of deep soil is about 20 000 Bq m$^{-3}$ (corresponding to a soil with about 25 Bq kg$^{-1}$ thorium, porosity 50%, density 1.5 gm cm$^{-3}$, and an emanation coefficient of 0.3).

Compared with the situation for radon, groundwater is usually of small importance for transporting thoron due to thoron's short half-life and the slow movement of groundwater. However, pore water can have an effect on the concentration of thoron in soil gas by displacement of air by water. If $r$ is the solubility coefficient for thoron in water (volumetric thoron concentration in water divided by volumetric thoron concentration in air) and $f_w$ is the fraction of the pore space filled with water then the thoron concentration in soil gas $C_g$ is

$$C_g = \frac{C_{g0}}{1 + f_w(r - 1)} \tag{2}$$

where $C_{g0}$ is the concentration in the absence of water. Since $r$ is less than one for all temperatures of liquid water, the effect of increased water in the pore space will be to increase the concentration of thoron in the soil gas. Equation 2 neglects the time needed for diffusion of thoron into water, which could be important for water in larger pores.

## 4.2.    Transport of Thoron

Thoron must be transported through porous media by diffusion or by the flow of soil gas to reach open air space. Diffusion is typically the dominant mechanism for release of thoron from soil to the atmosphere. Given the short half-life of thoron, only a thin layer of soil (typically the top few centimeters) can effectively release thoron. The diffusion is not necessarily simple three-dimensional molecular diffusion through the air of the open pore space. In dry soils, there can be significant molecular sorption retarding pore diffusion by temporarily trapping atoms upon the surface of grains (Schery and Whittlestone, 1989).

The combined effects of diffusion and pressure-induced flow (if present) can be estimated for semi-infinite homogeneous soil under steady-state conditions with (Schery, 1989)

$$J = \gamma\varepsilon\phi\left(\frac{D_e}{\lambda}\right)^{1/2} \tag{3}$$

$$\gamma = \frac{v}{2\varepsilon(\lambda D_e)^{1/2}} + \left(\frac{v^2}{4\varepsilon^2\lambda D_e} + 1\right)^{1/2} \tag{4}$$

where $J$ = flux density from the soil (atoms or activity per unit bulk area per unit time)

$\phi$ = source term (atoms or activity released to the interconnected pore space per unit volume of interconnected pore space per unit time)

$\varepsilon$ = porosity of the interconnected pore space

$\lambda$ = decay constant

$D_e$ = effective interstitial diffusion coefficient of Fick's law ($J = -\varepsilon D_e \nabla C$, $C$ = concentration in pore air, $J$ bulk flux density; see Appendix)

$v$ = Darcian velocity (volume per unit bulk area per unit time) of the soil gas (positive upward)

Equation 3 assumes that the thoron content of the air is negligible compared with that of the deeper soil gas. The effect of flow is contained in the factor $\gamma$. The equations for radon transport would be the same except radon's (smaller) decay constant is used. Hence, because of the decay constant in the denominator of the right side, equation 4 shows that soil gas flow causes a smaller enhancement of thoron transport than is the case for radon transport.

If flow can be neglected, equation 3 simplifies to

$$J = \varepsilon\phi\left(\frac{D_e}{\lambda}\right)^{1/2} \tag{5}$$

The diffusion length, defined as the distance traveled by thoron in its mean life (see Appendix), is given by

$$l = \left(\frac{D_e}{\lambda}\right)^{1/2} \tag{6}$$

A typical diffusion coefficient for soil is 0.02 cm$^2$ s$^{-1}$ (Sextro et al., 1987), which corresponds to a diffusion length of about 1.3 cm. Very diffuse soils (porous sands and gravel) might have lengths of several centimeters, whereas clays or soils with high water content might have lengths of only a few millimeters.

The arithmetic mean flux density of thoron from the continental lands (excluding the polar ice caps) is probably about 1.5 Bq m$^{-2}$ s$^{-1}$ (Schery et al., 1989; Schery, 1986). The flux from soil will normally be reduced in moist climates or during the rainy season by blocking of pores with water. The thoron flux from the oceans is below the limit of direct detection. Rock and earth materials form a major component of nonwood building materials, so they tend to have a thorium content comparable to that of soils and rocks. A median value for bare building materials such as concrete and limestone is probably of the order of 0.05 Bq m$^{-2}$ s$^{-1}$ (Folkerts et al., 1984). This value is small compared with soils because of reduced coefficients for emanation and diffusion. Certain building materials with elevated thorium content such as pumice stone can have much higher values (see Table 3).

## 5.  THORON IN THE OUTDOOR ATMOSPHERE

Diffusion from the top few centimeters of soil is the dominant process by which thoron is released to the outdoor atmosphere. Essentially no significant thoron is released from the oceans, and the direct release through plants (transpiration, etc.) is at most a few percent. After its release at the soil's surface, transport of thoron is primarily by eddy diffusion for the vertical direction and by wind for the horizontal direction. Typically, thoron moves vertically only a few tens of meters and laterally only several hundreds of meters before decay. As a result, airborne thoron is found in a relatively thin surface layer over continental land. In contrast, an important progeny of thoron, $^{212}$Pb, has a long enough half-life that it will be found higher in the atmosphere (several kilometers or more) and farther out to sea.

The vertical transport of thoron can be estimated with a simple one-dimensional eddy diffusion model (Grumm et al., 1990). If $z$ is the height above the surface of the earth, $J$ is the thoron flux density from the soil, and $K$ is the average eddy diffusion coefficient near the earth's surface, then in the steady-stage limit the thoron concentration at height $z$ is

$$C(z) = \frac{J}{\sqrt{\lambda K}} e^{-z\sqrt{\lambda/K}} \tag{7}$$

Using typical values for temperate climates of $J = 1.5$ Bq m$^{-2}$ s$^{-1}$ and $K = 8 \times 10^3$ cm$^2$ s$^{-1}$ (winter) or $K = 5 \times 10^4$ cm$^2$ s$^{-1}$ (summer) (Liu et al., 1984) gives for the thoron concentration at 1 m height 15 Bq m$^{-3}$ (winter) and 6 Bq m$^{-3}$ (summer). These values can be compared with measurements in six U.S. states over different seasons that gave an average value of 16 Bq m$^{-3}$ (Schery, 1986) and measurements reported by Fontan et al. (1964) lying in the range 2–10 Bq m$^{-3}$. Luxin et al. (1988) report an average value of 3.8 Bq m$^{-3}$ for areas with soils of average thorium content. Analysis of seasonal and diurnal effects in the above data suggest about 10 Bq m$^{-3}$ as a representative average value for temperate climates.

Figure 2 shows measurements of thoron averaged over 2-h intervals for a height of 1 m during the winter season at Socorro, NM (Grumm et al., 1990). Soils near this site have typical thorium concentrations but tend to be drier than average with slightly elevated diffusion coefficients, leading to above average thoron exhalation and concentration. As is usually the case, the major factor causing variation in the thoron concentration is variation in the vertical eddy diffusion mixing. Mixing is normally greatest in the late afternoon (high ground temperature) and smallest in the early morning (low ground temperature). Wind and rain are the two other factors most likely to affect outdoor thoron concentrations. High winds tend to decrease thoron due to greater vertical mixing. Heavy rains (heavy enough to fill pores in soil) can reduce thoron by significantly reducing the flux of thoron from soil. The influence of rain is greater on thoron than radon since thoron will come from nearby surface layers of soil, a zone easily influenced by local rain (Grumm et al., 1990). Light rains can sometimes enhance thoron flux if they are not strong enough to block pore space. The mechanism can be either an increase in the emanation fraction or moisture-induced desorption if the soil is initially dry.

The only important removal process for thoron gas in the atmosphere is decay. However, dry (and occasionally wet) deposition can be important for

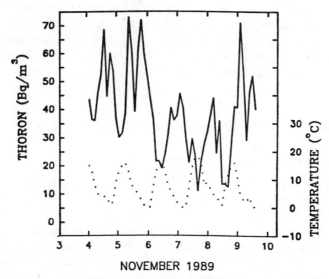

**Figure 2.** Outdoor thoron concentration (—) and temperature (---) at Socorro, NM, 1 m above ground for 6 days starting midnight 11 April 1989. For low wind conditions thoron typically reaches a maximum in the early morning when vertical mixing is at a minimum.

thoron progeny, so their distribution in the atmosphere is more complicated. Because of its very short half-life there is not much time for $^{216}$Po to separate from thoron. The $^{216}$Po concentration will normally follow that of thoron, except perhaps a few centimeters above surfaces. The progeny $^{212}$Pb has a relatively long half-life, so its atmospheric distribution will depend upon a combination of production, mixing, decay, and dry deposition. It normally does not escape directly from soil since it is chemically reactive and attaches to soil particles. Although produced by decay of thoron in the atmosphere, $^{212}$Pb is typically in strong disequilibrium with thoron. Its concentration will be lowest just above the soil, will gradually rise to a maximum at a height of tens or hundreds of meters, then will drop at greater heights (well above where thoron is found) because of decay (Jacobi and André, 1963). At one meter elevation over continental land in temperate climates an average value for $^{212}$Pb is about 0.07 Bq m$^{-3}$ (Schery, 1986).

Near the earth's surface the airborne radioactive gases affect the electrical properties of the atmosphere because of the ions they produce. Ions are important for trace chemical processes of the atmosphere as well as fog and smog production. At one meter height, the major mechanism of ion production from thoron and its progeny will normally be alpha-particle decays from thoron and $^{216}$Po. For example, assuming equilibrium between thoron and $^{216}$Po at a concentration of 10 Bq m$^{-3}$, and an energy of 35 eV to create an ion pair, the data from Table 1 indicate a charged pair production rate of about 4 ion pairs cm$^{-3}$ s$^{-1}$. This compares with a total rate of about 10–25 ion pairs cm$^{-3}$ s$^{-1}$ over land, which results in positive and negative ion densities averaging of the order of 1000 cm$^{-3}$ (Hoppel et al., 1986). The deposition of thoron (and radon) progeny on soil and plants leads to a surface deposit of radioactivity, which provides an even higher production of positive and negative ions in the few centimeters just above their surfaces. There is considerable variation in these ion densities due to changes in the concentration of the airborne nuclides.

## 6.  THORON AND ITS PROGENY IN ENCLOSED AIR SPACES

### 6.1.  Thoron in Mines and Factories

Thoron and its progeny can be significant in underground mines. Thorium itself is usually mined from open-air surface deposits. However, thorium minerals are commonly associated with uranium minerals, so the thoron exhalation from the ores of uranium mines is often significant. In addition, because of the possibility of restricted ventilation and proximity to bare soil and rock, any underground mine can have significant concentrations of both radon and thoron. If ventilation is not present, underground enclosures can be expected to have thoron concentrations approaching the high values of soil gas. Hence, ventilation, whether natural or manmade, is the key to the absolute concentrations of thoron and its progeny. Very few published data are available for thoron gas. The focus has been on thoron progeny since they are usually the more significant contributors to dose.

Bigu (1981) reviews thoron and thoron progeny in uranium mines. His experimental data indicate a median ratio for PAEC (Tn)/PAEC (Rn) of 0.65 with a range of about 0.4 to over 1.5. Bigu modeled the effect of mine ventilation and showed that this ratio tended to increase with increasing ventilation (the *absolute* concentration of all progeny of course decreased with increased ventilation). The PAEC (Tn) followed the concentration of $^{212}$Pb very well, as would be expected given that $^{212}$Pb is the major contributor to PAEC (Tn). The mines studied by Bigu had a ratio for the activity of $^{238}$U relative to $^{232}$Th in the rock in the range of 0.4–4.

Stranden (1985) and Dixon et al. (1985), have reported measurements on a variety of underground mines in Norway and the United Kingdom. The ratio PAEC (Tn)/PAEC (Rn) was usually in the range of 0.1–1. For mines with an activity concentration ratio $^{232}$Th/$^{226}$Ra $\sim 1$, Stranden found that the PAEC (Tn)/PAEC (Rn) was typically between 0.5 and 1. The PAEC (Tn)/ PAEC (Rn) ratio tended to be higher in sedimentary rock, although the absolute value of PAEC (Tn) tended to be highest in igneous rocks.

On average, it appears underground mines might be expected to have a PAEC (Tn)/PAEC (Rn) ratio on the order of one-half, with the absolute value PAEC (Tn) quite variable depending upon ventilation. Occupied mines will normally have PAEC (Tn) well under $2 \times 10^{-5}$ J m$^{-3}$ as an incidental effect of meeting health standards for radon. Unoccupied mines and natural underground caves will tend to have higher values of PAEC (Tn), but lower values for the ratio PAEC (Tn)/PAEC (Rn), due to the generally poorer ventilation of natural convection. In temperate climates this natural convection will tend to be greatest in the winter. Then the subsurface air is warmer than the outside air, leading to greater convective exchange.

A few measurements have been reported for thoron gas in above-ground occupational settings, such as factories and processing plants. Values from 1100 to 11 000 Bq m$^{-3}$ for thoron were common in a United Kingdom gas mantle factory, while values of 1800–18 000 Bq m$^{-3}$ were reported as typical for the indoor air of two thorium processing plants in India (Duggan, 1973).

## 6.2.  Thoron in Houses and Buildings

The 1980s saw a growing interest in possible pollution effects from airborne radon isotopes and their progeny in everyday housing. Estimates indicate that these airborne isotopes are usually the single largest source of effective dose from all sources of ionizing radiation in the nonoccupational setting (UNSCEAR, 1988; Committee on Biological Effects of Ionizing Radiation, 1990). Radon and its progeny have received the major attention, but there has been some study of thoron and its progeny (UNSCEAR, 1982; Schery, 1989, 1990).

From a physical science point of view, thoron progeny are a significant source of ionizing radiation in the indoor environment. Table 4 shows the ratio PAEC (Tn)/PAEC (Rn) for various locations, summarizing much of the available information. A strong correlation between PAEC (Tn) and PAEC (Rn) seems fairly pervasive over a range of housing and locations, although evidence indicates the relation is not linear. A study in France (Rannou, 1987) found that the phenomenological relation (in rounded numbers)

$$PAEC \ (Tn) \propto [PAEC \ (Rn)]^{0.4} \tag{8}$$

agreed well with the indoor data on thoron and radon progeny. Houses with high levels of radon progeny will thus tend to have proportionally less thoron progeny, as seen in Table 4. Houses with higher concentrations of radon are more likely to be studied, so there is some bias in the measurements of Table 4 to locations with higher PAEC (Rn). Weighted by a full cross section of all housing, the average ratio is probably about 0.5, or even a little higher.

Thoron gas is more difficult to measure so there are fewer data available for it. In a study of 25 indoor locations in the United States, Schery (1985) found an arithmetic mean value of 7.0 Bq m$^{-3}$ for measurements 1 m above the floor in rooms in which occupants spent a lot of time. Data were biased toward the winter season and southern latitudes. In a study of 24 measurements at 1 m above the floor in living spaces in Australia, Schery (1989) obtained an arithmetic mean thoron concentration of 16 Bq m$^{-3}$. These measurements were in the winter season and distributed over all states of Australia. The highest value was 207 Bq m$^{-3}$ for an atypical location of a cellar apartment with a dirt floor. In a study of 10 predominantly brick buildings in Innsbruck, Austria, Steinhausler (1975) found the time-averaged indoor concentrations ranged from 3.7 to 41 Bq m$^{-3}$. In rural villages of areas with average $^{232}$Th content in the soil (about 30 Bq kg$^{-1}$) in Southern China (near Enping County) Luxin et al. (1988) report a mean thoron gas concentration of 17.5 Bq m$^{-3}$. Luxin et al. also report an average indoor thoron concentration of 168 Bq m$^{-3}$ for nearby high background radiation areas where the thorium content of soil averages about 220 Bq kg$^{-1}$. These constitute some of the few presently identified areas that might qualify as indoor thoron radioactive "hot spots."

Based on these thoron measurements, and modeling of the more common progeny measurements, the average ground level indoor thoron concentration 1 m above the floor is probably within a factor of two of 10 Bq m$^{-3}$ for single-family housing in the United States. Another rule of thumb that appears to synthesize some of the U.S. data is that in one-story houses coupled to the soil the average thoron tends to be about one-fifth of the radon. However, the ratio appears to be higher in Australian housing, perhaps due to higher average ventilation and overall lower indoor radon levels. Because of its short half-life, thoron usually will not be well mixed within a house. Basements

**Table 4 The Ratio of the Potential Alpha-Particle Energy for the Major Thoron Progeny to That of the Major Radon Progeny for Various Locations**

| Location | PAEC (Tn)[a] / PAEC (Rn) | Comments |
|---|---|---|
| Italy (Latium) | 1.3 | Anomalous (volcanic) area, 50 dwellings, poor ventilation |
| Canada (Elliott Lake) | 0.3 | Samples at 95 dwellings, source activity $^{238}$U/$^{232}$Th ~ 1 |
| Hungary | ~0.5 | 22 dwellings |
| Norway | 0.5 | 22 dwellings, source activity $^{238}$U/$^{232}$Th ~ 1 |
| Federal Republic of Germany (Western Part) | 0.5 | 150 measurements spread over a year |
| Federal Republic of Germany (Southwestern) | ~0.8 | 95 dwellings |
| Federal Republic of Germany | ~0.5 | 27 houses |
| United States | 0.6 | 68 measurements in 20 states, mean PAEC (Rn) = 44 nJ m$^{-3}$ |
| China (Hubei Province) | 0.4 | 37 measurements, $^{238}$U/$^{232}$Th ~ 0.6 |
| France (Finistere) | 0.3 | 219 homes, mean PAEC (Rn) = 298 nJ m$^{-3}$ |
| Hong Kong | 0.8 | 10 indoor sites, atypical tropical coast |
| Austria | 0.7 | 12 dwellings |
| United Kingdom | 0.14 | 8 dwellings |
| United States | 0.3 | 53 measurements in eight southeastern cities on main floor, mean PAEC (Rn) = 280 nJ m$^{-3}$ |

[a]Rounded values of means or medians.

*Sources.* UNSCEAR (1988), Schery (1990), Dudney et al. (1990); T. Ren (Chinese Ministry of Public Health, personal communication, 1990).

can be expected to have higher concentrations than upper stories. Even within a single room thoron is likely to be unevenly mixed.

### 6.3.   Modeling Indoor Thoron

The concentration of thoron and thoron progeny in the indoor environment will be controlled by a number of factors, including entry rate of thoron, ventilation, and deposition of progeny upon indoor surfaces. A simplified model described by Porstendorfer et al. (1987) is useful for estimating average concentrations. Let $V$ be the volume of a room and $E$ the rate of input of thoron into the room from building materials or the soil. Assume an outdoor thoron concentration $C_{o0}$ and a ventilation rate $u$. For steady-state conditions averaged over the room volume, the thoron concentration $C_o$ is given by

$$C_o = \frac{E/V + uC_{o0}}{\lambda + u} \tag{9}$$

Common ventilation rates are in the range 0.1–5 h$^{-1}$, so usually $\lambda \gg u$ and $u$ in the denominator can be neglected. As a result, for cases where outdoor air is not an important source of thoron, the indoor thoron concentration should be less sensitive to ventilation than is the radon concentration. There is evidence that this may be the case (Zarcone et al., 1986).

Indoor air will normally contain a mixture of thoron and its progeny. (In fact, given that several progeny have half-lives significantly longer than thoron, it is possible to have progeny present without thoron.) Normally the progeny originate only from decay of thoron after it enters the indoor air space. Even when present in air entering indoor air space, progeny are susceptible to plateout during passage through small openings such as joints around pipes and cracks in the foundation. However, if the entrance route is large openings, such as open doors or windows, progeny may be able to enter without plateout loss.

The model described by Porstendorfer et al. (1987) is also useful for estimating indoor thoron progeny. Depositional (plateout) losses to the walls are assumed to fall into two classes: the more readily deposited "unattached" particles (particles too small to be measured by a condensation nuclei counter) and the slower-to-deposit "attached" particles (progeny attached to aerosol particles detectable by a condensation nuclei counter). Other processes included in the model are the attachment rate for progeny to aerosol particles by collision, and detachment from aerosols due to recoil of a heavy nucleus upon alpha decay. Under steady-state conditions and assuming no direct source of progeny from outdoor air, the concentrations of the unattached ("free") $C_i^f$ and attached $C_i^a$ particles of the $i$th progeny are given by

$$C_i^f = \frac{(\lambda_i C_{i-1}^f + r_{i-1} \lambda_i C_{i-1}^a)}{(u + \lambda_i + X + q^f)} \tag{10}$$

$$C_i^a = \frac{(1 - r_{i-1}) \lambda_i C_{i-1}^a + X C_i^f}{(u + \lambda_i + q^a)} \tag{11}$$

where $X$ = attachment rate to aerosol ($h^{-1}$)
$q^f$ = deposition rate of unattached particles ($h^{-1}$)
$q^a$ = deposition rate of attached particles ($h^{-1}$)
$r_i$ = fraction of the progeny that become detached from aerosol due to the recoil process (dimensionless)
$\lambda_i$ = decay constant for the $i$th isotope ($h^{-1}$)
$C_o$ = average thoron concentration from equation 9
$C_o^f = C_o$
$C_o^a = 0$

The index $i$ runs from 0 to 3, with $i$ = 0, 1, 2, and 3 corresponding with $^{220}$Rn, $^{216}$Po, $^{212}$Pb, and $^{212}$Bi. The concentration of the other progeny are either in equilibrium with members of this set or are not important for dosimetry calculations. The attachment rate $X$ is assumed to be proportional to the aerosol concentration $Z$:

$$X = \beta Z \tag{12}$$

where $\beta$ is the attachment coefficient. The recoil fraction is negligible for beta decay, so the only nonzero $r_i$ is $r_1$. Only two classes of particles are assumed instead of a more realistic spectrum of particle sizes. Hence, $q^a$, $q^f$, and $X$ are all average or effective values that can vary with such factors as the size spectrum of particles and the rate of mixing of the indoor air.

Figure 3 shows representative calculations with equations 9–12, assuming that the input of thoron from outside air is negligible ($C_{o0} = 0$). The concentration of radioisotopes is plotted as a function of the ventilation rate for a representative aerosol concentration of 20 000 particles $cm^{-3}$. Parameters used in these calculations are $q^a = 0.2\ h^{-1}$, $q^f = 40\ h^{-1}$, $\beta = 5 \times 10^{-3}\ cm^3$ $h^{-1}$, $r_1 = 0.5$, $u = 0.75\ h^{-1}$, and $E/V = 400\ Bq\ m^{-3}\ h^{-1}$. These might be representative values for single-family homes in the United States. With the exception of $r_1$, a factor of two uncertainty in any parameter is probably representative.

Figure 3 shows that higher ventilation has little effect on the thoron concentration over the range of ventilations covered. On the other hand, the attached progeny of $^{212}$Pb and $^{212}$Bi are significantly reduced by increased ventilation. Since thoron progeny are usually the major cause of radiation dose, the extent of disequilibrium among the isotopes is an important consideration. This disequilibrium can be quantified by the equilibrium factor $F$, which is defined as the PAEC (Tn) due to the progeny actually present divided by the PAEC (Tn) that would be present if complete equilibrium existed with thoron (see equation A3 of the Appendix). The equilibrium factor is plotted as a dotted line in Fig. 3. It decreases rapidly with increasing ventilation rate.

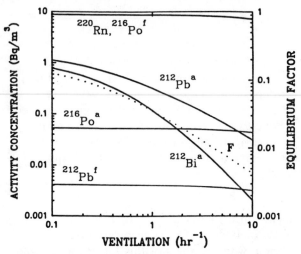

**Figure 3.** Model calculations for thoron and thoron progeny as a function of ventilation for an indoor location such as a one-story house in a temperate climate assuming a condensation nuclei concentration of 20 000 cm$^{-3}$.

The equilibrium factor is also sensitive to the aerosol concentration. Lower aerosol concentration will cause increased disequilibrium because plateout losses of unattached progeny to the walls are then greater.

With the exception of $^{216}$Po, the unattached progeny are normally a small fraction of the total progeny. Nevertheless, they are potentially important to dosimetry calculations since they are more readily deposited on tissue surfaces during the breathing process because of high deposition velocities. The majority of these unattached particles are electrically neutral, but a small fraction (perhaps 10% (Shimo et al., 1985), or a little higher, is representative) will be found in the positive charge state. The exact value will depend on such factors as the presence of neutralizing agents and the concentration of aerosol particles. Many other physicochemical properties of attached and unattached progeny are likely to be important for dosimetry considerations. Examples include the solubility of particles, their diffusion coefficients, and their geometrical shape. This is an area of current research and a review of some recent developments can be found in Hopke (1988).

### 6.4.    Source of Thoron for Indoor Air

An interesting issue not yet fully understood is the source of the thoron in buildings. When indoor radon was first studied, an important discovery was the major role played by soil as a source of radon. Houses in temperate or colder climates typically have a small underpressure of several Pascals at their foundations due to outside wind and convective effects of heating. This underpressure is sufficient to draw significant amounts of radon through the cracks and openings of foundations. Traditionally, it has been assumed that passage times are too long for the short-lived thoron to enter by this mechanism, so only building materials and outside air were believed the source of thoron (UNSCEAR, 1982). However, careful estimates of transport times show that they are frequently short enough to permit direct entry of thoron (Schery, 1986). Other evidence that soil can be a major source of thoron has accumulated (Schery, 1986). A site-specific study of a house of common construction and materials showed that the primary mode of entry of thoron was through cracks in the foundation (Zarcone et al., 1986). As previously discussed, houses with higher radon concentrations tend to have higher concentrations of thoron and its progeny. If thoron is coming from a different source, it is difficult to explain this correlation, given the importance of soil for controlling variation in indoor radon (Sextro et al., 1987). Finally, concentrations of thoron and its progeny in buildings often appear to be lower in upper stories if in terms of the ventilation system they are more isolated from soil. The recent data of Dudney et al. (1990) for basements and main floors of houses in the southeastern United States are particularly convincing in this respect.

On the other hand, it is quite feasible that under appropriate conditions building materials can provide significant input of thoron. A median value for the flux density $J$ for uncoated masonry materials such as concrete is of the order of $45 \times 10^{-3}$ Bq m$^{-2}$ s$^{-1}$ (Folkerts et al., 1984). Neglecting the (probably small) effects of outside air and ventilation, equation 9 becomes

$$C_o = \frac{JA}{\lambda V} \tag{13}$$

where $A$ is the internal surface area of the room and $E = JA$. Using a representative ratio of $A/V = 2$ m$^{-1}$, equation 13 gives $C_o = 7.1$ Bq m$^{-3}$, which is of the order of the concentration common in indoor air. Wood products typically have a much smaller thorium content and thoron exhalation. So critical issues in determining the importance of building materials are the extent to which masonry materials are present and the type of coverings upon their surfaces (paint, sealants, carpet, etc.). Further research will be needed to determine the relative importance of soil and building materials as sources of indoor thoron. Present evidence suggests that over the range of housing types, building materials will be relatively more important than is the case with radon, but that for single-story family housing of wood frame construction, soil is a major source.

## 7. HEALTH EFFECTS FROM THORON AND THORON PROGENY

### 7.1. Introduction

Historically, a beneficial effect was associated with the inhalation of radon isotopes and progeny. Springs and spas were popular attractions in pre-war

Europe and there still exists some contemporary support for these activities (Serdyukova and Kapitanov, 1969). Concern about possible adverse health effects associated with breathing radon isotopes and their progeny became widespread after World War II in connection with occupational exposure, particularly in uranium mines. Epidemiological data for miners in Colorado, Czechoslovakia, and elsewhere demonstrated fairly convincing evidence that exposure to high levels of radon progeny over long periods of time correlated with an increased incidence of lung cancer. The primary concern was radon progeny, but some consideration was given to thoron and thoron progeny (Evans and Goodman, 1940; Duggan, 1973). The major problem identified was radiation dose to the respiratory tract caused by alpha-particle decays. Radon and thoron progeny deliver dose by the same mechanism. However, the overall problem from thoron in a mixed atmosphere of thoron and radon is usually less due to typically lower PAEC for thoron progeny, and dosimetry details related to the different half-lives and alpha-particle energies.

For thoron and thoron progeny, there exist no direct epidemiological data and essentially no comprehensive laboratory studies with animals. Hence, estimates of health effects are based on dosimetry models and risk estimates from studies of other isotopes, such as radon. However, the similarity of the dose delivery mechanisms for radon and thoron progeny make such projections less difficult than they could be otherwise. Probably the uncertainties in projecting risk from thoron using risk estimates for radon are less than the uncertainties inherent in the initial risk estimates for radon.

## 7.2.    Mechanisms for Delivery of Radiation Dose

The primary health hazard associated with thoron and its progeny is lung cancer due to alpha-particle ionization damage to cells of the lung and bronchial epithelium. Densely ionizing alpha-particle radiation is believed to be an effective agent for both mutagenesis and carcinogenesis. It can cause double-strand breaks of DNA, which are difficult for a cell to repair. The progeny of thoron, as nonvolatile particles (either attached or unattached to aerosols), are readily deposited on respiratory tract tissues, particularly the bronchial epithelium. Figure 4 shows estimates of the weighted dose equivalence per unit PAEC (Tn) as a function of the activity median aerosol diameter adapted from James (1985). It illustrates that smaller-sized progeny generally deliver higher dose. However, uncertainties are large in the dosimetry calculations. Although $^{212}Pb$ itself is not an alpha-particle emitter, it is the critical isotope in air controlling PAEC (Tn) and alpha-particle dose. Compared with radon progeny, the dose delivered by $^{212}Pb$ is somewhat mitigated by its longer half-life and greater probability of being inhaled in the attached state. Before inhalation, $^{212}Pb$ has a longer time to attach to aerosol particles. The deposition velocity to surfaces of these larger (attached) particles is less, so there is a greater chance of $^{212}Pb$ being expelled from the

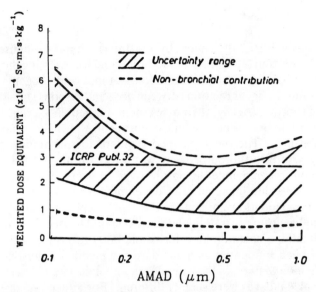

**Figure 4.** Weighted dose equivalent from thoron progeny per unit PAEC (Tn) as a function of activity median aerosol diameter (AMAD). The label ICRP Pub. 32 indicates the estimate from ICRP (1981). (Adapted from James (1985).)

lung without deposition. Current estimates are that the effective dose equivalent delivered per unit of PAEC (Tn) is about one-third to one-fifth of that delivered by PAEC (Rn) (ICRP, 1981; James, 1988). Given the frequently great disequilibrium between thoron and its progeny, it is theoretically possible for thoron itself to deliver a significant dose to the lung. For example, using the effective dose-equivalent conversion recommendations from ICRP (1981) and the parameters used for Fig. 3, the effective dose from thoron relative to that from its progeny would range from about 7% at $u = 1 \text{ h}^{-1}$ up to about 58% at $u = 10 \text{ h}^{-1}$.

As a consequence of the long half-life of $^{212}\text{Pb}$, a large fraction of the $^{212}\text{Pb}$ activity deposited in the alveolar–interstitium region of the lung is absorbed into the blood and transported to other organs of the body. Activity deposited in these organs can result in irradiation of their tissue by alpha particles when the progeny $^{212}\text{Bi}$ decays. Approximately 10% of the inhaled $^{212}\text{Pb}$ activity is deposited in the alveolar–interstitium (James, 1988), and the ICRP estimates that 30% of this deposited activity is absorbed by other tissues. ICRP models (James, 1988) estimate the majority of the resulting total effective dose equivalent per unit exposure to be due to a redeposition in the skeleton of $5 \times 10^{-6} \text{ Sv m}^3 \text{ J}^{-1} \text{ s}^{-1}$. This is only a few percent of the corresponding value for the bronchial epithelium, so these estimates suggest the net effect of redeposition of $^{212}\text{Pb}$ is to lower the overall dose.

## 7.3.   Risk Assessment

As of the early 1980s, the report by the International Commission on Radiological Protection (ICRP, 1981) was influential for assessing the lung cancer risk due to inhalation of airborne radon isotopes and progeny. The Commission assumed a linear relation between dose and response at all dose levels. In terms of thoron progeny, their assessment implied an average lifetime risk coefficient for lung cancer per unit exposure in the range of about 0.014–0.04 per J·h m$^{-3}$. The recommended standard for maximum annual occupational exposure to thoron progeny was 0.050 J·h m$^{-3}$. A miner receiving this maximum recommended exposure over 20 years would be expected to have a 1.4–4.3% lifetime chance of developing lung cancer.

The recommendation implied by the ICRP report for maximum annual occupational exposure was designed to correspond to an effective dose equivalent of 50 mSv, a common limit for whole-body exposure. This recommendation has been influential as a standard for member countries of the European Common Market. In the United States as of the late 1980s, recommended occupational standards were slightly different. For miners, no distinction was made between radon and thoron progeny, so the annual thoron progeny exposure limit was lower, 0.014 J·h m$^{-3}$. For industrial settings other than mining (covered by the Occupational Safety and Health Administration) a standard existed for maximum thoron concentration: 11 100 Bq m$^{-3}$. The most recent trends in regulatory exposure limits have been toward an upper effective dose limit for the combined contributions of all radiations, rather than separate standards for each agent. Furthermore, exposures are supposed to be kept as low as practical even when below the suggested limit.

The details of these risk assessment models are complicated and undergoing change. Factors that must be considered include age and sex of the subject, synergistic effects with other pollutants (such as cigarette smoke), and the rate, as well as integrated amount, of exposure. Especially difficult is the assessment of risk at lower levels of exposure. The lowest levels exhibiting a clear lung cancer effect from radon progeny in the miner data are cumulative exposures of the order of 0.4 J·h m$^{-3}$. When a dose-equivalence conversion factor is used that is one-third smaller for thoron progeny, this corresponds to continuous exposure at a PAEC level of about 650 nJ m$^{-3}$ over a lifetime of 70 years. Below this level, further assumptions are required to assess risk. The prevailing view is that there is some adverse health effect no matter how small the dose. A common practice is the assumption of a no-threshold linear dose–response relation down to zero dose. However, some scientists who have studied this issue feel that a threshold exists below which exposure to ionizing radiation poses no significant risk (Wachsmann, 1989). In fact, some scientists advance arguments why exposure to small amounts of ionizing radiation actually carries a beneficial effect. For one thing, humans have evolved on an earth with trace amounts of thoron pervasive in its atmosphere. The beneficial effects could be in ways unrelated to cancer and overlooked in studies focusing on cancer. The arguments for the strongly ionizing alpha-particle radiations of radon and thoron progeny do not seem as strong as the arguments for more weakly ionizing radiations (beta and gamma), but they nevertheless exist (Wachsmann, 1989; Serdyukova and Kapitanov, 1969). Such hormetic effects have been found to be true of a number of environmental agents from sunlight to trace metals in our diet.

The combination of limited data for thoron levels in environmental air, and the uncertainties in the dosimetry models, makes health projections for specific occupational and nonoccupational situations difficult. With this qualification, we can mention some of the limited assessments available. For atmospheres containing both radon and thoron, usually the dose equivalence from thoron progeny is smaller, but not necessarily insignificant. Dixon et al. (1985) estimated the median annual equivalent dose from thoron progeny over a cross section of eight mines in the United Kingdom. They found the equivalent dose ranged from 1 to 30% of the dose from radon progeny. The highest proportional dose from thoron progeny was in mines in sedimentary rocks. The mines studied were conventional operations producing ores such as gypsum, potash, and tin (but not uranium or thorium). Nonoccupational exposure to thoron progeny in everyday housing has received some study. In temperate climates of North America and Europe, current estimates put the effective dose equivalence from thoron progeny to be about 10–20% of that from radon progeny (UNSCEAR, 1982; Schery, 1990). Contingent on the validity of the prevailing risk assessments for radon, Schery (1990) projected up to 4000 deaths per year in the United States due to indoor thoron. This is a risk worthy of investigation, being comparable to other safety hazards of concern for the population, such as accidental deaths due to poisoning or firearms. It is much greater than projections of the effects of radioactivity from sources such as weapons testing fallout or emissions from power reactors.

There has probably been fairly adequate assessment of thoron as a pollutant in occupational settings, at least relative to the other major pollutants. For the common situation of a mixture of isotopes where radon progeny dominate dose, it is reasonable to give priority to assessment and control of radon. The few cases where thoron progeny dominate seem to have been identified. More complete evaluation of their health hazard involves basic dosimetry questions common to many types of ionizing radiation. The priority here could still be basic dosimetry studies of such things as the effect of alpha-particle dose rather than specific study of thoron.

In the area of nonoccupational exposure there is much to learn about thoron and its progeny. There is the general issue for all ionizing radiation of the effects of low-level exposure. In addition, specific to thoron, is the question of the relative importance of sources (soil, building materials, etc.) for supplying indoor air. The answer is particularly important should high levels be found and it becomes necessary to reduce them. Presumably, many of the procedures effective for thoron will also be effective for radon, but there is uncertainty on this point. Another issue is whether there exist many locales with unusually high levels of indoor thoron. This is perhaps the most important present uncertainty. It was the discovery of residences with high levels of indoor radon in the United States that gave impetus to much of the increased concern about radon as an indoor air pollutant. So far, China (Luxin et al., 1988) is one of the few countries that has identified radioactive "hot spots" involving thoron gas.

One apparent distinction between radon and thoron is that due to controlling physical factors the variation in thoron levels found among houses is such that the fraction of houses with very high levels of thoron appears to be less than the corresponding fraction for radon. Therefore, there are relatively fewer striking cases of high risk identified for individual houses. This lack of striking cases may influence society's perception of the importance of thoron as a pollutant more than the data for the significant integrated dose being delivered to the general population.

At present there exist no specific regulations for nonoccupational maximum indoor concentrations of thoron and thoron progeny. The United States Environmental Protection Agency's guideline of 420 nJ m$^{-3}$ for radon progeny would correspond with 1260 nJ m$^{-3}$ for thoron progeny, assuming a one-third smaller effective dose equivalence conversion factor. If one applies the rule of thumb of allowing one-tenth of the limit allowed for occupational exposure, then the result would be about 570 nJ m$^{-3}$ based on ICRP (1981) assuming continuous exposure. For a representative equilibrium factor of 0.02 (Schery, 1989) this corresponds to a thoron concentration of about 370 Bq m$^{-3}$.

## 8. EXPERIMENTAL MEASUREMENT OF THORON AND THORON PROGENY

### 8.1. Measurement of Thoron Gas

Direct measurement of thoron gas in air is usually more difficult than direct measurement of radon gas. The reasons have to do with small atomic concentrations of thoron and interfering signals from the radon gas, which is usually concurrently present. Radon and thoron decays must be separated by some technique, and there is likely to be a smaller number of decays from thoron and its progeny. Otherwise, the underlying approaches to radon and thoron gas measurement are similar. The demand for commercial measurement of thoron gas has been limited, so most instruments that have been built are specialized designs intended for research applications.

For higher concentrations of thoron, such as found in soil gas, it may be possible to determine thoron concentration from the alpha-particle decay rate in simple one-unit devices such as scintillation cells. The gas must be introduced and counted quickly so that the change in the alpha-particle count rate due to thoron decay can be monitored to distinguish it from the background due to radon. The device must be cleared and the gas reintroduced for each measurement. This technique is impractical for lower levels, such as outside air. For example, a 1-L cell containing a mixture of radon and thoron at about 10 Bq m$^{-3}$ each would produce only several counts total from thoron over its mean life above a comparable background due to radon. However, Crozier (1969) was able to use a variation of this technique with a larger volume ionization chamber to measure thoron emanations from soil.

Perhaps the most ubiquitous approach to measurement of thoron gas at lower levels such as found outdoors is the two-filter method. The concept is to pass the air stream through a chamber with a filter at its entrance and exit. The first filter removes all progeny from the air so that any progeny collected on the second filter will be due to thoron in the chamber. For a fixed sampling protocol, counts from thoron progeny on the second filter will be proportional to the thoron concentration of the entering air. The second filter need not actually be a filter but any device capable of measuring decay of thoron progeny such as an ionization detector or even a passive device such as an ionization-sensitive film or thermoluminescient chip. Since radon gas will also cause progeny on the second filter, procedures will normally be required to distinguish thoron progeny from radon progeny. The sensitivity of the two-filter technique is limited by the volume of the chamber between the filters. Up to a point, higher flow rates produce more thoron progeny on the second filter. However, when the concentration of thoron exiting the chamber becomes about the same as that entering the chamber, the only way to increase the progeny collected on the second filter is to enlarge the volume of the chamber so that the total rate of progeny production in the chamber is larger. Hence, the most sensitive detectors require large volumes, often running into many hundreds of liters.

Another problem with the two-filter technique is that all progeny produced in the chamber may not actually end up on the second filter. Some may be deposited on the walls of the chamber. This would be less of a problem if the fraction lost remained constant. However, it is likely to vary with trace properties of air, such as moisture, that affect the deposition velocity of the progeny. Various schemes have been devised to minimize these "plateout" losses. A simple expedient is to use a very high flow rate. Sometimes aerosols are injected into the chamber to provide a competing surface area to which progeny can attach for later collection on the final filter. This latter approach may require submicrometer aerosols at concentrations of $10^5$ cm$^{-3}$ or higher. A common approach to distinguish thoron progeny from radon progeny on the second filter is to employ alpha spectroscopy. Different counting intervals can also be used to take advantage of the different decay rates for radon and thoron progeny. Finally, a second two-filter system can be employed that delays the entering air long enough to remove all thoron. Its response then comes only from radon and can be subtracted from that of the primary system to deduce the thoron signal.

A number of systems using the two-filter principle for measuring thoron gas have been reported in the literature (Israel et al., 1968; Mark, 1969; Schery, 1985). Figure 5 shows a photograph of a system in use at New Mexico

**Figure 5.** Photograph of the New Mexico Tech 79-L continuous two-filter system for measurement of thoron gas. Progeny are deposited on a strip filter at the top of the apparatus after entrance and initial filtration at the bottom inlet hose. A microcomputer logs alpha counts and controls advance of the exit filter.

Tech (Grumm et al., 1990) capable of unattended measurement of outdoor thoron gas. The chamber volume is 79 L, and the exit filter is a strip 5.7 cm wide made from 1.2-$\mu$m-pore acrylic copolymer film (Versapor by Gelman). A measurement cycle involves deposition on the filter followed by movement to a 600-mm$^2$ silicon detector where a gate is set on the 8.8-MeV alpha particle from $^{212}$Po. For a flow rate of 65 L min$^{-1}$, and a 2-h sample and counting time, the sensitivity is about 2 alpha counts h$^{-1}$ Bq$^{-1}$ m$^3$ thoron.

There remains a need for more portable systems capable of measuring thoron gas and devices able to give an integrated response over a period of time. At intermediate concentrations, such as found in mine air, a pair of flow-through scintillation cells connected in series, separated by tubing or a chamber to allow decay of thoron, should be able to give a time-dependent response to both radon and thoron. Wang et al. (1985) describe a portable system for mine air that uses activated charcoal as the second filter. They use total alpha counting and separate the thoron progeny from the radon progeny using sequential counting periods. For a flow rate of 10 L min$^{-1}$ they report a lower limit of detection of about 1000 Bq m$^{-3}$ for thoron. It should be possible to improve the sensitivity of this design by increasing the volume of charcoal and using gamma spectroscopy to separate the progeny.

Another approach to thoron measurement is to use alpha spectroscopy with flow-through pulse-sensitive ionization chambers using air as the carrier gas. With careful design it may be possible to separate the thoron (6.3 MeV) or $^{216}$Po (6.8 MeV) from the radon (5.5 MeV) decays. In principle, integrated measurements of thoron gas over intervals of weeks or months should be possible with minor modification of the track-etch cups used for integrated radon measurement. For example, two cups can be exposed simultaneously with one cup having a larger diffusion barrier allowing entry of radon but not thoron. Subtraction of tracks from the two cups would provide the measure for thoron. Alternatively, a thin sheet could be placed over the track-etch film allowing passage of the 8.8-MeV alpha particle from the thoron progeny $^{212}$Po, but absorbing the lower-energy alphas from radon progeny. Despite preliminary development of these techniques, as of this writing no convincing routine operational systems have been reported in the literature.

An important use of thoron detectors is measurement of thoron exhalation rates from porous media such as soil or building materials. One technique is to place an "accumulator," an open-bottomed container, on a surface of the material and monitor the buildup of thoron within the container. The principles are the same as with measurement of radon exhalation except that the thoron concentration will reach steady-state more quickly and there is less chance of a high concentration in the container altering the transfer rate from the emanating material. To measure the exhalation rate to the open atmosphere, the air parameters inside the accumulator (pressure, temperature, etc.) should match those of the outside air as closely as possible. Pressure differences as small as one pascal can be important. The exhalation of thoron from soil is typically high enough that measurement of the thoron gas is not particularly difficult. An effective system is one designed by Stewart Whittlestone of the Australian Nuclear Science and Technology Organisation (Schery et al., 1989). This system utilizes 1-L flow-through scintillation cells and has a detection limit for thoron exhalation of about 0.02 Bq m$^{-2}$ s$^{-1}$. Keller et al. (1982) describe an accumulator system that collects thoron progeny directly on a surface barrier detector by electrostatic deposition. They quote a lower limit of detection of about $6 \times 10^{-3}$ Bq m$^{-2}$ s$^{-1}$. Systems such as these employing electrostatic deposition are subject to error due to variation of the mobility and collection efficiency for progeny with changing trace properties of air (e.g., moisture, aerosols, or neutralizing gases).

Techniques for calibration of thoron detectors have not been as refined as they are for radon detectors, where disagreement still exists among major labs at a level of around 5%. In principle, reference solutions of salts such as [228]Th could be de-aerated in analogy to a common technique for radon. However, the short half-life of thoron makes it more difficult to assure complete transfer of thoron. Another approach is to infer thoron calibration of a detector from its radon calibration. A traditional two-filter system employing injection of aerosol particles at high concentrations should have essentially equal collection efficiencies for radon and thoron progeny. If the exit filter is counted for alpha-particle decays, any difference in counting efficiency due to the different energies of the alpha particles can usually be made minimal. Inference of the thoron calibration of a scintillation cell from its radon calibration requires careful consideration of the geometrical pattern of the progeny. The thoron progeny [216]Po is likely to be suspended in air at the moment of decay (due to its short half-life), whereas the corresponding radon progeny [218]Po is more likely to have been deposited on the walls where the detection efficiency is different.

### 8.2.   Measurement of Thoron Progeny

Measurement of airborne thoron progeny is easier than measurement of thoron gas itself. Standard filter techniques for radon progeny work well for thoron progeny. One of the easiest measurements to make is that for [212]Pb. A known volume of air is passed through a filter. If initially the atmosphere contains roughly equal proportions of radon and thoron progeny on an activity basis, a delay of 6 h between collection and counting is sufficient for decay of the radon progeny and for establishment of equilibrium between [212]Pb and [212]Po. Total alpha-particle counting of the filter will then give the activity concentration $C_{Pb}$ of [212]Pb (in Bq m$^{-3}$):

$$C_{Pb} = \frac{N_\alpha \lambda^2}{f\dot{V}} (1 - e^{-\lambda T_1})^{-1} (e^{-\lambda T_2} - e^{-\lambda T_3})^{-1} \qquad (14)$$

where   $T_1$ = time from beginning of collection to end of collection (s)
$T_2$ = time from beginning of collection to beginning of counting (s)
$T_3$ = time from beginning of collection to end of counting (s)
$N_\alpha$ = total alpha-particle counts (dimensionless)
$\dot{V}$ = volumetric flow rate through the filter (m$^3$ s$^{-1}$)
$f$ = fraction of alpha-particle decays detected (dimensionless)
$\lambda$ = [212]Pb decay constant (s$^{-1}$).

Equation 14 assumes complete collection of [212]Pb on the filter.

Although mathematically more complex, protocols that provide all the major longer-lived radon and thoron progeny ([218]Po, [214]Pb, [214]Bi, [212]Pb, [212]Bi) are straightforward and available in the literature. If only total alpha counts are available, it is necessary to employ five separate counting intervals to obtain the five longer-lived progeny. Khan et al. (1982) recommend a collection time of 10 min, followed by counting intervals in minutes of (12, 14), (15, 30), (40, 70), (150, 210), and (280, 330) measured from the start of the collection interval. If $C$ is measured in Bq m$^{-3}$ and $\dot{V}$ in m$^3$ s$^{-1}$ then

$$KC_{Po218} = 0.98002I_1 - 0.22888I_2 + 0.07430I_3$$
$$- 0.03471I_4 + 0.028672I_5 \tag{15}$$

$$KC_{Pb214} = -0.06092I_1 - 0.01929I_2 + 0.04828I_3$$
$$- 0.07140I_4 + 0.07140I_5 \tag{16}$$

$$KC_{Bi214} = -0.05371I_1 + 0.05327I_2 - 0.01270I_3$$
$$- 0.05395I_4 + 0.06548I_5 \tag{17}$$

$$KC_{Pb212} = 0.00068I_1 - 0.00049I_2 + 0.00068I_3$$
$$- 0.00284I_4 + 0.01054I_5 \tag{18}$$

$$KC_{Bi212} = -0.01917I_1 + 0.01437I_2 - 0.01837I_3$$
$$+ 0.07005I_4 - 0.07899I_5 \tag{19}$$

where $I_1$, $I_2$, $I_3$, $I_4$, and $I_5$ are the counts in the respective counting intervals and $K = 10^4 \, \dot{V}f$. Knutson (1989) describes protocols for simultaneous measurement of radon and thoron progeny based on sequential counting over a number of equally spaced intervals of adjustable number and length. If alpha spectroscopy is available, determination of the thoron progeny $^{212}$Bi is especially simple, since the alpha particle emitted immediately after its decay (8.8 MeV from $^{212}$Po) is usually easily resolvable from the lower-energy alphas that accompany the other radon and thoron progeny. Yang and Tang (1978) provide formulas for arbitrary collection and counting times useful for measurements with alpha-particle spectroscopy.

## 9.  REDUCTION OF THORON AND THORON PROGENY IN ENCLOSED AIR SPACES

There has been little practical experience with techniques directed at reducing thoron in enclosed air spaces. The following comments will therefore be brief, primarily based on our understanding of mechanisms of thoron transport and projections from experience with radon.

For purposes of reduction in enclosed air spaces the only important difference between radon and thoron is the shorter half-life of thoron. In general this should make the reduction of thoron easier, since only a short delay in air delivery can lead to significant removal by decay. Large charcoal traps, which have been used in industrial settings to remove radon, should be even more effective with thoron since decay is more probable before breakthrough exhaust of thoron. Comparatively thin coats of paints, sealants, and coatings applied to source materials, which are ineffective for retarding radon, can be quite effective for reducing thoron exhalation. Perhaps the only technique less effective for thoron reduction is ventilation of the air space. Modest increases in ventilation effective for reducing radon may have little or no effect on thoron due to its short half-life (see equation 9). However, the technique will usually be effective for the important thoron progeny, which in practice are the objective.

A common technique for reducing radon in housing is to artificially create an underpressure beneath the foundation by suitable blowers and piping. The resulting pressure gradient reverses the flow of gas through cracks and open-

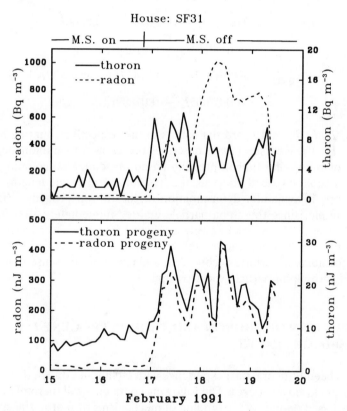

**Figure 6.** Measurements of thoron, radon, and their progeny in the bedroom of a one-story house in New Mexico showing the effectiveness of a sub-slab mitigation system for reducing indoor thoron for a case where soil is the primary source of thoron. When the mitigation system (M.S.) is on, producing downward airflow through the concrete slab that separates the house from the soil, the indoor concentrations of both radon and thoron are reduced.

ings in the foundation, causing a reduction of radon transport from the soil. As indicated by Fig. 6, this technique appears also to work well for thoron when soil is the primary source of thoron (Schery et al., 1991). However, the technique for housing with a crawl space where ventilation of the crawl space is increased by side vents or fans should be much less effective for thoron due to its insensitivity to ventilation. Thorough sealing of passages from soil to indoor air can reduce radon entry. This technique should be modestly more effective for thoron since even a small delay in passage is useful. Something as simple as closing a basement door could be helpful for reducing entry of thoron.

If direct reduction of thoron progeny is the object, both increased ventilation and filtration should be effective. Filtration should be easier for thoron progeny since most of the PAEC (Tn) is associated with $^{212}$Pb, which has a long half-life. Lower filtration rates should be possible. Personal respirators for industrial applications designed to remove radon progeny will work with thoron progeny. The thoron progeny $^{216}$Po builds up too quickly from thoron to be removed effectively by filters, but, unlike the corresponding radon progeny, $^{218}$Po normally carries insignificant dose (see equation 1) so this issue is mute. Ventilation of crawl spaces under houses will effectively reduce thoron progeny there, and would be useful in situations where there is direct entry of progeny from crawl spaces to indoor air.

## APPENDIX: NOMENCLATURE AND UNITS

The airborne concentration of the thoron decay products is often expressed as the potential alpha energy concentration (PAEC). This quantity incorporates the cumulative alpha energy density released into the air as the daughters decay to the stable isotope $^{208}$Pb. For radon progeny, the last decay product is the long-lived isotope $^{210}$Pb, and the PAEC (Rn) is (in units of nJ $m^{-3}$)

$$\text{PAEC (Rn)} = 0.579C(^{218}\text{Po}) + 2.86C(^{214}\text{Pb}) + 2.10C(^{214}\text{Bi}) \quad \text{(A1)}$$

where the concentrations $C$ are in Bq $m^{-3}$. For thoron progeny, the corresponding PAEC (Tn) is

$$\text{PAEC (Tn)} = 69.1C(^{212}\text{Pb}) + 6.56C(^{212}\text{Bi}) + 5.32 \times 10^{-4}C(^{216}\text{Po}) \quad \text{(A2)}$$

The equilibrium factor $F$, a measure of the disequilibrium between thoron and its progeny, can be calculated with

$$F = \frac{9.12 \times 10^{-1}C(^{212}\text{Pb}) + 8.66 \times 10^{-2}C(^{212}\text{Bi}) + 7.02 \times 10^{-6}C(^{216}\text{Po})}{C(^{220}\text{Rn})}$$

$$\text{(A3)}$$

Complete equilibrium (disequilibrium) corresponds to an $F$ value of 1(0). Another term used in reference to PAEC is equilibrium equivalent concentration of thoron, EEC (Tn), which is the concentration of thoron that would produce the observed PAEC (Tn) if there was complete equilibrium between thoron and its progeny ($F = 1$). If EEC (Tn) is in Bq m$^{-3}$ and PAEC (Tn) is in J m$^{-3}$, then EEC (Tn) = $1.32 \times 10^7$ PAEC (Tn).

The alpha-particle dose is the amount of energy absorbed by the lung tissue per unit mass. The international unit for dose is the gray (Gy), with one gray equal to one joule absorbed per kilogram of tissue. Because the biological effectiveness of an absorbed dose depends on the type of radiation, the modified quantity "dose equivalent" is used. The dose equivalent is determined by multiplying the absorbed dose by a dimensionless quality factor, which is about 20 for alpha particles, and unity for beta particles, gamma particles, and X-rays. The unit for dose equivalent is the sievert (Sv). The total health effect to the body is due to the exposure of all the individual tissues. This "effective dose equivalent" is the sum of all the body tissues' dose equivalents, each multiplied by a weighting factor to reflect different sensitivity to radiation damage. As an extreme example, alpha-particle dose to external hair or dead skin has much less effect than equal dose to living lung tissue or bone marrow which carry a higher weighting factor.

An older system of units is still in common use in parts of the English-speaking world. Here radiation activity is measured in curies (Ci) with 1 Ci = $3.7 \times 10^{10}$ Bq. A common unit for activity concentration is picocurie per liter (pCi/L) where 1 pCi/L = 37 Bq m$^{-3}$. For thoron only, 1 Bq m$^{-3}$ will correspond to 80 thoron atoms m$^{-3}$. Potential alpha energy concentration is measured in working levels (WL) with 1 WL = $2.08 \times 10^{-5}$ J m$^{-3}$. Accumulated exposure is measured in working level months (WLM) with 1 WLM = $3.51 \times 10^{-3}$ J·h m$^{-3}$. RAD and REM are used for dose and dose equivalence with 1 RAD = 0.01 Gy and 1 REM = 0.01 Sv. Thorium content of rocks and soils is often reported in parts per million (ppm). One ppm equals 4.03 Bq kg$^{-1}$. Further information on conversion factors can be found in UNSCEAR (1988).

For porous media transport this chapter uses the convention that the symbol $D_e$ stands for the effective interstitial diffusion coefficient satisfying Fick's law expressed in the form

$$J = -\varepsilon D_e \nabla C \tag{A4}$$

where $\varepsilon$ is the porosity, $C$ is the interstitial radon concentration, and $J$ is the bulk flux density. This convention is the same as used by Sextro et al. (1987) and, when the symbol $D_e$ is replaced by $k_e{}^*$, by Nero and Nazaroff (1984). With this convention, the diffusion length $l$, given by

$$l = \sqrt{\frac{D_e}{\lambda}} \tag{A5}$$

approximately represents the macroscopic straightline displacement traveled through the porous medium in a mean life. Another common convention uses a bulk diffusion coefficient, for example, given by the symbol $D$, related to the present coefficient by $D = \varepsilon D_e$ (Clements and Wilkening, 1974). Due to the complexities of porous media transport and the heuristic nature of Fick's law, neither convention is entirely satisfactory and without ambiguity, if an effort is made to interpret the physical significance at the molecular level. The tortuosity of the pore structure influences the microscopic path for diffusion, and the presence of sorption can make diffusion a combination of pore air, surface, and solution processes. Usually all that can be done is to choose a convention and try to make macroscopic measurements of the parameters consistent with that choice. For example, the porosity in equation A4 would ideally be determined from experiments involving surface flux density in combination with measurements of concentration gradients or diffusion lengths within a medium. The resulting value could be different from porosities determined by other methods. Realistically, such discrepancies are likely to be smaller than the usual sampling errors found in environmental measurements.

*Acknowledgments*

*Our research on thoron has received support from the U.S. Department of Energy and the U.S. Environmental Protection Agency. The Australian Nuclear Science and Technology Organisation hosted one of us (S. D. Schery) as a visiting scholar during much of the writing and editing process. Thanks to Pam Norton for timely typing of the manuscript.*

# REFERENCES

Bigu, J. (1981). "Mine models and the thoron problem in underground uranium mines." *Radiation Hazards in Mining*. Society of Mining Engineers, New York.

Burchfield, L. A., Akridge, J. D., and Kuroda, P. K. (1983). "Temporal distributions of radiostrontium isotopes and radon daughters in rainwater during a thunderstorm." *J. Geophys. Res.*, **88**, 8579–8584.

Clements, W. E., and Wilkening, M. H. (1974). "Atmospheric pressure effects on $^{222}$Rn transport across the earth-air interface." *J. Geophys. Res.*, **79**, 5025–5029.

Committee on Biological Effects of Ionizing Radiation (1990). *Health Effects of Exposure to Low Levels of Ionizing Radiation*. National Academy Press, Washington, DC.

Crozier, W. D. (1969). "Direct measurements of radon-220 (thoron) exhalation from the ground." *J. Geophys. Res.*, **74**, 4199–4205.

Dixon, D. W., James, A. C., Strong, J. C., and Wrixon, A. D. (1985). "A review of all sources of exposure to natural radiation in UK mines." In H. Stocker, ed., *Occupational Radiation Safety in Mining*. Canadian Nuclear Association, Toronto, pp. 241–247.

Dudney, C. S., Hawthorne, A. R., Wallace, R. G., and Reed, R. P. (1990). "Radon-222, $^{222}$Rn progeny, and $^{220}$Rn progeny levels in 70 houses." *Health Phys.* **58**, 297–311.

Duggan, M. J. (1973). "Some aspects of the hazard from airborne thoron and its daughter products." *Health Phys.* **24**, 301–310.

Durrance, E. M. (1986). *Radioactivity in Geology.* Halsted Press, New York.

Evans, R. D., and Goodman, C. (1940). "Determination of the thoron content of air and its bearing on lung cancer hazards in industry." *J. Ind. Hyg. Toxicol.* **22**, 89–99.

Folkerts, K. H., Keller, G., and Muth, H. (1984). "Experimental investigations on diffusion and exhalation of $^{222}$Rn and $^{220}$Rn from building materials." *Radiat. Prot. Dosim.* **7**, 41–44.

Fontan, J., Birot, A., and Blanc, D. (1964). "Variation de la concentration des gas radioactif naturels et de l'équilibre entre les gas et leurs descendants, dans l'air, au niveau de sol." *Geofis. Meteorol.* **13**, 80–87.

Grumm, D., Schery, S., and Whittlestone, S. (1990). "Two-filter continuous monitor for low levels of $^{220}$Rn and $^{222}$Rn." In *Proceedings of the 1990 International Symposium on Radon and Radon Reduction Technology*, EPA/600/9-90/0056. U.S. Environmental Protection Agency, Research Triangle Park, NC, Paper B-III-4.

Hopke, P. K. (1988). *The Initial Atmospheric Behavior of Radon Decay Products*, Report DOE/ER-0375. National Technical Information Service, Springfield, VA.

Hoppel, W. A., Anderson, R. V., and Willett, J. C. (1986). "Atmospheric electricity in the planetary boundary layer." *The Earth's Electrical Environment.* National Academy Press, Washington, DC.

International Commission on Radiological Protection (ICRP) (1981). *Limits for Inhalation of Radon Daughters by Workers*, Report 32. Pergamon Press, New York.

Israel, H. (1971). *Atmospheric Electricity*, Vol. 1. National Technical Information Service, Springfield, VA.

Israel, H. (1972). *Atmospheric Electricity*, Vol. 2. National Technical Information Service, Springfield, VA.

Israel, H., Horbert, M., and de La Riva, C. (1968). *Measurements of the Thoron Concentration of the Lower Atmosphere in Relation to the Exchange ("Austausch") in this Region*, Final Technical Report, Contract DAJA 37-67-C-0593. U.S. Army, APO New York.

Jacobi, W., and André, K. (1963). "The vertical distribution of radon 222, radon 220 and their decay products in the atmosphere." *J. Geophys. Res.* **68**, 3799–3813.

James, A. C. (1985). "Dosimetric assessment of risk from exposure to radioactivity in mine air." In H. Stocker, ed., *Occupational Radiation Safety in Mining.* Canadian Nuclear Association, Toronto, pp. 415–425.

James, A. C. (1988). "Lung dosimetry." In W. W. Nazaroff and A. V. Nero, eds., *Radon and Its Decay Products in Indoor Air.* Wiley, New York.

Keller, G., Folkerts, K. H., and Muth, H. (1982). "Method for the determination of $^{222}$Rn (radon)—and $^{220}$Rn (thoron)—exhalation rates using alpha-spectroscopy." *Radiat. Prot. Dosim.* **3**, 83–89.

Khan, A., Busigin, A., and Philips, C. R. (1982). "An optimized scheme for measurement of the concentrations of the decay products of radon and thoron." *Health Phys.* **42**, 809–826.

Knutson, E. O. (1989). *Personal Computer Programs for the Use in Radon/Thoron Progeny Measurements*, USDOE Report EML-517. Environmental Measurements Lab, New York.

Langmuir, D., and Herman, J. S. (1980). "The mobility of thorium in natural waters at low temperatures." *Geochim. Cosmochim. Acta* **44**, 1753–1766.

Liu, S. C., McAfee, J. R., and Cicerone, R. J. (1984). "Radon 222 and tropospheric vertical transport." *J. Geophys. Res.*, **89**, 7291–7297.

Luxin, W., Yongru, Z., Zufan, T., Weihu, H., Dequing, C., and Rongling, Y. (1988). "Cancer mortality study in high background radiation areas of Yangjiang, China." In *Epidemiological Investigations on the Health Effects of Ionizing Radiation*. Institut für Strahlenschutz, Germany, pp. 7–25.

Mark, T. D. (1969). "Eine Apparatur hoher Empfindlichkeit zur Messung der naturlichen Radioaktivitat der Atmosphäre." *Ber. Nat.-Med. Ver.* **57**, 71–93.

National Bureau of Standards (NBS) (1941). *Safe Handling of Radioactive Luminous Compound*, NBS Handbook H-27. U.S. Government Printing Office, Washington, DC, p. 3.

National Council on Radiation Protection and Measurements (NCRP) (1988). *Measurement of Radon and Radon Daughters in Air*, Report No. 97. NCRP, Bethesda, MD.

Nero, A. V., and Nazaroff, W. W. (1984). "Characterizing the source of radon indoors." *Radiat. Prot. Dosim.* **7**, 23–39.

Organization for Economic Co-operation and Development (OECD) (1965). *World Uranium and Thorium Resources*. European Nuclear Energy Agency, Paris.

Overstreet, W. (1967). *The Geologic Occurrence of Monazite*, Geol. Surv. Prof. Pap. 530. U.S. Government Printing Office, Washington, DC.

Porstendorfer, J., Reineking, A., and Becker, K. H. (1987). "Free fractions, attachment rates and plate-out rates of radon daughters in houses." In P. K. Hopke, ed., *Radon and Its Decay Products*, ACS Symp. Ser. 331. American Chemical Society, Washington, DC, pp. 285–300.

Rannou, A. (1987). *Contribution a l'étude du risque lie a la présence du radon 220 et du radon 222 dans l'atmosphere des habitations*, Rapp. CEA-R-5378. Commissariat a l'energie Atomic, Saclay, France.

Rutherford, E. (1901). "Einfluss der Temperatur auf die Emanationen radioactiver Substanzen." *Phys. Z.* **29**, 429–431.

Schery, S. D. (1985). "Measurements of airborne $^{212}$Pb and $^{220}$Rn at varied indoor locations within the United States." *Health Phys.* **49**, 1016–1067.

Schery, S. D. (1986). "Studies of thoron and thoron progeny: Implications for transport of airborne radioactivity from soil to indoor air. In *Indoor Radon*, SP-54. Air Pollution Control Association, Pittsburg, PA.

Schery, S. D. (1989). "Radon isotopes and their progeny in the indoor environment." In P. N. Cheremisinoff, ed., *Encyclopedia of Environmental Control Technology*, Vol. 2. Gulf Publishing, Houston, TX, pp. 897–920.

Schery, S. D. (1990). "Thoron in the environment." *J. Air Waste Manage. Assoc.* **40**, 493–497.

Schery, S. D., Li, Y., Grumm, D., and Turk, B. (1991). "Sources and mitigation of indoor thoron." In *Proceedings of the 84th Annual Meeting, Air and Waste Management Association*, 91-63.3, AWMA, Pittsburg, PA.

Schery, S. D., and Whittlestone, S. (1989). "Desorption of radon at the earth's surface. *J. Geophys. Res.*, **94**, 18297–18303.

Schery, S. D., Whittlestone, S., Hart, K. P., and Hill, S. E. (1989). "The flux of radon and thoron from Australian soils." *J. Geophys. Res.*, **94**, 8567–8576.

Serdyukova, A. S., and Kapitanov, Yu. T. (1969). *Radon Isotopes and Short-Lived Products of Their Disintegration in Nature*, Chapter 9. TT72-51014. National Technical Information Service, Springfield, VA.

Sextro, R. G., Moed, B. A., Nazaroff, W. W., Revzan, K. L., and Nero, A. V. (1987). "Investigations of soil as a source of indoor radon." In P. K. Hopke, ed., *Radon and Its Decay Products*, ACS Symp. Ser. 331. American Chemical Society, Washington, DC, pp. 10–29.

Shimo, M., Asano, Y., Hayashi, K., and Ikebe, L. (1985). "On some properties of $^{222}$Rn short-lived decay products in air." *Health Phys.* **48**, 75–86.

Staatz, M. H., Armbrustmacher, T. J., Olsen, J. C., Brownfield, I. K., Brock, M. R., Lemons, J. F., Coppa, L. V., and Clingan, B. V. (1979). *Principal Thorium Resources in the United States*, Geol. Surv. Circ. 805. U.S. Geological Survey, Arlington, VA.

Stein, L. (1987). "Chemical properties of radon." In P. K. Hopke, ed., *Radon and Its Decay Products*, ACS Symp. Ser. 331. American Chemical Society, Washington, DC, pp. 240–251.

Steinhausler, F. (1975). "Long-term measurements of $^{222}$Rn, $^{220}$Rn, $^{214}$Pb and $^{212}$Pb concentrations in the air of private and public buildings and their dependence on meteorological parameters." *Health Phys.* **29**, 705–713.

Stranden, E. (1985). "Thoron daughter to radon daughter ratios in mines." In H. Stocker, ed., *Occupational Radiation Safety in Mining*. Canadian Nuclear Association, Toronto, pp. 604–606.

Tanner, A. B. (1964). "Radon migration in the ground: A review." In J. Adams and W. Lowder, eds., *The Natural Radiation Environment*. University of Chicago Press, Chicago, IL, pp. 161–190.

United Nations Scientific Committee on the Effects of Atomic Radiation (UNSCEAR) (1982). *Ionizing Radiation: Sources and Biological Effects*, E.82.IX.8. United Nations, New York.

United Nations Scientific Committee on the Effects of Atomic Radiation (UNSCEAR) (1988). *Sources, Effects and Risks of Ionizing Radiation*, E.88.IX.7. United Nations, New York.

Wachsmann, F. (1989). "Are small doses really dangerous." *Health Phys. Soc. Newsl.* **17**(2), 1.

Wang Rui-Kai, Wang Ming-Chang, Li Hong-Quan, Wang Sheng-De, and Tian Meng-Xia (1985). "Activated carbon filter for measurement of radon and thoron." In H. Stocker, ed., *Occupational Radiation Safety in Mining*. Canadian Nuclear Association, Toronto, pp. 486–492.

Wang, Zuoyuan (1990). "Natural radiation in China: Level and distribution." *Jpn. Chem. Anal. Cent.* **18**, 44–57.

Wedepohl, K. H., ed. (1978). *Handbook of Geochemistry*, Vol. 11-5, Chapter 90. Springer-Verlag, Berlin.

Yang, F., and Tang, C. (1978). "A general formula for the measurement of concentrations of radon and thoron daughters in air." *Health Phys.* **34**, 501–503.

Zarcone, M. J., Schery, S. D., Wilkening, M. H., and McNamee, E. (1986). "A comparison of measurements of thoron, radon, and their daughters in a test house with model predictions." *Atmos. Environ.* **20**, 1273–2279.

# 11

# ADVANCED METHODS FOR THE EVALUATION OF ATMOSPHERIC POLLUTANTS RELEVANT TO PHOTOCHEMICAL SMOG AND DRY ACID DEPOSITION

*Paolo Ciccioli and Angelo Cecinato*

*Istituto sull'Inquinamento Atmosferico del CNR, Area della Ricerca di Roma, Monterotondo Scalo, Italy*

*Gaseous Pollutants: Characterization and Cycling,* Edited by Jerome O. Nriagu.
ISBN 0-471-54898-7   © 1992 John Wiley & Sons, Inc.

# 1. INTRODUCTION

A substantial portion of chemical compounds released into the low troposphere as a consequence of natural and man-made emission undergo a complex chain of physical and chemical transformations before they are deposited to the earth's surface (Schroeder and Lane, 1988). To describe their fate, it might be useful to imagine the troposphere in general, and particularly the atmospheric boundary layer (ABL), as a three-dimensional assembly of interconnected mixed-flow reactors, each one characterized by a well-defined chemical and physical equilibrium. Reagents are injected into the air parcel defining each reactor by direct emission. Mixing within each reactor and transfer to contiguous reactors mainly occurs by advection and diffusion caused by convective and turbulent motion. Mass transfer is also determined by wet and dry deposition to the ground.

Diffusion, impaction, and sedimentation are the basic mechanisms through which dry deposition occurs (Schroeder and Lane, 1988). If components are present in air as aerosols, their deposition velocity is a function of the size and shape of particles (McMahon and Denison, 1979). For gaseous components, physicochemical features such as molecular weight, polarity, solubility in water, and chemical reactivity (Schroeder and Lane, 1988) together with the chemical and physical nature of the recipient surface determine the deposition rates (Beilke and Gravenhorst, 1980; Hosker and Lindberg, 1982).

Wet deposition encompasses all processes by which chemicals are removed from the reactor in the aqueous form (i.e., rain, snow, and fog). Brownian capture, nucleation, and solution are the processes responsible for the incorporation of chemicals into droplets (Schroeder and Lane, 1988) and they can take place either within clouds or below them if precipitation occurs.

The size of the air parcel identifying each reactor is variable and depends on several meteorological factors. One of the most important is the height of the mixing layer, which can change from few tenths to thousands of meters above the ground during the day–night cycle (Smith and Hunt, 1978). The chemistry taking place in these imaginary reactors is also affected by the meteorology, being influenced by the temperature, pressure, solar radiation intensity, and relative humidity. The composition of the mixture resulting from these reactions is constantly changing, partly as a result of variable inputs, quality of emission, and differential deposition, and partly due to the different reactivity occurring in the reactor. If chemical and meteorological conditions promote high reactivity in the atmosphere, a substantial conversion of the total mass of reagents to yield various products is observed. Under steady-state conditions, the rate of production and that of removal by all processes determine the mean residence time of reagents and products in air and their aerometric concentrations (Beilke and Gravenhorst, 1980).

Since the rate of formation and that of removal by all processes strictly depend upon the physical state and chemical nature of individual constituents, residence times ranging from seconds (radicals) to several days and, in some instances, years (low reactive compounds) are observed in the atmosphere. Air mass circulation driven by local or global forces combined with diffusion provides dispersion of chemicals over different air parcels. Since the actual distance traveled depends on the amount of time a given species resides in the atmosphere (Schroeder and Lane, 1988), longer residence times correspond to longer distances traveled by the constituent before it is deposited on the ground. In spatial terms, distances of the order of thousands kilometers can be covered by chemical products characterized by a few days residence time (Hov et al., 1987).

Because new inputs of reagents and products can be injected into a moving air mass and because the emission and reactivity can occur in all the air parcels placed along its trajectory, quite large areas can be affected by dispersion and deposition of chemicals species. Under favorable conditions, compounds characterized by a few days residence time can be dispersed over an entire continent (synoptic-scale, long-range transport), whereas those having residence times of weeks or months can affect the whole troposphere and, by diffusing into the lower stratosphere, the entire atmosphere (global-scale, long-range transport) (Beilke and Gravenhorst, 1980; Hov et al., 1986). If direct emission and/or its chemical conversion gives rise to species potentially toxic to human beings or the environment, their dispersion and deposition changes the natural composition of air, soil, and water and can give rise to pollution.

It is widely recognized that a substantial part of the reactivity occurring in the ABL (and, hence, in our imaginary reactors) has to be attributed to photochemical reactions promoted and maintained by the UV light (Demerjan et al., 1974; Cox, 1979; Finlayson-Pitts and Pitts, 1986; Singh, 1987; Carlier and Mouvier, 1988). A complex chain of simultaneous and consecutive reactions, initiated when hydrocarbons ($HC_s$) and nitrogen oxides ($NO_x$) are released into the air and exposed to sunlight, converts these reagents (also called precursors) into secondary pollutants. Photochemical oxidants, whose most abundant and important member is ozone, represent the main but not the only family of secondary pollutants formed by photochemical reactions.

Oxidation of $NO_x$ and sulfur compounds (mainly $SO_2$) is also responsible for the production of acid components in air (Finlayson-Pitts and Pitts, 1977; Cox, 1979; Penkett et al., 1979; Atkinson and Lloyd, 1982; Cox and Penkett, 1983; Calvert and Stockwell, 1983; Calvert et al., 1985). These can be formed either by gas phase and heterogeneous reactions between gases and particles or by reactions taking place in solution after photochemical oxidants and primary pollutants are dissolved in or captured by water droplets. Formation of strong acids either within or below clouds results into high hydrogen ion levels in rain, snow, and fog. Because of the close coupling between gas and aqueous phase chemistry, photochemical pollution and acid deposition (wet and dry) can be regarded as different facets of the photochemical reactivity of the atmosphere (National Academy of Sciences (NAS), 1983) and have to be approached in a comprehensive way (Cox and Penkett, 1983; Calvert et al., 1985; Finlayson-Pitts and Pitts, 1986).

Photochemical pollution, often termed photochemical smog or "Los Angeles smog," from the area where it was first detected and investigated, was recognized to be a potential pollution problem in California as early as 1950 (Haagen-Smit, 1952; Haagen-Smit and Fox, 1956). Twenty years later, it was found to occur in many large cities in the United States (U.S. Department of Health, Education and Welfare (USDHEW), 1970), Europe (Guicherit and Van Dop, 1977), Japan, and Australia (Organization for Economic Co-operation and Development (OECD), 1975) because of the large amounts of $HC_s$ and $NO_x$ released into the atmosphere as a consequence of heavy traffic and industrial activities. High levels of ozone were reached during hot and sunny days when low circulation occurs in the ABL. In the mid-latitudes, such conditions are typical of the summer season and are associated with anticyclonic weather (Hov et al., 1986). Together with ozone, elevated levels of aldehydes, peroxyacyl nitrates ($PAN_s$), and fine particles were recorded (USDHEW, 1970; Altshuller, 1983b; Roberts, 1990).

Reduced visibility, due to absorption and scattering of light caused by fine particles, explains why the term *smog* was coined and, although misleading, is still widely used. However, the damage to plants was identified in the early 1950s as "gas-type injuries" and mainly attributed to high levels of ozone (Haagen-Smit, 1952). Since then, a large number of studies have been undertaken to assess all possible adverse effects caused by ozone. These studies have confirmed the observations made in the early 1950s and established that this pollutant can cause serious injuries to agricultural crops, natural plants, and forests, diseases to human beings, and weathering of materials.

An extensive review on the ozone effects can be found in the Air Quality Criteria for Ozone and Photochemical Oxidants issued by the EPA in 1986 (U.S. Environmental Protection Agency (USEPA), 1986). In the same volumes, phytotoxic properties of $PAN_s$, described by Taylor (1969), are reported together with the effects of all gaseous species involved in the photochemical cycle.

Concern about ozone effects arises from the fact that, in spite of control programs undertaken during the last 30 years, many of the major metropolitan areas in the United States and Europe still exceed national ambient air quality standards or national guidelines fixed by the various countries (Chock and Heuss, 1987; Grennfelt et al., 1988). In addition, levels of ozone in many rural and forest areas of the United States (USEPA, 1984) and Western Europe (Lubkert et al., 1984) are, by a wide margin, higher than the air quality limits fixed by the World Health Organization (WHO) (1986) for the protection of human health and vegetation.

Although part of tropospheric ozone found in sites located far away from densely populated areas can be attributed to injection from the stratosphere through tropopause folds (Johnson and Viezee, 1981; Viezee and Singh, 1982) and, to a minor extent, to reactions of naturally emitted compounds (Altshuller, 1983a), the largest contribution seems to come from long-range transport from heavily polluted areas (Cox et al., 1975; Hov et al., 1978; Schijoldager, 1984; Levy et al., 1985; Altshuller, 1986). Wolff et al. (1982) have reported elevated ozone levels in an area extending from the Midwest to the Gulf coast of the United States, whereas long-range transport of peroxyacetyl nitrate (PAN) and ozone were reported in northern Europe (Hov, 1984; Brice et al., 1984).

Long-range transport of ozone, photochemical oxidants, acidic species, and their precursors is possible because the residence time of these pollutants in the atmosphere (Hov et al., 1986) is comparable to the typical length of time necessary to reach a break-down situation in the ABL, which, in the mid-latitudes, ranges between 2 and 5 days (Smith and Hunt, 1978). When anticyclonic weather is established over parcels of air characterized by high emission of precursors, photochemical smog can rapidly develop and propagate over regions the size of some European countries or American states. It has been suggested that photochemical pollution can also be transported much farther away if intrusion into the low troposphere of $PAN_s$ accumulated in the upper troposphere takes place (Singh, 1987). Once injected in the ABL by seasonal mixing, $PAN_s$ can decompose to give $NO_2$, which, in turn, can react with natural $HC_s$ to produce ozone.

Acid deposition, which in the form of acid rain was reported more than 100 years ago in England (Cowling, 1982), was recognized as a long-range transport problem in Europe in the late 1970s when the OECD project (Ottar, 1976; OECD, 1977) showed that most countries in Europe received as much sulfur deposition from the surrounding countries as from their own sources. The impetus was derived from the work undertaken in the Scandinavian region (Odén, 1968; Odén, 1976; Bolin et al., 1972) where damage to the environment was associated with acid precipitation (Overrein, 1972; Overrein et al., 1980). Since that time, a co-operative program for the evaluation of the long-range transmission of air pollutants in Europe (called EMEP) has been in operation with the aim of providing governments with information of deposition and concentration of air pollutants as well as on the quantity and significance of pollutant fluxes across national boundaries. The potential danger that acid deposition might cause to ecosystems in some northern American regions (Likens et al., 1972; Likens, 1976; Schofield, 1976) was also recognized by the United States and Canada, who signed an agreement with the aim of mitigating transboundary fluxes and deposition of acid components.

Although acid rain contributes to acid deposition in a variable way depending upon latitude, geography, and season and does not always represent the main deposition process for acid components, it has been considered for some years the main acidification source for aquatic and forest ecosystems. Actually, some of the effects attributed to acid rain are common to dry acid deposition and in some instances it is not easy to distinguish these effects from those arising from primary and secondary gaseous pollutants. Adverse effects to aquatic life (Schofield, 1796; Overrein et al., 1980) can be definitely assigned to acid deposition (both wet and dry) only in watersheds where the geology is resistant to weathering, rainfall is heavy, and run-off water incapable of neutralizing acidity (Driscoll and Newton, 1985). For this reason, different critical loads of sulfate and nitrate can be established for watersheds having different sensitivity to acidification (Henriksen and Brakke, 1988). However, evidence must be collected that factors other than acid deposition do not play any important role in causing declines or extinction of fish before any definite conclusion is drawn (Lefohn and Brocksen, 1984).

To what extent forest decline in the United States, Canada, Germany, and Norway has to be attributed to acid deposition and particularly to acid rain is also controversial since the symptoms of many natural or man-made stresses are quite similar to those typical of atmospheric pollution (Lefohn and Brocksen, 1984; Woodman and Cowling, 1987). According to some authors (Gaffney et al., 1987) research in this field must widen and pollutants other than hydrogen ion, sulfate, and nitrate in precipitation should be considered to fully evaluate the potential detrimental effects on forest ecosystems. In particular, the dissolution of photochemical oxidants and oxidized organics derived by photochemical cycle into water droplets might result in the accumulation of phytotoxic compounds in rainwater, thus causing injuries to plants.

Although these aspects need further investigation, there is little doubt that acid deposition can be a source of stress for vegetation and it is conceivable that this might contribute to forest decline in combination with photochemical smog and dry deposition under certain conditions (McLaughlin, 1985).

The exceedingly complex mechanisms through which photochemical pollutants and acidic species can be formed, the different time scale, and the domain for their transport and deposition make an accurate apportionment of sources or source regions contributing to observed damage extremely difficult without the use of mathematical models capable of taking into account emission, reactivity, transport, and deposition processes of all chemical species involved in this environmental cycle. Models offer the advantage that source–receptor relationships can be understood, simulations under different scenarios can be performed, and the desired number of air parcels can be investigated. Through the use of models, long-term effects could be predicted and adequate control strategies adopted. This is a very important aspect to consider because the regional nature of photochemically originated pollution might require that different countries have to take concerted actions to prevent pollution.

Present limitations to the extensive use of mathematical models arise from the fact that a compromise must be reached between complexity and operationality to make models computationally efficient (Hov et al., 1987). In practical terms, this means that the more extensive the description of chemistry and physics, the shorter is the time period analyzed and the domain investigated. The number of variables to be considered is so high that several assumptions, which are only approximations of real processes, must be introduced in the mathematical description. The most advanced Eulerian and Lagrangian models available today for long-range transport can accurately describe only episodic events, and none of them can be used for all questions related to photochemical pollution and acid deposition (Hov et al., 1987). However, some specific aspects can be treated in great detail, and predictions based on models, although affected by uncertainty, are of great help in designing control strategies.

To reduce uncertainty, models can be validated by comparing predictions with observations. This approach can be used to decide whether a certain description used by a given model can be rejected, accepted, or parameterized. In any case, the mathematical treatment can be simplified and the model improved. Simplifications make possible long-term predictions and the investigation of larger domains without any increase in the computing time. However, validation of models is not a simple task since it requires a large number of observations, collected within the frame of a well-defined strategy (Derwent, 1990), including concentrations of the main pollutants measured over the area under investigation, emission rates of precursors released by all natural and man-made sources located in the same area, and deposition velocities of both precursors and secondary pollutants measured over different recipient surfaces. All these data must be provided by large, harmonized, monitoring networks giving careful consideration to spatial coverage, measurement accuracy, and siting (Derwent, 1990).

Collection of data is useful not only for verifying compliance of air quality standards and model validation, but also for checking whether and to what extent reactions observed in laboratory experiments really occur in the natural atmosphere. The possibility of providing reliable information on constituents whose composition is continuously changing as a consequence of variable inputs, reactivity, and transport is intrinsically difficult and requires dedicated instrumentation for sample collection and analysis. Generally speaking, sensitivity, precision and accuracy are the basic features critical for analytical methods used to measure pollutants relevant to photochemical smog and acid deposition. In many instances, these features must be combined with a short response time so that the evolution of pollutants in air or precipitation can be properly followed, whereas integration in space is often desired to reduce uncertainty derived by improper selection of the sampling site.

The aim of this chapter is to review the techniques that at various stage of development can be proposed as suitable candidates for the evaluation of species relevant to photochemical pollution and dry acid deposition. Their principle of operation is described, possibilities are analyzed, and examples of their application are presented.

## 2. REACTIONS CONTROLLING PHOTOCHEMICAL SMOG POLLUTION AND DRY ACID DEPOSITION IN THE LOWER TROPOSPHERE

A detailed presentation of the photochemistry occurring in the ABL is beyond the scope of this chapter. However, a short description of the main reactions controlling the formation of photochemical oxidants and acidic species relevant to dry deposition gives an idea of the number of chemicals involved in this complex environmental cycle and the problems connected with their determination.

Although the presence of light is necessary to start the photochemical cycle, its absence does not interrupt the reactivity of the atmospheric species. Many of the reactions occurring during the day cease at night, but new reactions occur. Some nighttime reactions act as a sink for photochemical oxidants and acidic species; others give rise to compounds that can restart the photochemical cycle at sunrise. In this case, products behave as temporary nighttime storage for reactive species.

Due to the substantial differences existing between day and night cycles it is useful to represent them with separate figures. Figure 1 is a schematic diagram of the main daytime reactions. It is not very detailed but gives an overview of how the cycle works. A detailed description of the mechanisms and rates of various reactions with adequate references can be found in the excellent book of Finlayson-Pitts and Pitts (1986) and in a recent review article by Carlier and Mouvier (1988). Different symbols are used to distinguish species derived from anthropogenic or natural emission (circles) from those formed during the photochemical reactions (squares and polygons). Squares with rounded sides indicate compounds that can be emitted and formed at the same time.

Further distinction is made between radicals and short-living compounds from species that, to some extent, behave as chain terminators and undergo accumulation in the atmosphere. Marked lines with arrows indicate reactions between compounds enclosed in boxes. Lines emerging from the boxes lead to reaction products. A dotted line is used for products that are pumped back into the cycle. Reagents that are part of the natural atmosphere ($O_2$, $H_2O$, and light) are indicated on the lines.

In many cases, lumped mechanisms have been used to draw Fig. 1 and no distinction is made between homogeneous and heterogeneous reactions. Moreover, some reactions that can occur during the day as well at night are

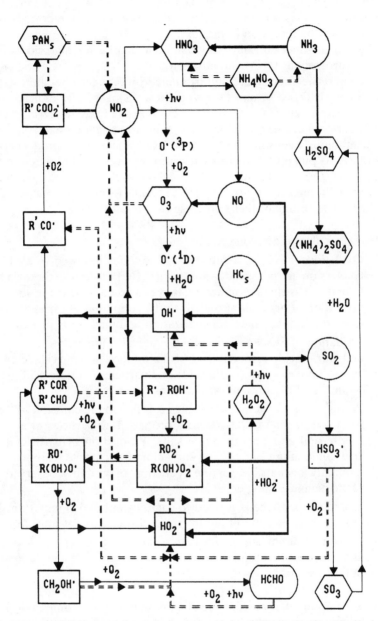

**Figure 1.** Schematic diagram of the main daytime reactions relevant to photochemical smog pollution and acid deposition. For explanation of the symbols see text.

reported in only one figure. For the sake of clarity, species derived from nighttime reactions entering into daytime cycle are not reported, because their role will be discussed in the text.

The working mechanism of daytime cycle can be summarized as follows: When hydrocarbons (HC$_s$) and nitrogen oxides (NO$_x$; i.e., NO and NO$_2$) are injected into the atmosphere and irradiated by sunlight, the photostationary state linking NO, NO$_2$, and O$_3$ is somehow changed. OH radicals, derived from ozone photolysis followed by reaction of the excited oxygen atom with atmospheric water, react with organics giving rise to additional radical species that continuously consume NO to form NO$_2$. Some of the products of these NO reactions pump OH radicals back into the cycle so that they can further react with NO. The real motor of the whole cycle is thus the OH radicals. They are continuously restored by HO$_2$ radicals formed by organic oxidation products. The results of these processes are the following:

1. Accumulation of ozone.
2. Oxidation of HC$_s$ to aldehydes and ketones followed by their partial conversion to peroxyacyl nitrates (PAN$_s$). Because these last compounds act as chain terminators, they can accumulate in the atmosphere. Oxidation of organics also leads to the formation of carbonaceous particles through complex polymerization and nucleation processes.
3. Oxidation of SO$_2$ and NO$_2$ by OH radicals ultimately leads to sulfuric and nitric acid, respectively. Acidic species are thus injected in the atmosphere as gases (nitric acid) and particles (sulfuric acid) as a consequence of photochemical reactions.
4. Partial neutralization of inorganic acids due to the emission of ammonia from soil or man-made activities produce fine particles of ammonium nitrate and ammonium sulfate salts. Production of particles (sulfuric acid, ammonium sulfate salts and ammonium nitrate, carbon particles) is mainly responsible for the reduction of visibility observed during photochemical smog episodes.
5. Hydrogen peroxide is formed to some extent and is important in promoting oxidation of SO$_2$ and various nitrogen oxides into droplets.
6. Dynamic equilibrium between emission, reaction, deposition processes, and transport determines aerometric concentrations and their variation during the day.

It is worth noting that reactions leading to ammonium nitrate and PAN$_s$ are equilibrium reactions whose constants highly depend on the temperature of the atmosphere. If the air parcel warms up, ammonium nitrate particles release nitric acid and ammonia into the atmosphere. In similar conditions, PAN$_s$ release NO$_2$ and peroxyacylradicals and thus act as a carrier for NO$_2$ in remote locations where levels of NO$_x$ are low (Singh, 1987).

Since several parameters control the reactions shown in Fig. 1, it is not easy to predict daily profiles of photochemical oxidants and acidic species in air. Different trends can be observed, depending upon the type of site (urban, suburban, rural, forest, remote) as well as the geography and meteorology. However, at short distances from the emission sources (suburban and semi-rural sites placed between 20 and 100 km downwind of large cities) the trend is usually characterized by maximum values of ozone, $PAN_s$, nitric acid, and sulfate particles centered between the middle of the day and late afternoon (5 p.m.) (Tuazon et al., 1981; Cox and Penkett, 1983). These maxima are concurrent with the consumption of $NO_x$ (Spicer, 1977).

The maximum levels of oxidants and acid compounds reached in air is a function of the ratio between $HC_s$ and $NO_x$. However, the quality of organic emission has a great influence in the formation of secondary pollutants. For instance, some organic compounds (such as many alkenes) have a greater potential for producing ozone and $PAN_s$ than other hydrocarbons that pump back more OH radicals into the cycle through formaldehyde formation (Atkinson, 1986; Derwent and Jenkin, 1990a,b).

A schematic view of possible nighttime reactions is shown in Fig. 2. The pathways are not so well established as those for daytime reactions because some of the products observed in laboratory experiments have not been detected yet in the atmosphere (Finlayson-Pitts and Pitts, 1986). It is believed that ozone is mainly consumed by reaction with NO to give $NO_2$. It is also possible that ozone reacts with alkenes to give rise to Griegee's intermediates and carbonyl compounds (Atkinson and Carter, 1984). Due to the different reactivity of such intermediates toward primary and secondary pollutants, it is hard to predict what the final products will be. It has been suggested that they can contribute to the formation of organic acids in air. Because the addition of ozone to the double bond is rather slow, such reactions can be observed only under favorable conditions. They might also occur to some extent during daytime. Reactions of both NO and $NO_2$ with peroxyalkyl and alkoxyl radicals formed during the day rapidly remove these species from the atmosphere (Atkinson et al., 1984).

Similar reactions occur with OH and $HO_2$ radicals. While alkyl nitrates can be quite stable during daytime and some of them adsorbed on particles, alkyl nitrites, nitrous acid, and peroxynitric acid are very quickly photolyzed and therefore can be considered an additional source for $NO_x$ and radicals when the sun rises (Finlayson-Pitts and Pitts, 1986). In this way they enter into the cycle described in Fig. 1. However, production of alkylnitrites, alkylnitrates, nitrous acid, nitric acid, and peroxynitric acid through the chain of reactions mentioned above is, in many instances, quite small, since it depends on the decomposition process of $PAN_s$ and the availability of NO.

More important pathways for producing alkylnitrates (especially alkyldinitrates), nitrous acid, and nitric acid appear to be linked to the reactions of $NO_2$ with ozone and water (Wayne et al., 1991). Ozone oxidizes $NO_2$ to form $NO_3$ radicals, which, in turn, can react either with organic compounds or again with $NO_2$. The former process leads to alkylnitrates (Japar and Niki,

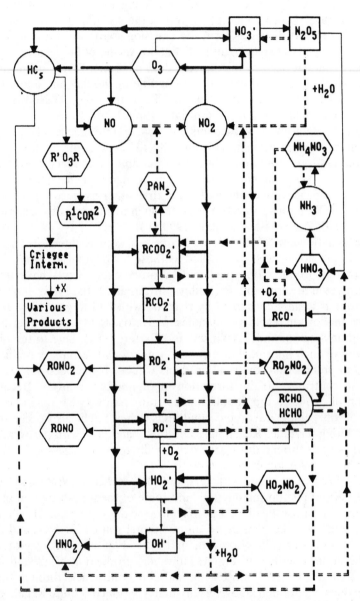

**Figure 2.** Schematic diagram of the nighttime reactions relevant to photochemical smog pollution and dry acid deposition.

**Table 1 Typical Range of Mixing Ratios of Chemical Species Relevant to Photochemical Smog Formation and Dry Acid Deposition Found in Areas Subjected to Direct Emission or Long-Range Transport**

| Species | Mixing Ratio |
|---|---|
| Gas phase[a] | |
| $SO_2$ | 1–200 ppb |
| NO | 0.05–1000 ppb |
| $NO_2$ | 1–500 ppb |
| $NO_3$ | 5–400 ppt |
| $HNO_2$ | 30 ppt–8 ppb |
| $HNO_3$ | 0.1–50 ppb |
| $NH_3$ | 1–20 ppb |
| PAN | 0.5–20 ppb |
| $O_3$ | 20–200 ppb |
| HCHO | 1–20 ppb |
| OH | 0.01–0.4 ppt |
| Particles[b] | |
| Sulfates | 0.2–40 ppb |
| Nitrates | 0.1–10 ppb |

[a]Selected values from Finlayson-Pitts and Pitts (1986, pp. 368–369)
[b]Allegrini and De Santis (1989).

1975; Atkinson et al., 1985), whereas the latter can produce $N_2O_5$, which, in turn, gives nitric acid by reaction with water (Tuazon et al., 1983). Water also reacts with $NO_2$ to produce nitrous and nitric acid (Heikes and Thompson, 1983; Pitts et al., 1984). This last reaction is believed to occur mainly on surfaces that release nitrous acid but retain nitric acid (Finlayson-Pitts and Pitts, 1986).

According to many authors (Wayne et al., 1991), $NO_3$ radicals play a role during the night similar to that of OH radicals during daytime. By taking into account both chemical and physical processes, the final result leads to a decrease of ozone, PAN, $HNO_3$, and NO combined with an increase of $NO_3$, nitrous acid, alkyl nitrates, and $NO_2$. Transition from a day to a night cycle is reflected in the diurnal trends followed by the various species during photochemical smog events. The range of concentrations of precursors, photochemical pollutants, and acidic species that can be found in air is reported in Table 1.

## 3. MONITORING APPROACHES

Although there is quite good agreement among the ways that compounds react, great uncertainties still remain in the precise assessment of the role that each reaction plays within the cycle under different environmental conditions. There are difficulties in defining the contribution of certain heterogeneous reactions and in evaluating deposition processes. The reactivity of $HC_s$ and their degradation products toward $NO_3$ and ozone, the final fate of Grieege's intermediates, and production of nitrous and nitric acids on surfaces are some of the issues that need further investigation. Moreover, deposition velocities for many gaseous compounds over different surfaces are not known or entail great uncertainty. Direct information on the working mechanism of the photochemical cycle could be gained by following the trends and detecting levels of as many species as possible participating in day/night reactions in different geographical regions under different meteorological conditions.

Since a large monitoring network equipped with the full set of instruments available for detecting pollutants is not possible because of the high costs and lack of trained people, two different monitoring approaches are presently followed. They represent the best compromise between what is desirable and what is really possible. One approach is aimed at looking at photochemical and acid deposition processes over a large scale in space and time. The number of stations required is very high, so the equipment is limited. Because it is not always possible to have trained personnel running these stations, simple apparatuses are preferred. Quality assurance is fundamental to provide meaningful data. Time resolutions from 1 to 12 h, depending upon the type of pollutant monitored, are accepted; automation of sampling and analysis is highly desirable but is not mandatory. Reliability with time failures of the order of few months is as important as high performances. If unattended instrumentation is used, automatic tuning and calibration is often requested. This type of monitoring is useful for identifying areas exposed to photochemical pollution and acid deposition, correlating their impacts with observed damage, verifying the usefulness of control strategies, studying long-range transport of pollutants, and validating models.

The other way to monitor secondary pollutants is to concentrate efforts on selected sites where it is expected that some of the chain reactions described in Figs. 1 and 2 can be better detected and studied. Selection of the type of site (urban, suburban, rural, forest, or remote) as well as that of season and latitude is based on the specific aspects to be investigated (emission, reactivity, transport, and deposition). More specific issues, such as the influence of man-made or natural emission on the day/night cycle of pollutants or the occurrence of specific mechanisms and pathways, can also be a matter of investigation. Duration of these monitoring campaigns can range from a few days to some weeks. Proper instrumentation is highly desirable to achieve better precision and accuracy together with a different resolution in time and space. Daily trends are recorded with a time resolution ranging from few seconds to few hours, whereas an integration in space from a single point to few kilometers can be reached. Efforts are made to detect radical species to gain better insight on day/night chemistry.

From an instrumental viewpoint, these campaigns represent a good opportunity to test new concepts in analytical chemistry based on environmental measurements. If different techniques are used to measure the same pollutant, it is possible to evaluate advantages and drawbacks associated with their adoption. Artifacts and interferences can be identified and possible ways to prevent them studied. The suitability of certain techniques for field measurements and their reliability can be also investigated.

## 4.  MONITORING TECHNIQUES

It is impossible to cover in few pages all the developments in monitoring techniques achieved in the last few years. Discussion, therefore, will be limited to those advanced methods that reasonably can be applied to detection of primary and secondary pollutants within realistic monitoring programs. Among sophisticated techniques, only those that are believed to provide unique information will be mentioned.

Techniques have been divided into three main groups according to the physical or chemical principle:

1. Spectroscopic techniques
2. Diffusion techniques based on denuders
3. Chromatographic techniques

Wet chemical methods are not discussed here because most of these should be well known to the reader since they have been used for many years. Adequate description of these methods can be found in the Air Quality Criteria issued by the USEPA in 1986.

### 4.1.  Spectroscopic Techniques

This section deals with techniques involving detection and quantitative evaluation of a given compound in air through the measure of radiation either emitted, reflected, or absorbed. Measurements of photochemical pollutants and species relevant to dry deposition in tropospheric air have been successfully performed with the following techniques:

- Differential optical absorption spectroscopy (DOAS) carried out in the UV/visible regions of the spectrum
- Fourier transform infrared spectroscopy (FTIR)
- Tunable diode laser absorption spectroscopy (TDLAS) operating in the mid-infrared
- Chemiluminescence
- Fluorescence

Flame photometry will be discussed in the next section in conjunction with diffusion and chromatographic techniques, whereas UV photometry, which has been in use for several years for the monitoring of ozone, does not need extensive description. The basic principle is UV absorption at the 254-nm mercury emission line using double-beam measurements (Bowman and Horak, 1972). The method is so specific that long-path UV photometers with improved electronics for accurate double-beam measurements of absorbancies are used as reference methods for calibrating ozone analyzers. UV photometry is one of the reference methods recommended by the USEPA for the monitoring of ozone (USEPA, 1986). It provides enough sensitivity to detect background levels of this pollutant in the troposphere.

### 4.1.1.  Differential Optical Absorption Spectroscopy (DOAS)

DOAS was developed and first applied to atmospheric studies in 1979 when it was used for the determination of $O_3$, HCHO, and $NO_2$ in air (Platt et al., 1979). Since that time, its application field has been widened to include several other chemicals participating in the day/night cycles shown in Figs. 1 and 2. In particular, the capability of monitoring OH, $NO_3$ radicals, and $HNO_2$ afforded by DOAS has made an outstanding contribution to the knowledge of chemical processes occurring in the atmosphere (Finlayson-Pitts and Pitts, 1986; Wayne et al., 1991).

DOAS is based on the measure of the intensity of absorption bands characteristic of a given pollutant recorded in the UV/visible regions. Through this measure, the column density of the compound along the light path is derived by knowing its differential absorption cross section. The light source can be high-pressure Xe lamps, incandescent quartz–iodine lamps, or laser systems. The light source is placed from a few hundred up to thousands of meters from the receiving point, depending upon the type of pollutant to be measured and its concentration in air. The light beam crossing the air parcel is received on a Newton-type telescope, which focuses the light into the entrance slit of a spectrograph (Platt and Perner, 1980). The grating disperses the light so that different wavelengths are projected across the exit slit. A small segment of the dispersed spectrum is rapidly scanned by a slotted disk scanning device consisting of a series of slits etched radially into a thin metal disk rotating in the focal plane. The slotted disk uses only one slit as an exit slit at any time. The light passing through the exit slit is amplified and signals are recorded, sent to a computer via an A/D converter, and stored in separate channels by using gated electronics. By accumulating a sufficient number of scans on several hundreds of channels, each one corresponding to intervals of about 0.2 nm, the absorption spectrum on a region of 20–40 nm is reconstructed.

The contribution of each pollutant to the spectrum is obtained by sequential subtraction. This is possible because small bandwidth features of the spectrum can be assigned to a given pollutant based on reference spectra previously recorded in the laboratory. Detection limits depend on the cross section of the compound to be detected, the light path used, and the number of scans accumulated.

The main advantages of DOAS with respect to other spectroscopic techniques are absence of wall losses, its specificity, the potential of real-time measurements, and simultaneous determination of several species during the same experiment. Integration in space and short-time resolution render DOAS extremely reliable for following daily trends. Potential handicaps include the dependence of signal from visibility conditions, particularly for measurements with very long path and wavelengths below 300 nm.

The only determination that is rather difficult to carry out in the field and poses several constraints on the measuring site is that of OH radicals (Hofzumahaus et al., 1990), because a two-laser system is necessary for generating UV light pulsed at 308 nm. To achieve definite identification of OH radicals, a resolution of 0.003 nm is required because two lines, shifted by 0.0055 nm, must be simultaneously monitored (Hubler et al., 1982). A path-length ranging from 3 to about 10 km is necessary to reach the sensitivity needed for aerometric determinations. Detection limits of 0.03 ppt can be achieved in this particular application of DOAS (Hubler et al., 1982, 1984). Although the uncertainty in the measurement of OH radicals carried out by DOAS is still quite high, distinct diurnal profiles have been reported in both polluted and rural sites (Hubler et al., 1984; Perner et al., 1987). The average levels as well as the maximum concentration occurring in the middle of the day were in agreement with model predictions, providing an important experimental support to theoretical studies (Perner et al., 1987).

Field determinations of pollutants other than OH radicals are much easier to perform with DOAS because high-pressure or incandescent lamps can be used as light sources. In this case, a very simple setup comprises a mirror, a lamp, and a power supply and can be easily installed in urban, rural, and remote areas.

The best example of the unique capability of DOAS for detecting pollutants that are impossible to evaluate by other techniques is the measurement of $NO_3$ radicals at ppt levels in air (Platt et al., 1980). By measuring the optical density of the absorption bands occurring at 623 and 662 nm, nighttime profiles of $NO_3$ radicals were determined in both polluted and unpolluted atmospheres (Platt et al., 1981; Perner et al., 1990a). Trends observed were characterized by a sharp maximum centered a few hours after sunset (Platt et al., 1980) or just before sunrise (Perner et al., 1990a). In some instances, formation of $NO_3$ radicals was concurrent with the decrease of ozone and $NO_2$ concentrations (Platt et al., 1980). The rapid removal of $NO_3$ radicals after 8 p.m. observed in polluted areas was attributed to heterogeneous reactions of both $NO_3$ and its reaction product, $N_2O_5$, with water droplets on surfaces to yield nitric acid and other compounds. This conclusion was reached because the decay of $NO_3$ radicals was found to be a function of the relative humidity and the content of suspended particles in air (Platt et al., 1980; Perner et al., 1990a).

Until diffusion techniques based on denuders were developed, DOAS and, with some limitations, FTIR were the only techniques capable of measuring $HNO_2$ in the atmosphere. As early as 1979, nighttime profiles of $HNO_2$ were determined in Julich by looking at two absorption bands occurring between 330 and 380 nm (Perner and Platt, 1979). If the most intense band is used (354 nm), detection limits of 0.1 ppb can be achieved over a light path of 10 km. The contribution that heterogeneous and homogeneous reactions can give to the nighttime formation of $HNO_2$ and the role that this pollutant can play in the production of OH radicals at sunrise are some of the issues that the development of DOAS has made possible to investigate (Finlayson-Pitts and Pitts, 1986).

The list of other species that can be evaluated by DOAS includes $SO_2$, $O_3$, HCHO, NO, and $NO_2$ (Platt and Perner, 1980). For many of them, the sensitivity of DOAS is sufficient to allow accurate determinations in slightly polluted areas and, in some instances, unpolluted areas (Finlayson-Pitts and Pitts, 1986). Recently, $NH_3$ has also been added to this list (Neftel et al., 1990; Perner et al., 1990b). Similarly to what happens with NO, the absorption bands of $NH_3$ occur in the far UV region (200–220 nm) where scattering of air molecules and absorption of ozone and oxygen is observed so that the usable light path is limited to 300 m with detection limits of about 3 ppb.

The capability of DOAS to provide detailed daily profiles of $O_3$, $SO_2$, $NO_2$, $HNO_2$, HCHO, NO, and $NH_3$ is well illustrated by the results shown in Figs. 3 and 4. All determinations were carried out by Perner et al. (1989, 1990b) during two intercomparison monitoring campaigns held in our Institute in September 1988 and May 1990, respectively. The first monitoring campaign was devoted to the evaluation of methods for the measure of $HNO_3$ and nitrate, the second was focused on the techniques used for $NH_3$.

Parallel determinations of some species were carried out and a direct comparison between different analytical approaches was possible. In Fig. 3a–d are reported daily profiles of $O_3$, $NO_2$, $SO_2$, and $HNO_2$ obtained by DOAS, UV photometry, chemiluminescence with luminol, and annular denuder methods (ADM), respectively (Ciccioli et al., 1990; Allegrini et al., 1990; De Santis, 1989). The principle of operation and performance of some of the techniques used in parallel with DOAS is discussed in the next sections. A fair agreement is found for $O_3$ and $HNO_2$, whereas discrepancies exist among the levels of $NO_2$ and $SO_2$ measured by different techniques. Larger deviations are observed at night. A detailed analysis of the various profiles and monitoring conditions suggested that these discrepancies were unlikely to arise from erroneous response of the instruments used. They can be better explained by the different types of sampling carried out during the experiments (Ciccioli et al., 1990). While DOAS was measuring pollutants over the Tiber valley with the beam partly crossing the emission sources, the other instruments (all single-point air monitors) were detecting only the portion reaching the edge

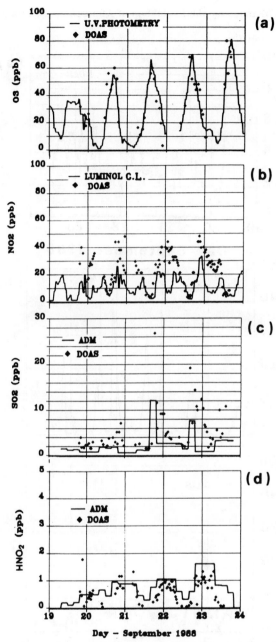

**Figure 3.** Daily profiles of some primary and secondary pollutants detected through the simultaneous use of optical and diffusional techniques. Data refer to the Intercomparison Monitoring Exercise on Nitric Acid and Nitrate Measurements held in a rural area near the city of Rome (Montelibretti, Italy) from 19–24 Sep 1988. Data with Differential Optical Absorption Spectriscopy (DOAS) were obtained by Perner et al. (1989). Data with the luminol-chemiluminescence analyzer from Unisearch (Canada) (NO2) and UV photometry from Dasibi (USA) (ozone) were obtained by Ciccioli et al. (1990). Values of $SO_2$ and $HNO_2$ by the annular denuder method (ADM) were collected by De Santis (1989).

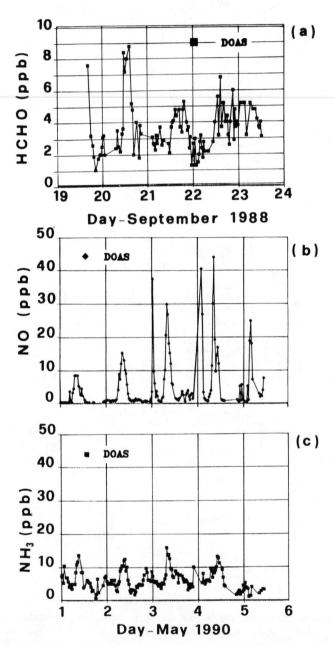

**Figure 4.** Diurnal profiles of HCHO, NO, and NH₃ recorded by DOAS during the Intercomparison Monitoring Exercise on Nitric Acid and Nitrate Measurements (19–24 Sep 1988) and the Intercomparison Monitoring Exercise on Ammonia (1–6 Sep 1990) held in Montelibretti, Italy. (From Perner et al., 1989, 1990b.)

of the valley where the monitoring site was located. Due to the corrugated shape of the ground and low mixing occurring at night, only part of the emission occurring within the valley was allowed to reach the sampling site during night hours. This explanation of observed differences is supported by the fact that during daytime, when the air masses were well mixed, closer values of $NO_2$ and $SO_2$ were recorded. This suggests that comparisons between techniques having different resolution in space and time should be made with care since large differences in response might arise from meteorological and geographical factors that are sometimes difficult to account for.

Quite interesting are daily profiles obtained for $HNO_2$, which establish the behavior of this pollutant in air during the development of a photochemical smog episode. Trends are consistent with its formation at night followed by decomposition induced by photolysis after sunrise. Results in Fig. 3d represent the first example of simultaneous detection of $HNO_2$ by DOAS and a diffusion technique based on denuders. Daily profiles of $NH_3$ shown in Fig. 4c seem to be consistent with both emission from soil and formation from thermal decomposition of ammonium nitrate. However, very little can be said about the accuracy of DOAS determinations because the evaluation of the results obtained during this intercomparison exercise is still in progress. The capability of DOAS for providing detailed information of the levels of HCHO is discussed in Sections 4.1.2, 4.1.3, and 4.1.5.

In spite of the great advantages offered by this optical technique, its use until now has been quite limited, mainly because no instruments of this type were commercially available. Today, DOAS systems are built in Europe and measurements of $NH_3$ carried out with commercial instruments in a rural site in Switzerland have already been reported (Neftel et al., 1990). Daily profiles of $HNO_2$, $SO_2$, and HCHO recorded in Milan, Florence, and Rome together with measurements of benzene, toluene, and xylene have been carried with the same instrument by the manufacturing company (Rancon s.r.l., 1990). Since during these campaigns no attempts were made to compare data collected by DOAS with those obtained by other methods, we cannot say very much about the reliability of these determinations. However, values of the mixing ratios and daily profiles seem to be in agreement with expectations based on the emission rates and reactivity of these pollutants in air. A novelty is represented by the monitoring of volatile aromatic compounds carried out in urban areas. They are detected at ppb levels using absorption bands in the UV region ranging between 230 and 270 nm.

If commercially available instruments could provide the same performance as research instruments and low maintenance could be combined with simplified procedures for the measurement of differential optical cross sections for reactive gases (such as $HNO_2$), DOAS would have a good chance of being widely adopted in large monitoring networks, because it can run unattended for long periods. The possibility of performing integrated determinations over quite long distances can be particularly helpful in those areas where, due to variable emission or geographical and meteorological factors, a large number of stations would be required to get values truly representative of the levels of pollution reached.

### 4.1.2. Fourier Transform Infrared Spectroscopy (FTIR)

The main advantages associated with the use of long-path FTIR in atmospheric investigations were outlined by Pitts et al. in 1977. Important atmospheric studies have been carried out with this technique in the Los Angeles area (Tuazon et al., 1981; Hanst et al., 1982; Biermann et al., 1987). Determinations of species playing a decisive role in photochemical pollution and acid generation in air have been determined. The list includes CO, NO, $NO_2$, $HNO_3$, $HNO_2$, PAN, $NH_3$, formaldehyde, formic acid, and nonmethane paraffinic hydrocarbons. The detection of organic nitrates in air (Hanst et al., 1982) is especially interesting. The simultaneous detection of $HNO_3$, PAN, $O_3$, $NH_3$, HCHO, and formic acid during a photochemical smog episode (Tuazon et al., 1981) shows clearly the close connection between photochemical smog formation and nitric acid generation in air. The information gained by making use of this technique has provided experimental evidence of the intimate connection between photochemical smog formation and the increased acidity of the atmosphere.

The analysis of trends obtained by long-path FTIR has stimulated research in several fields, including that focused on the development of new instrumentation (e.g., TDLAS) for the monitoring of acid compounds in air. The instrument is composed of a Michelson interferometer, an infrared light source, a multireflection White cell of about 25 m base path, two detectors, and a computer system for deconvolution of sinusoidal signals by Fourier transform. When the infrared light source crosses the interferometer the detector receives a signal (called interferogram) that is the result of constructive and destructive interferences produced by all wavelengths. When air flows through the chamber, compounds absorb the light and the interferogram is altered. Deconvolution of the interferogram by Fourier transform and its comparison with the empty cell permits the generation of a classical IR spectrum by computer.

The advantages of long-path FTIR over scanning systems can be summarized as follows (Finlayson Pitts and Pitts, 1986): 1. Sampling of different wavelengths is achieved at once. 2. All the source energy is used for generating the spectrum. 3. Resolution is constant over the entire range. Resolution of the order of $\frac{1}{16}$ cm$^{-1}$ is commonly achieved by commercial instruments. A unique feature, which is very useful in atmospheric studies, is that all compounds present in air are detected at once and recorded on the same spectrum.

The main limitation is that sensitivity for most noncriteria pollutants is not sufficient for ambient monitoring in moderately polluted or unpolluted areas (Finlayson-Pitts and Pitts, 1986). Another important limitation is that the instrumentation is complex and expensive, and cannot be easily transported. Although the volume to surface ratio is much higher than that of other optical instruments, wall effects might occur with "sticky" gases ($HNO_3$, $NH_3$).

Because of sensitivity limitations the use of long-path FTIR is restricted to urban or polluted areas where levels of pollutants are usually higher than detection limits. Nevertheless, long-path FTIR can be useful in intercomparison programs aimed at evaluating accuracy and precision of other techniques. A typical example of this application is shown in Fig. 5*a* where HCHO

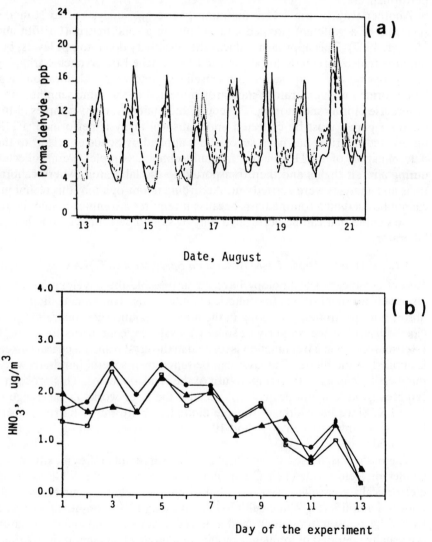

**Figure 5.** (*a*) Daily profiles of HCHO obtained by DOAS (– – –), FTIR (···), and TDLAS (---) during an Intercomparison held 13–21 Aug 1986 in the Los Angeles Area. (From Schiff and Mackay, 1989.) (*b*) Daily trends of $HNO_3$ recorded with TDLAS (▲), annular denuder method (ADM) (□) and filterpack method (FPM) (•) during a 13-day intercomparison held in North Carolina 29 Sep–12 Oct 1986. (From Schiff and Mackay, 1989.)

determinations carried out in the Los Angeles area by FTIR, TDLAS, and DOAS are reported (Schiff and Mackay, 1989). Trends as well as absolute values recorded by these three optical methods appear to be consistent in spite of the fact that the first two methods are not averaged over a long path. The fact that a given technique gives results in agreement with both DOAS and FTIR can be certainly regarded as a serious test for evaluating its performances.

Another way to use FTIR for atmospheric gas analysis is to scan the spectrum of a pollutant trapped and isolated in a solid matrix (Griffith and Shuster, 1987). This approach reduces the sensitivity down to ppt levels, but requires that compounds to be detected be enriched by cryogenic trapping on a glassy matrix composed of condensed $CO_2$ and water. Calibration must be performed with laboratory standards subjected to the same sampling procedure used for air monitoring. To reduce uncertainties, many samples of the same air parcel must be collected and statistical analysis performed. FTIR on isolated matrices has been successfully applied by Schuster (1990) to the determination of HCHO in the free atmosphere. Samples were collected during aircraft flights and then transported to the laboratory where calibrations and analyses were carried out. According to our opinion, this technique cannot be used on a routine basis because it requires a complex manipulation of the sample that can be performed only by skilled personnel in a dedicated laboratory.

### 4.1.3. Tunable Diode Laser Absorption Spectroscopy (TDLAS)

Recent advances in technology have made possible the development of instruments for monitoring atmospheric pollutants based on tunable diode laser absorption spectroscopy working in the mid-IR region. Their operation principle is based on the property of some lead-salt semiconductors to act as IR laser sources when a p-n junction is formed in the crystal and electrical current is applied to the diode. The laser can be tuned over several hundreds wavenumbers by changing the temperature of the semiconductor. This procedure requires good temperature control because the laser works in the range of 10–80 K. Since linewidths of tunable diode lasers are rather narrow, absorbancies ranging between $10^{-3}$ and $10^{-5}$ due to a single rotation line can be measured.

The basic design used for atmospheric measurements relies mostly on the studies of Reid et al. (1978). It is composed of a tunable diode laser, an electronically controlled cryocooler fed with liquid helium, a current control, a lock-in amplifier, a White cell (300 m path length) where the air sample is sucked by an aspirating pump, and a detection system. Tunable diode lasers are usually operated in frequency modulation mode at an amplitude 2.2 times larger than the peak-to-peak modulation amplitude. The detector output is also analyzed at 2 times the modulation frequency. By scanning the modulated radiation over an absorption line, a derivative-like signal is recorded.

The working principle, calibration procedures, and instrumental setup for tropospheric air monitoring were reviewed in 1983 (Schiff et al., 1983) and the first application to the determination of NO, $NO_2$, and $HNO_3$ can be found in the same article. Further improvements of the original design (Shiff and Mackay, 1989) resulted in a better sensitivity for $NO_2$ so that detection limits close to 10 ppt were reported. It is remarkable that such results were obtained by a TDLAS installed on an aircraft flying over the clean troposphere and in the free atmosphere.

Extensive evaluation of the whole system was made by NASA as a part of the global tropospheric experiment. TDLAS was proved to correlate well with laser-induced fluorescence and chemiluminescence (Schiff and Mackay, 1989). The suitability of TDLAS for monitoring HCHO at high levels is clearly shown in Fig. 5a (see Section 4.1.2, FTIR). Further support to this technique has been provided by Harris et al. (1990) who recently reported data collected during the 1988 *Polarstern* cruise. They have installed a TDLAS on board the ship. Results obtained by a wet method based on the reaction of HCHO with 2,4-dinitrophenylhydrazine (DNPH) seem to agree with figures given by TDLAS in the range of 50–200 ppt. Intercomparison with fluorometric methods performed in a rural area in North Carolina gave satisfactory results, whereas during the monitoring campaign carried out at Glendora agreement was found only during the first days (Shiff and Mackay, 1989).

As long as the intercomparison proceeded, large deviations among TDLAS and other methods were observed, which have been attributed to degradation of fluorometric instrumentation with the time (Schiff and Mackay, 1989). Figure 5b shows the consistency between results obtained by TDLAS, a diffusion technique based on annular denuder, and a filter pack collector in the monitoring of $HNO_3$. Detection limits are not reported, but they certainly fall in the range of 100–200 ppt.

TDLAS also makes possible sensitive determinations (20 ppt) of $H_2O_2$ in the atmosphere. Parallel determinations of this pollutant carried out by TDLAS and two fluorometric techniques again showed a satisfactory agreement in a rural site, whereas large discrepancies between levels recorded by the various instruments were observed in polluted areas (Schiff and Mackay, 1989). From these results it is hard to evaluate the real performance of TDLAS in the determination of $H_2O_2$ because, as is discussed in Section 4.1.5, responses given by some fluorometric techniques might be subjected to serious artifacts when high levels of photochemical oxidants are present in air.

Monitoring of $NH_3$ by TDLAS was carried out by Brassington (1989) in a background station of the northwest coast of England. The cell was pumped at high speed to prevent wall effects. No detection limits are reported, but they can be estimated to be in the range of 50 ppt from the trends shown.

Automated TDLAS has been in operation for unattended monitoring of NO, $NO_2$, and $SO_2$ in the Scauinsland surveying station located in the Black Forest (Schmidtke et al., 1989). While detection limits of 50 and 300 ppt were achieved for $NO_2$ and NO, respectively, the highest sensitivity reached for $SO_2$ was 1 ppb.

Based on the results reported in the technical literature, the main advantages of TDLAS for the monitoring of air pollutants can be summarized as follows: high time resolution and excellent specificity, high sensitivity for $NO_2$, and the capability to detect some important noncriteria pollutants such as $HNO_3$, $NH_3$, HCHO, and $H_2O_2$ at trace levels. Potential restrictions to the use of TDLAS are represented by wall effects common to all systems where sample flows through a cell for detection. They can be a possible source of artifacts in the determination of "sticky" gases ($HNO_3$ and $NH_3$) or pollutants soluble in water ($H_2O_2$, $HNO_3$, $NH_3$) when monitoring is carried out at values of relative humidity close to 90%. Wall effects can be minimized by using high sampling speeds and through a careful check of the zero levels between two consecutive determinations. However, the real influence of these effects in producing sampling artifacts can be assessed only through an intensive evaluation of TDLAS carried out under rather different conditions such as those experienced in long monitoring campaigns.

TDLAS instruments available on the market look quite promising for atmospheric studies because they can cover the range of pollutants not detected by DOAS and exhibit higher sensitivity than FTIR. While the adoption of TDLAS for special research monitoring programs, particularly those aimed at defining the role played by $H_2O_2$, is undisputed, its use in large monitoring networks finds severe limitations arising from the cost of operation, the limited number of species that can be determined, and the impossibility of monitoring different pollutants at the same time. However, research on TDLAS is a fast-growing area and it is possible that unique applications (such as the determination of organic peroxyradicals in air) can justify its acquisition on a larger scale.

### 4.1.4. Chemiluminescence

Chemiluminescence-based instruments have been widely used for many years to monitor $O_3$, $NO_2$, and NO in ambient air. They exploit the ability of certain pollutants to react with another chemical species, giving rise to an electronically excited product capable of emitting light in a broad range of wavelengths when it returns to the ground state. The measure of emitted light can be used for detection, since its intensity is proportional to pollutant concentration. Ozone chemiluminescence analyzers exploit the reaction with ethylene to give formaldehyde, whereas in the $NO_x$ instruments ozone is added to convert NO to $NO_2$. Therefore chemiluminescence analyzers must be equipped with bottles or photochemical flow reactors capable of supplying a constant input of chemical reagent.

The specificity of the reactions mentioned above is such that chemiluminescence is promulgated by the USEPA (1986) and other national agencies as the reference method for monitoring ozone and NO. Detection of $NO_2$ by chemiluminescence is possible if previous conversion to NO is accomplished.

In the past, this step was achieved by means of catalytic converters held at high temperature (ca. 300 °C). This was the working principle of $NO_x$ analyzers until it was found that other nitrogen-containing pollutants ($HNO_3$, $HNO_2$ and PAN) were also partly converted to NO so that measurements of $NO_2$ carried out with such analyzers were affected by serious errors (Winer et al., 1974).

The theoretical possibility of eliminating interferences in the determination of $NO_2$ combined with the fact that other instrumental methods have yet reached detection limits for NO lower than chemiluminescence-based analyzers has stimulated further investigations aimed at improving this spectroscopic technique. Two main approaches have been followed. The first has been to improve the conventional design so that both NO and $NO_2$ could be detected at ppt levels without interference. Both the reaction chamber and the electronics have been modified (Drummond et al., 1985). Dark currents and interfering chemiluminescence were measured in a zero-mode by passing air/ozone mixtures in a Teflon-coated relaxation chamber. The NO signal was calculated from the difference between the measure mode (no passage through the relaxation chamber) and the two closest zero modes. The photomultiplier was cooled at $-40$ °C and connected to a photon counter with two different reading times (30 and 180 s). The chamber was kept at reduced pressure for longer radiative lifetime. Detection limits of 10 ppt were reached with these modifications.

Today the preferred approach for detecting $NO_2$ by chemiluminescence is its conversion to NO in a photolytic chamber (580 mL) by irradiating the sample with a 300-W Xe lamp (Kley and McFarland, 1980). The light is filtered by cutting wavelengths below 350 nm, passed through a shutter, and reflected by a UV mirror capable of reflecting wavelengths between 350 and 375. The conversion efficiency (ca. 50%) can be evaluated by passing a known amount of $NO_2$ in synthetic air through the same chamber, and model calculations can be used to derive the actual concentration. Detection limits in same range as NO can be obtained with this system (Bruning and Rohrer, 1990). To this basic design a gold converter can be added for measuring $NO_y$, defined as all nitrogen-containing pollutants excluding $HNO_3$ (Fahey et al., 1985).

A similar setup was used by Bruning and Rohrer (1990) during the *Polarstern* cruise. They were able to detect changes of NO and $NO_2$ from 10 to 80 ppt when the ship was moving from northern to southern latitudes. Levels of $NO_y$ ranging between 50 and 200 ppt were detected.

Instruments for NO and $NO_2$ based on the same principles just described are presently available on the market. They can work either in the switching mode just described or in the parallel mode with two NO analyzers.

The other approach to improve chemiluminescence detection has been to investigate chemical species capable of reacting with $NO_2$ but not with other nitrogen containing pollutants. By improving the original method proposed by Maeda et al. (1980), a monitor for $NO_2$ where chemiluminescence is produced by reaction with luminol has been developed (Schiff et al., 1986).

Light is detected at 425 nm. Air flows through a fabric wick kept wet by luminol solution continuously fed by a micropump. A photomultiplier, placed very close to the reaction cell, measures the emitted light. The instrument is quite simple and compact. It offers the advantage that can be operated in the field by batteries and can detect $NO_2$ at levels of 10–20 ppt. A comparison made with TDLAS has given satisfactory results (Schiff et al., 1986).

Examples of daily trends obtained by measuring $NO_2$ with a commercially available luminol-chemiluminescence monitor are shown in Fig. 3$b$. Reasons explaining discrepancies with DOAS have already been discussed and cannot be attributed to its working principle. However, these $NO_2$ analyzers are affected by matrix interferences. Ozone was found to produce a positive response, although the extent (0.33%) is such that it can be considered negligible in many practical instances. A more serious interference is caused by PAN which contributes 25% of its content to the $NO_2$ signal (Kelly et al., 1990).

At present, second-generation chemiluminescence analyzers equipped with a photolysis cell seem to be the only ones that can compete with DOAS and TDLAS in terms of sensitivity and specificity. They can be used for field investigations as well as for the monitoring of NO and $NO_2$ in large networks, although the price is quite high in relation to the number of pollutants that can be detected.

Out-of-line chemiluminescence detection for the monitoring of $H_2O_2$ has been also reported (Jacob et al., 1986). Air is passed through a sampling tube kept at $-40$ °C where $H_2O_2$ is trapped together with water vapor. Determination is carried out after melting the ice by reaction with peroxyoxalate. The method is manual and does not discriminate between $H_2O_2$ present as gas and that dissolved in droplets. It might be also affected by artifacts due to trapping of ozone and other pollutants.

### 4.1.5.   Fluorescence

Fluorescence (i.e., emission of light from a pollutant returning to the ground state after being excited with an appropriate light source) is the common way for detecting $SO_2$ in polluted environment. $SO_2$ present in air samples is excited using zinc or cadmium lamps (213.8 and 228.8 nm, respectively) and fluorescence is monitored in the 220 to 400 nm range. Today, more sophisticated ways to exploit fluorescence have been investigated, based on the use of lasers as source of light. They permit detection of air pollutants at ppt levels.

Papenbrock and Stuhl (1990) developed an instrument for detecting $HNO_3$ in air. They found that photolysis of this compound with a ArF-laser yields excited OH(A) radicals in a two-photon, two-step mechanism. $HNO_3$ is thus detected by looking at the fluorescence emitted by excited OH (308.8 nm) when the excimer laser crosses the chamber where air containing $HNO_3$ is sampled. Detection is carried out at right angle with respect to the incoming light to minimize interferences from the laser beam. To improve detection (longer lifetime), the photolysis cell is kept at low pressure with a rotary pump. If signal is accumulated for 1 h, detection limits of 30 ppt can be obtained. After each measurement the chamber is evacuated at $5 \times 10^{-4}$ mbar to obtain the zero signal.

Data have been collected with this technique in several locations in Europe and the instrument was also used on the *Polarstern* cruise (Papenbrock and Stuhl, 1990). Daily trends have been reported showing the capability of this technique to provide reliable information on sites characterized by rather different pollution levels. These are shown in Fig. 6. Comparisons made with diffusion techniques have shown low deviations. A similar instrumental design (Bradshaw et al., 1985) has been used for the detection of $NO_2$ at ppt levels. As previously mentioned, good agreement was found with TDLAS and photolysis–chemiluminescence in rather unpolluted air samples (Schiff and Mackay, 1989).

Laser-induced fluorescence is probably the most promising way for measuring OH radicals in air (Hard et al., 1984). Air is sampled into a chamber where excitation of OH radicals is performed by a laser beam set at 282 nm. Emission is detected at 308 nm. Interference from photolysis of $O_3$ and water to give OH radicals together with difficulties in calibrating the instrument are great limiting factors to the application of this technique. They have been only partly rectified by reducing the pressure through expansion of the sample in the cell. However, it has been estimated that interference ranging between $10^5$ and $10^6$ radicals/cm$^3$ can affect determinations in air (Hard et al., 1989). Since concentrations range between 1 and $6 \times 10^6$ radicals/cm$^3$, uncertainty is quite high.

It has recently been proposed that changing the excitation wavelength from

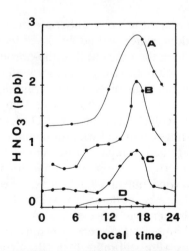

**Figure 6.** Diurnal variations of HNO$_3$ in polluted and unpolluted environments determined by laser-induced fluorescence. (*A*) Bochum (FRG), 13 Aug 1986; (*B*) Schauinsland, Black Forest (FRG), 14 Jul 1987; (*C*) Petten (The Netherlands), 12 Aug 1987; (*D*) *Polarstern* cruise (30 W, 19.8 to 14.4 N over the Atlantic Ocean), 26 Sep 1988. (From Papenbrock and Stuhl, 1990.)

282 to 308 nm might result in a much lower interference from $O_3$ and water and a more intense emission (Hofzumahaus et al., 1990). However, this choice poses severe detection problems because laser stray light and OH fluorescence have the same wavelength. These can be resolved by separating the two signals if the radiative lifetime of OH radicals is much longer than laser pulse duration. Detection is possible by working at low pressures and by making use of gated electronics for photon counting and off/on switching of the photomultiplier. Since such a system has not been tested in the field, theoretical calculations on detection limits and interferences must be verified.

Instruments of this type are complex, expensive, and often designed for a specific air pollutant. They suffer the same limitations as TDLAS and FTIR as far as wall effects are concerned. Their development is justified only by the fact that few alternatives exist for detecting certain pollutants. This is particularly true for OH radicals. For this reason, cheaper fluorometric monitors have been developed that exploit the reaction of a pollutant with another species to given intense fluorescent products. This is the working principle of continuous monitors for $H_2O_2$, $NH_3$, HCHO, and $SO_2$ in air.

Peroxydase-catalyzed dimerization of $p$-hydroxyphenylacetic acid followed by fluorescent detection has been proposed by Lazrus et al. (1986) for detecting $H_2O_2$. Air is passed through a coil and scrubbed with an aqueous solution. Peroxides are dissolved in the solution and continuously analyzed by fluorescence. Organic peroxides, which interfere in the $H_2O_2$ determination, are monitored on a separate channel after reaction with enzyme catalase has removed $H_2O_2$ from the aqueous solution. The net $H_2O_2$ concentration is determined by the difference between the two channels. The same chemistry is used in the system developed by Dasgupta et al. (1988), but sample collection is achieved with the use of microporous tubing through which the scrubbing solution flows.

The basic principles of a diffusion scrubber are the same as diffusion techniques based on denuders, which will be discussed in detail in Sections 4.2. The air stream flows along the outer wall of the porous tubing and has free access to the scrubbing solution. $H_2O_2$, together with other gaseous components, diffuses into the tube, dissolves in the scrubbing solution, and is determined by fluorescence after pH is raised to 9.5 using ammonia. Peroxides are removed by a $MnO_2$ column and the net $H_2O_2$ determined by the difference between two readings (with and without column).

These two systems are quite simple to operate but seem to be affected, to a different extent, by interference in photochemically polluted areas. During the intercomparison carried out at Glendora (see Section 4.1.3, TDLAS) continuous scrubbing fluorescence detection gave values of $H_2O_2$ mixing ratios that were three times higher than those of TDLAS in all the range of measurements. Diffusion scrubbing fluorescence results were closer to those of TDLAS but differences were still quite high. Better agreement among the three techniques were found in less polluted environment.

The reason for these differences in performance has been attributed mainly to the fact that fluorometric monitors respond with varying efficiency to peroxides. Since the net content of $H_2O_2$ is determined by the difference, errors can be quite large in polluted areas. In our opinion an additional source of discrepancy arises from the different sampling systems used. While continuous scrubbers efficiently sample gases and particles of a certain size, diffusion scrubbers are more selective for gases. This difference might explain why the continuous scrubber always gives higher values for $H_2O_2$ than the diffusion scrubber.

The different sampling capabilities of continuous and diffusion scrubbers have been exploited to detect $NH_3$ and $NH_4^+$ (Neftel et al., 1990) by adapting the fluorometric technique proposed by Genfa and Dasgupta (1989). One channel (diffusion scrubber) detects ammonia, whereas the other (continuous scrubber) collects ammonia and $NH_4^+$ present in small and large particles. Both solutions (0.001 M HCl) react with o-phtaldialdehyde and sulfite to give intensely fluorescent products. They are excited at 365 nm and detected at 450 nm. Higher sensitivity than commercial DOAS is reported. The authors correctly pointed out that they did not evaluate the sampling efficiency for particles, but, based on theoretical considerations, they believed that it was size dependent so that $NH_4^+$ determinations could not have been accurate.

A diffusion scrubber combined with fluorescence detection is also the principle of operation of automatic monitors for $SO_2$ and HCHO, having detection limits of 175 and 100 ppt, respectively (Dasgupta et al., 1988). An extensive description of the solution used for inducing fluorescence in the air sample, linearity ranges, and factors affecting detection is reported. A field evaluation of HCHO carried out in a rural area by using continuous scrubbing fluorometric detection, TDLAS, and 2,4-DNPH derivatization gave consistent results with good agreement between the various data sets (Schiff and Mackay, 1988).

By contrast, large deviations between continuous-scrubbing, fluorometric detection and TDLAS were observed during the last days of a monitoring campaign held in the Los Angeles area. These differences were attributed to degradation of the performance of the fluorometric instrument with the time (Schiff and Mackay, 1989). A similar effect was observed when TDLAS was compared with another fluorometric technique based on the reaction between formaldehyde dehydrogenase and $NAD^+$ (Lazrus et al., 1988). From the results of this campaign it is difficult to decide to what extent one can rely on presently available fluorometric techniques for the evaluation of HCHO, particularly because the observed deviations from TDLAS show opposite trends.

## 4.2.  Diffusion Techniques Based on Denuders

The ability to perform the monitoring of both gases and particles with the same device and the possibility of obtaining accurate information comparable to the more sophisticated optical techniques but at a lower cost and with simpler nonspecific devices that are affordable by any type of monitoring station are the main reasons why we have dedicated an entire section to diffusion techniques based on denuders. The use of denuders should be regarded not as another way to collect samples, but as a different approach to the monitoring of atmospheric pollutants. It has great potential for improvement if proper analytical techniques (including some optical methods already discussed) become a part of the method.

Denuders have been known for several years and were first employed by Durham et al. (1978) for removing $SO_2$ during sampling of particulate matter (which is why the term denuder was used). The tube was coated with lead dioxide, but no discussion on the efficiency was reported. In 1979, Ferm obtained quantitative removal of $NH_3$ from the air stream by optimizing the geometrical features of a denuder coated with oxalic acid. Through a detailed description of factors influencing denuder collection, he showed that efficiency higher than 99% can be reached under certain conditions.

A diffusion technique based on denuders was thus proposed for the interference-free monitoring of $NH_3$ and ammonium nitrate on particulate matter. Detection limits of 0.001 ppb for sampling periods of 24 h were obtained. This earlier geometry consisted of a cylindrical glass tube having an internal diameter of 0.3 cm and a length of 50 cm. The performances of such a denuder geometry are well described by the Gormley-Kennedy equation (Gormley and Kennedy, 1949):

$$\frac{Cz}{Co} = 0.819 \exp\left(\frac{-14.62\,DL}{4F}\right) \tag{1}$$

where Co is the concentration entering the tube, Cz the average concentration leaving the tube, $D$ the diffusion coefficient of the gas, $F$ is the flow rate through the tube, and $L$ is its length. Equation 1 is valid under the assumption that the walls of the tube act as a perfect sink for a given gas and collection is carried out under laminar flow conditions. Discrimination between gas and particles occurs because the lower diffusion coefficient prevents particles from reaching the walls. Of course, gravitational forces and turbulent motion, which can cause impaction of particles on the tube walls, must be prevented. For this reason, denuders are usually placed in vertical position, their sampling speed is maintained at Reynold's numbers smaller than 2000, and a small portion of the inlet tube is uncoated.

The possibilities afforded by denuders are twofold. First, collection of gases can be performed by avoiding interference from particles (which unavoidably occurs if impregnated filter or traps filled with liquids or solid sorbents are used for gas sampling). This requires that the gas deposited on the denuder be determined somehow. At present, the usual approach consists of coating the inner walls of the denuder with a chemical compound that strongly reacts with the pollutant to yield stable products that can be analyzed and quantitatively determined by chromatographic or colorimetric techniques. Analysis is performed after recovering the product by liquid extraction or thermal desorption. However, the possibility of using in situ or on-line detection in a way similar to that adopted with diffusion scrubbers already described (see Section 4.1.5) should not be disregarded.

The second possibility is to prevent reactions between particles and gases during collection of particulate matter. In some instances, such reactions can greatly affect the composition of the sample. This is often the case when the formation of a given pollutant proceeds through equilibrium reactions. Ammonium nitrate can be used as an example. If a such pollutant is retained on a filter and continuously exposed to gaseous species capable of reacting with particulate matter (i.e., strong acids), its decomposition can yield gaseous products ($HNO_3$ and $HN_3$) which can be released from the filter (Appel and Tokiwa, 1981).

On the other hand, gaseous pollutants such as $NO_2$ might be partly retained on particulate matter collected on filter and can release $HNO_3$ and $HNO_2$ by reaction with water. The former product can further react with $NH_3$ present in the air stream to form ammonium nitrate. If gases are not removed, particulate matter deposited on the filter can have a rather different composition than that of particles suspended in air.

This is just an example of a large variety of complex reactions that can take place between gases and particles during sampling. Discrimination between phases to prevent artifacts due to multiphase reactivity is the conceptual basis that renders diffusion techniques using denuders the ideal candidates for a correct sampling, although interference cannot always be prevented or completely evaluated.

In recent years, great improvement in the denuder technology has been achieved through the development of the annular geometry (Possanzini et al., 1983). This novel design makes it possible to overcome the main limitations of the tubular geometry for atmospheric monitoring. This arrangement, simulating an assembly of tubular denuders displaced in parallel according to a circular geometry, can be described by the following equation:

$$\frac{Cz}{Co} = 0.82 \, \exp\left(\frac{-22.53 \, DL}{4F}\right) \frac{d_1 + d_2}{d_1 - d_2} \qquad (2)$$

where $D$, $L$, and $F$ have the same meaning as equation 1, and $d_1$ and $d_2$ are the inner and outer diameters of the annulus (Possanzini et al., 1983).

It has been shown that through a judicious choice of the annulus, better working efficiencies (defined as $1 - Cz/Co$) than with a tubular denuder are achieved at $L/F$ ratios that are an order of magnitude lower than those needed by tubular denuders. This means that laminar flow conditions are reached and maintained with sampling flow rates as high as 30 L/min against the 2 L/min afforded by tubular geometries. High sampling flow rates and high efficiency are combined with large capacity (up to milligram levels in some instances). Such performances permit the collection of sufficient amounts of gas and particles for the analysis in a rather short time. The development of annular denuders explains why diffusion techniques have enjoyed increasing popularity in the last few years. In any further discussion we shall usually be referring to annular denuders when speaking about diffusion techniques.

As it happens for any enriching system, sampling times necessary for atmospheric monitoring cannot be determined a priori. They depend on several factors and can change from one pollutant to another. When the monitoring of pollutants is performed by diffusion techniques based on denuders the following factors must be taken into account:

1. Concentration of the pollutant in air
2. Sensitivity of the analytical method used for the analysis
3. Total capacity of the denuder
4. Monitoring strategy

The first three factors do not require further comment, but the fourth needs an explanation. Since denuders can be combined in series for the simultaneous collection of several pollutants (Allegrini and De Santis, 1989), different approaches can be followed depending upon the monitoring strategy. They can be summarized as follows:

1. Determination of a single component in the gas phase
2. Determination of several gaseous components at once using the same denuder or a sampling train of denuders placed in series
3. Determination of gases and particles during the same enrichment step
4. Determination of evaporation processes from the filter used for the collection of particulate matter.

If a single gaseous pollutant is to be monitored, only factors 1, 2, and 3 determine the sampling time. Due to the efficiency and capacity of annular denuders, sample collection can change from 30 min to 12 h. In the former case daily trends can easily be followed, whereas a 12-h sampling time might be necessary when background levels have to be detected. If multiple gas sampling or speciation of particles is required, the less concentrated component determines the minimum enrichment factor necessary for an accurate analysis. This explains why profiles of samples averaged over 3–4 h are often reported even though some of the pollutants monitored could have been determined with better time resolution if a single species was monitored. The same is true if evaporation processes are to be investigated, some additional denuders are inserted after the filter or filter-pack collectors for the evaluation of blow-off gases.

The common criticism that diffusion techniques require longer sampling times than some optical techniques is often unjustified and should be evaluated case by case. It is correct, in general, for tubular geometry, but might not be applicable to annular denuders, especially if determination of a single component is required. Simultaneous determination of different pollutants often leads to a decrease in time resolution; however, this effect is largely compensated for by the higher number and quality of information gained on the atmospheric processes.

Although, in principle, diffusion techniques can be applied to any chemical species, in practice there are some limitations. Assume, for instance, that a given pollutant A has the same reactivity as pollutant B toward the wall coating and that they give the same species C as final product. Determination of A in the presence of B is impossible if C is used for their detection since B produces a positive interference on the determination of A that cannot be evaluated. In this case, only coatings leading to different products can be employed. However, if the reactivity of A is higher than that of B, equations describing their penetration efficiencies give a different mass distribution profile along the denuder and differential methods can be applied (Febo et al., 1989). Two denuders placed in series can be used. Because the walls act as a perfect sink for A, this gaseous component will be mainly concentrated at the inlet of the first denuder with a mass distribution function characterized by a rapid exponential decay with the length. Pollutant B will be distributed in a quasi-homogeneous way between the two denuders; due to its low reactivity toward the wall coating, the denuder will behave as a chromatographic column working in frontal analysis. By subtracting the amount of C collected in the second denuder from the amount of C present in the first denuder we can evaluate the content of C resulting from the reaction of A with the wall coating. A full discussion of the physical and chemical considerations that make such an approach possible, together with the various cases in which it can be used, can be found in the papers recently published by Febo et al., (1989).

### 4.2.1. Instrumental Design Used with Diffusion Techniques

Denuders can be incorporated into different instruments according to the specific type of pollutant to be detected and different aspects of photochemical smog and dry acid deposition investigated. The simplest instrumental setup consists of an aspirating pump, a flowmeter, a flow controller, and an inlet to remove particles greater than about 5 μm. These components are connected to an annular or sometimes a tubular denuder coated with a thin film of absorbing material that is able to specifically react with the pollutant to be monitored. After sampling, the denuder is extracted by a proper solvent and analyzed for the product of the reaction between the pollutant and the wall coating.

Determinations can be carried out by any instrumental technique capable of detecting the product. Trains of annular denuders connected in series have been used for the simultaneous monitoring of pollutants present in air as gases and particles (Allegrini and De Santis, 1989). Such an instrumental setup is placed in a temperature-regulated room containing several parallel lines that are used in sequence for air monitoring. A microprocessor controls the sampling time of each line. It automatically connects the selected train to the aspirating pump and the inlet to the atmosphere. The same instrument can be used to determine other pollutants by simply changing the sequence and coatings of annular denuders.

Another type of instrumentation uses rotating annular denuders where walls are kept wet by a liquid solution acting as a perfect sink for the pollutant to be collected (Slanina et al., 1987). The gaseous component is removed from the air stream by dissolution in the liquid layer. After a prefixed time, denuder rotation and air sampling are stopped. The solution, containing the reaction product with the pollutant, is automatically removed from the annulus and transferred to a sequential collector where it is stored until analysis is performed. Then, a known amount of fresh solution is fed into the denuder and a new sampling sequence is initiated. Cycles of 30 min are usually adopted. Up to two different denuders can be independently operated. The sampling unit works automatically until the reservoir of fresh solution is consumed and the sequential collector filled. This method can be used for the simultaneous collection of pollutants if the species formed in the liquid layer are stable (Keuken et al., 1988).

A third type of instrument combines thermoanalytical techniques with denuders for the collection and analysis of pollutants (Slanina et al., 1981). Usually, it is applied to the speciation of components present in particulate matter that can be volatilized at a certain temperature. By heating a portion of the total denuder length, a pollutant present in air as a particle can be evaporated when it crosses the hot zone. In the gas phase the pollutant can reach the denuder walls by diffusion and can be selectively removed by the air stream if the surface of the walls behaves as a perfect sink for that pollutant. The principle can be extended to the speciation of particles composed of pollutants with different evaporation or dissociation temperatures. In this case, various zones, kept at different temperatures, serve for the collection. Each zone is maintained at a temperature allowing the evaporation (or dissociation) of a given pollutant. This diffusion approach is termed the thermodenuder method and can be combined on-line with optical, electrochemical, and chromatographic detection techniques.

### 4.2.2. Application of Denuder Techniques to the Monitoring of Pollutants

All three diffusion techniques described in the previous sections have been employed in the determination of $HNO_3$ and give detection limits ranging from 30 to 100 ppt, depending upon the sampling time chosen. A train consisting of sodium chloride and sodium carbonate denuders (Perrino et al., 1990) placed in series has been shown to provide reliable air measurements of $HNO_3$ and the possible formation of $NO_3^-$ ions due to the reaction of

$NO_2$ with water to give nitrous and nitric acid on the denuder can be evaluated by differential techniques. The use of a second denuder is necessary only in very polluted areas where mixing ratios of $NO_2$ higher than 70–80 ppb are present.

Determinations of nitrate ion in the extract are performed by ion chromatography equipped with a suppression column for enhanced sensitivity. A differential conductivity cell is connected to the analytical column for detection. If the train of denuders is combined with a denuder for ammonia evaporation processes of ammonium nitrate during sampling can be evaluated and the total content of nitrate in particles can be correctly measured (Allegrini et al., 1990). $HNO_3$ released by dissociation of ammonium nitrate deposited on the acetate filter is collected on a backup nylon filter, whereas $NH_3$ is collected on the last denuder. Mixing chambers are inserted between denuders and a cyclone is placed after the first denuder to avoid adsorption of $HNO_3$ on cyclone walls. The whole sampling unit is operated at a flow rate of 15 L/min.

Automated samplers are usually composed of six trains of denuders that cover a 24-h sampling cycle with a time resolution of 4 h. This setup is specifically designed for air monitoring in unpolluted areas where low concentrations of nitric acid are found. Time resolutions ranging between 1 and 2 h are possible in urban areas, but two automatic samplers are required to follow daily profiles.

Rotating denuders with a liquid layer made of a diluted solution of formic acid with a pH of 3.7 were used for simultaneous collection of $HNO_3$, HCl, and ammonia (Keuken et al., 1988). With a sampling flow rate of 30 L/min, detection limits of 30–50 ppt are obtainable with 30 min time resolution. Also, in this case, ion chromatography is the preferred technique for analyzing the extract.

Thermodenuders coated with $MgSO_4$ have been applied to the determination of $HNO_3$ and nitrate (Keuken et al., 1990). During sampling, only the first part (10 cm) of the denuder collects $HNO_3$; the second part (10 cm) adsorbs the volatile nitrogen-containing compounds. Both zones are kept at room temperature. The last portion (40 cm or more) is heated at 140 °C so that ammonium nitrate decomposes and $HNO_3$ formed is adsorbed on the walls by diffusion. After sampling, the various zones of the denuder are sequentially heated at 700 °C under a flow of nitrogen starting from the last one. Adsorbed components are evaporated and sent to a $NO_x$ chemiluminescence analyzer for detection. The profile obtained closely resembles that produced by a gas chromatograph. The first peak corresponds to ammonium nitrate, the second to all $NO_y$ less nitric acid, and the last peak is composed of $NO_y$ including $HNO_3$. Differential techniques are used for the evaluation of $HNO_3$. The time resolution is 30 min at a flow rate of 30 L/min.

In spite of the conceptual and instrumental differences, all these techniques yielded consistent results in the determination of $HNO_3$ performed during the intercomparison monitoring campaign held in our Institute in September

1988. Mixing ratios determined with various methods were found to agree within the experimental errors. These results, combined with the good agreement shown between annular denuders, TDLAS (see Fig. 5b), and laser induced fluorescence (see Section 4.1.5), suggest that diffusion techniques are accurate and sensitive for the monitoring of this pollutant. An example of the amount of information that can be simultaneously collected by a train of annular denuders is shown in Fig. 7b and c. Data of $HNO_3$, nitrate, sulfate and evolved nitrate refer to the intercomparison exercise on $HNO_3$ already mentioned (Allegrini et al., 1990).

Trends shown in Fig. 7b and c can be combined with those reported in Fig. 3a–d to get a comprehensive picture of what happens to photochemical pollutants and acidic species when pollution episodes develop over areas located downwind of a large city. Photochemical smog episodes of this type are quite common from March to late September when anticyclonic weather is established in central Italy. During the day, the supply of both primary and secondary pollutants is mainly provided by the urban area of Rome because of the sea breeze circulation. This meteorological situation, typical of central–southern Italy, is also common to many Mediterranean countries.

Daily trends reported in Fig. 7a suggest a second application of annular denuders: the determination of $NH_3$. Data reported in this figure were obtained by two rotating denuders filled with the same solution used for $HNO_3$ (Keuken et al., 1990). Solid and dotted lines refer to two denuders operated in parallel by the same sampling unit. The figure shows clearly the reproducibility and high time resolution afforded by this denuder technique. These determinations, carried out simultaneously with those of precursors and products presented in Figs. 3 and 7, highlight the important role played by this pollutant in neutralizing acidic species in air.

As we shall see in Section 4.3.1, there are two processes controlling the concentration of $HNO_3$ at the site: evaporation of ammonium nitrate and direct transport from the city. In the former case, $HNO_3$ formed in the urban area is deposited or neutralized by ammonia emitted in the rural area surrounding the city of Rome so that only ammonium nitrate particles reach the site. In the latter case, a portion of $HNO_3$ survives the deposition and neutralization processes and is directly transported over the site together with other photochemical pollutants.

Due to the high emission rate, serious artifacts can affect the determination of $HNO_3$ and nitrate if $NH_3$ is not removed by the air stream during sampling. With the rotating denuder system used in Fig. 7a, $NH_3$ is determined as ammonium ion by spectrophotometric techniques (blue indophenol method). Detection limits of about 130 ppt can be achieved. An alternative to liquid solutions is the use of organic acids deposited on denuder walls. All act as perfect sinks for $NH_3$, but citric acid is often preferred because of its low vapor pressure (Allegrini and De Santis, 1989). The same analytical procedure previously described is used for analyzing the extract. Detection of $NH_3$ can also be accomplished by HPLC combined with UV absorption at 230 nm if

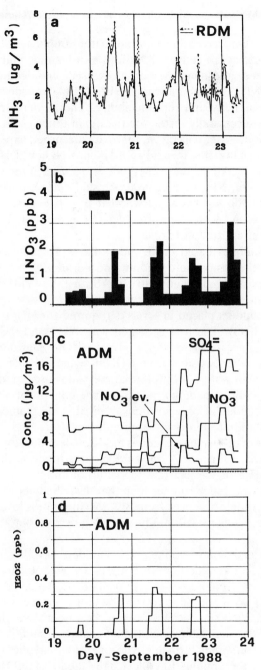

**Figure 7.** Daily trends of NH$_3$, HNO$_3$, nitrate, and sulfate in particulate matter, nitrate evolved from the filter, and gaseous H$_2$O$_2$ recorded during the Intercomparison Monitoring Exercise on Nitric Acid and Nitrate Measurements (19–24 Sep 1988) held in Montelibretti, Italy. (*a*) Data obtained with two rotating annular denuders (RDM). (From Keuken et al., 1990.) (*b* and *c*) Data obtained with a train of annular denuders (ADM). (From Allegrini et al., 1990.) (*d*) Data obtained with the annular denuder method (ADM). (From Ciccioli et al., 1990.)

phosphoric acid is used as coating material. After sampling, $m$-toluoyl chloride is added to the extract to convert $NH_3$ into $m$-toluamide and this product is injected, separated, and quantified on a reversed-phase column (Possanzini and Di Palo, 1989). The main advantage of HPLC over spectrophotometric methods is its simplicity and reproducibility. Thermodenuders coated with $V_2O_5$ have been successfully tested for the quantitative collection of $NH_3$ in air (Keuken, 1989). After collection, thermally desorbed vapors (700 °C) are passed through a converter to oxidize $NH_3$ into $NO_x$ which is detected by chemiluminescence. All of these methods were used during the intercomparison campaign on $NH_3$ held in 1990. As said before, data are still under evaluation and we do not know yet whether they give consistent results between themselves and with respect to the optical methods based on DOAS whose results are reported in Fig. 4.

Diffusion techniques based on denuders have provided a simple way to detect $HNO_2$ in the atmosphere (Sjödin and Ferm, 1985). So far, they represent the only possible alternative to DOAS for the monitoring of this pollutant in different environments. Recent studies have shown that a train of three annular denuders placed in series is required for $HNO_2$ determination (Perrino et al., 1990). The first is coated with NaCl and the other two are coated with sodium carbonate + 1% glycerol. The inlet denuder removes $HNO_3$, while the other two account for $HNO_2$ eventually formed by reaction of NO and $NO_2$ with water. A differential method is used to measure $HNO_2$. Efficiency evaluation is made by measuring the content of nitrite ion found in the last two denuders. Concentrations of nitrite are obtained by subjecting the extracted solution to ion chromatographic analysis.

The suitability of diffusion techniques based on denuders for the monitoring of $HNO_2$ is shown in Fig. 3$d$. A fair agreement is found during all 5 days of monitoring, in spite of the fact that the setup of annular denuders used for this investigation was much simpler than that described above and hence was partly subjected to interferences from $SO_2$. Mixing ratios lower than 100 ppt of $HNO_2$ can be detected with a 3-hr sampling carried out at a flow rate of 15 L/min.

Both rotating and well-coated annular denuders have been applied to the determination in air of another important product of photochemical reactions: $H_2O_2$. Interference from ozone was prevented or evaluated by differential techniques. $H_2O_2$ formation due to the reaction of ozone with water can be avoided by adding NO to the air stream. This approach, first adopted by Tanner et al. (1986), was later extended to rotating denuders (Keuken et al., 1988). Solutions containing formaldehyde and $p$-hydroxyphenylacetic acid are inserted between the annulus to collect and stabilize $H_2O_2$ and $SO_2$ in the absorption medium. Amperometric flow injection analysis is used for detection. With this technique, the sample is injected twice. The first injection is analyzed directly for $H_2O_2$ and peroxides, whereas in the second injection only peroxides are detected as $H_2O_2$ is decomposed over a Pt converter. The net content of $H_2O_2$ in the solution is determined by the difference of the two signals (Keuken et al., 1987). Detection limits of few ppt can be reached when the collection time of rotating denuder is 30 min and the flow rate is 32 L/min.

An easier way to determine $H_2O_2$ by annular denuders is to deposit a film of titanium(IV) oxalate–sulfuric acid on the walls (Possanzini et al., 1988). After sampling, the layer is extracted with xylenol orange and absorbance of the resulting complex is measured by spectrophotometry. The influence of ozone is evaluated by differential methods. Detection limits of about 10 ppt are reported for 3 h sampling at 15 L/min.

Examples of $H_2O_2$ determinations carried out during the intercomparison monitoring campaign on $HNO_3$, already mentioned, are shown in Fig. 7d. Unfortunately, no attempts have been made to compare the performances of denuder methods developed for $H_2O_2$ with those obtained by TDLAS and fluorometric-based instruments. Such a comparison would be of great help in establishing the real uncertainties in the determination of this important pollutant with available techniques.

At the present time, it is not clear whether large differences in the mixing ratios of $H_2O_2$ detected by different research groups can be attributed to the variability of its concentration in air or the inadequacy of monitoring systems. Certainly, $H_2O_2$ is a difficult pollutant to determine because of its multiphase reactivity and distribution, but confusion is added by the tendency to measure it without any previous careful evaluation of the limitations of the technique adopted. In particular, very little attention is paid to the content of ozone and $SO_2$ in air when $H_2O_2$ is monitored. In this respect, intercomparison monitoring campaigns restricted to one pollutant are almost useless in practice because they do not tell us whether the presence of other compounds is heavily affecting the results of the exercise.

The approach followed in the 1988 intercomparison of $HNO_3$ has revealed that it is almost impossible to critically evaluate the accuracy of different methods used for the determination of any pollutant participating in the photochemical cycle without having detailed information on what happens to its precursors and all compounds that somehow influence its concentration through the chain of reactions shown in Figs. 1 and 2. If large deviations occur between methods, only the analysis of daily profiles of possible interfering constituents can be used to identify the possible source of artifacts.

Low boiling aldehydes, ketones, and carboxylic acids can also be included in the list of photochemical pollutants that can be detected by diffusion techniques. The former two classes of organic components can be enriched on annular denuders coated with 2,4-DNPH (Possanzini et al., 1987). The reaction yields dinitrophenylhydrazones that are separated by reversed-phase liquid chromatography. Detection is accomplished by UV/visible absorption or amperometric techniques. Although electrochemical devices are more sensitive, care must be taken to achieve a constant reproducibility (Ciccioli et al., 1987).

Since 2,4-DNPH acts as a perfect sink only for HCHO, the method has some limitations in terms of sampling flow rates and collected volume if $C_2$–$C_4$ aldehydes and ketones have to be determined. Fifty ppt is the detection limit for HCHO that can be achieved with small denuders. If HPLC is not included in the laboratory facilities, HCHO can also be determined by colorimetric methods. In this case, the well-known chromotropic acid reaction can be adapted to diffusion techniques.

Determination of organic acids in air is probably one of the most interesting and promising applications of annular denuders. Formic and acetic acid are efficiently sampled on denuders coated with sodium or potassium carbonate at a flow rate of 10 L/min. After collection, the walls are washed with water and the solution is analyzed by ion exclusion chromatography. A conductimeter is used for detecting the ions. Detection limits of 50 and 300 ppt have been reported for formic and acetic acid respectively on 3-h samples. The method has been used in remote sites in Australia (Noller et al., 1986) and Central Europe (Puxbaum et al., 1988) and found to be free from interference.

The high reactivity of ozone toward olefins has been exploited by Possanzini and Di Palo (1988) to develop a method for detecting ozone. Two denuders connected in series are employed. The first is coated with eugenol, the second with 2,4-DNPH. Ozone is removed from the air stream by reaction with eugenol. Formaldehyde, which is the stoichiometric reaction product, is released in the gas phase and quantitatively collected on the denuder coated with 2,4-DNPH. Through the quantification of formaldehyde it is possible to evaluate the level of ozone in air. During sampling, HCHO is removed by placing a third denuder coated with 2,4-DNPH at the inlet of the sampling line. With this arrangement both HCHO and ozone can be monitored at once using the same analytical technique. In spite of its apparent complexity, the method is rather simple and gives results that are in good agreement with UV-photometry (see Fig. 8). It offers the possibility to measure ozone in remote areas where automatic instrumentations are often difficult to install and run.

The same field of application can be found for diffusion techniques developed for the monitoring of $NO_2$. The coating used in this case is a mixture of guaiacol and NaOH (Possanzini et al., 1990). The extract is analyzed for nitrate by ion chromatography using conductimetric detection. Since guaiacol acts as a perfect sink for $NO_2$ and this primary pollutant is always present at ppb levels, small denuders can be connected to a battery-operated pump for field monitoring in remote sites. Laboratory tests have been performed to evaluate the removal efficiencies of denuders for PAN and encouraging results have been reported (Parmar and Grosjean, 1990). It is possible that very soon denuder sampling will be also applied to PAN determination.

$SO_2$ can be simply collected by sodium carbonate/glycerin-coated annular denuders and quantified by analyzing the extract for sulfate using ion chromatography (Allegrini et al., 1987). Continuous methods for $SO_2$ (such as by fluorescence) are often not sensitive enough for background monitoring, and optical detectors (such as flame photometry) respond to all sulfur compounds.

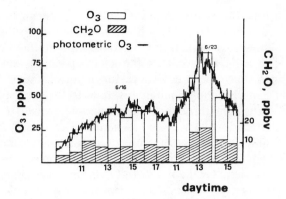

**Figure 8.** Simultaneous determination of ozone and HCHO by using a train of annular denuders and UV photometry. For the denuder coatings see text. (From Possanzini and Di Palo, 1988.)

Diffusion methods therefore provide a reliable technique for routine determination of this primary pollutant in large monitoring networks. Possible interference arising from reduced sulfur compounds can be easily evaluated with differential procedures. Detection limits in the range of 100 ppt have been reported for 12-h sampling periods. Better time resolution is obtained if $SO_2$ mixing ratios fall in the ppb range, as is well illustrated by the results shown in Fig. 3c. The same rotating denuder system used for the collection of $H_2O_2$ permits the determination of $SO_2$ in air samples (Keuken et al., 1988).

Flow injection systems make it possible to quantify the amount collected by flame photometry. However, this detection technique is more often combined with denuders for discriminating gas from particles. It is based on the property of sulfur-containing compounds to produce excited species that emit light at a wavelength of 394 nm when they are burned on a hydrogen flame. The light emitted is detected by a photomultiplier after being filtered. Flame photometry has been developed and is mainly used in combination with gas chromatography because it is sensitive to ppt levels of sulfur compounds eluted from the column. It can still be used for the determination of volatile sulfur compounds in polluted areas where mixing ratios higher than 20 ppb of each pollutant are present in air. An important feature of flame photometry is that it is equally sensitive to gas and particles once the $H_2$/air ratio of the flame is properly adjusted.

The idea of removing sulfur gases by diffusion for real-time measurement of sulfate aerosols by flame photometry has been followed by several research groups (Durham et al., 1978; Garber et al., 1983). Tubular denuders for $SO_2$ and $H_2S$ are inserted between the sampling line and the flame photometric detector connected to an aspirating pump. Temperature must be controlled to avoid volatilization of sulfuric acid. Sulfur particles are evaluated by subtraction techniques. The main limitations of these types of monitors is that no discrimination is made between sulfuric acid and other sulfate aerosols.

Because the knowledge of their relative concentrations is of key importance in atmospheric investigations concerning acid deposition, automated thermodenuder systems have been developed for the quantitative determination of sulfuric acid particles and ammonium sulfate in the atmosphere (Slanina et al., 1985). Air is sent through a line composed of two tubular copper/ copper oxide-coated denuders placed in series. The first is maintained at 120 °C so that sulfuric acid aerosols are evaporated and deposited on the denuder walls, whereas ammonium sulfate particles pass through it unaffected. The second denuder, kept at 300 °C, decomposes ammonium sulfate aerosols into sulfuric acid, which is then deposited on the walls by diffusion. The $SO_2$ and reduced sulfur compounds are removed from the air stream by means of another set of denuders coated with copper/copper oxide and charcoal that are placed at the inlet of the sampling line. After collection, copper sulfate is decomposed by sequentially heating the train of denuders at 800 °C. $SO_2$ evolved is sent to the flame photometric detector in a stream of $N_2$. $SO_2$, sulfuric acid, and ammonium sulfate are determined by measuring the area of the corresponding peaks. With 1-h sampling, detection limits as low as 3 ppt can be attained.

A semicontinuous analyzer for sulfuric acid and sulfur aerosols with time resolution of 30 min has been recently described by Bohm and Israel (1990). A thin-walled, uncoated, stainless steel thermodenuder kept at 135 °C removes sulfuric acid from the sample, whereas other sulfur aerosols pass through it unaffected and reach the flame photometric detector. An inlet denuder removes sulfur gases from the incoming air, while a cyclone removes particles larger than 10 μm. After 15 min of sampling, air free from particles and sulfur gases is passed through the denuder. Temperature is raised to 700 °C and sulfuric acid is vaporized and sent to the photometric detector. After every cycle, air free from sulfur compounds is passed through the flame photometric detector to check the baseline. The system is controlled by a personal computer and can be run unattended for one month. It has been used for $SO_2$, sulfuric acid, and sulfate aerosols determinations in polluted areas in Germany. Daily profiles have been reported.

The last advanced application of denuder techniques that is worth mentioning deals with the evaluation of organic matter in carbon particles. This information is of key importance when the amount of organic and elemental carbon is used for source apportionment (Chow et al., 1991) or where photochemical reactions lead to an increase of organic matter on particles (Zielinska et al., 1989; Ciccioli et al., 1991). Since the organic material can be either blown off from the filter or adsorbed on particles during sampling, thermograms obtained by heating particles will show a level of organic carbon that will be different as a function of sampling duration and the concentration of organic species in air.

To evaluate these two effects, sampling of carbonaceous particles has been accomplished by inserting denuders made of quartz filters impregnated with carbon materials in front and after the main filter assembly (Eathough et al., 1991). The front denuder prevents deposition of volatile organics on carbon particles and the latter allows the determination of the amount of material volatilized from filter. The correct amount of organic carbon present in particles can then be evaluated by subjecting the filter and backup denuder to the same thermoanalytical process. The results obtained during a field monitoring campaign carried out in a deserted area in Utah confirmed the usefulness of denuder techniques for a correct sampling of carbonaceous particulate matter.

### 4.3.  Chromatographic Techniques

The simultaneous presence in air of different pollutants often requires separation steps for their unambiguous identification and quantitative determination at trace levels. Chromatography, in its various forms, can be used to perform this separation efficiently by exploiting the ability of a solid sorbent or a liquid coating to retain molecules having different vapor pressure, molecular size, solubility, and heat of adsorption.

Separation is performed by injecting the mixture (gaseous or liquid) into a column filled with a solid material whose surface can be coated with liquid stationary phases. A liquid or a gas is passed through the column so that a dynamic equilibrium of the components between the stationary and mobile phase is reached. The different affinity of the compounds present in the mixture toward the stationary phase results in a differential retention within the column so that individual components reach the column outlet at different times. Depending upon the performances of the column used, complete separation of the various components is obtained.

Detection of eluted compounds is possible if physical or chemical devices, capable of sensing the molecules emerging from the column in the presence of the mobile phase, are placed at the column outlet. Efficiency and capability depend upon the geometrical features of the column (column diameter and length, particle size). Selectivity is strictly related to the nature of the stationary phase used for the separation. In gas chromatography, the role of the mobile phase is negligible if compared to that of the liquid or solid sorbent. In liquid chromatography, both mobile and stationary phases are equally important in determining the separation process.

A third element, acting in practice as a multiplication factor for discriminating between the compounds present in complex mixtures, is the detection system. If the detector is sensitive and selective for a given compound, the determination is possible even in the case of incomplete separation. This is particularly true for gas chromatography (GC) where a large variety of selective and sensitive detectors are available. Among them are the flame ionization detection, electron capture, flame photometry, alkali flame, and photoionization detection. While the first is of general use because it is equally sensitive to all organic compounds, the others can be used to selectively detect compounds where specific functional groups or atoms are present in the molecule.

The combination of GC with mass spectrometry (GC-MS) allows unambiguous identification of eluted compounds through their mass spectra. If used in a selected ion detection (SID) system, GC-MS is extremely specific and sensitive. It requires, however, that the mass spectra of the compounds to be determined be known.

With the few exceptions already discussed, HPLC (high-performance liquid chromatography) is applied mostly to the analysis of high boiling components or to the cleanup of the soluble organic fraction (SOF) extracted from particles. Both reversed and normal phase liquid chromatography are suitable for this type of application. Ion liquid chromatography is used mostly in combination with sampling techniques based on denuders, impregnated filters, and liquid solutions on impingers.

Some of the applications of liquid chromatography have already been discussed. Applications of chromatographic techniques to the analysis of environmental samples have recently been reviewed (Liberti and Ciccioli, 1985, 1986) and details can be found in these articles. In this chapter, discussion will be restricted to the following items:

1. Improvement in the GC determination of $PAN_s$ in air
2. Evaluation of hydrocarbons in air by GC
3. Identification of nitrated polyaromatic hydrocarbons (nitro-PAH) formed by photochemical reactions

In our opinion, they represent the most interesting applications of chromatographic techniques for the analysis of air pollutants relevant to photochemical smog pollution and dry acid deposition.

### 4.3.1.  Determination of $PAN_s$ by Gas Chromatography

Gas chromatography with electron capture detection (ECD) has been the preferred method for the monitoring of peroxyacetyl nitrate (PAN) and its upper homologue, peroxypropyonyl nitrate (PPN), since the early 1960s (Darley et al., 1963). Until recently, it has represented the only possible alternative to long-path FTIR. The original method is simple, sensitive, and selective. It has been refined and adapted by numerous practitioners over the ensuing years although the essential features have been maintained to a large extent. Many of these modifications have been described in a recent review on the atmospheric chemistry of organic nitrates (Roberts, 1990).

Analysis is performed by direct injection of a known volume of air (typically 3–5 cm³) into a short column (50 cm to 1 m) packed with an inert support (usually Chromosorb) coated with 5–10% of polyethylene glycols of various types. The method can be automated and run unattended by connecting the column inlet to a sampling device composed of a multiport valve equipped with a calibrated loop, an aspirating pump, and a timer controlling sample transfer into the column. Elution of PAN and PPN is accomplished in 2–4 min, depending on the column temperature and flow rate. The column provides a good separation with respect to other electron capturing species (mainly chlorofluorocarbons and nitroalkanes) present in air. The resolution is not sufficient, however, for a base-to-base separation of PAN and PPN. Analysis is carried out at temperatures ranging between 25 and 40 °C to reduce $PAN_s$ decomposition occurring within the column. A time resolution varying from 12 to 30 min can be achieved depending upon the relative humidity of the sample as water is sensed by the ECD. It gives a broad, saturated peak with a positive/negative response showing severe tailing so that the sample can be injected only when the signal returns to the baseline. Although water can be eliminated by heart-cut techniques, this procedure is not adopted because it seems that water somehow prevents partial decomposition of PAN (Watanabe and Stephens, 1978).

The ECD can be operated in both constant current and constant frequency modes by using $N_2$ or mixtures of argon/methane for generating thermal electrons necessary for the electron capture process. In constant current the sensitivity is lower than constant frequency, but linearity and dynamical range are wider (Ciccioli, 1989). The choice of operating conditions is thus dependent on the levels of $PAn_s$ to be detected. Detection limits of the order of 50–100 ppt can be reached in constant frequency mode, under very clean conditions and with detector temperatures of 25–30 °C.

The importance of the detector operating conditions is shown in Fig. 9 where signals of the same $PAN_s$ analyzer working in constant current and constant frequency modes are compared (Ciccioli, 1989). PAN and PPN monitors based on electron capture detection are commercially available in Europe and Canada. They can be connected to electronic integration systems for $PAN_s$ quantification and data can be stored on personal computers by using software developed for chromatography. We have direct experience of a commercial $PAN_s$ analyzer produced in our country whose performance has been compared with that of a home-made analyzer built in our Institute. The former showed better sensitivity, precision, and accuracy although both gave consistent results in moderately polluted areas.

Figure 10a shows diurnal variations of PAN and PPN recorded during the intercomparison monitoring campaign on $HNO_3$ mentioned in the previous sections (Ciccioli et al., 1990). Data reported in this figure refer to the commercial instrument, but results obtained with the home-made analyzer were comparable within the experimental errors (5%).

The type of information gained through $PAN_s$ monitoring can be better understood by comparing daily profiles shown in Fig. 10a with those reported in Figs. 3a and 7b where trends of ozone and $HNO_3$ are shown. Since, unlike ozone, $PAN_s$ are mainly produced by man-made sources that, in our case, are concentrated in the urban area of Rome located 30 km away from the

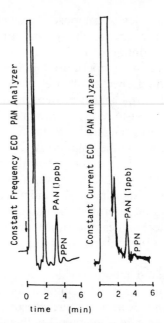

**Figure 9.** Comparison between the signal response of a commercially available PAN$_s$ analyzer (Carlo Erba, Italy) working in the constant current and constant frequency modes. (From Ciccioli, 1989.)

sampling site, they can be taken as good indicators of direct transport of photochemical pollution from the city. During the photochemical episode investigated, transport from Rome mainly occurred during the second and the last day of monitoring. It was detected by the sharp increase in the PAN and PPN levels in the middle of the day between 2 and 4 p.m. when strong vertical and horizontal mixing took place (Ciccioli et al., 1989a). Trends of this type are typical of our site when photochemical smog develops in the urban area of Rome. A comparison with daily profiles reported in Fig. 3a suggests that local sources of natural origin were contributing more than 50% of the ozone measured. During this monitoring campaign, it was not possible to distinguish direct transport from the city from local production simply by looking at ozone profiles. However, in some episodes, transport of ozone is clearly detected by a sharp peak concurrent with that of PAN with maximum ozone levels exceeding 100–120 ppb are recorded in the middle of the day (Ciccioli et al., 1989a).

PAN$_s$ monitoring is also useful to identify the different pathways leading to HNO$_3$ formation. If we compare Figs. 10a and 7b we see that the increase of HNO$_3$ observed on the third and the fourth day of monitoring cannot be explained in terms of concurrent transport of these pollutants from the city as PAN$_s$ and HNO$_3$ show different trends. Because local sources could not justify the levels of atmospheric acidity measured, the only possible expla-

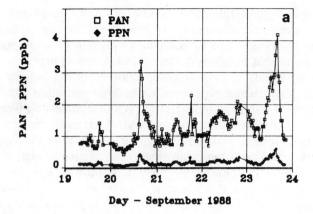

**Day — September 1988**

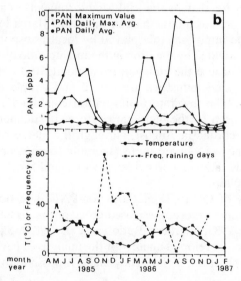

**Figure 10.** (*a*) Daily trends of PAN and PPN recorded during the Intercomparison Monitoring Exercise on Nitric Acid and Nitrate Measurement held in Montelibretti, 19–24 Sep 1988. Data obtained with a GC-ECD analyzer. (From Ciccioli et al., 1990.) (*b*) Seasonal variations of PAN in the Tiber valley from 1985 to 1987 recorded by using a GC-ECD analyzer. (From Ciccioli et al., 1989a.)

nation was that in those two days $HNO_3$ was generated by evaporation of ammonium nitrate particles left aloft from the previous days. This hypothesis was strongly supported by the measurement of evolved nitrate made possible by the use of annular denuders (see Section 4.2.2). Figure 7*c* shows that the maximum value of evolved nitrate (which is an indirect indication of evaporation processes) was reached on the same days in which direct transport from the city was negligible and was observed during the warmest hours of the day. This example illustrates well how the simultaneous monitoring of photochemical oxidants can be combined with that of acidic species to investigate equilibrium reactions occurring in the atmosphere.

While daily trends of $PAN_s$ provide information on long-range or local transport of photochemical pollutants, seasonal variations can be used to establish to what extent photochemical pollution affects a given area or if intrusion of $PAN_s$ takes place due to seasonal exchange between the lower and upper troposphere. Figure 10b reports monthly values of PAN obtained in our Institute during a two-year monitoring campaign carried out with a GC-ECD analyzer (Ciccioli et al., 1989a). A positive correlation is found with the temperature, and a negative correlation is observed with the frequency of rain events. The former is used as an approximate indicator of the solar radiation intensity, while the latter is an indirect way to detect when weather conditions were unfavorable to photochemical smog formation. Seasonal trends indicate that in the mid-Mediterranean areas, photochemical pollution takes place for a large part of the year (March to late October).

Evidence of the important role played by local transport from the urban area of Rome is provided by the drop in the mixing ratio observed in August. It occurs mainly because the emission from mobile and stationary sources is curtailed. Schools and industries are closed and people leave the city for summer holidays. The low rate of formation combined with high thermal decomposition caused by the high temperatures in summer leads to the low values of PAN recorded in August, although the meteorological conditions in the Mediterranean basin are quite favorable to photochemical smog pollution. During this campaign no evidence was found for PAN intrusion from the upper troposphere.

The sensitivity of GC-ECD methods for PAN detection can be further improved if enrichment procedures based on cryogenic techniques are adopted (Singh and Salas, 1983; Vierkorn-Rudolph et al., 1985). Sampling devices of this type have been developed mainly for the analysis of PAN in remote areas (Vierkorn-Rudolph et al., 1985). They make it possible to evaluate PAN at ppt levels.

An instrument of this type was at work during the *Polarstern* cruise (Muller et al., 1990), and was able to detect the small differences in PAN concentration existing between the northern and southern hemispheres. Measurements were carried out on board a ship moving over the Atlantic ocean. High-resolution capillary chromatography, suggested by us as a possible way to improve sensitivity of presently available $PAN_s$ analyzers (Liberti and Ciccioli, 1986), was shown to provide a good separation of all volatile organic nitrates (Roberts et al., 1989). However, the only practical result reported was obtained on less efficient wide-bore capillary columns internally coated with Carbowax 400 (Helmig et al., 1989). The sensitivity is slightly improved with respect to packed columns but the separation between PAN and PPN is basically unchanged.

Although highly efficient narrow-bore columns are advantageous in preventing PAN decomposition (Roumelis and Glavas, 1989), the amount of air that can be directly injected into these columns is so small that only concentrated samples can be analyzed. Cryogenic focusing is possible but thermal desorption can take place during the vaporization step necessary to transfer the sample into the column. These problems, combined with increased sophistication of the instrumentation required with capillary GC, explain why packed columns are still preferred.

The main limitation associated with the use of GC-ECD based $PAN_s$ analyzers is the need for frequent calibrations because the ECD response is highly sensitive to impurities present in the carrier gas. Although pure gases are often combined with adsorption traps to eliminate gases affecting the ECD response, the risk of contamination still exists, especially when the instrumentation is frequently moved for field measurements and the carrier gas disconnected. It is thus a good practice to perform frequent calibrations so that possible changes in sensitivity can be detected and proper response factors for evaluating $PAN_s$ concentration can be applied.

Unfortunately, calibration systems for PAN are more complicated than the PAN analyzer itself. The simplest one, developed by Grosjean et al. (1984), involves a photochemical reactor, permeation tubes filled with $NO_2$ and alkenes for PAN production. Analysis by ion liquid chromatography after alkaline hydrolysis of PAN is required for quantification. Permeation tubes containing PAN dissolved in high boiling solvents have also been employed in automated instrument calibration (Muller et al., 1990). This approach requires, however, that synthesis and purification of PAN be performed and permeation rates be checked by alkaline hydrolysis.

Because more severe problems are encountered in the calibration of PPN, the same response factor is often assumed for PAN and PPN. Although large differences in the electron capture cross section are unlikely to be observed between these two pollutants, the assumption that PAN and PPN give the same response can be an additional source of uncertainty in comparing values obtained by various research groups, especially if different detector geometry or working conditions are chosen.

To overcome the limitations associated with electron capture detection, Burkhardt et al. (1988) have proposed a GC method whereby PAN is determined by fluorescence induced by luminol after its catalytic conversion into $NO_2$. The advantage is that the instrument reading is directly expressed in ppb and signal response can also be checked by a $NO_2$ source. Moreover, air can be used as carrier gas. In principle, this type of monitor should allow a better comparison between data sets obtained by different research groups because it appears easier to use and the calibration is less rigorous. Performances of GC-ECD and GC-fluorescence PAN analyzers have been recently compared (Blanchard et al., 1990) and good agreement was found during a 24-day intercomparison study carried out in Canada. Adoption of GC analyzers equipped with chemiluminescence detection would also eliminate the need of calibration systems for PPN.

The availability of commercial GC-ECD and GC-chemiluminescence automated analyzers for $PAN_s$ seems to indicate that many of the problems connected with the monitoring of these photochemical oxidants, including the availability of reliable standards for field calibration, will soon be resolved and GC-based instruments will be included in large monitoring networks. This is highly desirable because data presently available are not sufficient to assess the role played by $PAN_s$ in the cycling of nitrogen-containing pollutants, to evaluate long-range transport on regional scale, to verify possible accumulation in the upper troposphere, or to measure the possible exchange occurring with the lower troposphere.

### 4.3.2.  Determination of Hydrocarbons in Air by Gas Chromatography

There are several reasons justifying the quantitative determination of individual organic compounds in air and emission sources. The main reason is that the reactivity of hydrocarbons toward OH radicals (Atkinson, 1986), ozone (Atkinson et al., 1984), and $NO_3$ radicals (Atkinson et al., 1985) is quite variable depending upon the chain length, structure, number of double and triple bonds, and presence of benzene rings and functional groups in the molecule. Differences in reactivity and products formed mean that different amounts of photochemical oxidants and acidic species can be generated during a pollution episode. Equal amounts of different organic mixtures lead to different ozone and $PAN_s$ production (Derwent and Jenkin, 1990a), and hence, levels of acidic species.

Another reason is that identification of reaction products (aldehydes, ketones, Greege's intermediates, alcohols, organic acids) makes it possible to establish to what extent reaction pathways and mechanisms postulated through smog chamber experiments actually occur in the atmosphere. This is of great help for improving models because it would result in a better parameterization of hydrocarbon chemistry.

The speciation of hydrocarbons in air is probably the only way to evaluate the relative contribution of natural and man-made hydrocarbons to photochemical smog formation. This is a fundamental prerequisite if realistic control strategies aimed at reducing photochemical smog pollution are to be developed. Moreover, the availability of emission rates for individual hydrocarbons is crucial for a serious validation of regional models developed for predicting photochemical pollution and acid deposition.

In practice, the determination of hydrocarbons is made difficult by the fact that a method capable of providing separation, identification, and quantitative determination of all organic compounds that can be released or formed in air does not exist. The common use of the level of total hydrocarbons less methane, promulgated by various national agencies as a suitable index for expressing hydrocarbon content in air and emission sources, certainly has not contributed to the promotion of research in this field.

Difficulties in the speciation of organic compounds arise mainly from the high complexity of the mixture, low levels present in air, and different physical and chemical properties of individual hydrocarbons. An accurate quantification would require separation of all the compounds. In practice, this is not possible because the resolution afforded by the best columns currently available is not sufficient for complete separation.

Although specific multidetection systems can help to discriminate between some co-eluted components, some information is lost anyway. In addition, the amount of matter needed for detection and/or identification largely exceeds what can be transferred into the column by direct injection. Consequently, enriching procedures must be adopted for collecting hydrocarbons. There are three common methods for sample enrichment: condensation of hydrocarbons by passing air through a tube kept at very low temperatures (liquid argon or oxygen), adsorption on traps filled with solid sorbents capable of retaining hydrocarbons at ambient temperature, and a combination of these two methods (i.e, absorption on trapping materials maintained at low temperatures). None of these is capable of enriching all hydrocarbons with carbon numbers ranging from $C_2$ to $C_{16}$ at the levels required for individual detection. Methane is excluded from the list of hydrocarbons requiring preconcentration because it reaches ABL levels (1–2 ppm) easily detectable by direct injection. Compounds with carbon numbers above $C_{16}$ are present mainly as aerosols and are collected on filters.

When cryogenic enrichment on empty tubes or cooled traps is employed, the main limiting factor is condensation of water and $CO_2$ in the collection system. After a certain volume of air is passed through the enriching device, lines or traps get plugged and no more air can be passed through them. The presence of high amounts of these gases in the sample also poses severe limitations to the chromatographic and detection system if certain columns are used for the separation.

When collection is made on traps maintained at ambient temperature other limiting factors can also affect the enriching process. If the absorbent is strong enough to retain low boiling compounds (this usually occurs when the specific surface area of the adsorbent is much larger than about 200 m²/g), it is almost impossible to desorb high boiling components without reaching temperatures higher than 300 °C. This unavoidably leads to sample losses due to decomposition of thermally unstable compounds. If the adsorbent is light (specific surface area between 20 and 100 m²/g) its capability to retain compounds is limited so that volatile compounds are lost when enriching factors sufficient for the detection of high boiling compounds are reached.

The need to find a satisfactory compromise between enrichment and recovery combined with the inadequacy of columns to separate all hydrocarbons are the fundamental reasons why different approaches are followed in the determination of organic compounds. Methods are usually identified by the range of carbon number to be quantified. The combination of the enrichment procedure and chromatographic column adopted defines the carbon range and the way the sampling is performed.

With cryogenic techniques, the storage of samples on empty tubes is difficult, so the enriching unit is preferentially combined with the GC system (McClenny et al., 1984) through a multiport valve. After collection, the tube containing sample is connected to the column inlet by switching the valve. The frozen sample is then heated, vaporized, and injected into the column. Water and $CO_2$ can be removed by means of diffusion systems based on ion exchange membranes (McClenny et al., 1984). This process does not affect nonpolar components but can selectively remove species containing reactive functional groups. By using a personal computer for driving the various operations, systems of this type can be automated and run unattended for long periods of time (McClenny et al., 1984).

In remote areas, collection on electropolished canisters is the preferred way of sampling (Schmidbauer and Oehme, 1986). The canister is taken to the laboratory and an aliquot of the sample is injected into the GC column after being dried and cryo-focused with procedures similar to those just described. Water can be removed by traps filled with desiccants (Schmidbauer and Oehme, 1986). Enrichment on adsorbing materials offers the advantages over canisters in that enrichment and sampling are carried out at once and the sample can be averaged over a longer time. Traps are often preferred because they are easier to handle and can be cleaned during the desorption step.

A wide variety of solid sorbents, known since the mid-1970s, can be used for collecting hydrocarbons. The list includes porous polymers, such as Tenax (Pellizzari et al., 1976; Knoeppel et al., 1980; Arnts, 1985; Yokouchi et al., 1986, 1990; Kanakidou et al., 1988; Clement et al., 1990) and Chromosorb 102 (Louw et al., 1977), graphitic carbons with different surface areas (Bruner et al., 1974; Ciccioli et al., 1976, 1986; Bloemen et al., 1990), and carbon molecular sieves (Rudolph et al., 1981; Bloemen et al., 1990). Traps filled with different adsorbents have sometimes been used to widen the range of carbon numbers investigated (Rudolph et al., 1981; Bloemen et al., 1990). Sampling is often carried out at low temperatures to improve the adsorption efficiency of hydrocarbons with a certain carbon number (Bruner et al., 1974; Kanakidou et al., 1988; Bloemen et al., 1990).

Traps have been combined with a large variety of capillary and packed columns (Liberti and Ciccioli, 1985). Capillary chromatography has been shown to provide a better resolution but requires cyrofocusing of the organic material at the column inlet. This additional step is needed because desorption of traps takes time and requires flow rates (10–30 mL/min) higher than those normally used by narrow-bore capillary columns (1–5 mL/min). Cryofocusing is performed with the same enriching techniques used for cryogenic sampling in air (Yokouchi et al., 1986; Kanakidou et al., 1988) or by modifying the injection system (Arnts, 1985; Bloemen et al., 1990).

All methods intended for the determination of hydrocarbons of importance in photochemical studies fall into two main categories regardless of the enriching procedure chosen. The first category includes all methods using capillary columns internally coated with aluminum oxide/KCl (Schneider et al.,

1978) or any other having a similar performance. They allow the elution and complete separation of all alkanes, alkenes, alkines, and arenes in the $C_1$–$C_7$ carbon range, whereas polar compounds and hydrocarbons with molecular weight higher than $C_7$ so retained cannot be detected. Positive identification of eluted compounds can be achieved through the measurement of the retention index, so that flame ionization is preferred for detection and quantification. To preserve the column performance and extend its lifetime, water is removed from the sample using dessicants or ion exchange membranes. Their use should not affect the sample quality since polar compounds, which are more selectively adsorbed by these devices, are not analyzed by the column.

Determinations of hydrocarbons with this technique have been carried out in polluted and remote areas (Schmidbauer and Oehme, 1986; Kanakidou et al., 1988; Bloemen et al., 1990). Detection limits ranging between tenths of ppt to several ppb of individual components were demonstrated. An example of the GC profile recorded in a moderately polluted area is reported in Fig 11 (Bloemen et al., 1990). In this case, collection was carried out on traps

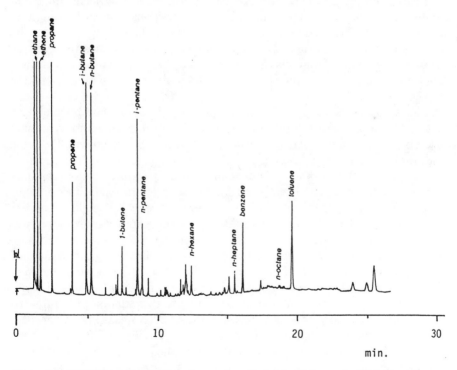

**Figure 11.** GC-FID profile of a 2-L air sample collected in Bilthoven in 1990. Trapping was carried out using tubes filled with graphitic carbons and carbon molecular sieves. A Poraplot U fused-silica column (0.53 mm i.d., 20 μm coating) was used for the analysis. Water was removed by Nafion tubes (Perma Pure). (From Bloemen et al., 1990.)

filled with two graphitic carbons and carbon molecular sieves kept at $-10$ °C. A drying system comprising two diffusion membranes set in series was used. The instrument was completely automated and was run unattended.

Although a large number of reactive hydrocarbons from natural and anthropogenic origin can be detected and quantified through this method, some important information is missing in the chromatogram of Fig. 11. Trimethyl and tetramethylbenzenes, exhibiting high potential for producing ozone (Derwent and Jenkin, 1990a) and oxygenated compounds formed by photochemical reactions or emitted by man-made and natural sources are not detected. This is why methods falling in the second category have been developed. They make use of capillary columns internally coated with thick films of methylphenylsilicones which offer the advantage that all components up to $C_{16}$ are eluted. With these columns water does not represent a serious problem and drying systems can be avoided in many instances. Collection is preferentially carried out on traps kept at ambient temperature so that water and $CO_2$ are only partly retained. If 1- to 2-L samples are collected, components lower than $C_5$ are partly lost (Knoeppel et al., 1980; Arnts, 1985; Yokouchi et al., 1990; Clement et al., 1990).

In polluted areas, where 300–500 cm$^3$ of sample are sufficient for quantitative determination, determination of hydrocarbons with carbon number equal to 4 may be possible. Separation of low boiling compounds can be achieved with GC instruments having subambient capabilities by selecting proper temperature programs (Yokouchi et al., 1986). If hydrocarbons starting from $C_3$ are to be evaluated, the upper limit of the carbon range is limited to $C_{13}$.

Figure 12 shows a typical chromatographic profile obtained by analyzing a 2-L air sample collected in a forest area of Central Italy with the method used in our laboratory. Air was enriched on glass tubes filled with the same adsorbents used for generating the chromatogram shown in the previous figure, but sampling was carried out at ambient temperature. A 60-m capillary column coated with DB-1 was used for the separation. Quantitative determination of all hydrocarbons with retention times equal or higher than isoprene was possible. Positive identification of the various constituents was carried out by repeated analysis using the mass spectrometer as detection system. It is interesting to note the presence in the chromatogram of nonanal and decanal, which could not have been detected with the previous method. These compounds were also found in a rural area in Japan (Yokouchi et al., 1990) and their presence in air was attributed to degradation of natural hydrocarbons. With low-polar columns, the use of mass spectrometry for identification purposes is almost mandatory since co-elution of different components is often observed and compounds with different polarity are analyzed by the column.

The increasing demand for emission and aerometric data of individual hydrocarbons can be satisfactorily fulfilled by the two chromatographic methods just described. While the method based on aluminum oxide/KCl columns

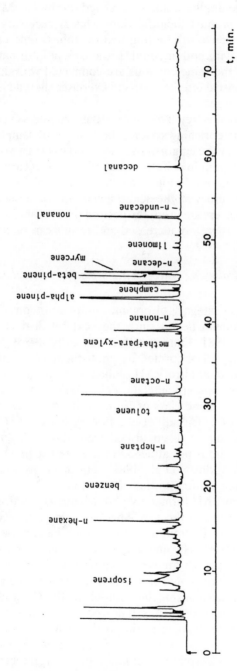

**Figure 12.** GC-FID profile of a 2-L air sample collected in the Regional Park of the Monti Cimini (Viterbo, Italy) in Sep 1990. The trapping material was the same as that used for generating Fig. 11. A 60-m narrow- bore column (0.32 mm) coated with DB-1 (1.2 μm film thickness) was used for the analysis. Identification was carried out by mass spectrometry. Original data from the authors (unpublished).

may already be installed in large monitoring networks, the other might not be so easy to run because duplicate analysis carried out by GC-MS might be frequently required for product identification. This is necessary especially when many sources are active in emitting hydrocarbons. One of the main problems that still exists with both methods is the lack of information on the accuracy obtained when the same columns are connected with different enriching procedures. Extensive intercomparison exercises should be promoted to standardize the procedures.

Another serious problem is time resolution. If automated sampling is performed by using the same enriching system, the number of samples that can be run each day is limited by the length of the chromatogram. This corresponds to about one analysis each hour in polluted areas or less in rural areas. Because the sample is not averaged over a long time, is not possible to follow daily profiles with the same accuracy as the acidic species or some photochemical pollutants. If traps or canisters are used, the desired time resolution is reached but the amount of work is greatly increased and trends can be followed for only a few days.

### 4.3.3. *Methods for Evaluating Nitro-PAH Formed by Photochemical Reactions*

Nitro-PAH belongs to the category of organic compounds considered to be potentially harmful to human health since they exhibit distinct mutagenic activity (Schuetzle et al., 1981; Finlayson-Pitts and Pitts, 1986). Nitro-PAH can be either directly emitted or formed by photochemical reactions. 1-Nitropyrene is the most abundant nitro-PAH of man-made origin. It is released mainly by autovehicular emission (Schuetzle et al., 1981) where it is associated with other members of the same class or isomers that are present at lower levels (Schuetzle et al., 1981; Ciccioli et al., 1989b). Nitro-PAH formed by photochemical reactions can be distinguished from those that are emitted because the functional group is not attached to the site (or sites) having the highest electron density (Atkinson et al., 1987). This happens because of the differences in the reactions leading to their formation.

The occurrence of nitro-PAH formed by photochemical reactions was first detected in samples collected in the Los Angeles basin (Pitts et al., 1985). The unexpected presence of fluoranthene and pyrene having the functional group in position 2, not justified by any of the known sources present in that area, suggested that a mechanism similar to that proposed for the conversion of toluene into *m*-nitrotoluene could be invoked to explain their formation in air. According to this mechanism, the addition of the OH radical to the site with the highest electron density is followed by the attack of $NO_2$ in position 2, with loss of water leading to the final product. This sequence of reactions occurs in the gas phase and only after any nitroarenes formed are deposited on particles by adsorption. Formation of 2-nitrofluoranthene/pyrene isomers should then be observed when temperature is high and levels of photochemical oxidants suggest high atmospheric reactivity.

Later studies (Arey et al., 1986; Atkinson et al., 1987; Zielinska et al., 1989) have led to a better understanding of nitroarenes formation in air. They have shown that the mechanism involving OH radicals and $NO_2$ is common to many volatile arenes (particularly interesting is the formation of 3-nitro-biphenyl from byphenyl) and that other pathways exist for the formation of 2-nitrofluoroanthene.

Compelling evidence has been provided that fluoranthene in the gas phase can react with $N_2O_5$ to yield the 2-nitro derivative. According to the cycle shown in Fig. 2, this reaction is expected to be the dominant pathway at night whenever high photochemical activity develops in the preceding day. In any case, occurrence of nitro-PAH should be ubiquitous since it is not linked to a specific source. Observations made in Europe (Ramdahl et al., 1986; Liberti and Ciccioli, 1986; Ciccioli et al., 1989b) have shown clearly that 2-nitro-fluoranthene/pyrene isomers are actually present in dust collected during enhanced photochemical activity. An analysis of various emission sources also revealed that only one particular industrial process able to release components of this type has been identified so far. This occurs because conditions simu-lating nighttime situations (emission of fluoranthene in the gas phase, presence of ppm levels of $NO_2$, ozone formation by electrostatic discharge) are some-how obtained (Ciccioli et al., 1989b). However, few sources of this type exist around the world and their emission cannot justify the ubiquitous occurrence of 2-nitrofluoranthene/pyrene isomers. It is thus safe to say that their presence in air can be attributed almost exclusively to photochemical reactivity.

All methods for the evaluation of photochemical originated nitro-PAH (Pitts et al., 1985; Ramdahl et al., 1985; Liberti and Ciccioli, 1986; Ciccioli et al., 1989b; Zielinska et al., 1989) have the same basic features in common: high-volume collection of particles lower than 10 $\mu$m, solvent extraction of the SOF, cleanup of the extract on columns filled with alumina or silica particles, isolation of the fraction containing nitro-PAH, followed by GC-MS analysis for the identification and quantification of species present. If volatile components are to be determined simultaneously, collection can be performed on polyurethane foam placed after the filter (Atkinson et al., 1987). After sampling, the backup collector is extracted and subjected to the same puri-fication and analytical steps used for dust. Capillary chromatography is the preferred method for separating nitro-PAH because of its high efficiency and low column activity. Either long columns (60 m) coated with low-polar liquid phases (DB-5) (Pitts et al., 1985; Ramdahl et al., 1985; Liberti and Ciccioli, 1986; Zielinska et al., 1989) or slightly polar (DB-17), short capillaries (17 m) (Ciccioli et al., 1989b) can accomplish a satisfactory separation of nitro-PAH. While the latter column provides a better separation of nitrofluoran-thene/pyrene isomers, the former makes it possible to investigate nitro-PAH over a wider range of carbon numbers. The high complexity of the fraction to be analyzed requires the combined use of retention index and GC-MS in SID for positive identification. Specific ions that can be used for detection are $[M^+]$, $[M\text{-}NO]^+$, $[M\text{-}NO_2]^+$, $[M\text{-}HNO_2]^+$, and $[M\text{-}NO\text{-}CO]^+$.

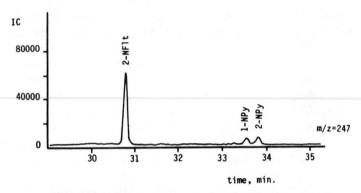

**Figure 13.** GC-MS profile of nitroarenes with molecular weight 247 obtained by analyzing the organic fraction extracted from a 600-m³ ambient particulate sample collected in the center of Milan (Italy) in Mar 1990. A 17-m capillary column coated with DB-17 was used for the separation. (From Ciccioli et al., 1991.)

Figure 13 shows a typical profile of nitrofluoranthene/pyrene isomers recorded during the GC-MS-SID analysis of a dust sample collected in Milan (Ciccioli et al., 1991). It is quite different from those found in the urban area of Rome (Ciccioli et al., 1989b), but closely resembles those found in the Los Angeles area (Zielinska et al., 1989). The presence of photochemically originated nitroarenes can be clearly detected and the relative abundance of the man-made nitro-PAH (1-nitropyrene) evaluated. Figure 14 reports day/night concentrations of 2-nitrofluoranthene in Milan (Ciccioli et al., 1991). The

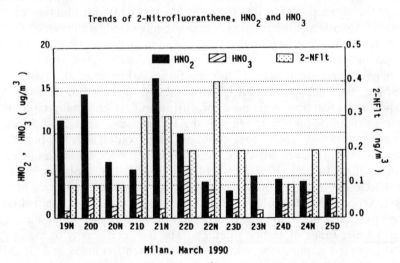

**Figure 14.** Day/night variations of 2-nitrofluoranthene observed in downtown Milan. (From Ciccioli et al., 1991.)

occurrence of higher values recorded at night from 21 to 24 March seems consistent with the existence of the nighttime pathway involving $N_2O_5$.

Although monitoring of nitro-PAH is difficult, time-consuming, and affordable only in laboratories where instrumentation and dedicated personnel are available, the basic structure needed is the same as that required for the monitoring of volatile hydrocarbons, so the high price can be justified by the amount of information that can be collected. We have intentionally given equal importance to the monitoring of volatile organics and nitroarenes because we believe that if the former are important for a better understanding of the reactions leading to photochemical oxidants, the latter are more strictly linked to the adverse effects that photochemical pollution can have on human health.

While great attention has been devoted to the effect of ozone, $PAN_s$, and acidic species, very little has been said about the mutagenic, carcinogenic, or teratogenic risk associated with photochemical pollution. It has to be stressed that among the fraction containing nitroarenes that exhibits a high specific direct mutagenic activity (revertants/g of extract), the highest number of revertants is often observed in the most polar fraction whose composition is virtually unknown and may be influenced by the photochemical reactivity of the atmosphere. So far, we know that at least nitroarenes are formed during the day/night cycle and that they have effects comparable to the most toxic organic compounds (such as polychlorinated dibenzodioxins) identified in the atmospheric environment.

## 5. CONCLUSIONS

Many tools for investigating the cycling of compounds relevant to photochemical pollution and dry acid deposition are now available to the environmental chemist. Some of them look adequate in terms of sensitivity and precision. In many instances desired time resolution and accuracy can be conjugated with simplicity of operation. Gaps still exist in certain areas. Some of these gaps could be filled by promoting repeated intercomparison exercises covering not only the species that are the main subject of the experiment, but also those components that are related to them through the chain of reactions shown in Figs. 1 and 2. Other gaps can be only filled through decisive advances in instrumentation and technology. Aspects more closely related to health effects should be incorporated in the monitoring strategies because they may have a more direct impact than that currently being attributed to some primary pollutants.

## REFERENCES

Allegrini, I., and De Santis, F. (1989). "Measurement of atmospheric pollutants relevant to dry deposition." *CRC Crit. Rev. Anal. Chem.* **21**, 237–254.

Allegrini, I., De Santis, F., Di Palo, V., Febo, A., Perrino, C., Possanzini, M., and Liberti, A. (1987). "Annular denuder method for sampling reactive gases and aerosols in the atmosphere." *Sci. Total Environ.* **67**, 1–16.

Allegrini, I., Febo, A., and Perrino, C. (1990). "Field intercomparison exercise on nitric acid and nitrate measurement: Critical evaluation of the results." In G. Restelli and G. Angeletti, eds., *Physico-chemical Behavior of Atmospheric Pollutants*, Proc. 5th Eur. Symp., Varese, Italy, 25–28 September 1989. Kluwer Academic Publishers, Dordrecht, pp. 69–74.

Altshuller, A. P. (1983a). "Natural volatile organic substances and their effect on air quality in the United States." *Atmos. Environ.* **17**, 2131–2165.

Altshuller, A. P. (1983b). "Measurements of the products of atmospheric photochemical reactions in laboratory studies and in ambient air—relationships between ozone and other products." *Atmos. Environ.* **17**, 2383–2427.

Altshuller, A. P. (1986). "Review paper: The role of nitrogen oxides in nonurban ozone formation in the planetary boundary layer over N America, W Europe and adjacent areas of ocean." *Atmos. Environ.* **20**, 245–268.

Appel, B. R., and Tokiwa, Y. (1981). "Atmospheric particulate nitrate sampling errors due to reactions with particulate and strong acids." *Atmos. Environ.* **15**, 1087–1089.

Arey, J., Zielinska, B., Atkinson, R., Winer, A. M. Ramdahl, T., and Pitts, J. N., Jr. (1986). "The formation of nitro-PAH from the gas-phase reaction of fluoranthene and pyrene with the OH radical in the presence of NOx." *Atmos. Environ.* **20**, 2339–2345.

Arnts, R. R. (1985). "Precolumn sample enrichment device for analysis of ambient volatile organics by GC-MS." *J. Chromatogr.* **329**, 399–405.

Atkinson, R. (1986). "Kinetics and mechanisms of the gas-phase reaction of the hydroxyl radical with organic compounds under atmospheric conditions." *Chem. Rev.* **88**, 69–201.

Atkinson, R., and Carter, W. P. L. (1984). "Kinetics and mechanisms of the gas-phase reactions of ozone with organic compounds under atmospheric conditions." *Chem. Rev.* **84**, 69–201.

Atkinson, R., and Lloyd, A. C. (1982). "An updated chemical mechanism for hydrocarbons/NOx/SO2 photoxidation suitable for inclusion in atmospheric simulation models." *Atmos. Environ.* **16**, 1341–1355.

Atkinson, R., Ashmann, S. M., Carter, W. P., Winer, A. M., and Pitts, J. N., Jr. (1984). "Formation of alkylnitrates from the reaction of branced and alkyl peroxy radicals with NO." *Int. J. Chem. Kinet.* **16**, 1085–1101.

Atkinson, R., Ashmann, S. M., Winer, A. M., and Pitts, J. N., Jr. (1985). "Kinetics and atmospheric implications of the gas-phase reactions of NO3 radicals with a series of monoterpenes and selected organics at 294 ± 2 °K." *Environ. Sci. Technol.* **19**, 159–163.

Atkinson, R., Arey, J., Zielinska, B., Winer, A. M., and Pitts, J. N., Jr. (1987). "The formation of nitropolycyclic hydrocarbons and their contribution to the mutagenicity of ambient air. In S. S. Sandhu, D. M. De Marini, M. J. Mass, M. M. Moore, and J. L. Mumford, eds., *Short-term Bioassays in the Analysis of Complex Environmental Mixtures*. Plenum, New York, Vol. V, pp. 291–309.

Beilke, S., and Gravenhorst, G. (1980). "Cycles of pollutants in the troposphere." In B. Versino and H. Ott, eds., *Physico-chemical Behavior of Atmospheric Pollutants*, Proc. Eur. Symp. Ispra, Italy 16–19 October 1979. CEC, Brussels, EUR 6621 DE, EN, FR, pp. 331–353.

Biermann, H. W. Tuazon, E. C., and Winer, A. M. (1987). *In-situ Long Pathlength FT-IR and DOAS Measurements of Gaseous Species During the Citrus College Study*. Paper presented at the Third International Conference on Carbonaceous Particles in the Atmosphere, Berkeley, CA, October 5–8, 1987, Abstr. No. B6, p. 18.

Blanchard, P., Shepson, P. B., So, K. W., Schiff, H. I., Bottenheim, J. W. Gallant, A. J., Drummond, J. W., and Wong, P. (1990). "A comparison of calibration and measurement techniques for gas chromatographic determination of atmospheric peroxyacetyl nitrate (PAN)." *Atmos. Environ.* **24A**, 2839–2846.

Bloemen, H. J., T., Bos, H. P., and Dooper, H. P. M. (1990). "Measuring VOC for the study of atmospheric processes." *Int. Lab.* **20**, Sept., 23–26.

Bohm, R., and Israel, G. W. (1990). "A new analyzer for sulfuric acid and sulfur aerosol in ambient air: Development and measurements." In G. Restelli and G. Angeletti, eds., *Physico-chemical Behaviour of Atmospheric Pollutants*, Proc. 5th Eur. Symp. Varese, 25–28 September 1989. Kluwer Academic Publishers, Dordrecht, pp. 145–151.

Bolin, B., Granat, L., Ingelstrom, L., Joannesson, Mattsson, Odén, S., Rodhe, H., and Tamm, C. O. (1972). *Sweden's Case Study for the United Nations Conference on the Human Environment: Air Pollution Across National Boundaries. The Impact on the Environment of Sulfur in Air and Precipitation*. Norstadt and Sons, Stockholm, pp. 1–97.

Bowman, L. D., and Horak, R. F. (1972). "A continuous ultraviolet absorption ozone photometer." *Anal. Instrum.* **10**, 103–108.

Bradshaw, J., Rodgers, M. O., Sandholm, S. T., Kesheng, S., and Davis, D. D. (1985). "A two-photon laser-induced fluorescence field instrument for ground-based and airborne measurements of atmospheric $NO_2$." *J. Geophys. Res.* **90**, 12861–12873.

Brassington, D. J. (1989). "Measurements of atmospheric HCl and NH3 with a mobile tunable diode laser." In R. Grisar, G. Schmidtke, M. Tacke, and G. Restelli, eds., *Monitoring of Gaseous Pollutants by Tunable Diode Laser*, Proc. Int. Symp. Freiburg, F.R.G., 17–18 October 1988. Kluwer Academic Publishers, Dordrecht, pp. 16–24.

Brice, K. A., Penkett, S. A., Atkins, D. A. F., Sandalls, F. J., Bamber, D. J., Tuck, A. F., and Vaugham, G. (1984). "Atmospheric measurements of peroxyacetyl nitrate (PAN) in rural south-east England: Seasonal variations winter photochemistry and long-range transport." *Atmos. Environ.* **18**, 2691–2702.

Bruner, F., Ciccioli P., and Di Nardo, F. (1974). "Use of graphitised carbon black in environmental analysis." *J. Chromatogr.* **99**, 661–672.

Bruning, D., and Rohrer, F., (1990). "Surface NO and $NO_2$ mixing ratios measured between 30 °N and 30 °S in the Atlantic region." In G. Restelli and G. Angeletti, eds., *Physico-chemical Behavior of Atmospheric Pollutants*, Proc. 5th Eur. Symp. Varese, 25–28 September 1989. Kluwer Academic Publishers, Dordrecht, pp. 699–704.

Burkhardt, M. R., Maniga, N. I., Stedman, D. H., and Paur, R. J. (1988). "Gas chromatographic method for measuring nitrogen dioxide and peroxyacetyl nitrate in air without compressed gas cylinders." *Anal. Chem.* **60**, 816–819.

Calvert, J. C., and Stockwell, W. R. (1983). "Acid generation in the troposphere by gas-phase chemistry," *Environ. Sci. Technol.* **17**, 428A–443A.

Calvert, J. C., Lazrus, A. J., Kok, G. L., Heikes, B. G., Walega, J. G., Lind, J., and Cantrell, C. A. (1985). "Chemical mechanisms of acid generation in the troposphere." *Nature (London)* **317**, 27–35.

Carlier, P., and Mouvier, G. (1988). "Initiation à la physico-chimie de la basse troposphère." *Pollut. Atmos.* **117**, 12–23.

Chock, D. P., and Heuss, J. M. (1987). "Urban ozone and its precursors." *Environ. Sci. Technol.* **21**, 1146–1153.

Chow, J. C., Lowenthal, D. H., Pritchett, L. C., Frazier, C. A., Watson, J. G., Neuroth, G. R., and Robbins, R. (1991). *Differences in Carbon Thermograms for Diesel and Gasoline Powered Vehicles.* Paper presented at the Fourth Conference on Carbonaceous Particles in the Atmosphere, Vienna, Austria, 3–5, April, 1991, p. 14.

Ciccioli, P. (1989). "Determination of peroxyacetyl nitrate (PAN) and peroxypropionyl nitrate (PPN)." In I. Allegrini, A. Febo, and C. Perrino, eds., *Field Intercomparison Exercise on Nitric acid and Nitrate Measurements—Methods and Data.* CEC, Brussels, Air Pollut. 22, pp. 194–201.

Ciccioli, P., Bertoni, G., Brancaleoni, E., Fratarcangeli, R., and Bruner, F. (1976). "Evaluation of organic pollutants in the open air and atmospheres in industrial sites using graphitized carbon black traps and GC-MS analysis with specific detectors." *J. Chromatogr.* **126**, 757–770.

Ciccioli, P., Brancaleoni E., Cecinato, A., Di Palo, C., Brachetti, A., and Liberti, A. (1986). "Gas chromatographic evaluation of the organic components present in the atmosphere at trace levels with the aid of Carbopack B for the preconcentration of the sample." *J. Chromatogr.* **351**, 433–449.

Ciccioli, P., Draisci, R., Cecinato, A., and Liberti, A. (1987). "Sampling of aldehydes and carbonyl compounds in air and their determination by liquid chromatographic techniques." In G. Angeletti and G. Restelli, eds., *Physico-chemical Behaviour of Atmospheric Pollutants*, Proc. 4th Eur. Symp. Stresa, Italy, 23–25 September 1986. Reidel, Dordrecht, pp. 131–141.

Ciccioli, P., Brancaleoni, E., Cecinato, A., and Brachetti, A. (1989a). "Diurnal and seasonal variations of peroxyacetyl nitrate (PAN) in a semirural area of Central Italy." In L. J. Brasser and W. C. Mulder, eds., *Proceedings of the Eighth Clean Air Congress.* Elsevier, Amsterdam, Vol. 3, pp. 497–502.

Ciccioli, P., Cecinato, A., Brancaleoni, E., Draisci, R., and Liberti, A. (1989b). "Evaluation of nitrated polycyclic aromatic hydrocarbons in anthropogenic emission and air samples. A possible means of detecting reactions of carbonaceous particles in the atmosphere." *Aerosol Sci. Technol.* **10**, 296–310.

Ciccioli, P., Possanzini, M., Cecinato, A., Di Palo, V., Brancaleoni, E., Brachetti, A., Mura, S., Perner, D., Parchatka, U., Karbach, H., and Eslick, I. C. (1990). "Trends of photochemical oxidants observed during the intercomparison field exercise on nitric acid and nitrate measurements held in Montelibretti." In G. Restelli and G. Angeletti, eds., *Physico-chemical Behaviour of Atmospheric Pollutants,*

Proc. 5th Eur. Symp., Varese, Italy, 25–28 September 1989. Kluwer Academic Publishers, Dordrecht, pp. 75–82.

Ciccioli, P., Cecinato, A., Brancaleoni, E., Cabella, R., and Buttini, p. (1991). *Nighttime Formation of 2-Nitrofluoranthene in an Urban Area.* Paper presented at the Fourth Conference on Carbonaceous Particles in the Atmosphere, Vienna, Austria, 3–5, April, 1991, p. 15.

Clement, B., Riba, M. L., Leduc, R., Haziza, M., and Torres, L. (1990). "Concentration of monoterpenes in a maple forest in Quebec." *Atmos. Environ.* **24A**, 2513–2516.

Cowling, E. B. (1982). "An historical resume of progress in scientific and public understanding of acid precipitation and its biological consequences." In F. M. D'Itri, ed., *Acid Precipitation: Effects on Ecological Systems.* Ann Arbor Science Publishers, Collingwood, MI, pp. 43–83.

Cox, R. A. (1979). "Rates, reactivity and mechanism of homogeneous atmospheric oxidation reactions." In B. Versino and H. Ott, eds., *Physico-chemical Behaviour of Atmospheric Pollutants,* Proc. 1st Eur. Symp. Ispra, Italy, 16–18 October 1979. CEC, Brussels, EUR 6621 DE, EN, FR, pp. 91–109.

Cox, R. A., and Penkett, S. A. (1983). "Formation of atmospheric acidity." In S. Beilke and A. Elshout, eds., *Proceedings of a CEC Workshop on Acid Deposition, Berlin (Reichstag).* Reidel, Dordrecht, pp. 56–81.

Cox, R. A., Eggleton, A. E. J., Derwent, R. G., Lovelock, J. E., and Pack, D. H. (1975). "Long-range transport of photochemical ozone in Northern-Western Europe." *Nature (London)* **255**, 118–121.

Darley, E. F., Kettner, K. A., and Stephens, E. R. (1963). "Analysis of peroxyacyl nitrates by gas chromatography with electron capture detection." *Anal. Chem.* **35**, 389–391.

Dasgupta, P. K., Dong, S., Hwang, H., Yang, H. C., and Genfa, Z. (1988). "Continuous liquid-phase fluorometry coupled to a diffusion for the real-time determination of atmospheric formaldehyde, hydrogen peroxide and sulfur dioxide." *Atmos. Environ.* **22**, 949–963.

Demerjan, K. L., Kerr, J. A., and Calvert, J. G. (1974). "The mechanism of photochemical smog formation." *Adv. Environ. Sci. Technol.* 1–263.

Derwent, R. G. (1990). "Evaluation of a number of chemical mechanisms for their application in models describing the formation of photochemical ozone in Europe." *Atmos. Environ.* **24A**, 2615–2624.

Derwent, R. G., and Jenkin, M. E. (1990a). "Hydrocarbon involvement in photochemical ozone formation in Europe." *U.K., At. Energy Res. Estab., Rep. AERE* **R13736**.

Derwent, R. G., and Jenkin, M. E. (1990b). "Hydrocarbons and the long range transport of ozone and PAN across Europe." *U.K., At. Energy Res. Estab., Rep. AERE* **R13816**.

De Santis, F. (1989). "Methods for the collection and the analysis of HNO2, NH3, SO2 and related anions in particles." In I. Allegrini, A. Febo, and C. Perrino, eds., *Field Intercomparison Exercise on Nitric Acid and Nitrate Measurement— Methods and Data.* CEC, Brussels, Air Pollut. Res. Rep. 22, pp. 202–206.

Driscoll, C. T., and Newton, R. M. (1985). "Chemical characteristics of Adirondack lakes." *Environ. Sci. Technol.* **19**, 1018–1024.

Drummond, J. W., Ehhalt, D. H., and Volz, A. (1985). "An optimized chemilumi-nescence detector for tropospheric NO measurements." *J. Amos. Chem.* **2**, 287–306.

Durham, J. L., Wilson, W. E., and Bailey, E B. (1978). "Application of a SO2 denuder for continuous measurements of sulfur in submicrometric aerosols." *Atmos. Environ.* **12**, 883–886.

Eathough, D. J., Wadsworth, A., Eathough, D. A., Crowford, J., Hansen, L. D., and Lewis, E. A. (1991). *A Multiple-System, Multichannel Diffusion Denuder Sampler for the Determination of Fine-Particulate Organic Material in the Atmosphere.* Paper presented at the Fourth Conference on Carbonaceous Particles in the Atmosphere, Vienna, Austria, 3–5, April, 1991, p. 19.

Fahey, D. W., Eubank, C. S., Hubler, G., and Fehsenfeld, F. C. (1985). "Evaluation of a catalytic reduction technique for the measurement of total reactiv odd-nitrogen NOy." *J. Atmos. Chem.* **3**, 435–468.

Febo, A., De Santis, F., Perrino, C., and Giusto, M. (1989). "Evaluation of laboratory and field performance of denuder tubes: A theoretical approach." *Atmos. Environ.* **23**, 1517–1530.

Ferm, M. (1979). "Method for the determination of atmospheric ammonia." *Atmos. Environ.* **13**, 1385–1393.

Finlayson-Pitts, B. J., and Pitts, J. N., Jr. (1977). "The chemical basis of air quality: kinetics and mechanisms of photochemical air pollution and application to control strategies." *Adv. Environ. Sci. Technol.* **7**, 75–162.

Finlayson-Pitts, B. J., and Pitts, J. N., Jr. (1986). "Atmospheric Chemistry: Fundamentals and Experimental Techniques." New York.

Gaffney, J. S., Streit, G. E., Spale, W. D., and Hall, J. H. (1987). "Beyond acid rain: Do soluble oxidants and toxins interact with SO2 and NOx to increase ecosystem effects?" *Environ. Sci. Technol.* **21**, 519–524.

Garber, R. W., Daum, P. H., Doering, R. F., D'Ottavio, T., and Tanner, R. L. (1983). "Determination of ambient aerosol and gaseous sulfur using a continuous FPD. III. Design and characterization of a monitor for airborne application." *Atmos. Environ.* **17**, 1381–1385.

Genfa, Z., and Dasgupta, P. K. (1989). "Fluorometric measurements of aqueous ammonium ion in a flow injection system." *Anal. Chem.* **61**, 408–412.

Gormley, P. G., and Kennedy, M. (1949). "Diffusion from a stream flowing through a cylindrical tube." *Proc. R. Ir. Acad., Sect. A* **52**, 163–169.

Grennfelt, P., Saltbones, J., and Schjoldager, J. (1988). *Oxidant Data Collection in OECD Europe 1985–1987 (OXIDATE)* Report on Ozone, Nitrogen, Dioxide and Peroxyacetyl Nitrate—October 1985–March, April, September 1986. Norwegian Institute for Air Research, Lillestrom, NILU Rep. OR 31/88.

Griffith, D. W. T., and Schuster, G. (1987). "Atmospheric gas analysis using matrix isolation-Fourier transform infrared spectroscopy." *J. Atmos. Chem.* **5**, 59–81.

Grosjean, D., Fung, K., Collins, J., Harrison, J., and Breitung, E. (1984). "Portable generator for on-site calibration of peroxyacetyl nitrate and analyzers." *Anal. Chem.* **56**, 569–573.

Guicherit, R., and Van Dop, H. (1977). "Photochemical production of ozone in Western Europe (1971–1975) and its relation to meterology." *Atmos. Environ.* **11**, 145–155.

Haagen-Smit, A. J. (1952). "Chemistry and physiology of Los Angeles smog." *Ind. Eng. Chem.* **44**, 1342–1346.

Haagen-Smit, A. J., and Fox, M. M. (1956). "Ozone formation in photochemical oxidation of organic substances." *Ind. Eng. Chem.* **48**, 1484–1487.

Hanst, P. L., Wong, N. W., and Bragin, J. (1982). "A long path infrared study of Los Angeles smog." *Atmos. Environ.* **16**, 969–981.

Hard, T. M., O'Brien, R. J., Chan, C. Y., and Mehrabzadeh, A. A. (1984). "Tropospheric free radical determination by FAGE." *Environ. Sci. Technol.* **18**, 768–777.

Hard, T. M., Chan, C. Y., Mehrabzadeh, A. A., and O'Brien, R. J. (1989). "Pressure dependence of ozone interference in the laser fluorescence measurement of OH in the atmosphere: Comment." *Appl. Opt.* **28**, 26–27.

Harris, G. W., Klemp, D., Zenker, T., Burrows, J. P., Mattieu, B., and Jacob, P. (1990). "Polarstern 1988: Measurements of trace gases using tunable diode lasers and intercomparison with other methods." In G. Restelli and G. Angeletti, eds., *Physico-chemical Behaviour of Atmospheric Pollutants*, Proc. 5th Eur. Symp., Varese, Italy, 25–28 September, 1989. Kluwer Academic Publishers, Dordrecht, pp. 644–640.

Heikes, B. G., and Thompson, A. M. (1983). "Effects of heterogeneous processes on NO3, HNO2 and HNO3 chemistry in the troposphere." *J. Geophys. Res.* **88**, 10883–10895.

Helmig, D., Muller, J., and Klein, W. (1989). "Improvements in the analysis of atmospheric peroxyacetyl nitrate (PAN)." *Atmos. Environ.* **23**, 2187–2192.

Henriksen, A., and Brakke, D. F. (1988). "Sulfate deposition to surface waters." *Environ. Sci. Technol.* **22**, 8–14.

Hofzumahaus, A., Dorn, H. P., and Platt, U. (1990). "Tropospheric OH radicals measurement techniques: Recent developments." In G. Restelli and G. Angeletti, eds., *Physico-chemical Behavior of Atmospheric Pollutants*, Proc. Eur. Symp., Varese, Italy, 25–28 September, 1989. Kluwer Academic Publishers, Dordrecht, pp. 103–108.

Hosker, R. P., Jr., and Lindberg, S. E. (1982). "Atmospheric deposition and plant assimilation of gases and particles." *Atmos. Environ.* **16**, 889–910.

Hov, O. (1984). "Modeling of the long range transport of peroxyacetyl nitrate to Scandinavia." *J. Atmos. Chem.* **1**, 187–202.

Hov, O., Hesstvedt, E., and Isaksen, I. S. A. (1978). "Long range transport of tropospheric ozone." *Nature (London)* **273**, 341–344.

Hov, O., Becker, K. H., Builtjes, P., Cox, R. A., and Kley, D. (1986). *Evaluation of Photooxidants-Precursor Relationship in Europe*. CEC, Brussels, AP/60/87. Air Pollut. Res. Rep. 1.

Hov, O., Allegrini, I., Beilke, S., Cox, R. A., Eliassen, A., Elshout, A. J., Gravenorst, G., Penkett, S. A., and Stern, R. (1987). *Evaluation of Atmospheric Processes Leading to Acid Deposition in Europe*. CEC, Brussels, Air Pollut. Res. Rep. 10, CEC Rep. EUR 11441.

Hubler, G., Ehhalt, D. H., Patz, H. W., Perner, D., Platt, U., Schoeder, J., and Toennissen, A. (1982). "Determination of ground level OH concentrations by a long-path laser absorption technique." In B. Versino and H. Ott, eds., *Physico-*

*chemical Behaviour of Atmospheric Pollutants*, Proc. 2nd Eur. Symp., Varese, Italy, 29 September–1 October, 1981. Riedel, Dordrecht, pp. 10–19.

Hubler, G., Perner, D., Platt, U., Toennissen, A., and Ehhalt, D. H. (1984). "Ground level OH radical concentration; new measurements by optical absorption." *J. Geophys. Res.* **89**, 1309–1319.

Jacob, P., Tavares, T. M., and Klockow, D. (1986). "Methodology for the determination of gaseous hydrogen peroxide in ambient air." *Fresenius Z. Anal. Chem.* **325**, 359–365.

Japar, S. M., and Niki, H. (1975). "Gas-phase reactions of the nitrate radical with olefins." *J. Phys. Chem.* **79**, 1629–1652.

Johnson, W. B., and Viezee, W. (1981). "Stratospheric ozone in the lower troposphere. I. Presentation and interpretation of aircraft measurements." *Atmos. Environ.* **15**, 1309–1323.

Kanakidou, M., Bonsang, B., and Lambert, G. (1988). "Light hydrocarbons vertical profiles and fluxes in a French rural area." *Atmos. Environ.* **23**, 921–927.

Kelly, T. J., Spicer, C. W., and Ward, G. F. (1990). "An assessment of the luminol chemiluminescence technique for measurement of NO2 in ambient air." *Atmos. Environ.* **24A**, 2397–2403.

Keuken, M. P. (1989). "The determination of acid deposition-related compounds in the lower atmosphere." Ph.D. thesis discussed at the Vrije Universiteit te Amsterdam, ECN/1989.

Keuken, M. P., Bakker, F. P., Lingerak, W. A., and Slanina, J. (1987). "Flow injection analysis of hydrogen peroxide, sulfite, formaldehyde and hydroxymethanesulfonic acid in precipitation samples." *Int. J. Environ. Anal. Chem.* **31**, 263–279.

Keuken, M. P., Schoonebeek, C. A. M., Van Wensveen-Louter, A., and Slanina, J. (1988). "Simultaneous sampling of NH3, HNO3, HCl, SO2 and H2O2 in ambient air by a wet annular denuder system." *Atmos. Environ.* **22**, 2541–2548.

Keuken, M. P., Wayers-Ijpelaan, A., Otjes, R. P., and Slanina, J. (1990). "Determination of NH3, HNO3 and NH4NO3 in ambient air by automated thermodenuders system and a wet annular denuder system." In G. Restelli and G. Angeletti, eds., *Physico-chemical Behaviour of Atmospheric Pollutants*, Proc. 5th Eur. Symp., Varese, Italy, 25–28 September, 1989. Kluwer Academic Publishers, Dordrecht, pp. 6–10.

Kley, D., and McFarland, M. (1980). "Chemiluminescence detector for NO and NO2." *Atmos. Technol.* **12**, 63–69.

Knoeppel, H., Versino, B., Schlitt, H., Peil, A., Schauenburg, H., and Vissers, H. (1980). "Organics in air, sampling and identification." In B. Versino and H. Ott, eds., *Physico-chemical Behaviour of Atmospheric Pollutants*, Proc. 1st Eur. Symp., Ispra, Italy, 16–18 October 1979. CEC, Brussels, EUR 6621, DE, EN, FR, pp. 25–40.

Lazrus, A. L., Kok, G. L., Lind, J. A., Gitlin, S. N., Heikes, B. G., and Shetter, R. E. (1986). "Automated fluorometric method of hydrogen peroxide in air." *Anal. Chem.* **58**, 594–597.

Lazrus, A. L., Fong, K. L., and Lind, J. A. (1988). "Automated fluorometric determination of HCHO in air." *Anal. Chem.* **60**, 1074–1077.

Lefohn, A. S., and Brocksen, R. W. (1984). "Acid rain effects research—A status report." *J. Air Pollut. Control Assoc.* **34**, 1005–1013.

Levy, H., Mahlman, J. D., and Moxim, W. J. (1985). "Tropospheric ozone: the role of transport." *J. Geophys. Res.* **90**, 3753–3772.

Liberti, A., and Ciccioli, P. (1985). "Chromatography for the evaluation of the atmospheric environment." In F. Bruner, ed., *The Science of Chromatography*, J. Chromatogr. Libr. Elsevier, Amsterdam, Vol. 32, pp. 219–256.

Liberti, A., and Ciccioli, P. (1986). "High resolution chromatographic techniques for the evaluation of atmospheric pollutants." *HRC CC, J. High Resolut. Chromatogr. Chromatogr. Commun.* **9**, 492–501.

Likens, G. E. (1976). "Acid precipitation." *Chem. Eng. News* **54**, 29–44.

Likens, G. E., Bormann, F. H., and Johnson, N. M. (1972). "Acid rain." *Environment* **14**, 33–40.

Louw, C. W., Richards, J. F., and Faure, P. K. (1977). "The determination of volatile organic compounds in city air by gas chromatography combined with standard addition, selective substraction infrared spectrometry and mass spectrometry." *Atmos. Environ.* **11**, 703–717.

Lubkert, B., Lieben, P., Grosch, W., Jost, D., and Weber, E. (1984). *Oxidant Monitoring Networks—Oxidant Monitoring Data*. Report from an International Workshop, 23–25 October 1984. Schauinsland, Federal Republic of Germany/OECD Environment Directorate, Ministry of the Interior of the Federal Republic of Germany, Unweltbundesamt, Berlin.

Maeda, Y. K., Aoki, K., and Munemori, M. (1980). "Chemiluminescence method for the determination of nitrogen dioxide." *Anal. Chem.* **52**, 307–311.

McClenny, W. A., Pleil, J. D., Oliver, K. D., and Holdren, M. W. (1984). "Automated cyrogenic preconcentration and gas chromatographic determination of volatile organic compounds in air." *Anal. Chem.* **56**, 2947–2951.

McLaughlin, S. B. (1985). "Effects of air pollution on forests—A critical review." *J. Air Pollut. Control Assoc.* **3**, 512–534.

McMahon, T. A., and Denison, P. J. (1979). "Empirical atmospheric deposition parameters." *Atmos. Environ.* **13**, 571–585.

Muller, K. P., Rudolph, J., and Wohlfart, K. (1990). "Measurements of peroxyacetyl nitrate in the marine atmosphere." In G. Restelli and G. Angeletti, eds., *Physico-chemical Behaviour of Atmospheric Pollutants*, Proc. 5th Eur. Symp., Varese, Italy, 25–28 September, 1989. Kluwer Academic Publishers, Dordrecht, pp. 705–710.

National Academy of Sciences (NAS). (1983). *Acid Deposition, Atmospheric Processes in Eastern North America*. NAS, Washington, DC.

Neftel, A., Blatter, A., and Staffelbach, T. (1990). "Gas-phase measurements of NH3 and NH4 + with differential optical absorption spectroscopy and gas stripping scrubber in combination with flow-injection analysis." In G. Restelli and G. Angeletti, eds., *Physico-chemical Behaviour of Atmospheric Pollutants*, Proc. 5th Eur. Symp., Varese, Italy, 25–28 September, 1989. Kluwer Academic Publishers, Dordrecht, pp. 83–91.

Noller, B. N., Currey, N. A., Ayers, G. P., and Gillett, R. (1986). "Naturally acidic rainwater at site in Northern Australia." In H. F. Hartmann, ed., *Proceedings of the Seventh World Clean Air Congress.* Clean Air Society of Australia and New Zealand, Vol. 5, pp. 189–190.

Odén, S. (1968). *The Acidification of Air and Precipitation and its Consequences in the Natural Environment*, Ecol. Comm. Bull. No. 1. Swedish National Research Council, Stockholm, pp. 1–117.

Odén, S. (1976). "The acidity problem—An outline of concepts." *Water, Air, Soil Pollut.* **6**, 137–166.

Organization for Economic Cooperation and Development (OECD). (1975). *Photochemical Oxidant Air Pollution. A Report of the Air Management Sector Group on the Problem of Photochemical Oxidants and Their Precursors in the Atmosphere.* OECD, Paris.

Organization for Economic Cooperation and Development (OECD). (1977). *The OECD Programme on Long Range Transport of Air Pollutants.* OECD, Paris.

Ottar, B. (1976). "Monitoring long range transport of air pollutants: The OECD study." *Ambio* **5**, 203–206.

Overrein, L. N. (1972). "Sulfur pollution patterns observed: Leaching of calcium in forest determined." *Ambio* **1**, 143–145.

Overrein, L. N., Seip, H. M., and Tollan, A. (1980). *Acid Precipitation—Effects on Forest and Fish.* Final Report of the SNSF Project, 1972–1980. RECLAMO, Oslo, Norway.

Papenbrock, T., and Stuhl, F. (1990). "Detection of nitric acid in air by laser-photolysis-fragment-fluorescence (LPFF) method." In G. Restelli and G. Angeletti, eds., *Physico-chemical Behavior of Atmospheric Pollutants*, Proc. 5th Eur. Symp., Varese, Italy, 25–28 September, 1989. Kluwer Academic Publishers, Dordrecht, pp. 651–656.

Parmar, S. S., and Grosjean, D. (1990). "Laboratory tests of KI and alkaline annular denuders." *Atmos. Environ.* **24A**, 2695–2698.

Pellizzari, E. D., Bunch, J. E., Berkley, R. E., and McRae, J. (1976). "Determination of trace hazardous organic vapor pollutants in ambient atmospheres by gas chromatography-mass spectrometry-computer." *Anal. Chem.* **48**, 803–807.

Penkett, S. A., Jones, B. M. R., Brice, K. A, and Eggleton, A. E. J. (1979). "The importance of atmospheric ozone and hydrogen peroxide in oxidizing sulfur dioxide in cloud and rainwater." *Atmos. Environ.* **13**, 123–137.

Perner, D., and Platt, U. (1979). "Detection of nitrous acid in the atmosphere by differential optical absorption." *Geophys. Res. Lett.* **6**, 917–920.

Perner, D., Platt, U., Trainer, M., Hubler, G., Drummond, J., Junkermann, W., Rudolph, J., Shubert, B., Volz, A., Ehhalt, D. H., Rumpel, K. J., and Helas, G. (1987). "Measurements of tropospheric OH concentrations: A comparison of field data with model predictions." *J. Atmos. Chem.* **5**, 185–216.

Perner, D., Parchatka, U., Karbach, H. J., and Eslick, I. C. (1989). "Determination of atmospheric trace gases by long-path absorption measurements." In I. Allegrini, A. Febo, and C. Perrino, eds., *Field Intercomparison Exercise on Nitric Acid and Nitrate Measurement—Methods and Data.* CEC, Brussels, Air Pollut. Res. Rep. 22, pp. 208–223.

Perner, D., Karbach, H. J., and Eslick, I. C. (1990a). *CEC Air Quality Progress Report on A Field Measurement Study of the Nighttime Oxidation of Naturally Emitted Hydrocarbons by NO3*, CEC, Brussels, Rep. XII/392/90, pp. 119–126.

Perner, D., Ladstatter-Weissenmayer, A., and Gall, R. (1990b). *Determination of NH3, SO2, NO and NO2 by Long-path UV Absorption During the Intercomparison of Monterotondo, April 29 through May 5, 1990*. Preliminary Report to the Intercomparison Co-ordinator. To be published.

Perrino, C., De Santis, F., and Febo, A. (1990). "Criteria for the choice of a denuder sampling technique devoted to the measurement of atmospheric nitrous and nitric acids." *Atmos. Environ.* **24A**: 617–626.

Pitts, J. N., Jr., Finlayson-Pitts, B. J., and Winer, A. M. (1977). "Optical systems unravel smog chemistry." *Environ. Sci. Technol.* **10**, 787–790.

Pitts, J. N., Jr., Biermann, H. W., Atkinson, R., and Winer, A. M. (1984). "An investigation of the dark formation of nitrous acid in environmental chambers." *Int. J. Chem. Kinet.* **16**, 919–939.

Pitts, J. N., Jr., Sweetman, J. A., Zielinska, B., Winer, A. M., and Atkinson, R. (1985). "Determination of 2-nitrofluoranthene and 2-nitropyrene in ambient particulate matter: Evidence of atmospheric reactions." *Atmos. Environ.* **19**, 1601–1608.

Platt, U., and Perner, D. (1980). "Direct measurements of atmospheric HCHO, HNO2, O3, NO2 and SO2 by differential optical absorption in the near UV." *J. Geophys. Res.* **85**, 7453–7458.

Platt, U., Perner, D., and Patz, H. W. (1979). "Simultaneous measurement of atmospheric HCHO, O3 and NO2 by differential optical absorption." *J. Geophys. Res.* **84**, 6329–6335.

Platt, U., Perner, D., Winer, A. M., Harris, G. W., and Pitts, J. N., Jr. (1980). "Detection of NO3 in the polluted troposphere by differential optical absorption." *Geophys. Res. Lett.* **7**, 89–92.

Platt, U., Perner, D., Shroeder, H., Kessler, G., and Toenissen, A. (1981). "The diurnal variation of NO3." *J. Geophys. Res.* **86**, 11965–11970.

Possanzini, M., and Di Palo, V. (1988). "Simultaneous measurements of formaldehyde and ozone in air by annular denuder-HPLC techniques." *Chromatographia* **25**, 895–898.

Possanzini, M., and Di Palo, V. (1989). "Determination of atmospheric ammonia as a m-toluamide by denuder sampling and HPLC-UV detection." *Chromatographia* **28**, 27–30.

Possanzini, M., Febo, A., and Liberti, A. (1983). "New design of a high performance denuder for the sampling of atmospheric pollutants." *Atmos. Environ.* **17**, 2605–2610.

Possanzini, M., Ciccioli, P., Di Palo, V., and Draisci, R. (1987). "Determination of low boiling aldehydes in air and exhaust gases by using annular denuders combined with HPLC techniques." *Chromatographia* **23**, 829–834.

Possanzini, M., Di Palo, V., and Liberti, A. (1988). "Annular denuder method for determination of H2O2 in the ambient atmosphere." *Sci. Total Environ.* **77**, 203–214.

Possanzini, M., Di Palo, V., and Petricca, M. (1990). "Trace determination of NO2 and SO2 in air with annular denuder and ion chromatographic techniques." *Proc. Eur. Conf. Anal. Chem.—EUROANALYSIS VII, Vienna, Austria, August, 1990*, Paper No. A1,1, P-Mo-20.

Puxbaum, H., Rosenberg, C., Gregori, M., Lanzerstorfer, C., Ober, E., and Winiwarter, W. (1988). "Atmospheric concentrations of formic and acetic acid and related compounds in Eastern and Northern Austria." *Atmos. Environ.* **22**, 2841–2850.

Ramdahl, T., Sweetman, J. A., Zielinska, B., Atkinson, R., Winer, A. M., and Pitts, J. N., Jr. (1985). "Analysis of mononitro-isomers of fluoranthene and pyrene by high resolution capillary gas chromatography mass spectrometry." *HRC CC, J. High Resolut. Chromatogr. Chromatogr. Commun.* **8**, 849–852.

Ramdahl, T., Zielinska, B., Arey, J., Atkinson, R., Winer, A. M., and Pitts, J. N., Jr. (1986). "Ubiquitous occurrence of 2-nitrofluoranthene and 2-nitropyrene in air." *Nature (London)* **321**, 425–427.

Rancon S.r.l (1990). *Determinazione di inquinanti atmosferici con il sistema OPSIS.* Information sheets from the Italian OPSIS representative, Milan, Italy.

Reid, J., Shewchun, J., Garside, B. K., and Ballik, E. A. (1978). "High sensitivity pollution detection employing tunable diode lasers." *Appl. Opt.* **17**, 300–310.

Roberts, J. M. (1990). "The atmospheric chemistry of organic nitrates." *Atmos. Environ.* **24A**, 243–287.

Roberts, J. M., Fajer, R. W., and Springston, S. R. (1989). "The capillary gas chromatographic separation of alkylnitrates and peroxycarboxylic nitric anhydrides." *Anal. Chem.* **61**, 771–772.

Roumelis, N., and Glavas, S. (1989). "Decomposition of peroxyacetyl nitrate and peroxypropionyl nitrate during gas chromatographic determination with a wide bore capillary and two packed columns." *Anal. Chem.* **61**, 2731–2734.

Rudolph, J., Ehhalt, O. H., Khedim, A., and Jebsen, C. (1981). "Determination of C2–C5 hydrocarbons in the atmosphere at low parts per 10 to the 9 to high parts of 10 to the 12 levels." *J. Chromatogr.* **247**, 301–310.

Schiff, H. I., and Mackay, G. I. (1989). "Tunable diode laser absorption spectrometry as reference method for tropospheric measurements." In R. Grisar, G. Schmidtke, M. Tacke, and G. Restelli, eds., *Monitoring of Gaseous Pollutants by Tunable Diode Lasers*, Proc. Int. Symp., Freiburg, F.R.G., 17–18 October 1988. Kluwer Academic Publishers, Dordrecht, pp. 36–45.

Schiff, H. I., Hastie, D. R., Mackay, G., Iguchi, T., and Ridley, B. A. (1983). "Tunable diode laser systems for measuring trace gases in tropospheric air." *Environ. Sci. Technol.* **17**, 352A–364A.

Schiff, H. I., Mackay, G. I., Castledine, C., Harris, G. W., and Tran, G. (1986). "Atmospheric measurements of nitrogen oxide with a sensitive luminol instrument." *Water, Air, Soil Pollut.* **30**, 105–114.

Schijoldager, J. (1984). "Observations of photochemical oxidants in connection with long- range transport." In P. Greenfelt, ed., *The Evaluation and Assessment of the Effects of Photochemical Oxidants on Human Health, Agricultural Crops, Forestry, Materials and Visibility*, Proc. Int. Workshop, Goteborg, Sweden, February 29–March 2, 1984. Swedish Environmental Research Institute (IVL), Goteborg, IVL-EM 1850, ISBN 91-7810-153-0, pp. 1–7.

Schmidbauer, N., and Oehme, M. (1986). "Improvement of a cryogenic preconcentration unit for C2-C6 hydrocarbons in ambient air at ppt levels." *HRC CC, J. High Resolut. Chromatogr. Chromatogr. Commun.* **9**, 502–505.

Schmidtke, G., Klocke, U., Knothe, M., Kohn, W., Riedel, W. J., and Wolf, H.

(1989). "Atmospheric trace gas measurements on the Scauinsland surveying station." In R. Grisar, G. Schmidtke, M. Tacke, and G. Restelli, eds., *Monitoring of Gaseous Pollutants by Tunable Diode Lasers*, Proc. Int. Symp. Freiburg, F.R.G., 17–18 October 1988. Kluwer Academic Publishers, Dordrecht, pp. 25–35.

Schneider, W., Frohne, J. C., and Bruderreck, H. (1978). "Determination of hydrocarbons in the parts per 10 to the 9 range using glass capillary columns coated with aluminum oxide." *J. Chromatogr.* **155**, 311–327.

Schofield, C. L. (1976). "Effects of acid precipitation on fish." *Ambio* **5**, 228–230.

Schroeder, W. H., and Lane, A. D. (1988). "The fate of toxic airborne pollutants." *Environ. Sci. Technol.* **22**, 240–246.

Schuetzle, D., Lee, F. S. C., Prater, T. J., and Tejada, S. B. (1981). "The identification of polynuclear aromatic hydrocarbon (PAH) derivatives in mutagenic fractions of Diesel particulate extracts." *Int. J. Environ. Anal. Chem.* **9**, 93–144.

Schuster, G., Wilson, S., and Helas, G. (1990). "Formaldehyde measurements in the free troposphere." In G. Restelli and G. Angeletti, eds., *Physico-chemical Behaviour of Atmospheric Pollutants*, Proc. 5th Eur. Symp., Varese, Italy, 25–28 September, 1989. Kluwer Academic Publishers, Dordrecht, pp. 555–557.

Singh, H. B. (1987). "Reactive nitrogen in the troposphere. Chemistry and transport of $NO_x$ and PAN." *Environ. Sci. Technol.* **21**, 320–327.

Singh, H. B., and Salas, L. J. (1983). "Methodology for the analysis of peroxyacetyl nitrate (PAN) in the unpolluted atmosphere." *Atmos. Environ.* **17**, 1507–1516.

Sjödin, A., and Ferm, M. (1985). "Measurements of nitrous acid in an urban area." *Atmos. Environ.* **19**, 763–767.

Slanina, J., Van Lamoen-Doornenbaal, L., Lingerak, W. A., Meilof, W., Klockow, D., and Neissner, R. (1981). "Application of a thermodenuder analyzer to the determination of H2SO4, HNO3 and NH3 in air." *Int. J. Environ. Anal. Chem.* **9**, 59–70.

Slanina, J., Schoonebeek, C. A. M., Klockow, D., and Niessner, R. (1985). "Determination of sulfuric acid and ammonium sulfate by means of a computer controlled thermodenuder system." *Anal. Chem.* **58**, 1955–1960.

Slanina, J., Van Weusveen, A. M., Schoonebeek, C. A. M., and Voors, P. I. (1987). "Automated Denuder Systems." In G. Angeletti and G. Restelli, eds., *Physico-chemical Behaviour of Atmospheric Pollutants*, Proc. 4th Eur. Symp., Stresa, Italy, 23–25 September 1986. Reidel, Dordrecht, pp. 25–32.

Smith, F. B., and Hunt, R. D. (1978). "Meteorological aspects of the transport and pollution over long distances." *Atmos. Environ.* **12**, 461–478.

Spicer, C. W. (1977). "The fate of nitrogen oxides in the atmosphere." *Adv. Environ. Sci. Technol.* **7**, 163–261.

Tanner, R. L., Markovits, G. Y., Ferreri, E. A., and Kelly, T. J. (1986). "Sampling and determination of gas-phase hydrogen peroxide following removal of ozone by gas-phase reaction with nitric oxide." *Anal. Chem.* **58**, 1857–1865.

Taylor, O. C. (1969). "Importance of peroxyacetyl nitrate (PAN) as a phytotoxic air pollutant." *J. Air Pollut. Control Assoc.* **19**, 347–351.

Tuazon, E. C., Winer, A. M., and Pitts, J. N., Jr. (1981). "Trace pollutant concentrations in a multiday smog episode in the California South Coast Air Basin by long path length Fourier transform infrared spectrometry." *Environ. Sci. Technol.* **15**, 1232–1237.

Tuazon, E. C., Atkinson, R., Plum, C. N., Winer, A. M., and Pitts, J. N., Jr. (1983). "The reaction of gas-phase N2O5 with water vapor." *Geophys. Res. Lett.* **10**, 953–956.

U.S. Department of Health, Education and Welfare (USDHEW) (1970). *Air Quality Criteria for Photochemical Oxidants*, NAPCA Rep. No. AP-63. National Air Pollution Control Administration, Washington, DC.

U.S. Environmental Protection Agency (USEPA) (1984). *National Air Quality and Emission Trend Report—1983,* EPA Rep. No. EPA-450/4-84-029. U.S. Environmental Protection Agency, Office of Air Quality Planning and Standards, Research Triangle Park, NC.

U.S. Environmental Protection Agency (USEPA) (1986). *Air Quality Criteria for Photochemical Oxidants*, Vols. III–V. U.S. Environmental Protection Agency, Environmental Criteria and Assessment Office, Research Triangle Park, NC, Publ. No. EPA/600/8-84/020 d, e and f.

Vierkorn-Rudolph, B., Rudolph, J., and Diederich, S. (1985). "Determination of peroxyacetyl nitrate (PAN) in unpolluted areas." *Int. J. Environ. Anal. Chem.* **20**, 131–148.

Viezee, W., and Singh, H. B. (1982). *Contribution of Stratospheric Ozone to Ground Level Ozone Concentrations—A Scientific Review of Existing Evidence.* Final Report of the EPA Grant CR 809330010. SRI International, Menlo Park, CA.

Watanabe, I., and Stephens, E. R. (1978). "Reexamination of moisture anomaly in analysis of peroxyacetyl nitrate." *Environ. Sci. Technol.* **12**, 222–223.

Wayne, R. P., Barnes, I., Biggs, P., Burrows, J. P., Canosa-Mas, C. E., Hjorth, J., Le Bras, G., Moortgat, G. K., Perner, D., Poulet, G., Restelli, G., and Sidebottom, H. (1991). "The nitrate radical: Physics, chemistry and the atmosphere." *Atmos. Environ.* **25A**, 1–203.

Winer, A. M., Peters, J. M., Smith, J. P., and Pitts, J. N., Jr. (1974). "Response of commercial NO–NO2 analyzers to other nitrogen containing compounds." *Environ. Sci. Technol.* **8**, 1118–1121.

Wolff, G. T., Kelly, N. A., and Ferman, M. A. (1982). "Source regions of summertime ozone and haze episodes in the eastern United States." *Water, Air, Soil Pollut.* **18**, 65–81.

Woodman, J. N., and Cowling, E. B. (1987). "Airborne chemicals and forest health." *Environ. Sci. Technol.* **21**, 120–126.

World Health Organization (WHO) (1986). *Air Quality Guidelines, Review Draft.* WHO, Regional Office of Europe, Copenhagen.

Yokouchi, Y., Ambe, Y., and Maeda, T. (1986). "Automated analysis of C3–C13 hydrocarbons in the atmosphere by capillary gas chromatography with a cryogenic preconcentration." *Anal. Sci.* **2**, 571–575.

Yokouchi, Y., Mukai, H., Nakajima, K., and Ambe, Y. (1990). "Semi-volatile aldehydes as predominant organic gases in remote areas." *Atmos. Environ.* **24A**, 439–442.

Zielinska, B., Arey, J., Atkinson, R., and Winer, A. M. (1989). "The nitroarenes of molecular weight 247 in ambient particulate samples collected in Southern California." *Atmos. Environ.* **23**, 223–229.

# INDEX

**535**